Musculoskeletal Imaging

MRI, CT, Nuclear Medicine, and Ultrasound in Clinical Practice

Musculoskeletal Imaging
MRI, CT, Nuclear Medicine, and Ultrasound in Clinical Practice

1991

EDITED BY

John A. Markisz, M.D., Ph.D.

Associate Professor of Radiology, and Chief,
Division of Magnetic Resonance, Cornell University Medical College;
Associate Attending Radiologist, The New York Hospital, New York

FOREWORD BY

Joseph P. Whalen, M.D.

Professor and Chairman, Department of Radiology,
Cornell University Medical College;
Radiologist-in-Chief, The New York Hospital, New York

LITTLE, BROWN AND COMPANY
Boston/Toronto/London

First Edition

Library of Congress Catalog Card No. 91-61163

ISBN 0-316-54613-5

Printed in the United States of America
MV

To my father:

Hey Dad, this book's for you.

Contents

Foreword

During the past two decades, the perspective of clinicians confronting musculoskeletal problems has been fundamentally altered by the introduction and continued development of new and innovative diagnostic methods. The introduction and continued improvements in CT scanning, ultrasonography, and nuclear medicine have expanded and refined the information available from each of these modalities, and each has subsequently established a role for itself in the diagnosis of various musculoskeletal pathologies.

The appearance of magnetic resonance imaging early in the 1980s promised to open new vistas in imaging. Structures heretofore "invisible" to any imaging techniques became potential targets of investigation. It became very clear to those medical centers that first installed MRI systems that the initial applications of this exciting new modality would be in the central nervous system. As clinical experience accumulated and image quality improved significantly, especially with the practical development of surface coils, it became apparent that musculoskeletal applications would be extremely important, even rivaling those of the CNS in its clinical utilization. Experience in the many centers that now have MRI capabilities, along with the ever-increasing literature on musculoskeletal MRI, verifies its distinctive effectiveness.

This book clearly and authoritatively reviews the applications of these recently developed imaging modalities—CT, ultrasonography, nuclear medicine, and MRI—along with discussions of their appropriate use in musculoskeletal disease. MRI, the newest and least familiar of these techniques, is presented in considerably more detail than the others, constituting approximately one half of the text. The common theme throughout the chapters involves how and when various techniques should be used in various clinical settings. Although the emphasis is on clinical applications, anatomic and pathologic correlation is of prime importance. Multiple images of normal joint anatomy are presented in different planes along with a detailed delineation of the various component structures. Each chapter contains multiple examples of pathologic entities, along with a description of how that modality is applied in order to obtain a diagnosis.

The authors are experts in their fields, the information is current, and each chapter contains an extensive bibliography of recent medical literature. Dr. Markisz and his colleagues are to be congratulated for their timely review of the applications of these newer imaging modalities to musculoskeletal disease.

Joseph P. Whalen, M.D.

Preface

Major technological innovations during the past twenty years have provided physicians with new and very different methods of detecting pathologic changes in bone, surrounding soft tissue structures, and joints with increasing accuracy and sophistication. These new disciplines—nuclear medicine, ultrasonography, computed axial tomography (CT), and magnetic resonance imaging (MRI)—each provide important diagnostic information not previously available in a non-invasive manner. Even with the extraordinary advances now available from modern imaging procedures, the routine radiograph remains the initial examination of choice whenever a musculoskeletal problem is encountered. It is after the first radiographs are obtained that the choice of subsequent imaging procedures is not always clear. Will it be a scintigram, a sonography, or a tomogram? CT or MRI? With multiple new techniques now routinely available, when necessary, it becomes increasingly important to realize what information is available from the different types of examinations. The question this text attempts to answer is, "What is the most effective tool that we have available for the evaluation of a particular problem?"

These new modalities use inherently different principles, often yielding information unique to that imaging technique. The selection of the particular type of examination that should be requested depends on the specific clinical question to be answered. An understanding of the underlying principles and capabilities of each of these methods will permit an appreciation of the advantages and limitations of each and will help to ensure their proper clinical use. In this text we discuss the clinical applications of computed tomography, ultrasonography, nuclear medicine, and magnetic resonance imaging relative to the musculoskeletal system. Each is unique in the physical principles involved, and each will be most fruitful when applied in the proper clinical setting. The clinical applications are presented in order to demonstrate how and when each should be used, in order to maximize their utility. Detailed discussions of the physical principles employed by CT, nuclear medicine, and ultrasonography are only touched on here; they are well described in many standard texts. MRI, the newest and most complex of the technological advances being considered, is explained by Dr. Knowles in Chapter 8.

J. A .M.

With sincere appreciation and gratitude for their inspiration, encouragement, and support

Mary Ann Payne, M.D., Ph.D.

Joseph D. Gettler, Ph.D.

Elias Kazam, M.D.

Joseph P. Whalen, M.D.

Acknowledgments

Very few educational endeavors can be accomplished without the assistance of others, and this work is certainly no exception. I am deeply indebted to my co-authors for their exceptional contributions to this text. I appreciate the time and energies that went into producing their outstanding efforts, and I am sure readers will, too.

I owe thanks to all of my many colleagues at Cornell University Medical College and The New York Hospital, for their assistance, comments, and suggestions, which allowed me to complete this project. In particular, I would like to thank Drs. Ralph Marcove, Joseph Lane, Morton Coleman, Mark Pasmanteir, David Wolf, and David Altchek for their constructive assistance in obtaining valuable clinical correlations and insights.

I am especially grateful to the following individuals for their special assistance:

Dr. Mary Ann Payne, without whose support and encouragement this project would not have progressed beyond its rudimentary stages;

Dr. Patrick T. Cahill, for sharing his knowledge and experience with me and for encouraging my interest in the musculoskeletal system;

Drs. Marvin Eisenberg, J. Bruce Kneeland, Jack Barie, Tom Glendening, Ken Rapoport, and Joseph Yacavone, for contributing images used in this text;

Michael Aquilia, Richard Fischer, Edna Hong, May Navarro, Sang Joon Park, Steven Pantaleone, Pauline Soeffing, and Robert Harkinish, for obtaining the clinical MRI studies and aiding with the photography;

James Zaput of General Electric Medical Systems, for performance above and beyond the call of duty;

Edna Hong, Sang Joon Park, Michael Aquilia, and my brother Bob, each of whom spent countless hours aiding in the acquisition of normal anatomic images;

Jane Licht and the editorial staff of Little, Brown and Company, for providing their technical expertise and assistance during all phases of this project; and

Geri Davis and Martha Morong of Quadrata, Inc., for their outstanding effort in the design and production of this text.

Finally, sincere thanks must also go to my family, who volunteered untold hours of assistance and were, as always, supportive and understanding throughout this project.

Contributing Authors

Carol B. Benson, M.D.
Assistant Professor of Radiology, Harvard Medical School; Assistant Attending Radiologist, Brigham and Women's Hospital, Boston, Massachusetts

William D. Johns, M.D.
Assistant Clinical Professor of Nuclear Medicine, University of Connecticut School of Medicine, Farmington; Assistant Medical Director—Nuclear Medicine, Danbury Hospital, Danbury, Connecticut

R. James R. Knowles, Ph.D.
Associate Professor of Physics in Radiology, Cornell University Medical College; Senior Medical Physicist, The New York Hospital, New York, New York

Peter A. Kosovsky, M.D.
Assistant Professor of Radiology, Cornell University Medical College; Assistant Attending Radiologist, The New York Hospital, New York, New York

John A. Markisz, M.D., Ph.D.
Associate Professor of Radiology, and Chief, Division of Magnetic Resonance, Cornell University Medical College; Associate Attending Radiologist, The New York Hospital, New York, New York

Barbara J. McNeil, M.D., Ph.D.
Chairman, Department of Health Care Policy, Harvard Medical School; Attending Radiologist, Brigham and Women's Hospital, Boston, Massachusetts

Richard B. Rafal, M.D.
Assistant Professor of Radiology, Cornell University Medical College; Assistant Attending Radiologist, The New York Hospital, New York, New York

Robert Schneider, M.D.
Associate Professor of Radiology, Cornell University Medical College; Attending Radiologist, Hospital for Special Surgery, New York, New York

David A. Summerville, M.D., Ph.D.
Formerly Assistant Professor of Radiology, Harvard Medical School; Staff Physician, Nuclear Medicine, The Children's Hospital, Boston, Massachusetts

S. Ted Treves, M.D.
Professor of Radiology, Harvard Medical School; Chief, Division of Nuclear Medicine, The Children's Hospital, Boston, Massachusetts

Musculoskeletal Imaging
MRI, CT, Nuclear Medicine, and Ultrasound in Clinical Practice

1

Introduction

JOHN A. MARKISZ

The basis of early diagnostic imaging was established on November 8, 1895. This was the date that Wilhelm Röntgen discovered x-rays emanating from a Crookes tube by observing their effects on fluorescent salts and on photographic plates. On December 22, 1895, he first visualized the bones in his wife's hand as an image on a photographic plate, thus creating the field of radiology. He was honored for his efforts with the first Nobel Prize in physics 6 years later. News of this innovation spread quickly and within a year x-ray examinations were an accepted part of a medical workup, although the first radiologists included photographers and electricians, as well as physicians. Ever since, the conventional radiograph has been the fundamental technique used for the initial examination of abnormalities of the musculoskeletal system. This new technique was not universally acclaimed initially, as expressed in an editorial in the *London Pall Mall Gazette* [1]:

> We are sick of the roentgen ray: you can see other people's bones with the naked eye, and also see through eight inches of solid wood. On the revolting side of this indecency of this

> there is no need to dwell. . . . [W]e seriously put before the attention of the Government . . . that it will call for legislative restriction of the severest kind. Perhaps the best thing would be for all civilized nations to combine to burn all works on the roentgen rays, to execute all the discoverers and to corner all the tungstate in the world and whelm it in the middle of the ocean.

Although legislative restraints were imposed, the field of radiology persevered. In the ensuing years, with radiology now approaching its centenary, volumes have been written on the use of plain films in locating and characterizing various types of musculoskeletal lesions. Most skeletal abnormalities can now be so well identified on routine x-rays that no further workup may be necessary.

Generating a routine radiograph is dependent on two physical principles: X-rays will fog photographic film (the greater the concentration of radiation, the darker the area on the developed film); and different tissues will attenuate x-rays to different degrees, so that many will penetrate air, fewer will get through fat, fewer still through soft tissues, and very few through bone. Variations on the basic x-ray technique have increased the utility of the basic radiograph. With the technique of injecting contrast medium (air was used as early as 1905) into the joint space, the horizons of radiology expanded to develop arthrography, enabling accurate assessment of multiple problems within joints, including tendon and ligament damage. In areas other than the musculoskeletal system, variations on the same theme provided additional diagnostic information. The intravenous injection of contrast material allowed both the morphology and the function of the kidneys and urinary tract to be studied. A more invasive use of contrast medium led to the development of angiography and direct visualization of the vascular system. This also permitted an evaluation of the vascularity of musculoskeletal lesions. Orally administered contrast medium was responsible for the establishment of gastrointestinal radiology and visualization of the GI tract.

The regular radiograph is a two-dimensional image of three-dimensional structures superimposed on one another. The interpretive problem caused by multiple densities being placed on top of one another was partially reduced by tomography. Tomography (linear, circular, small-angle, etc.) permitted better visualization of an internal structure. The x-ray source and photographic plate were moved synchronously in a geometric pattern, which had the effect of focusing on a particular plane and blurring out anatomic parts not in the plane of interest. All of these methods, however, are derived from the standard radiograph.

In 1972, a new modality utilizing x-rays was developed by Hounsfield, who described the first feasible computer-assisted tomographic (CT) unit, which he was able to put into clinical use in 1973. The images obtained were cross-sectional views, clearly demonstrating internal anatomy and eliminating the problem of interpreting representations of superimposed structures. Six years later, he, like Röntgen, also received the Nobel prize in physics for his efforts. This new system was capable of directly obtaining high-resolution tomographic images that not only distinguished between bone and soft tissue, but also discriminated between much more subtle differences in tissue as well.

The earliest CT scanners employed an x-ray source in a round gantry, directly opposite a radiation detector, with the patient in the center. After the emission of a highly collimated beam of x-rays, the transmission through the patient is measured by the detector and recorded on a computer. The radiation passing through the patient is attenuated by everything in its path (fat, bone, soft tissue, air, etc.). The source and the detector then rotate through a small angle, and transmission data for this new position are similarly recorded by the computer. After multiple measurements, different types of reconstruction techniques (back projection, integral equations, Fourier transforms, or series expansion) can be used to generate a cross-sectional image. Rapid and seemingly unending technologic developments have continued to improve image quality. The synchronously rotating source and detector have been replaced by a circular array of detectors and a moving x-ray generator in today's fourth-generation units. The 4.5-minute acquisition time necessary for the first-generation scanners was decreased to the current 1 to 2 seconds. Motion, the source of severe artifacts on earlier studies because of the prolonged acquisition times, no longer is an impediment to quality examinations. Radiation exposure has also been significantly decreased with the improved technology. The resolution afforded by CT is excellent, and since it is ultimately dependent on x-rays for the production of an image, it can utilize orally administered contrast media to highlight the GI system and intravenously administered contrast material to delineate vascular structures and to determine the vascularity of tumors. It is also the only one of the four modalities we are discussing that can definitively identify calcifications, an important part of radiologic diagnoses.

The turn of the century was an extremely active time in experimental physics, with an unprecedented variety of new phenomena being studied. Just 3 months after Röntgen detected the emission of x-rays from a Crookes tube, on March 1, 1896, Henri Becquerel discovered natural radioactivity from uranium salts, which eventually led to the development of nuclear medicine. His Nobel prize was awarded 9 years later and was shared by Pierre and Marie Curie. Madame Curie was a graduate student of Becquerel, whose work so interested her husband Pierre that he left his research on piezoelectric crystals (later the basis for ultrasonography) to join her in this new and exciting field. With the fundamental principles established in nuclear physics, three different disciplines were combined in order that nuclear medicine could evolve to its present state: radiopharmacology, physiology, and electronics.

The ability to incorporate radioactive elements into various molecules that will be metabolized by, transported to, or eliminated by different organs or tissues is the basis for all nuclear scans. These radioisotopes emit radiation, which can be electronically counted or imaged. Although resolution is not as detailed as in other techniques, abnormalities are detected on the basis of physiologic function of the organ or system under investigation, rather than on anatomic changes. Radiolabeled red blood cells, for example, can provide information about cardiac function, organ perfusion, or gastrointestinal bleeding sites as well as help to discriminate between liver metastases and hepatic hemangiomata. Recent advances in labeling monoclonal antibodies with radioactive atoms promise to significantly increase the versatility and utility of this field. Electronic innovations have led to the improvement of resolution by the development of the gamma camera, to succeed the rectilinear scanner, and single-photon emission computed tomography (SPECT), permitting tomographic sections of areas of interest to be obtained. Advances in computer techniques now permit the routine evaluation of such physiologic parameters as cardiac output and glomerular filtration rate, as well as permit cine imaging of cardiac wall motion and tissue perfusion.

The field of ultrasonography has developed without directly providing a Nobel prize for any of the many scientists who shared in its development. Although Pierre Curie *was* investigating the piezoelectric effect, his attentions were shifted to the exploration of radioactivity, and he became a Nobel laureate for his work in the latter field. Imaging applications of the principles of sound propagation evolved only relatively recently, following many years of development of such diverse applications as underwater communication, submarine detection, ocean floor mapping, and flaw detection in metals. The word ultrasound simply refers to sound with frequencies above the upper limit of the range of hearing, approximately 20 kHz. Frequencies considerably higher than this, 1–20 MHz, are employed for diagnostic applications.

The piezoelectric effect involves the interconversion of mechanical and electrical energy, and is the basis for the production of ultrasonic transducers. When high pressure is applied to certain crystals, a voltage difference is developed on opposing faces of the crystal. This pressure can be provided by an ultrasonic wave, with the resulting voltage determined by the strength of the ultrasound. The converse is also true. If a voltage is applied across an appropriate crystal, there will be a distortion of the crystalline structure, which can be utilized to produce an ultrasound beam. The frequency of an ultrasound transducer is an intrinsic property of the crystal within it. The higher the frequency (with correspondingly smaller wavelengths), the greater the resolution. Associated with increasing frequency, however, is increased tissue absorption, resulting in decreased penetrating power, so that an adjustment must be made, depending on the clinical situation. Higher frequencies, above 5 MHz, are usually used for small, superficial structures, while 1–2-MHz transducers are necessary to penetrate bone.

At interfaces between tissues of different densities, reflections of sound waves, or echoes, are produced. Employing the properties of piezoelectric crystals, the transducer acts to both produce sound waves and receive echoes. A momentary high voltage is first applied to the crystal, causing an ultrasonic wave front to be emitted, which is then reflected by tissue interfaces. The crystal is then distorted by the returning echo, producing a small voltage that can be amplified and recorded. The crystal is therefore both the transmitter and the receiver of the ultrasound.

Images are obtained by scanning a sector and measuring the strength of the echoes as a function of time after the initial pulse. The stronger the echo, the brighter the point on the display screen. The distance of the point from the transducer is determined by the amount of time the echo takes to return after the initial pulsation. In this way, a series of lines is obtained for each sector, with each line composed of a series of points of varying intensity, thus producing the image.

The field of magnetic resonance imaging (MRI) survived both a name and initial change from its parent field, nuclear magnetic resonance (NMR). Although MRI is more descriptive and less offensive (to some), it cannot be denied that the procedure has directly descended from NMR. In 1945, Felix Bloch of Stanford University and Edward Mills Purcell of Harvard University independently demonstrated that magnetic nuclei could interact with electromagnetic radiation of discrete frequencies, establishing the field of NMR spectroscopy and providing the theoretic framework for the later development of MRI. Together, in 1952, they shared a Nobel prize for their efforts. NMR spectroscopy has long been an important tool of chemists for determining molecular structure. In 1972, Paul Lauterbur was able to utilize the principles of NMR to produce an image, calling the process zeugmatography, from the Greek term meaning "to join together." This was a reference to the coupling of the magnetic and radiofrequency fields by the magnetic nuclei.

Diseases of the musculoskeletal system can be organized into several broad categories—neoplastic (malignant and benign), infectious, inflammatory, traumatic, metabolic, and mechanical. Many different anatomic and physiologic changes occur and are reflected in patterns or signals that are distinctive for each modality utilized. None of the modern modalities can evaluate everything. Some are inherently better in certain situations because of the principles that allow the image to be obtained, or by the way in which they are acquired. Now we move on to examine these modalities so that we can get a better appreciation of the applications of these principles and the proper utilization of the methods themselves.

REFERENCE

1. Cade S. Radiation induced cancer in man. Br J Radiol 1957; 393–402.

BIBLIOGRAPHY

Ambrose J. Computerized transverse axial scanning (tomography): Part 2. Clinical applications. Br J Radiol 1973; 46:1023–47.

Edeiken J. Roentgen diagnosis of diseases of bone. Baltimore: Williams & Wilkins, 1981.

Hendee WR. Medical radiation physics. 2nd ed. Chicago: Year Book Medical Publishers, 1979.

Hounsfield GN. Computerized transverse axial scanning (tomography): Part 1. Description of the system. Br J Radiol 1973; 46:1016–22.

Johns EH, Cunningham JR. The physics of radiology. 3rd ed. Springfield: Charles C. Thomas, 1977.

Partain CL, James, Jr, AE, Rollo FD, Price RR: Nuclear magnetic resonance imaging. Philadelphia: Saunders, 1983:542.

Rocha AFG, Harbert JC. Textbook of nuclear medicine: Basic science. Philadelphia: Lea & Febiger, 1978.

Rollo FD. Nuclear medicine physics, instrumentation and agents. St. Louis: C. V. Mosby, 1977.

2

Anatomy of the Synovial Joints

JOHN A. MARKISZ

An understanding of normal joint anatomy and function is a fundamental prerequisite for the analysis and interpretation of any image of the musculoskeletal system. Since clinical applications of the tomographic capabilities of CT, MRI, SPECT, and ultrasonography are steadily increasing, an appreciation of the multiplanar relationships of the various components of the musculoskeletal system has become essential. This is especially true for the complex articulations comprising the various joints. In this section, MRIs of the major synovial joints are presented in transverse (axial), coronal, sagittal, and oblique planes to demonstrate their normal spatial relationships, in order that pathologic conditions can be more easily identified and understood.

Synovial joints are complex structures that involve freely movable articulations. The bony surfaces involved are in contact but are not continuous as they are, for example, in a symphysis, where only limited motion is permitted, or a gomphosis, a peg-and-socket system (e.g., teeth set into the mandible or maxilla) that will not allow any joint motion at all. The synovial joints all have certain features in common. There are articular surfaces, covered with articular cartilage, which is a compressible and elastic material that minimizes wear and tear at the points of contact and provides an essentially friction-free, lubricated surface. The joints are contained within a fibrous capsule composed of interwoven fibers of connective tissue. Tendons and ligaments often blend into the capsule to give it, and the joint, additional support. A synovial membrane lines the nonarticular segments of the joints, including the fibrous capsule, synovial bursae, and synovial tendon sheaths. Movement of one tissue plane over another is greatly facilitated by synovial fluid, a viscous liquid secreted by the synovial membrane. Under normal circumstances, very little fluid is found within a joint; less than 0.5 ml can be obtained from within a normal knee, the largest joint in the body.

Synovial joints are classified according to their shape and function. *Ball-and-socket* joints are formed by the spherical head of one bone inserted into a deep recess of another bone, such as is found in the shoulder and the hip. This type of joint is multiaxial, with the capability of movement around three axes. *Condylar* joints allow a principal movement to occur in one plane, and as such are essentially uniaxial, but also have a limited amount of rotation in a perpendicular plane. This type of motion can be found in the knee and the temporomandibular joints (TMJs). *Ellipsoid* joints are biaxial, being composed of an oval, convex surface on one side and an elliptical, concave surface on the other. The radiocarpal and metacarpophalangeal joints are examples of ellipsoid joints. *Hinge* joints are uniaxial joints that restrict movement to one plane, as occurs in the interphalangeal joints. *Pivot* joints are uniaxial with motion restricted to rotation around a fixed axis, as found in the proximal radioulnar joint during pronation and supination. *Plane* joints allow a sliding motion to occur and are formed by two relatively flat surfaces that slide over each other. The intermetacarpal joints are examples of this type of articulation. The *saddle* joint is composed of two bones, each of which has two articular surfaces, one concave surface that is at right angles to a convex surface. For each surface of this biaxial joint, the concavity of one bone is apposed to the convexity of the other, allowing motion in two planes at right angles to each other. The carpometacarpal joint of the thumb illustrates this type of motion. Joints that use one type of articulation are termed *simple* joints, whereas those that employ more than one type are referred to as *compound* joints.

The Temporomandibular Joint

Although not usually thought of as part of the musculoskeletal system, the TMJ is a complex, mobile structure that allows the mandible to be depressed, elevated, protruded, retracted, and even moved laterally. It is a condylar type of joint, and while usually referred to in the singular, both TMJs must operate simultaneously (but not necessarily symmetrically) for any movement to occur. The bony structures involved are the articular tubercle and the anterior portion of the mandibular fossa of the temporal bone, superiorly, and the condyle of the mandible, inferiorly. The surfaces of all of the osseous components are covered with articular cartilage. An articular disc separates the bony structures.

The articular disc is a complex ovoid structure, expanded at each end (with the posterior band larger than the anterior band) and a concave inferior surface that normally remains in contact with the articular surface of the condyle. The posterior band and the anterior band are separated by the thinner, concave intermediate zone. Tapering regions, the posterior bilaminar band and the anterior extension, project posteriorly and anteriorly, respectively. The entire circumference of the disc is connected to a fibrous capsule, and the anterior extension is attached to the pterygoideus lateralis tendon. Additional fibrous bands ensure close proximity of the articular disc to the condyle.

The fibrous capsule around the joint fits tightly inferiorly to the articular disc, where it attaches to the neck of the mandible, below the condyle. It forms a rather loose enclosure above the disc, being fastened to the articular tubercle anteriorly and the roof of the mandibular fossa and the squamotympanic fissure posteriorly. The capsule is lined with a synovial membrane. The lateral ligament extends from the zygoma to the posterolateral surface of the neck of the mandible, and of the many ligaments involving the mandible, it is the only one directly associated with the capsule, providing increased stability.

When the mouth is in a closed position, the condyle of the mandible is centered within the mandibular fossa, in contact with the concave surface of the intermediate zone. The posterior band of the articular disc lies directly superior to the condyle. When the mouth is open, both the condyle and the disc move anteriorly, and the disc rotates so that the condyle maintains contact with the intermediate zone. The condyle should then lie directly inferior to the articular tubercle, with the disc assuming an almost horizontal position. This anatomy is demonstrated for both open and closed mouths in normal sagittal planes (Figures 2–1A and B).

The Shoulder

The shoulder is a multiaxial ball-and-socket joint and is the most mobile joint in the body. The rounded head of the humerus fits into the socket formed by the glenoid cavity of the scapula. Because the glenoid is quite shallow, a relatively large range of motion is possible, but this configuration also reduces the integrity of the joint. The primary source of mechanical stabilization of the joint is found in the muscles that completely surround it: the supraspinatus superiorly; the infraspinatus and teres minor inferiorly; the subscapularis anteriorly; the deltoid anteriorly, posteriorly, and laterally; as well as the superior encasement of the coracoid process, the acromion, and the coracoacromial ligament.

The humeral head is covered with articular cartilage. It has two tuberosities, the greater tuberosity laterally, to which the tendons of the supraspinatus, infraspinatus, and teres minor muscles are attached, and the more anterior lesser tuberosity, to which the tendon of the subscapularis muscle is attached. Between these tuberosities lies the bicipital groove in which the tendon of the biceps brachii muscle resides. The articular surface of the glenoid cavity is lined with a thin fibrocartilage rim around it (the glenoid labrum). This protects the articular surface and serves to further stabilize the joint by increasing the size of the concavity.

The very loose-fitting fibrous capsule around the joint is lined with a synovial membrane. The capsule is loose enough to allow a significant range of motion. It is connected to the anatomic neck of the humerus laterally, the humeral neck and edge of the glenoid cavity medially, and the coracoid process superiorly. It includes the tendon of the long head of the biceps brachii. The capsule is reinforced by the merging into it of the tendons of the supraspinatus, infraspinatus, teres minor, and subscapularis muscles to form the rotator cuff. The capsule is strengthened anteriorly by the glenohumeral ligaments, which extend from the medial (anterior) edge of the glenoid cavity to above, on, and below the lesser tubercle. Additional anterior support is obtained from extensions of the tendons of the pectoralis major and teres major muscles. Multiple bursae surround the joint. Quite important to the normal operation and evaluation of the joint are the subacromial bursa and the subscapularis bursa. The subacromial bursa has subacromial and subdeltoid segments and does not normally communicate with the joint space if the rotator cuff mechanism is intact. It is located inferior to the acromion and deltoideus muscle, and superior and lateral to the rotator cuff. The subscapularis bursa does communicate with the joint space, and is located between the subscapularis tendon and the scapula.

The anatomy of the shoulder is illustrated in Figures 2–2 (axial), 2–3 (paracoronal), and 2–4 (parasagittal). The paracoronal images are obtained parallel to the supraspinatus muscle and tendon, while the parasagittal images are acquired perpendicular to them, being parallel to the glenoid labrum. The oblique planes were chosen because they best demonstrate the interrelationships of the various structures.

The Elbow

The elbow is a complex junction of the humerus, ulna, and radius and consists of two hinge articulations and a separate uniaxial pivot-type joint. One of the hinge articulations occurs between the articular surfaces of the trochlea of the humerus and the trochlear notch of the ulna, and the second occurs between the articular surfaces of the capitellum of the humerus and the head of the radius. The proximal radioulnar joint is a synovial pivot joint that couples the radius and ulna. A single fibrous capsule envelops all of the articulations, and a complex synovial membrane completely lines the capsule.

A folding of synovium between the radius and ulna partially divides the region into the humeroulnar and humeroradial sections. This fold is roughly triangular in shape, and contains extrasynovial fat. Three other fat pads are found between the synovial membrane and the fibrous capsule. The largest of the three overlies the olecranon fossa, into which it is forced by the triceps tendon during flexion. Fat pads over the radial fossa and coronoid fossa are pushed by the brachialis tendon during extension.

The elbow is stabilized by the collateral ligaments. The radial collateral ligament extends from the lateral epicondyle of the humerus to the annular ligament around the head of the radius. The ulnar collateral ligament is composed of two triangular portions, the anterior part extending from the medial epicondyle of the humerus to the medial aspect of the coronoid process, and the posterior part, bridging the inferomedial portion of the medial epicondyle to the medial aspect of the olecranon.

The proximal radioulnar joint is a pivot formed by the head of the radius rotating within the radial notch of the ulna. The annular ligament encircles the head of the radius, while both ends (anterior and posterior margins of the radial notch) are attached to the ulna, keeping it in contact with the radial notch. The internal surface of the annular ligament is in contact with the head of the radius and is covered with a thin layer of cartilage. The superior surface is continuous with the radial collateral ligament and the inferior surface is lined with synovial membrane. Normal axial, sagittal, and coronal anatomy is illustrated in Figures 2–5, 2–6, and 2–7, respectively.

The Wrist

The wrist is a compound joint that actually incorporates the distal radioulnar joint, a uniaxial pivot joint, and the radiocarpal joint, an ellipsoid joint. A triangular articular disc is part of both joints and aids in binding the radius and ulna together. The disc is attached to a depression between the head of the ulna and the styloid process and to the edge of the radius. The outer perimeter of the disc is affixed to the ligaments surrounding the joint. The disc is a bioconcave structure, with the proximal surface articulating with the head of the ulna and the distal surface with the lunate and triquetrum. The distal articular surface of the radius is in continuity with the distal surface of the disc, and together they form the concave articular surface necessary for their articulation with the convex surfaces of the scaphoid, lunate, and triquetrum bones, to form the radiocarpal joint. The fibrous articular capsule is lined with synovium and strengthened by the many strong ligaments bridging the joint.

The flexor retinaculum is a strong fibrous structure extending across the palmar aspect of the wrist from the pisiform and the hook of the hamate, to the tubercles of the scaphoid and trapezium, superficially, as well as deep to the medial lip of the trapezium. A small, enclosed canal, the carpal tunnel, is thus formed by the flexor retinaculum near the palmar surface and by the carpal bones dorsally. Through this tunnel pass the flexor tendons of the fingers and the median nerve and vessels. The median nerve is just deep to the flexor retinaculum and is best visualized as an ovoid structure on transverse images. The ulnar nerve and superficial branches of the median nerve course superficial to the flexor retinaculum. Distal to the carpal tunnel, the flexor retinaculum becomes continuous with the palmar aponeurosis. The extensor tendons are observed on the back of the wrist, covered by the superficial extensor retinaculum. This fibrous band stretches obliquely from the anterior portion of the radius to the triquetrum and pisiform bones. The normal anatomy of the wrist and hand is depicted in Figures 2–8 (axial), 2–9 (sagittal), and 2–10 (coronal).

The Hip

The hip is a ball-and-socket joint capable of relatively restricted motion, with the ball approximated by the femoral head and the socket by the acetabulum. The acetabulum is a complex structure in which the roof is formed by the ilium, the anteromedial aspect is formed by the pubic bone, and the posteromedial portion is formed by the ischium, and all are joined at the triradiate cartilage. The acetabular labrum is a fibrocartilaginous addition affixed to the rim of the acetabulum. It serves both to protect the acetabulum by shielding it and to increase the stability of the joint by deepening the acetabular cavity.

There is a strong fibrous capsule, thicker superiorly and anteriorly, where resistance is greatest. It is further reinforced in this region by the iliofemoral, ischiofemoral, and pubofemoral ligaments. Superiorly the capsule is attached above the acetabular labrum to the transverse acetabular ligaments and the obturator foramen. Inferiorly it surrounds the femoral neck, attaching at the trochanteric rim anteriorly and just above the base of the lesser trochanter posteriorly. The synovial membrane is quite large. It lines the capsule from the femoral neck to the articular cartilage of the femoral head, covering both surfaces of the acetabular labrum, the ligament of the head of the femur, and the mass of fat contained within the acetabular fossa.

The iliofemoral (iliotrochanteric) ligament consists of strong, thick medial and lateral bands, with a thin, weak central band. It extends from the anterior inferior iliac spine to the trochanteric line of the femur. The ischiofemoral ligament attaches to the ischium, inferior and posterior to the acetabulum, and extends over and around the neck of the femur. The pubofemoral ligament stretches from the superior pubic ramus to merge with the fibrous capsule and the medial band of the iliofemoral ligament. The transverse acetabular ligament is a fibrous extension of the acetabular labrum, covering the incomplete inferior portion of the acetabulum. It is through here that vessels and nerves that maintain the hip joint can enter. The ligament of the head of the femur consists of two bands extending from the fovea capitis of the femoral head to either side of the acetabular notch. The normal anatomy of the hip is depicted in Figures 2–11 (axial), 2–12 (coronal), and 2–13 (sagittal).

The Knee

The knee is a complex and essentially double condylar joint. Although motion is basically restricted to a hinge-type motion in one plane, some rotation is allowed. The bony structures involved are the articular surfaces of the femoral condyles, the tibial plateaus, and the patella. The femur and tibia are separated by two crescentic cartilaginous menisci, which cover approximately two-thirds of each tibial plateau. The patella is secured anteriorly by the quadriceps tendon superiorly, which blends into the patella tendon inferiorly.

The medial meniscus is C-shaped, thicker posteriorly than anteriorly, and firmly anchored to the medial tibial plateau. The lateral meniscus is more nearly a complete ring and more uniform in thickness. The posterior portion of the lateral meniscus is attached to the inferior fascicle of the popliteus tendon (which interposes between the meniscus and the fibular collateral ligament) and the ligament of Wrisberg (inserting on the lateral aspect of the medial femoral condyle). It is not firmly anchored to the tibial plateau, permitting a small amount of rotational mobility. Both menisci are connected anteriorly by the transverse ligament.

The posterior cruciate ligament is stronger and thicker than its anterior counterpart, extending from the lateral surface of the medial femoral condyle to the posterior intercondylar area of the tibia. The anterior cruciate ligament is a series of fibers connecting the medial part of the anterior intercondylar aspect of the tibia with the medial surface of the lateral femoral condyle. These ligaments help to maintain the integrity of the joint by keeping the femur and tibia in relatively close proximity. There is a bursa between the two ligaments.

The medial and lateral collateral ligaments aid in maintaining lateral stability of the knee. The medial collateral ligament is composed of two parts, the more superficial tibial collateral ligament, a broad band extending from the medial epicondyle of the femur to the medial condyle and shaft of the tibia, and the deeper medial capsular ligament, which is a thickening of the fibrous capsule, extending from the medial femoral epicondyle to the border of the medial meniscus. The two structures are separated by bursae. The lateral collateral ligament is a complex collection of fibrous structures, most importantly the iliotibial band and the fibular collateral ligament. The iliotibial band is an extension of the deep fascia separating the muscles of the leg and inserts on the anterolateral aspect of the tibia, blending with the fibrous capsule. The fibular collateral ligament, found posteriorly, is an extracapsular structure extending from the apex of the head of the fibula to the lateral femoral epicondyle.

The course of the synovial membrane is quite complex. From the superior border of the patella, it forms the suprapatellar bursa, just posterior to the quadriceps tendon, and is actually a part of the joint space. From the inferior part of the patella, it lines the Hoffa fat pad (infrapatellar fat pad). The free surfaces of the menisci are not covered by synovium. Multiple bursae surround the joint. Normal anatomy of the knee is represented in the axial plane in Figure 2–14, in the sagittal plane in Figure 2–15, and in the coronal plane in Figure 2–16.

The Ankle

The ankle is an extremely complex area. It technically is the talocrural joint; however, actually it is intimately related to several distinct joints: the inferior tibiofibular joint, the talocalcanceal or subtalar joint, and the talocalcaneonavicular joint. The exact separation between ankle and foot is not always clear, especially since most of the structures of interest—tendons, nerves, and vessels—traverse both regions.

Anterior to the actual joint pass (medial to lateral) the saphenous nerve, great saphenous vein, tibialis anterior tendon, extensor hallucis longus tendon, anterior tibial vessels, superficial and deep peroneal nerves, and the tendons of the extensor digitorum longus and peroneus tertius. Posterior to the medial malleolus lies the tibialis posterior tendon, the flexor digitorum tendon, posterior tibial vessels, tibial nerve, and flexor hallucis longus tendon. Posterior to the lateral malleolus extend the peroneus longus and brevis tendons. These posterior tendons course under the respective malleoli and pass anteriorly in the foot to insert as follows: The tibialis posterior insertions are multiple and varied and include the anterior part of the calcaneus, the navicular, the cuboid, and the second, third, and fourth metatarsals; the flexor digitorum continues across the sole of the foot to insert on the distal phalanx of the fifth toe, and contributes to the tendons inserting on the second, third, and fourth metatarsals; the flexor hallucis longus inserts on the base of the distal phalanx of the great toe; the peroneus brevis inserts on the base of the fifth metatarsal; and the peroneus longus inserts on the base of the first metatarsal. The Achilles tendon extends from the gastrocnemius and soleus muscles to insert on the posterior aspect of the calcaneus.

The true ankle joint, or talocrural joint, is a uniaxial hinge-type joint. The distal part of the tibia and medial malleolus and distal part of the fibula and lateral malleolus form a concavity into which fits the talus, the proximal end of which is convex. The bony surfaces involved are covered by articular cartilage and are connected by the medial (deltoid) and lateral (calcaneofibular and anterior and posterior talofibular) collateral ligaments.

The deltoid ligament is a thick, multicomponent structure that attaches along the posterior and anterior surfaces of the medial malleolus and extends to insert on the navicular, talus, and calcaneus. The tibialis posterior and flexor digitorum longus tendons cross medial to the deltoid ligament. The anterior and posterior talofibular ligaments extend from their respective surfaces on the lateral malleolus to the lateral articular facet (anteriorly) and the posterior process of the talus. The calcaneofibular ligament extends from the posterior portion of the lateral malleolus posteriorly and inferiorly to the posterolateral surface of the calcaneus. The peroneus longus and brevis tendons course superficial to it. There is a thin fibrous capsule that is strengthened by the medial and lateral collateral ligaments. A synovial membrane lines the fibrous capsule and extends a short distance between the tibia and fibula.

The inferior tibiofibular joint is a fibrous joint, or syndesmosis, in which the bones are held closely together by an interosseous ligament. Only minimal movement is permitted—a slight lateral rotation of the fibula when the ankle is dorsiflexed. The normal anatomy of the ankle and foot is illustrated in Figures 2–17 (sagittal) and 2–18 (axial).

BIBLIOGRAPHY

Clemente CD. Anatomy: a regional atlas of the human body. Baltimore: Urban & Schwarzenberg, 1981.

Grant JCB. Grant's atlas of anatomy. New York: Williams & Wilkins, 1972.

Koritke JG, Sick H. Atlas of sectional human anatomy. Baltimore: Urban & Schwarzenberg, 1988.

Snell RS. Clinical anatomy for medical students. Boston: Little, Brown, 1973.

Warwick R, Williams PL. Gray's anatomy. 35th British ed. Philadelphia: Saunders, 1973.

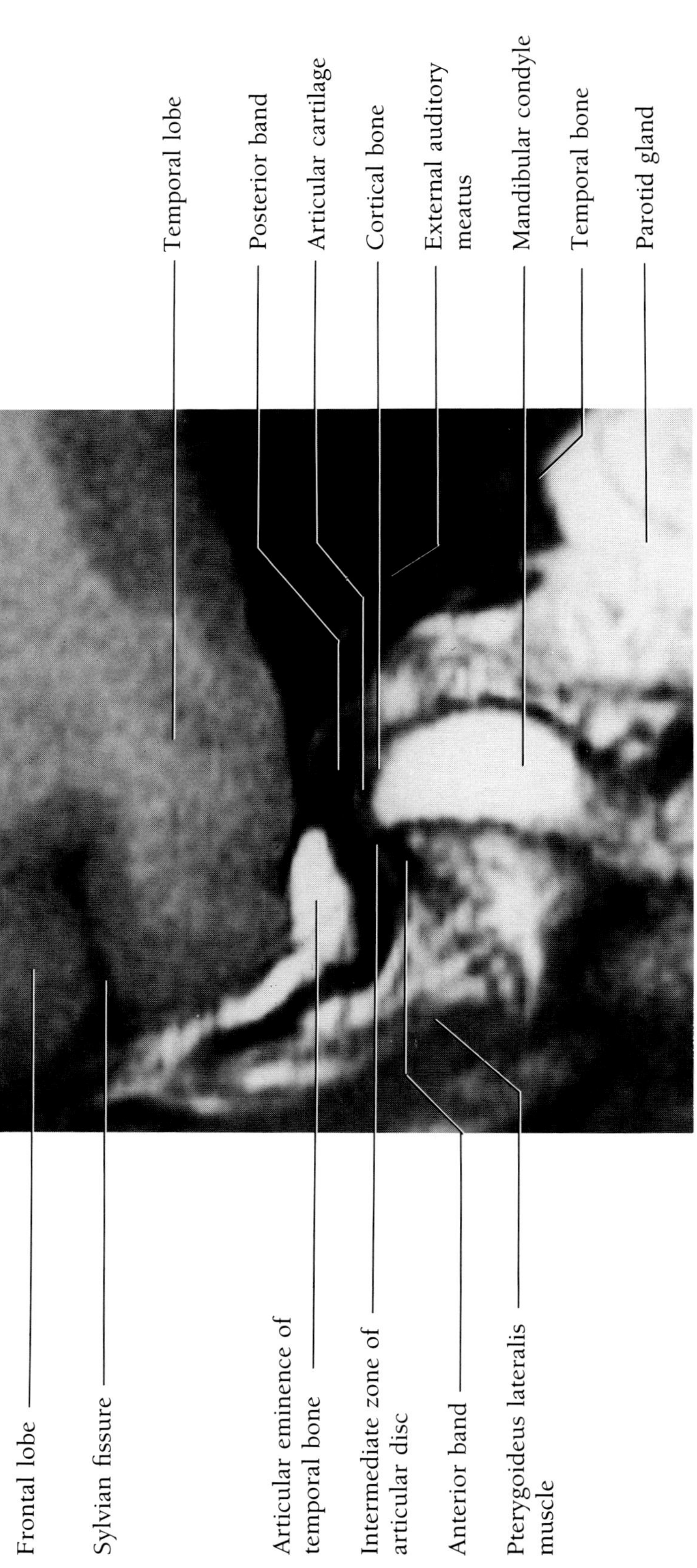

FIGURE 2–1(A)
The normal temporomandibular joint, viewed in sagittal section: Closed mouth view.

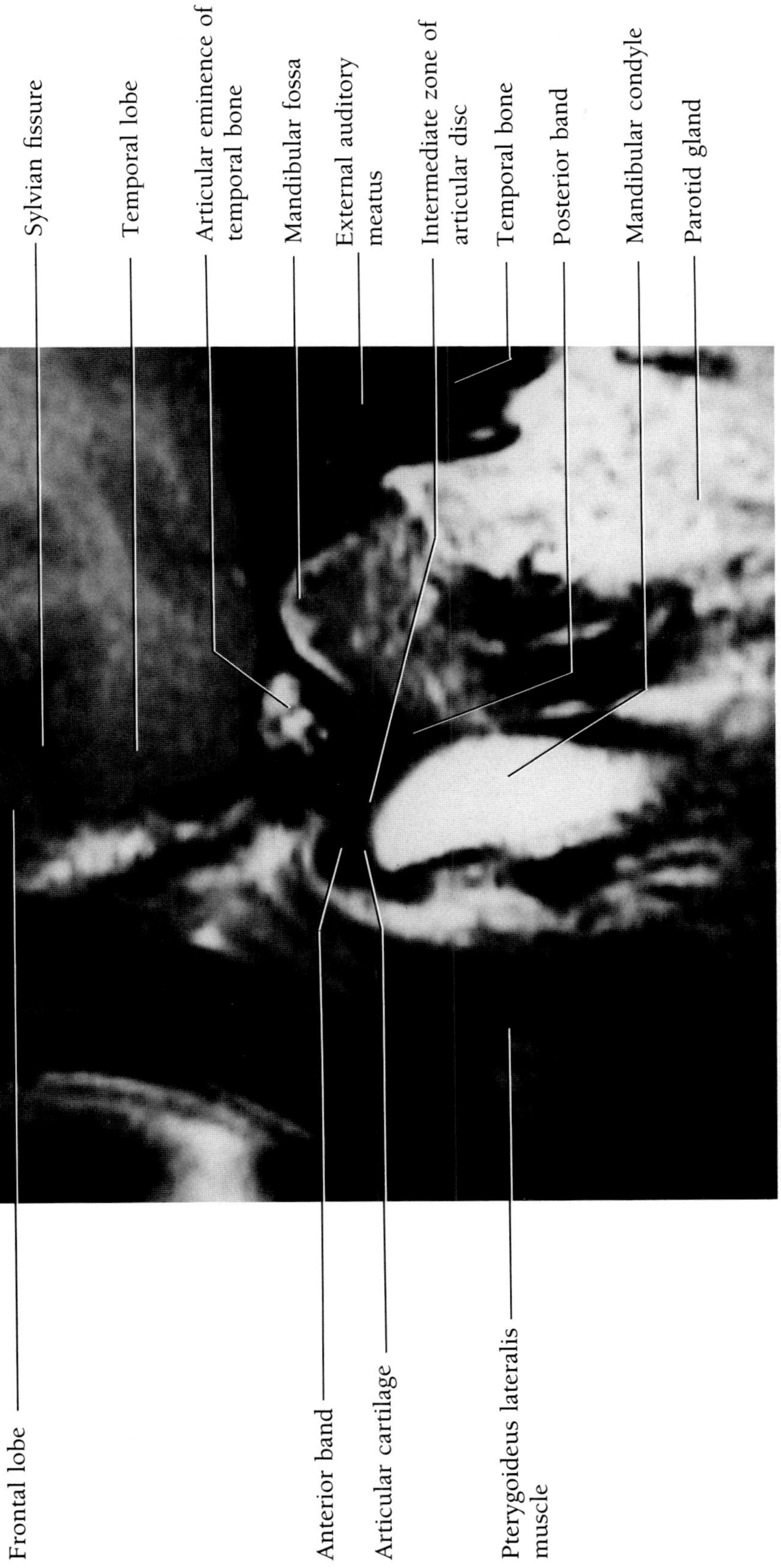

FIGURE 2–1(B)
The normal temporomandibular joint, viewed in sagittal section: Open mouth view.

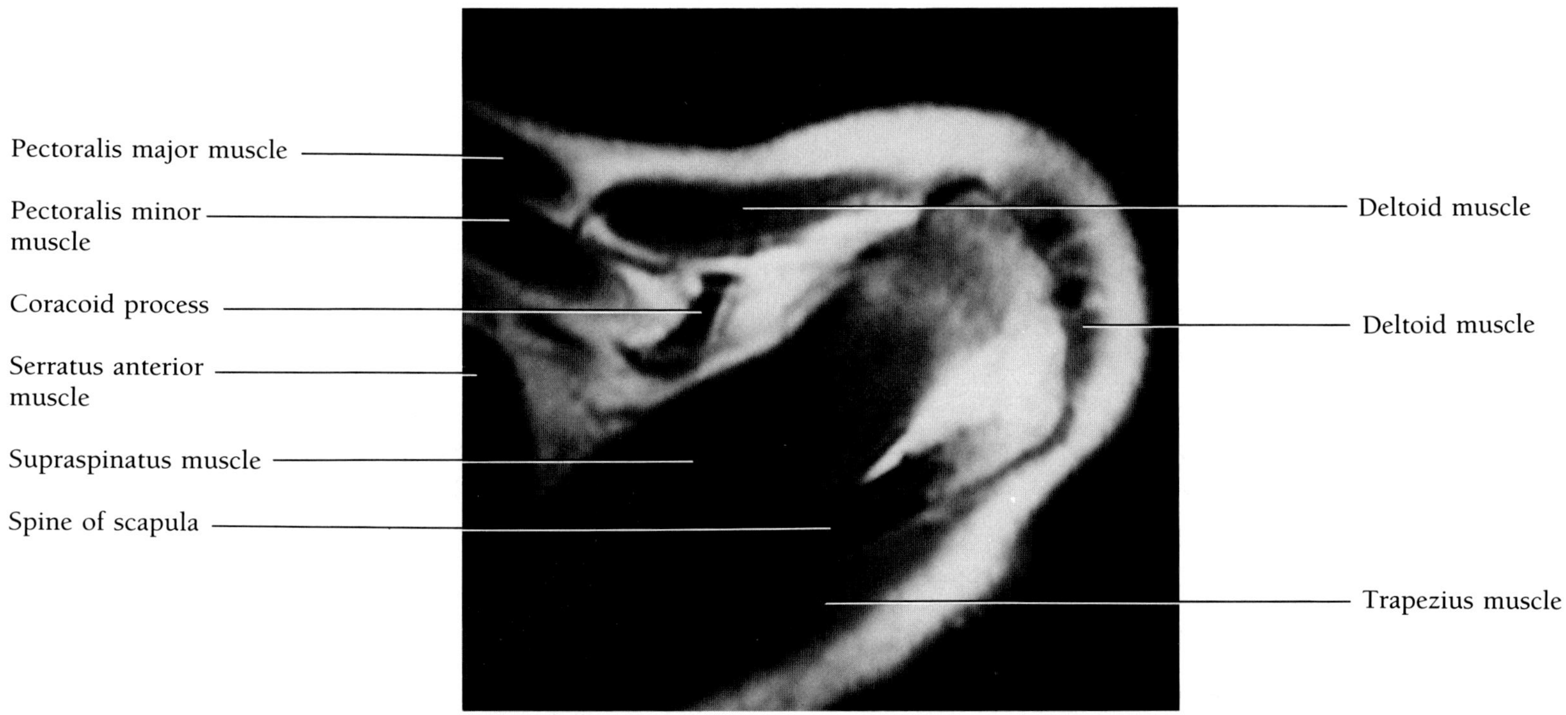

A

FIGURE 2–2
Axial images of the normal shoulder. A–E. *Through the plane of the supraspinatus muscle, successively inferior through the joint to the proximal part of the humeral shaft.*

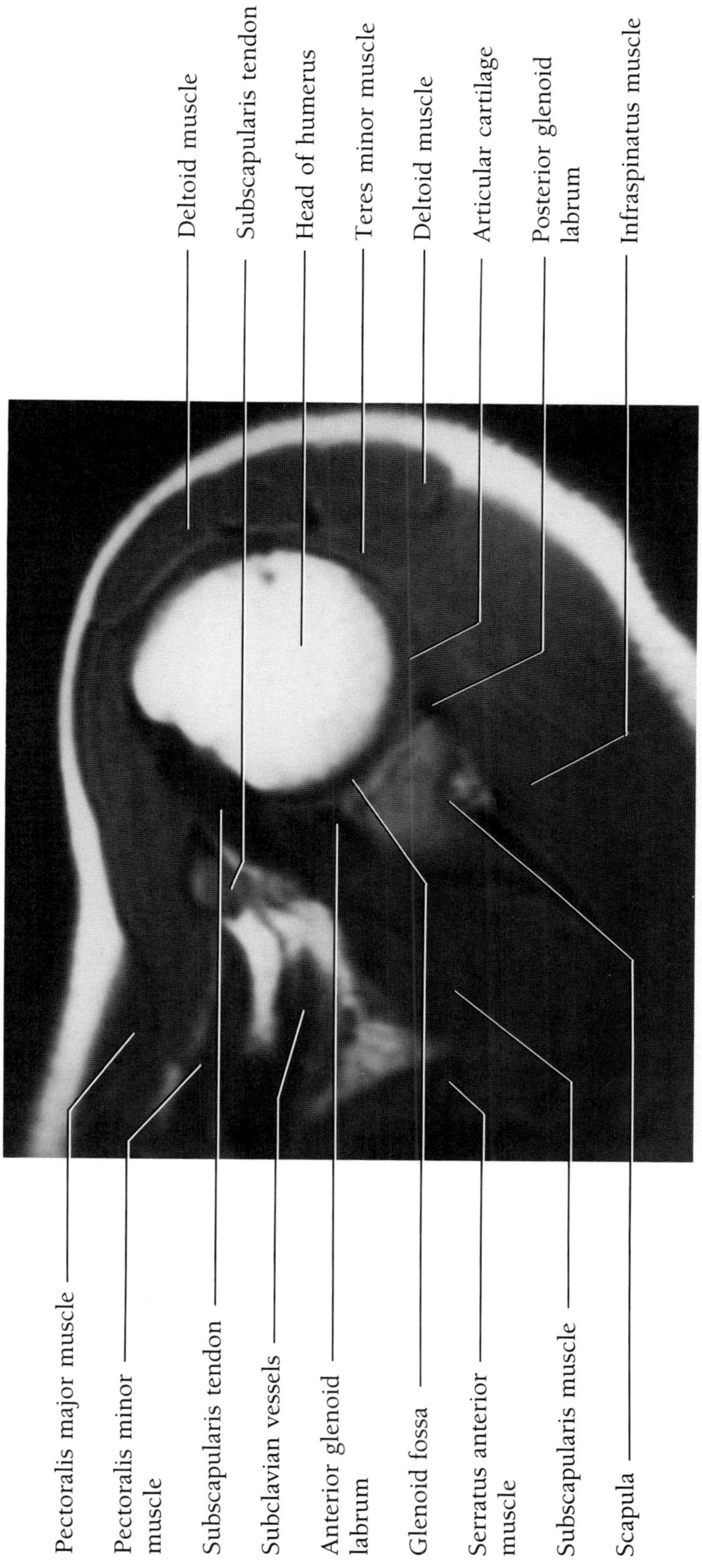
Pectoralis major muscle
Pectoralis minor muscle
Subscapularis tendon
Subclavian vessels
Anterior glenoid labrum
Glenoid fossa
Serratus anterior muscle
Subscapularis muscle
Scapula
Deltoid muscle
Subscapularis tendon
Head of humerus
Teres minor muscle
Deltoid muscle
Articular cartilage
Posterior glenoid labrum
Infraspinatus muscle

FIGURE 2–2(B)

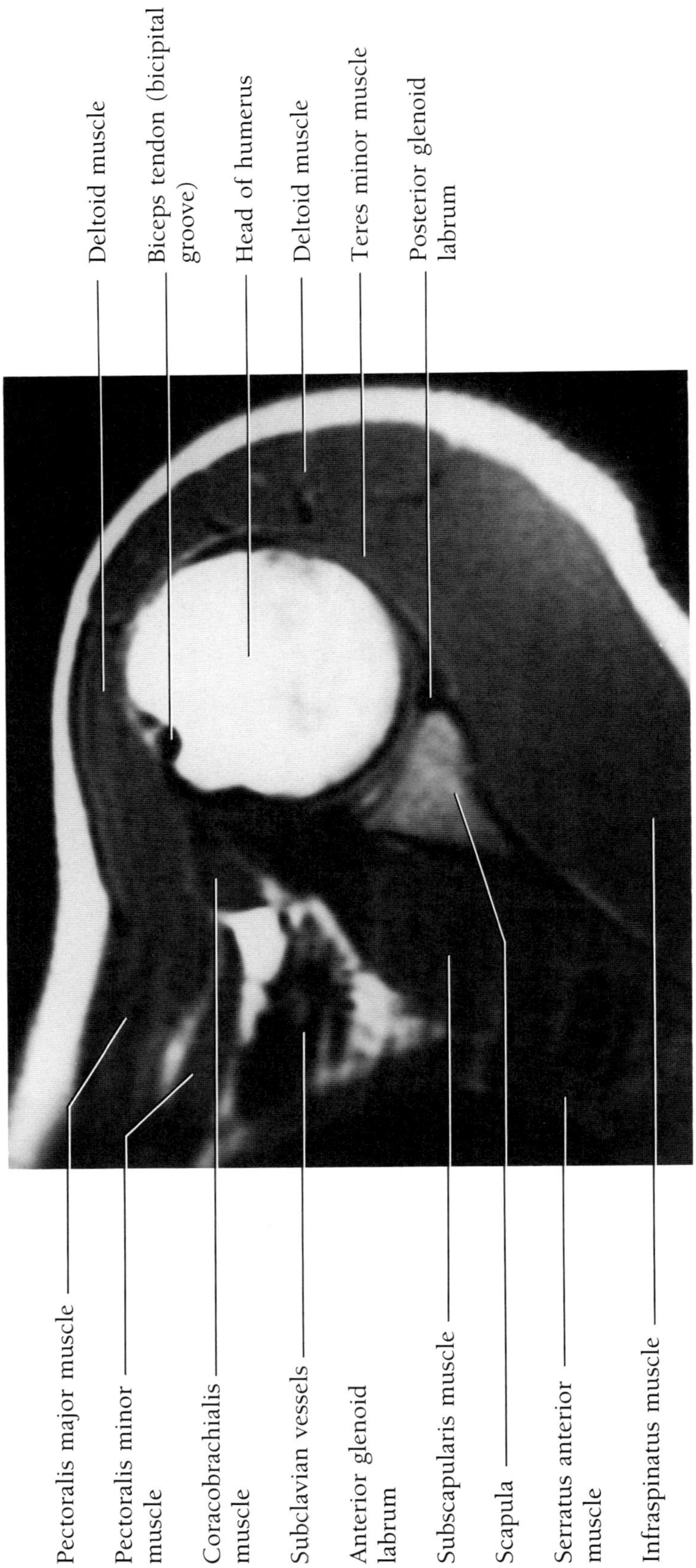
Deltoid muscle
Biceps tendon (bicipital groove)
Head of humerus
Deltoid muscle
Teres minor muscle
Posterior glenoid labrum
Pectoralis major muscle
Pectoralis minor muscle
Coracobrachialis muscle
Subclavian vessels
Anterior glenoid labrum
Subscapularis muscle
Scapula
Serratus anterior muscle
Infraspinatus muscle

FIGURE 2–2(C)

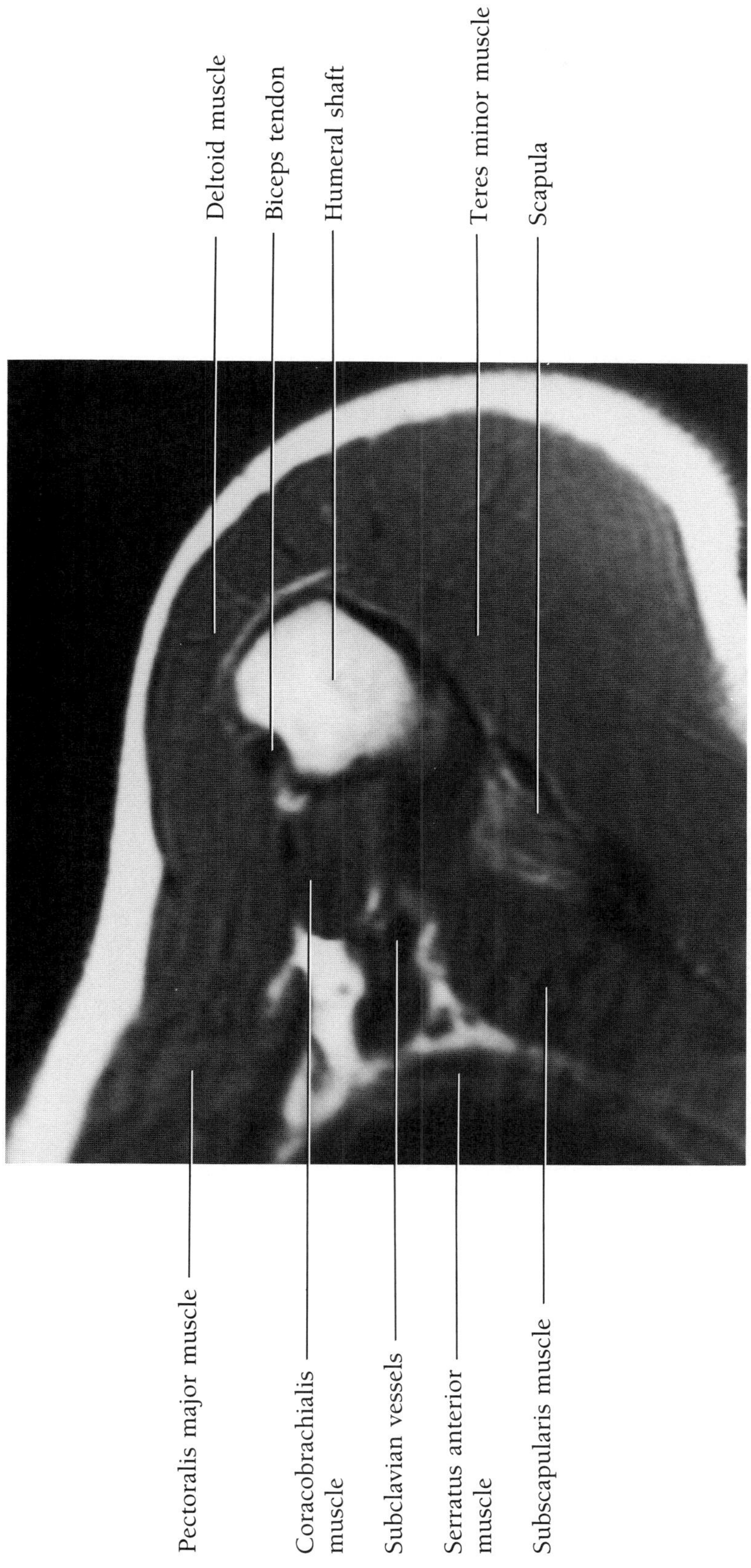
Deltoid muscle
Biceps tendon
Humeral shaft
Teres minor muscle
Scapula
Pectoralis major muscle
Coracobrachialis muscle
Subclavian vessels
Serratus anterior muscle
Subscapularis muscle

FIGURE 2–2(D)

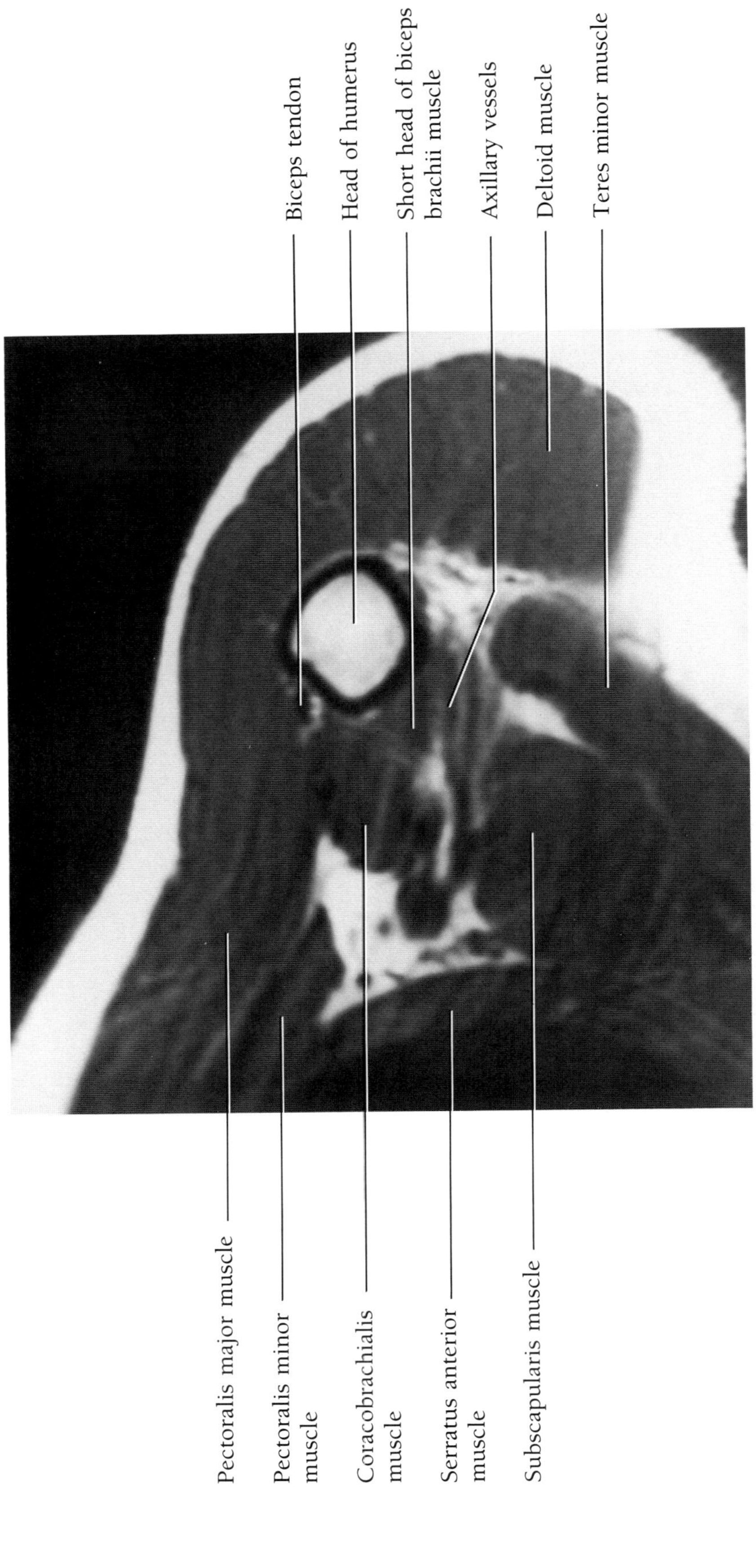
Biceps tendon
Head of humerus
Short head of biceps brachii muscle
Axillary vessels
Deltoid muscle
Teres minor muscle
Pectoralis major muscle
Pectoralis minor muscle
Coracobrachialis muscle
Serratus anterior muscle
Subscapularis muscle

FIGURE 2–2(E)

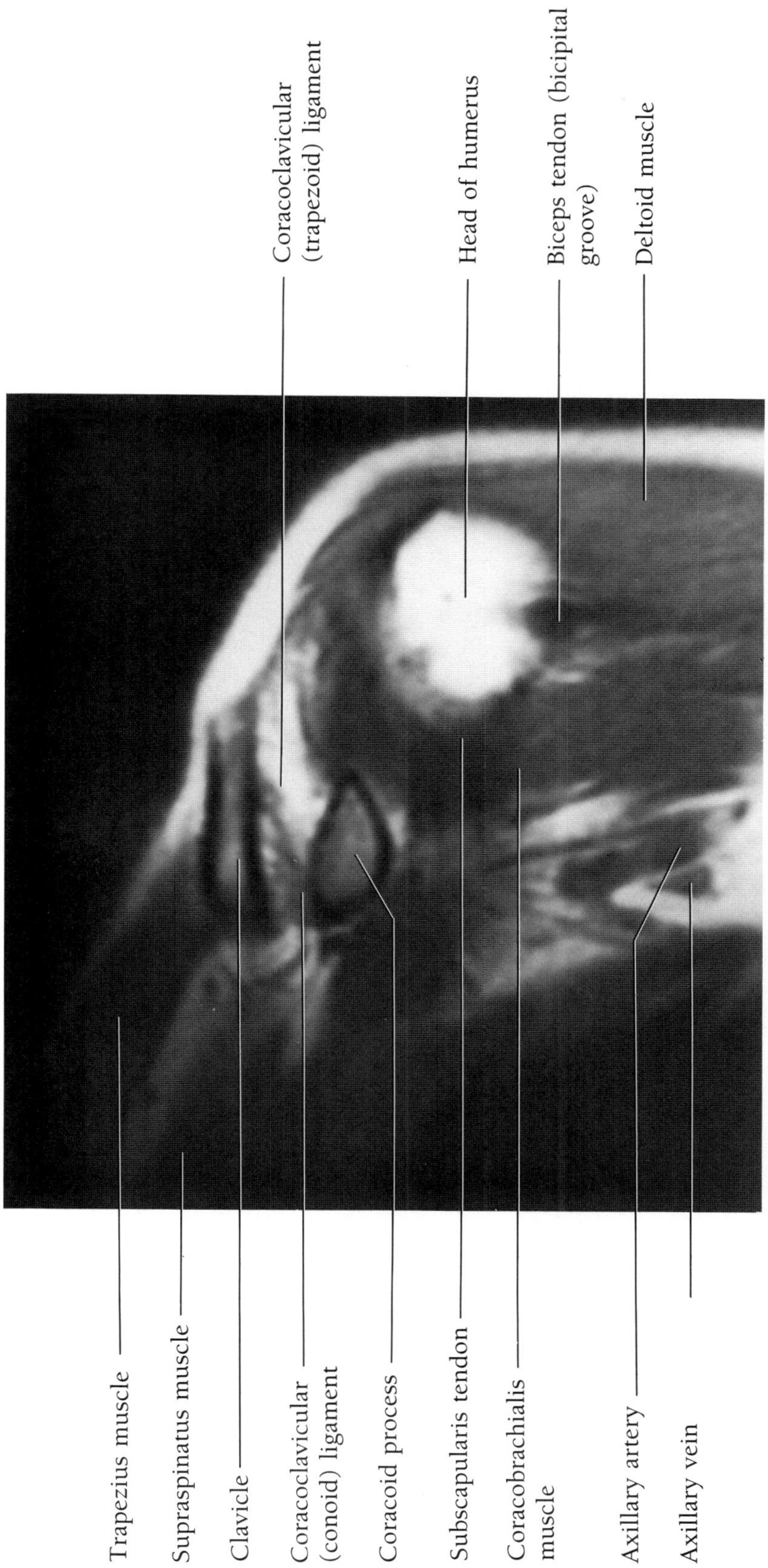

A

FIGURE 2–3
Paracoronal (oblique) images of the normal shoulder, oriented parallel to the supraspinatus muscle and tendon. A–E. *The most anterior section successively to the most posterior section.*

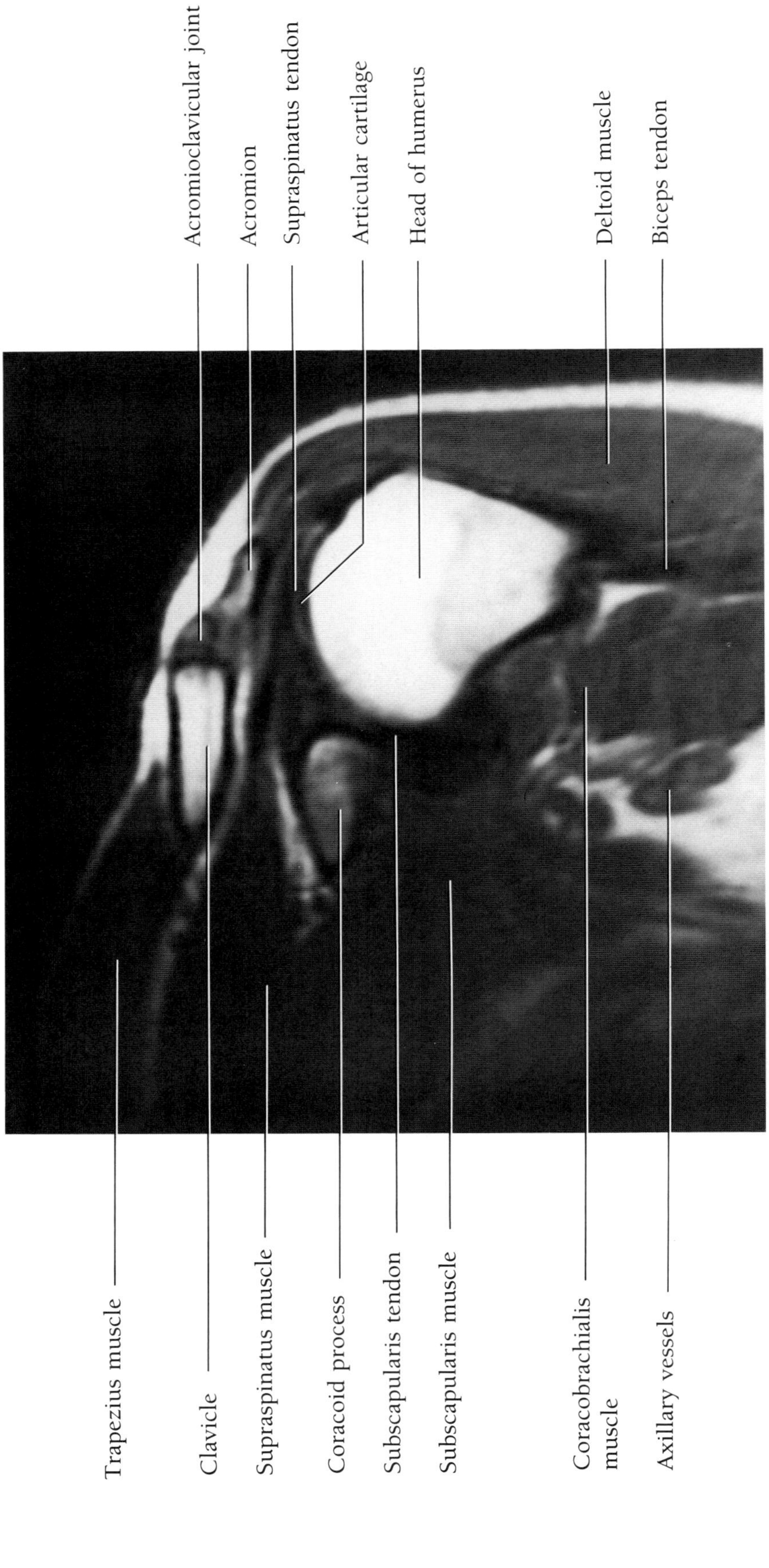
Acromioclavicular joint
Acromion
Supraspinatus tendon
Articular cartilage
Head of humerus
Deltoid muscle
Biceps tendon
Trapezius muscle
Clavicle
Supraspinatus muscle
Coracoid process
Subscapularis tendon
Subscapularis muscle
Coracobrachialis muscle
Axillary vessels

FIGURE 2–3(B)

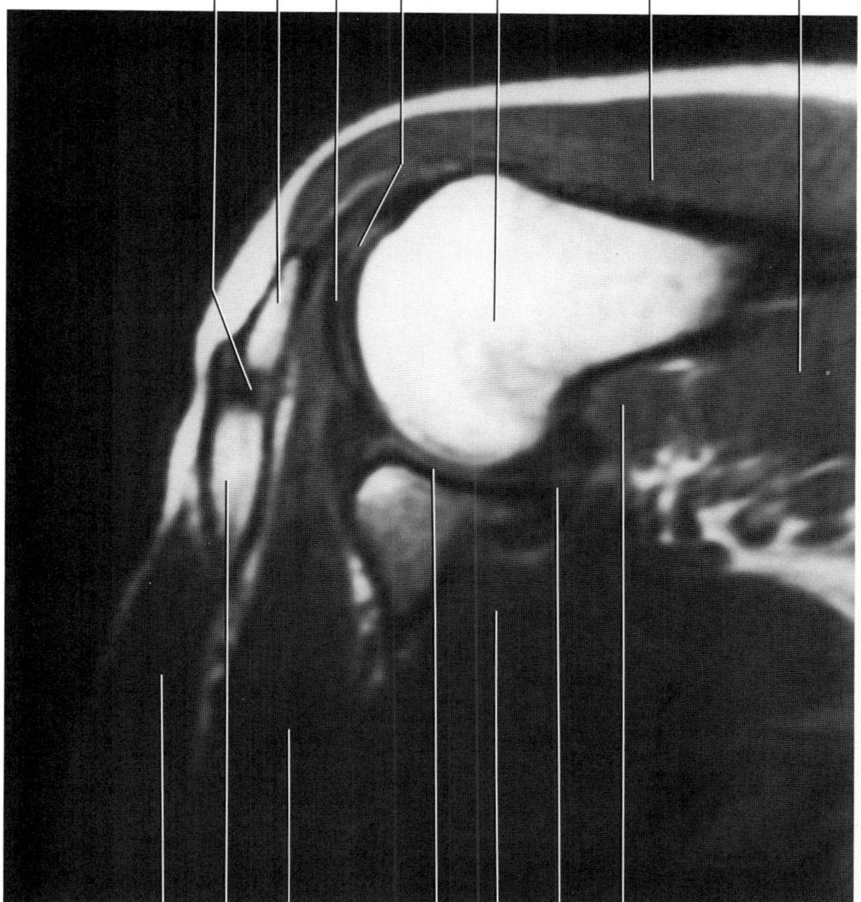
Trapezius muscle
Clavicle
Supraspinatus muscle
Glenoid fossa
Subscapularis muscle
Inferior labrum
Subscapularis muscle
Acromioclavicular joint
Acromion
Supraspinatus tendon
Articular cartilage
Head of humerus
Deltoid muscle
Coracobrachialis muscle

FIGURE 2–3(C)

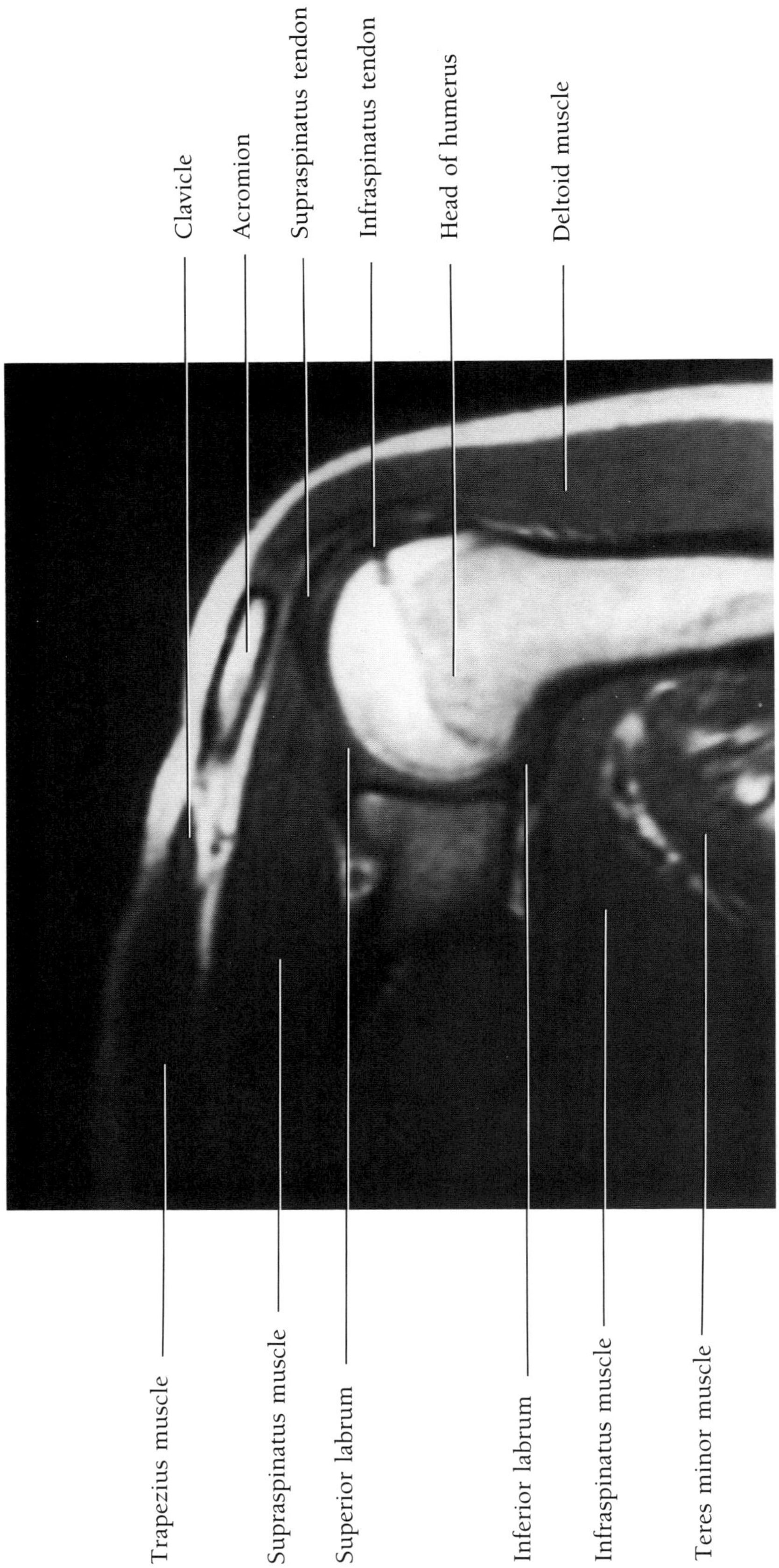
Clavicle
Acromion
Supraspinatus tendon
Infraspinatus tendon
Head of humerus
Deltoid muscle
Trapezius muscle
Supraspinatus muscle
Superior labrum
Inferior labrum
Infraspinatus muscle
Teres minor muscle

FIGURE 2–3(D)

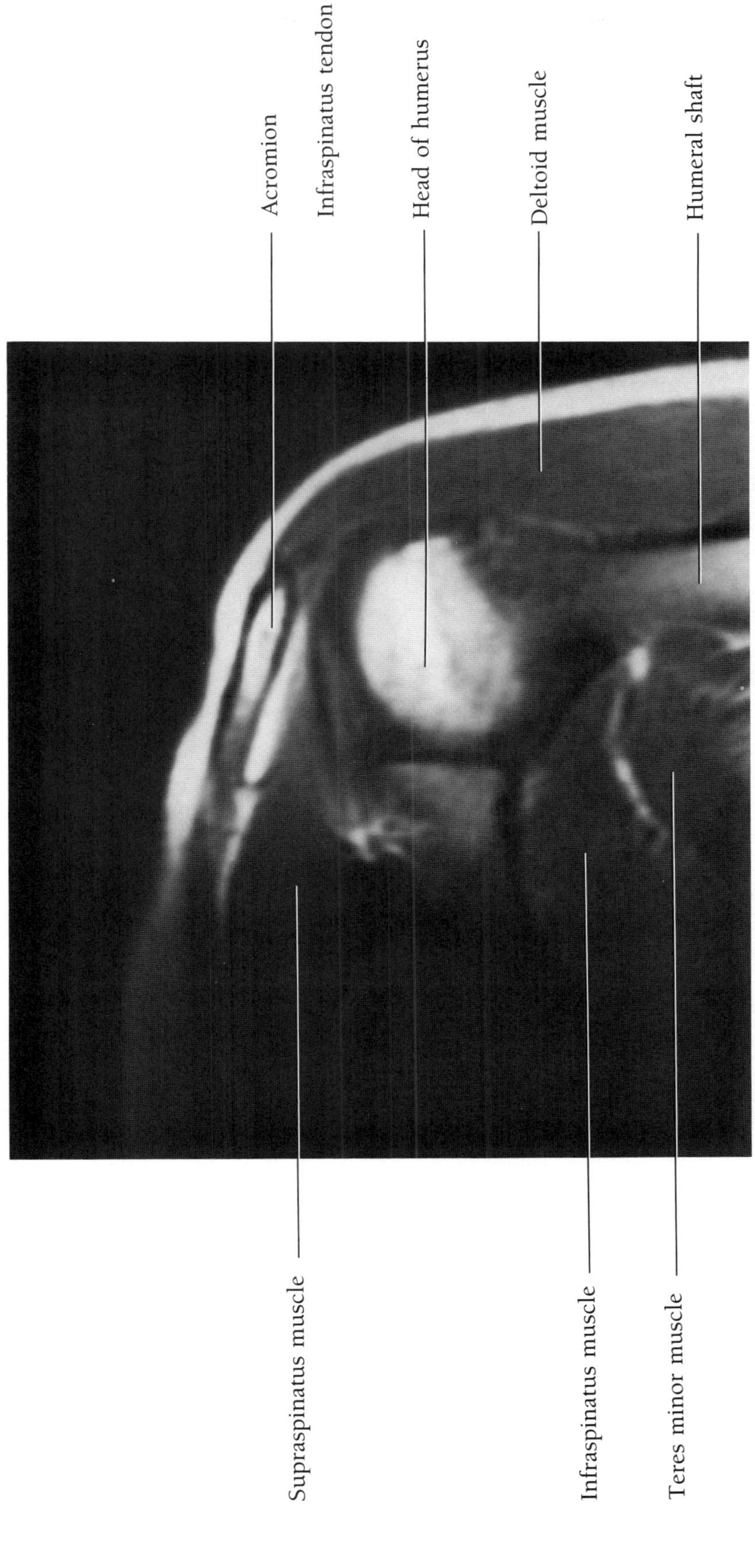
Acromion
Infraspinatus tendon
Head of humerus
Deltoid muscle
Humeral shaft
Supraspinatus muscle
Infraspinatus muscle
Teres minor muscle

FIGURE 2–3(E)

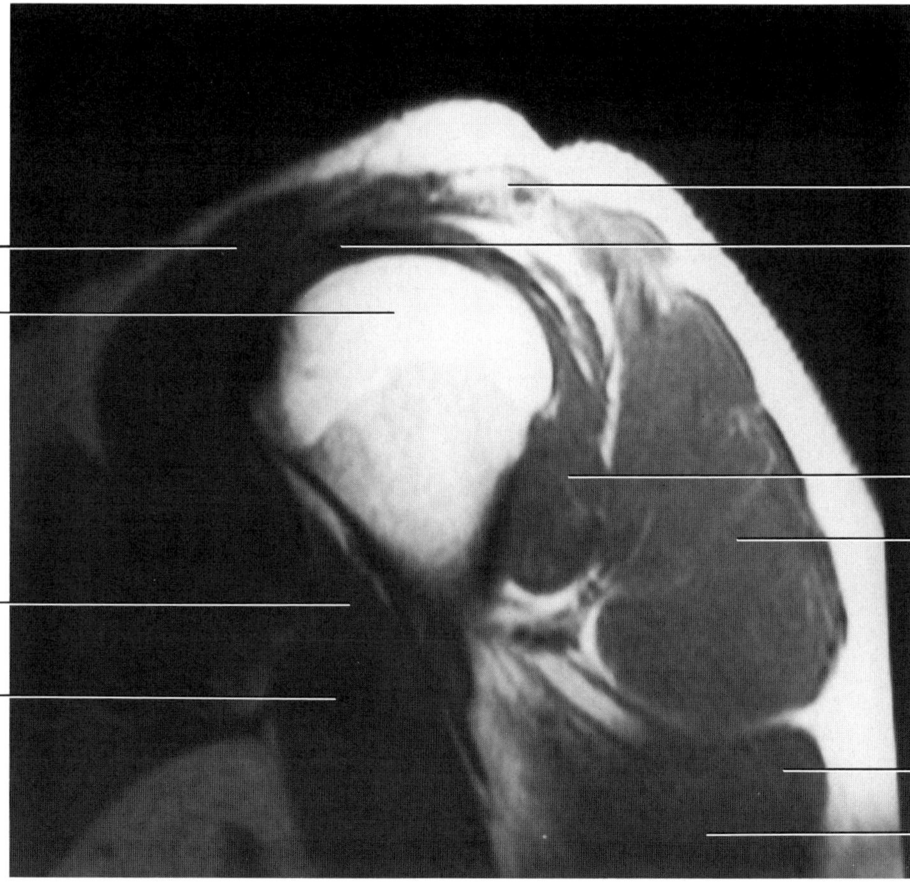

A

FIGURE 2–4
Parasagittal images of the normal shoulder, oriented perpendicular to the oblique images of Figure 2–3. A–F. *The most lateral section successively to the most medial section.*

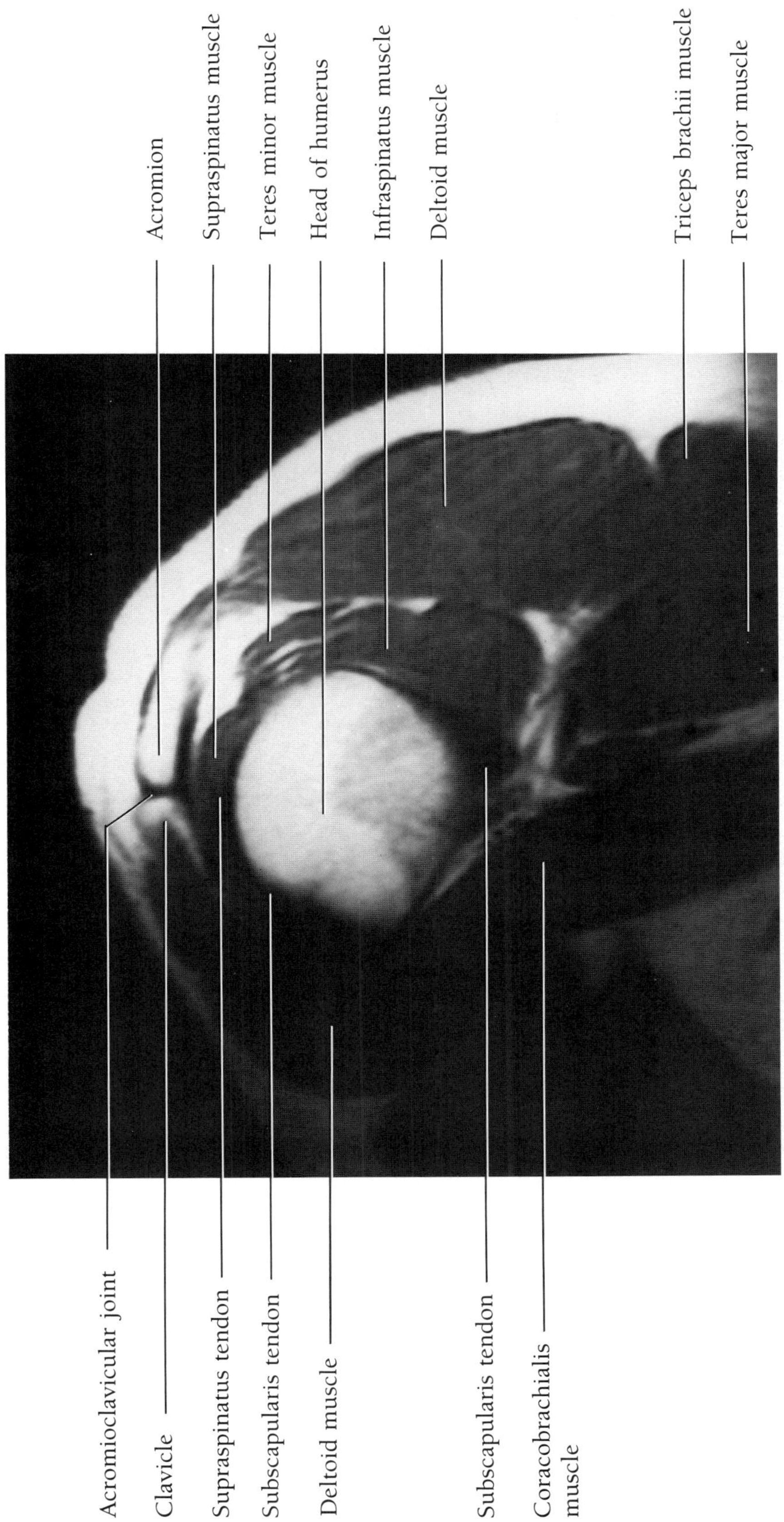
Acromion
Supraspinatus muscle
Teres minor muscle
Head of humerus
Infraspinatus muscle
Deltoid muscle
Triceps brachii muscle
Teres major muscle
Acromioclavicular joint
Clavicle
Supraspinatus tendon
Subscapularis tendon
Deltoid muscle
Subscapularis tendon
Coracobrachialis muscle

FIGURE 2–4(B)

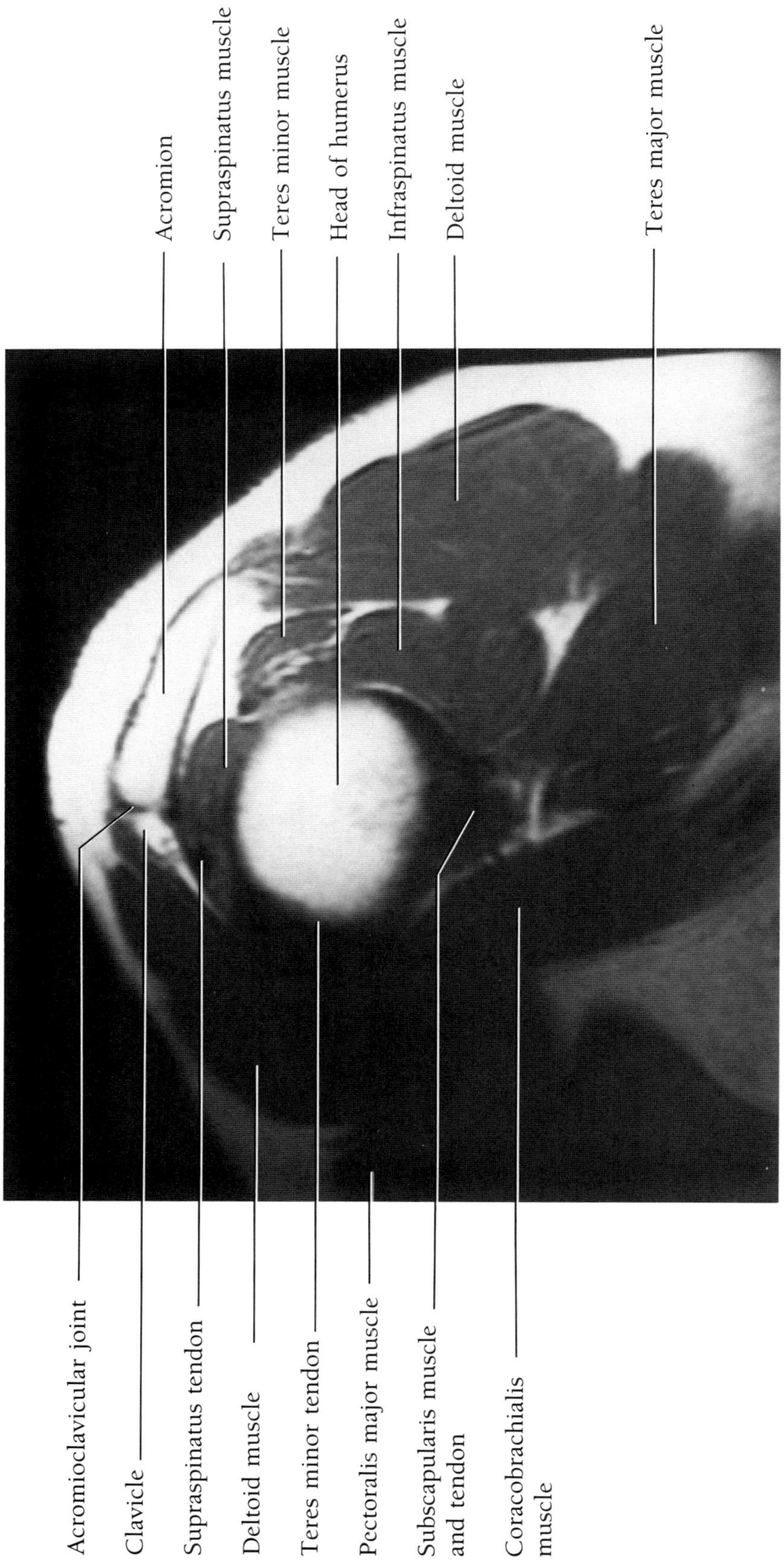
Acromion
Supraspinatus muscle
Teres minor muscle
Head of humerus
Infraspinatus muscle
Deltoid muscle
Teres major muscle
Acromioclavicular joint
Clavicle
Supraspinatus tendon
Deltoid muscle
Teres minor tendon
Pectoralis major muscle
Subscapularis muscle and tendon
Coracobrachialis muscle

FIGURE 2–4(C)

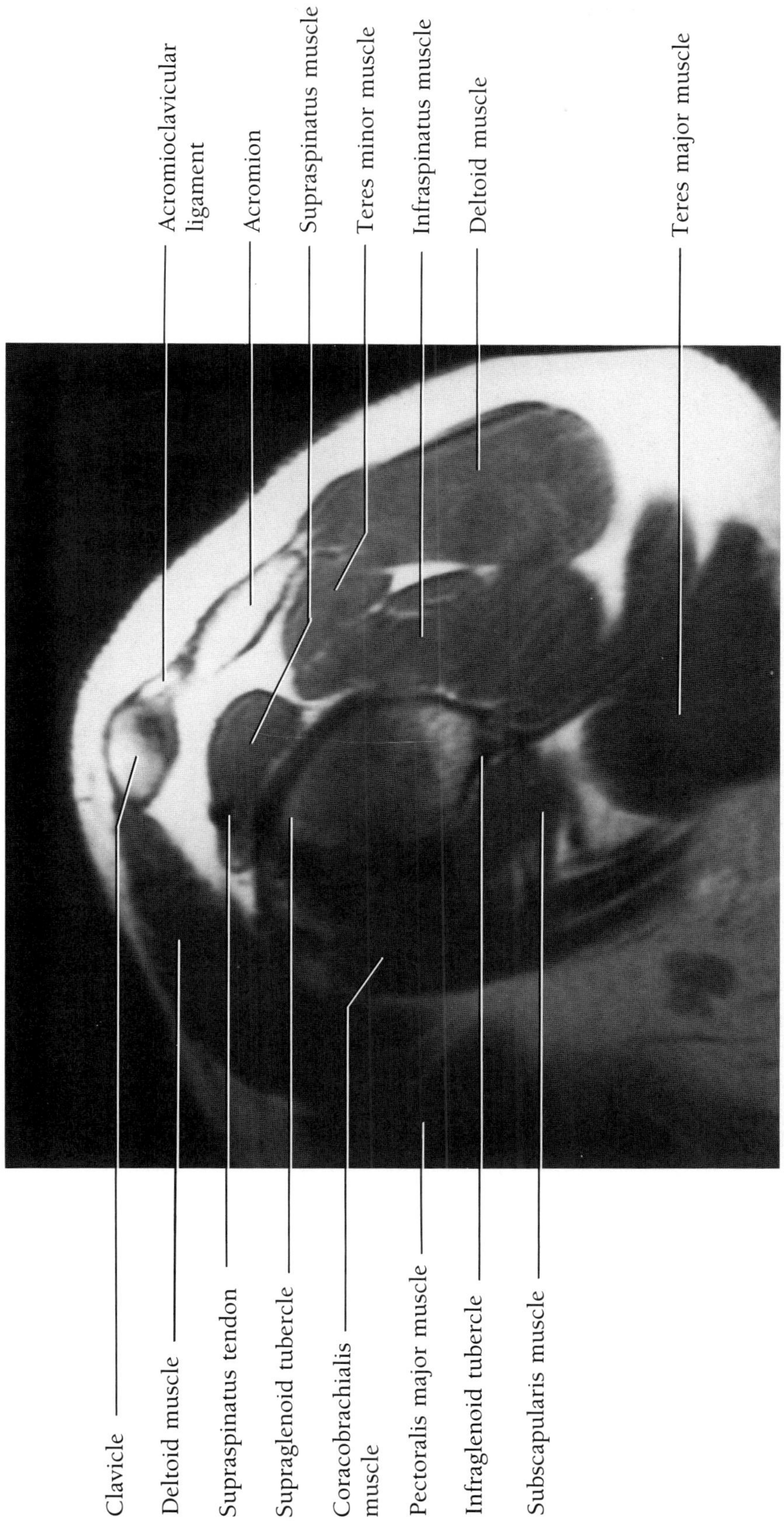
Acromioclavicular ligament
Acromion
Supraspinatus muscle
Teres minor muscle
Infraspinatus muscle
Deltoid muscle
Teres major muscle
Clavicle
Deltoid muscle
Supraspinatus tendon
Supraglenoid tubercle
Coracobrachialis muscle
Pectoralis major muscle
Infraglenoid tubercle
Subscapularis muscle

FIGURE 2–4(D)

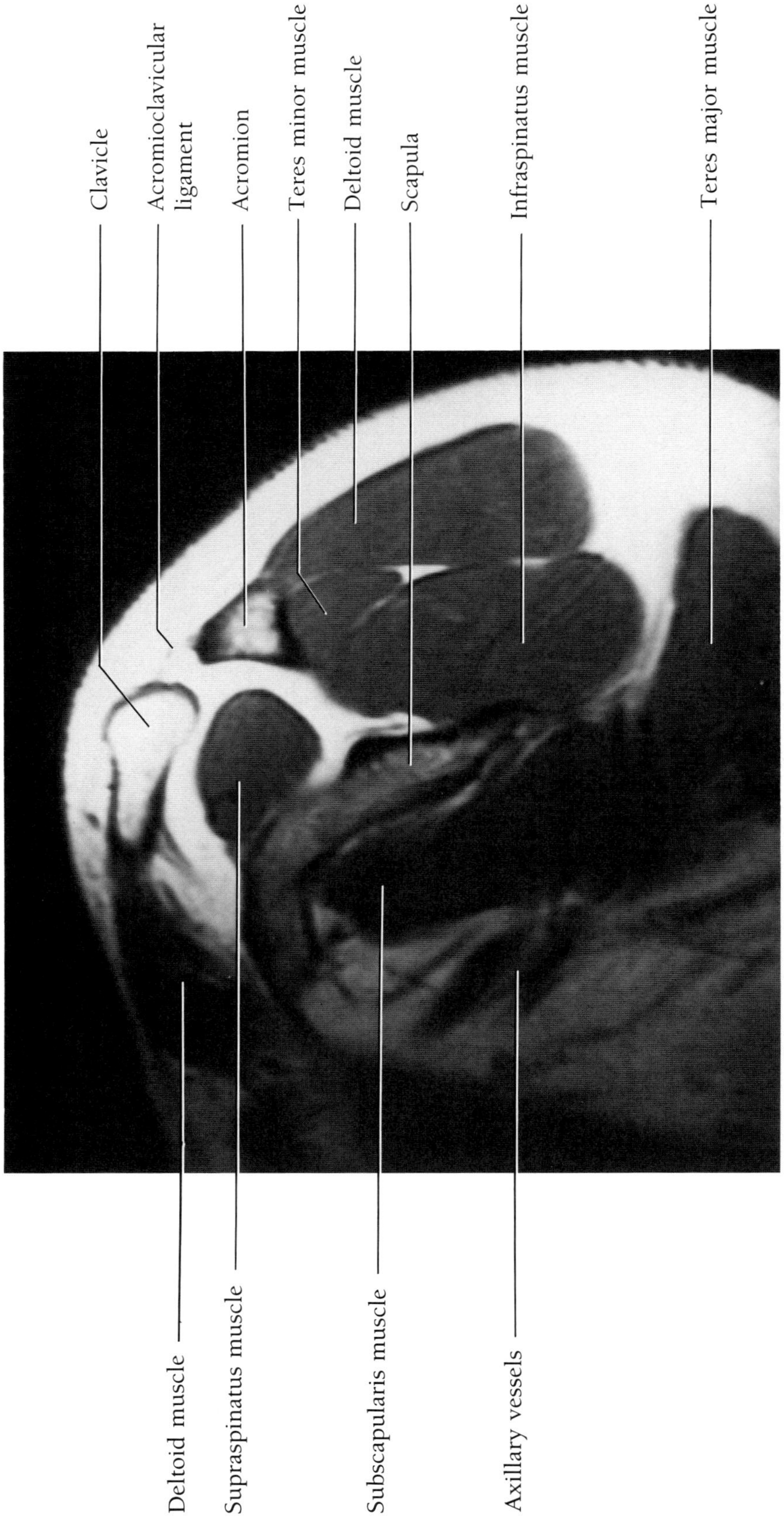
Clavicle
Acromioclavicular ligament
Acromion
Teres minor muscle
Deltoid muscle
Scapula
Infraspinatus muscle
Teres major muscle
Deltoid muscle
Supraspinatus muscle
Subscapularis muscle
Axillary vessels

FIGURE 2–4(E)

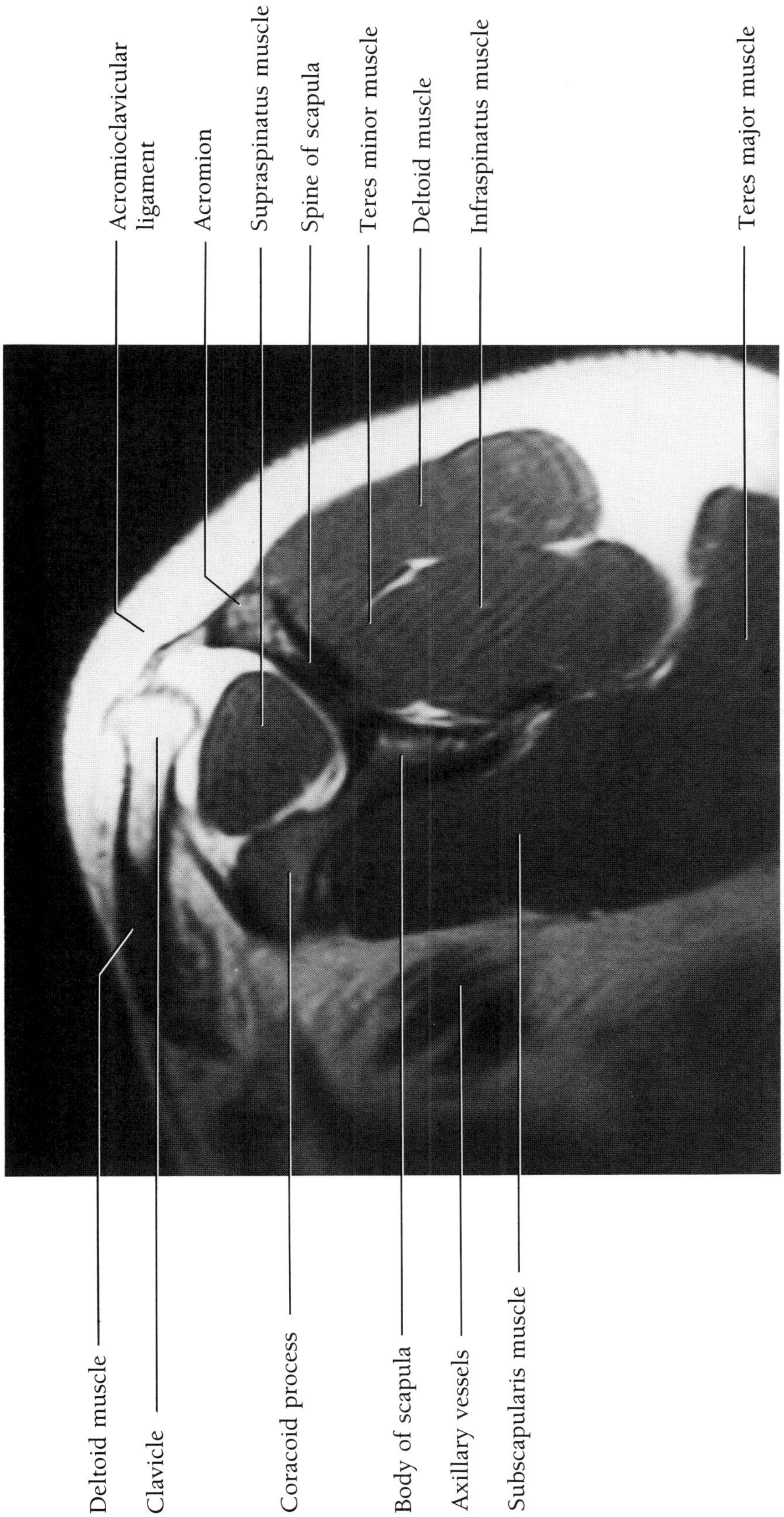
Acromioclavicular ligament
Acromion
Supraspinatus muscle
Spine of scapula
Teres minor muscle
Deltoid muscle
Infraspinatus muscle
Teres major muscle
Deltoid muscle
Clavicle
Coracoid process
Body of scapula
Axillary vessels
Subscapularis muscle

FIGURE 2–4(F)

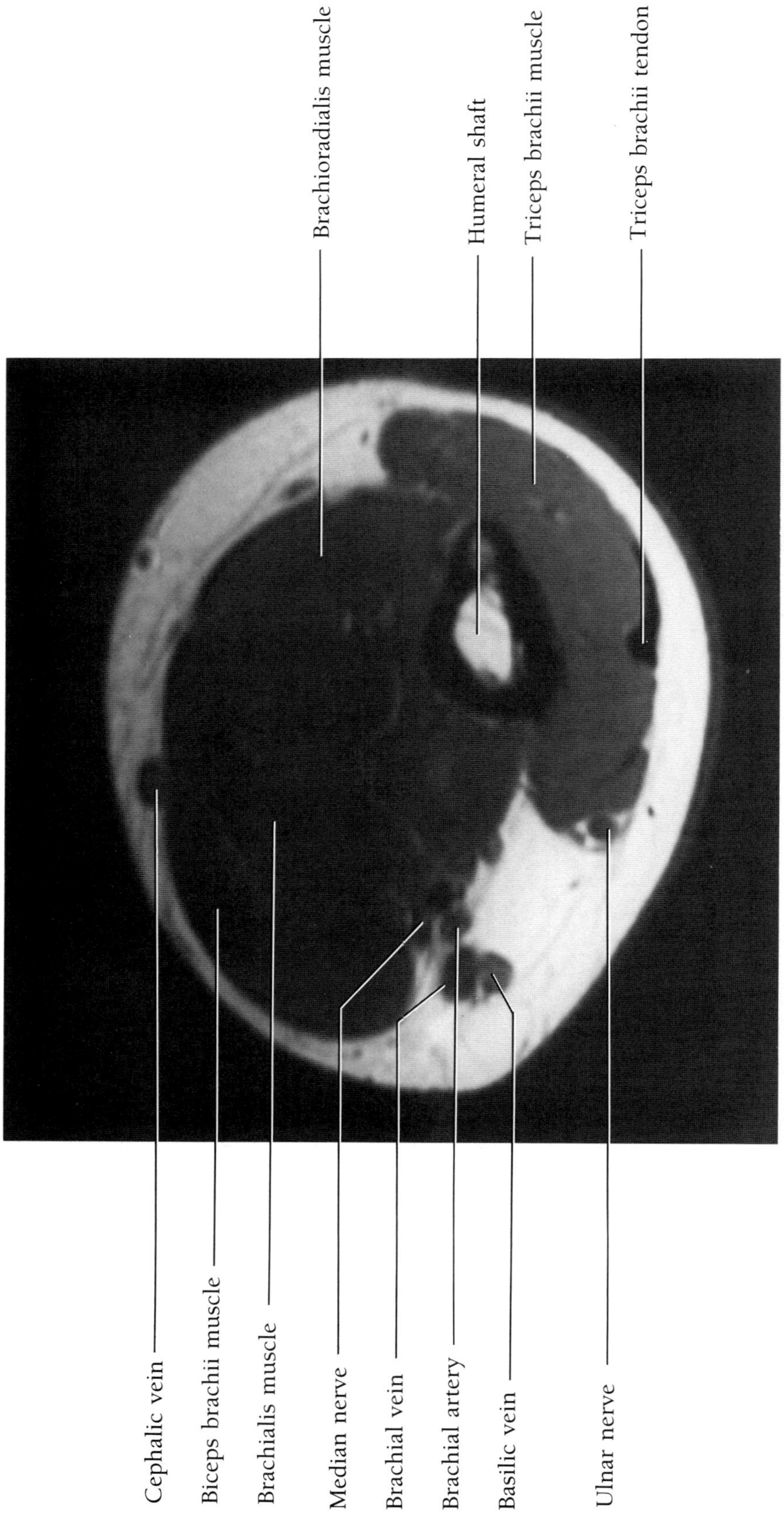

FIGURE 2–5(A)
Axial image of the normal elbow: Through the distal part of the humeral shaft.

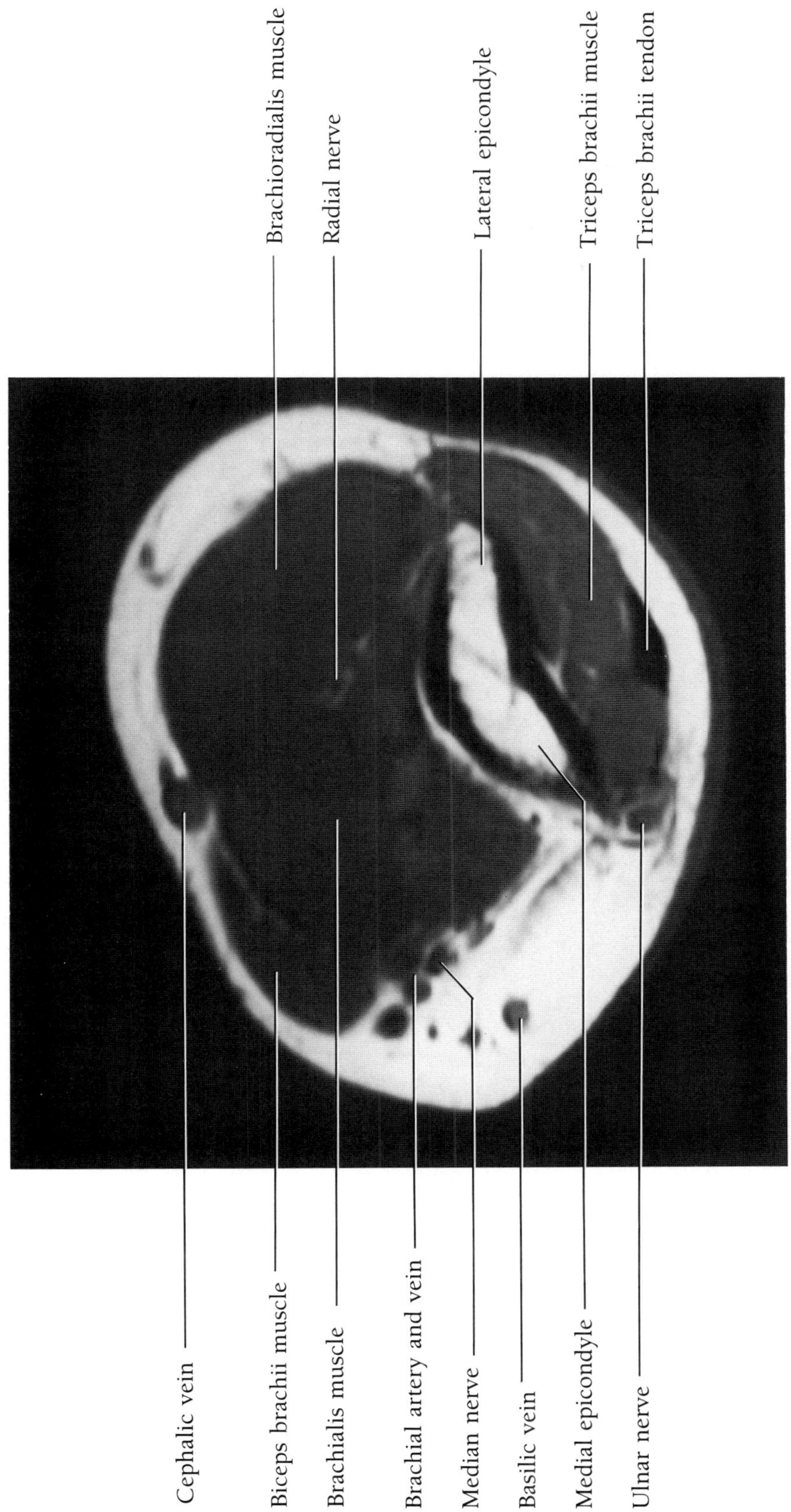

FIGURE 2–5(B)
Axial image of the normal elbow: The humeral epicondyles.

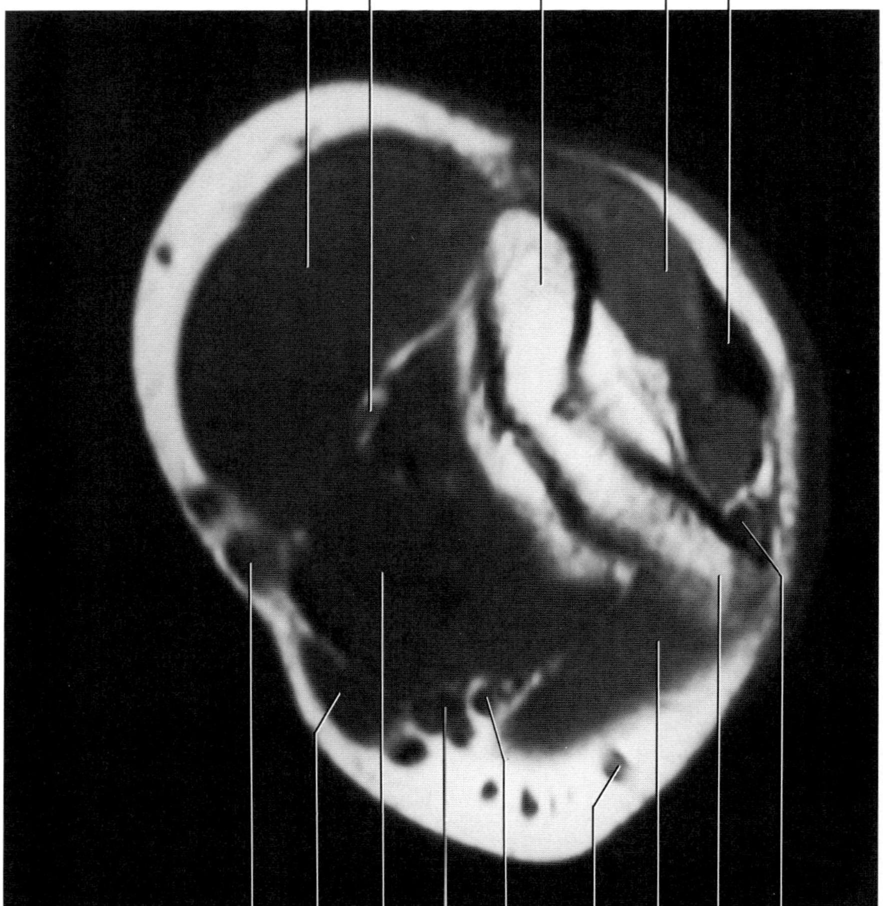

FIGURE 2–5(C)
Axial image of the normal elbow: The proximal olecranon process.

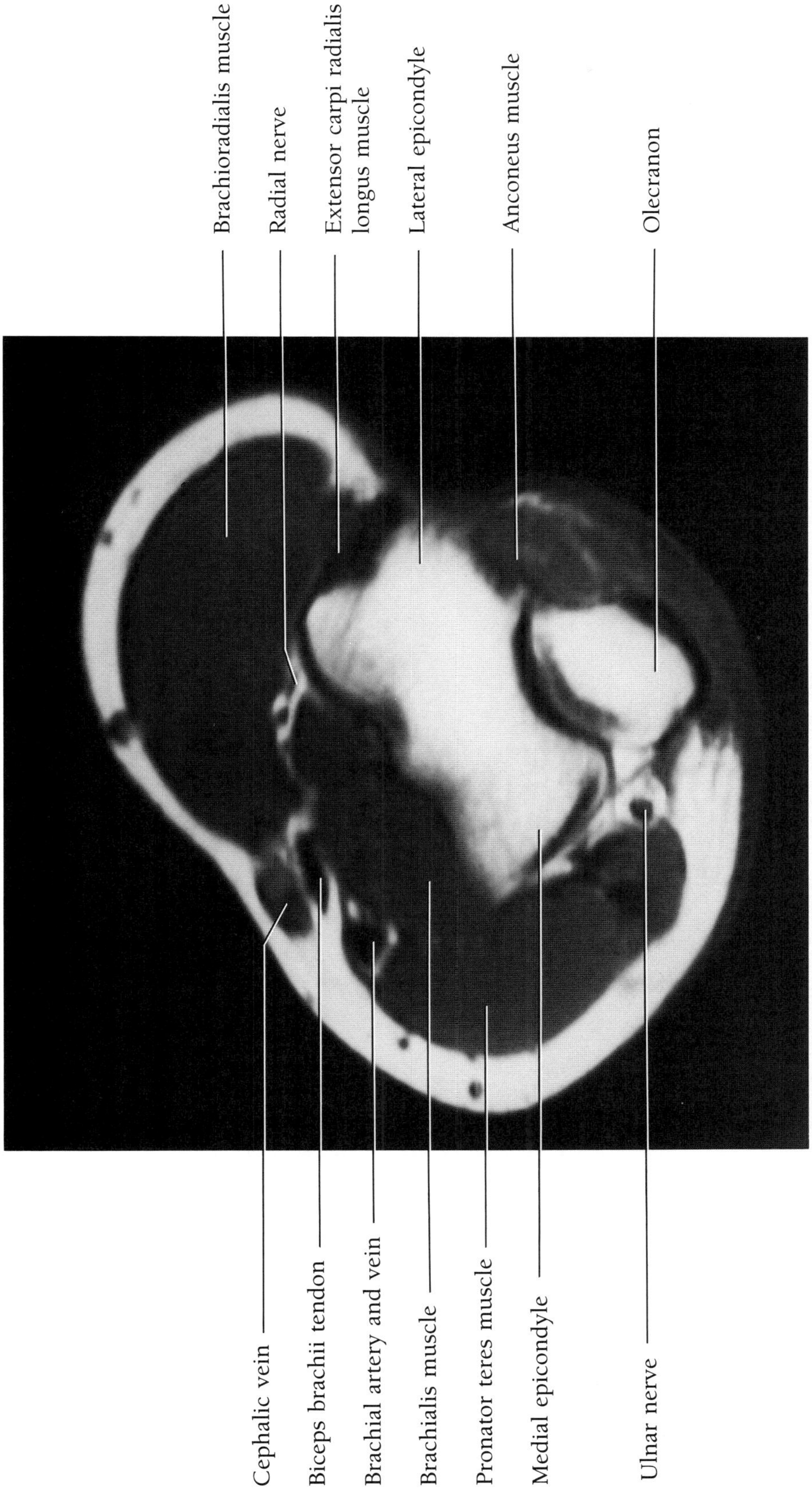

FIGURE 2–5(D)
Axial image of the normal elbow: The olecranon and distal epicondyles.

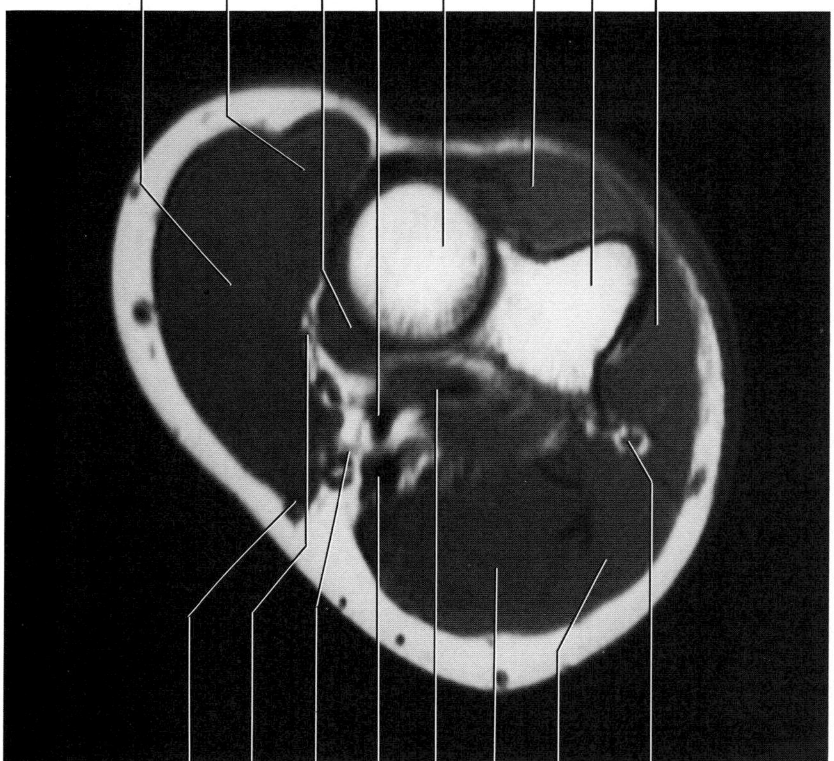

FIGURE 2–5(E)
Axial image of the normal elbow: The radioulnar joint.

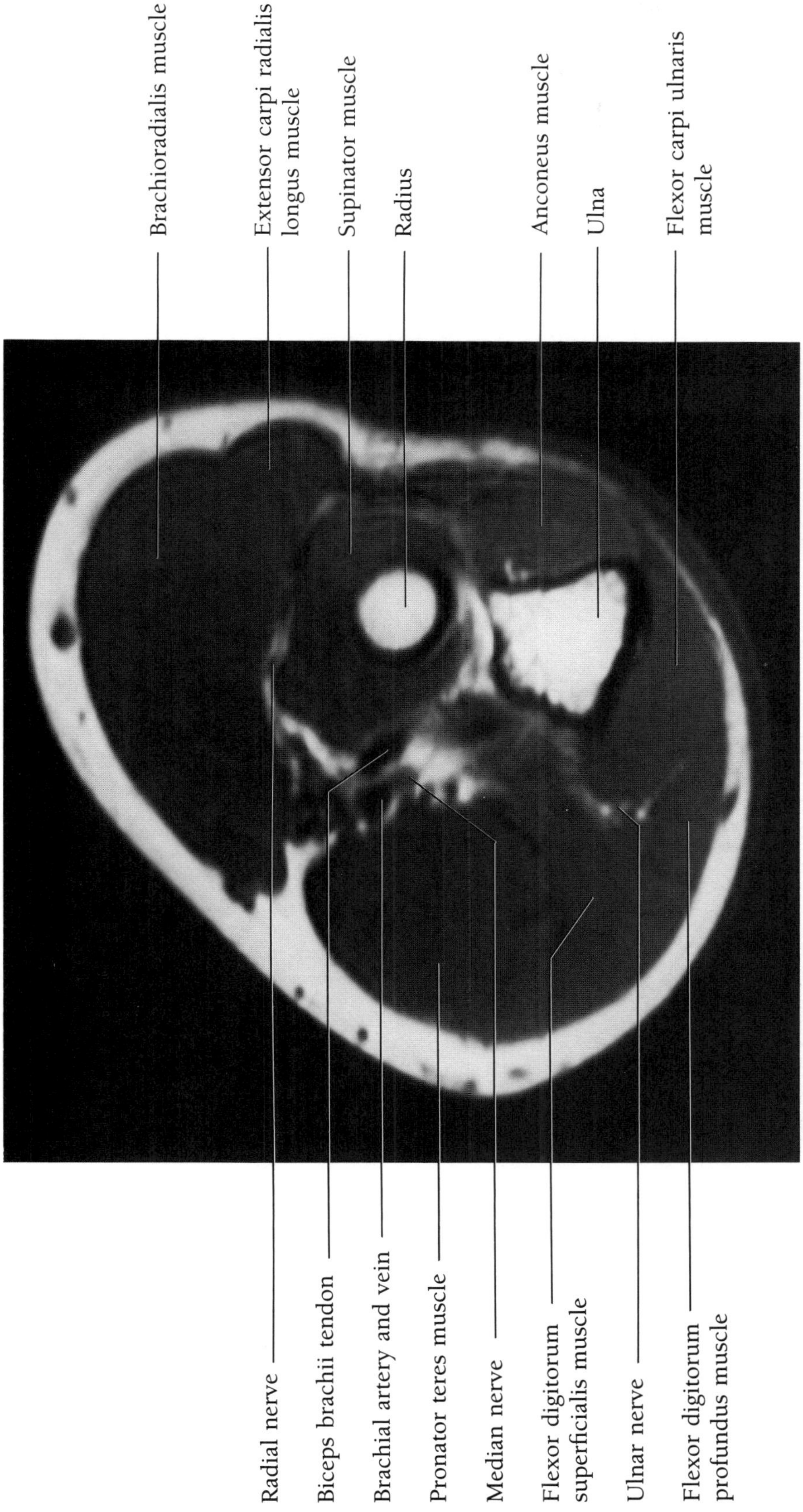

FIGURE 2–5(F)
Axial image of the normal elbow: Just distal to the proximal radioulnar joint.

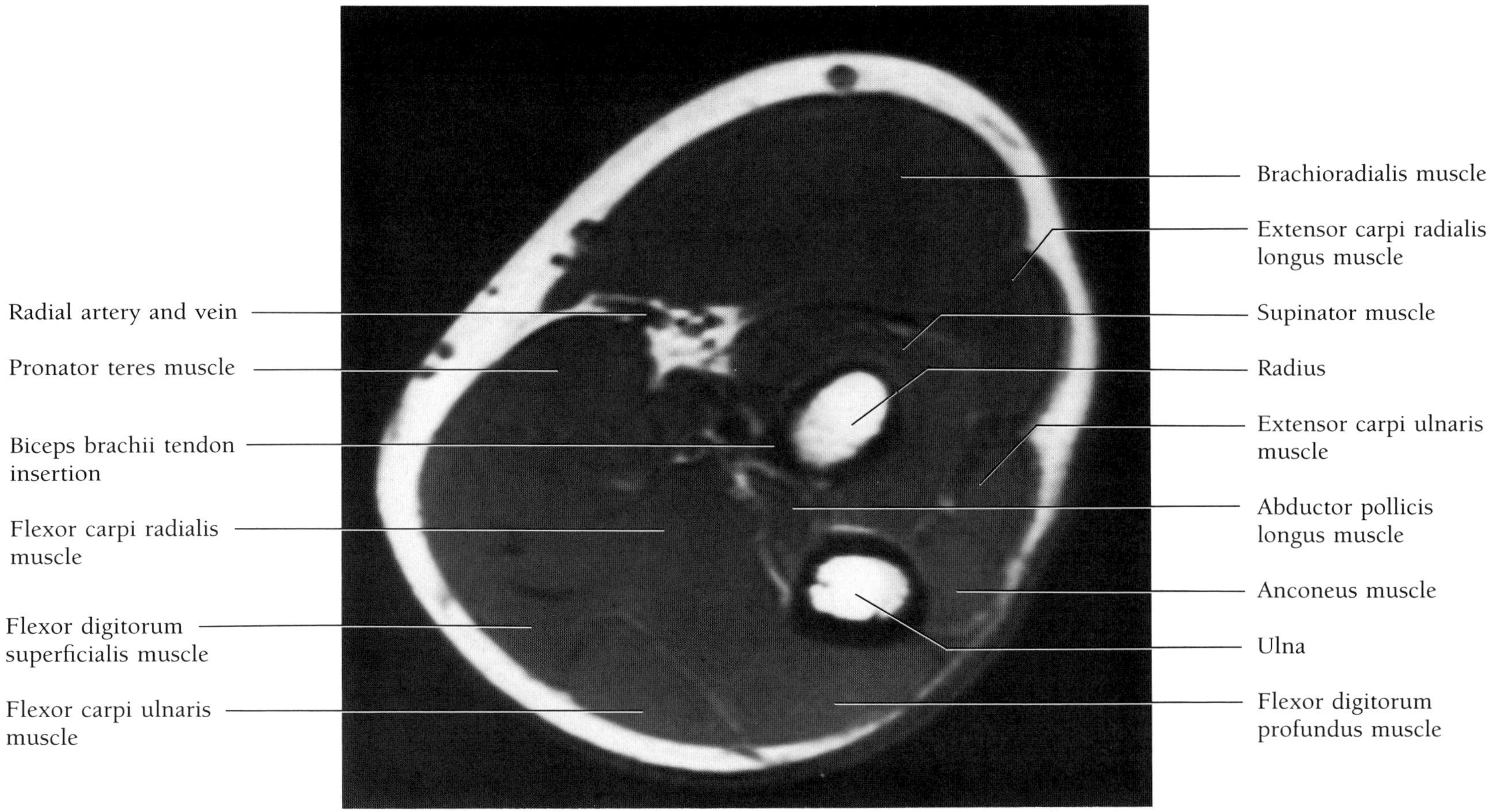

FIGURE 2–5(G)
Axial image of the normal elbow: The proximal part of the forearm.

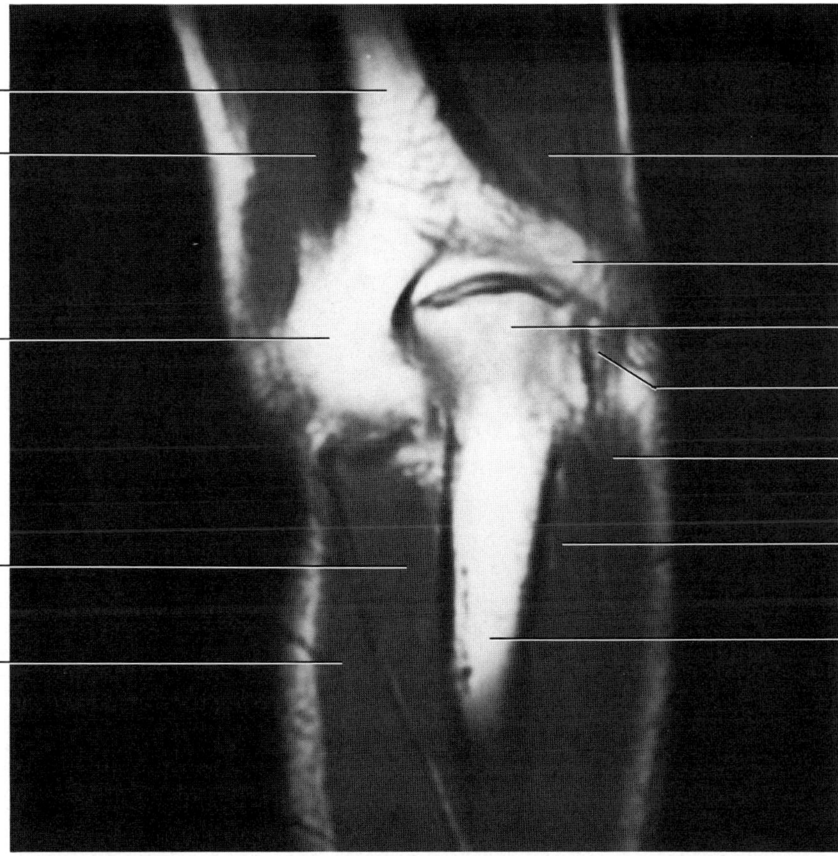

FIGURE 2–6(A)
Sagittal image of the normal elbow: Through the radius-capitellum articulation.

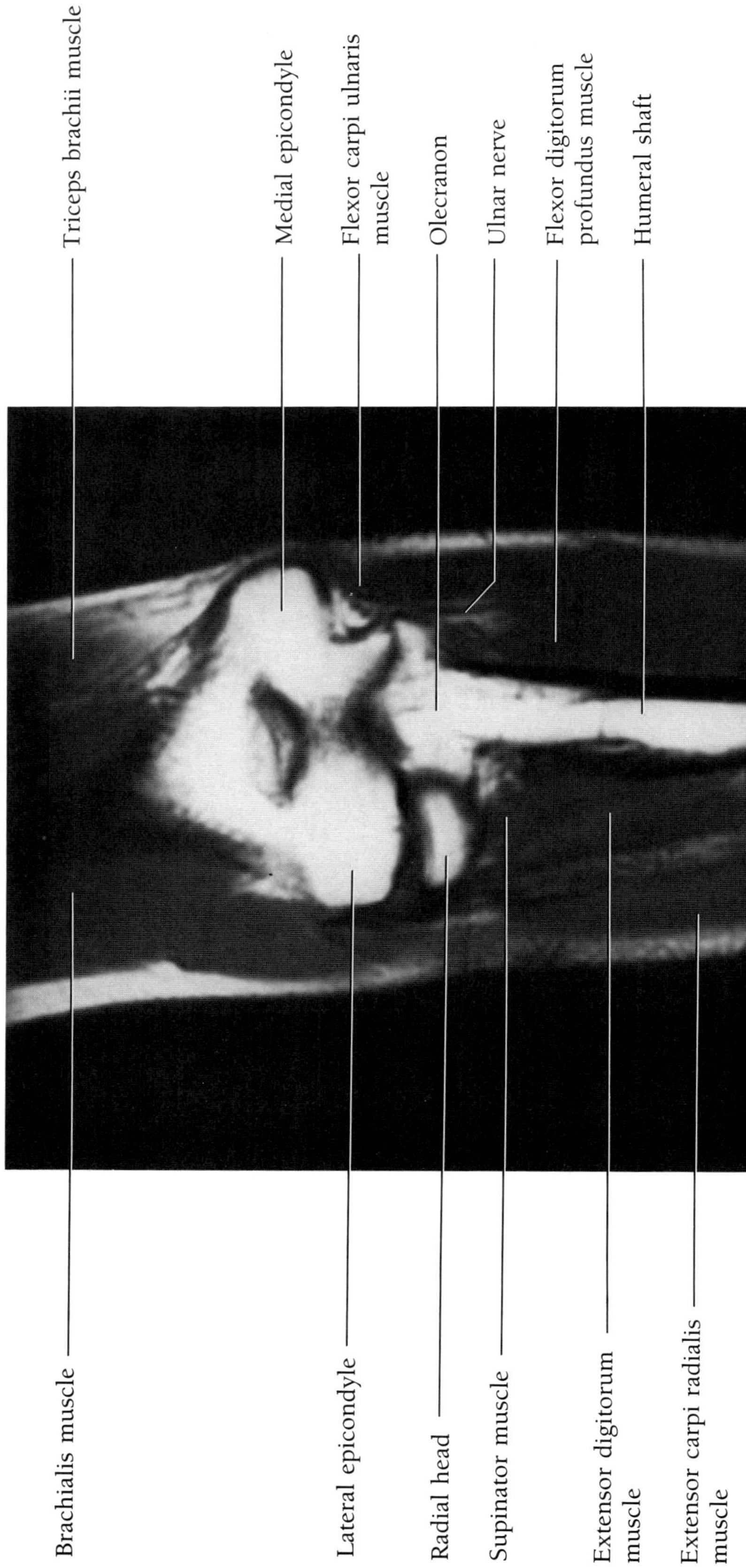

FIGURE 2–6(B)
Sagittal image of the normal elbow: Slightly medial to the section in A.

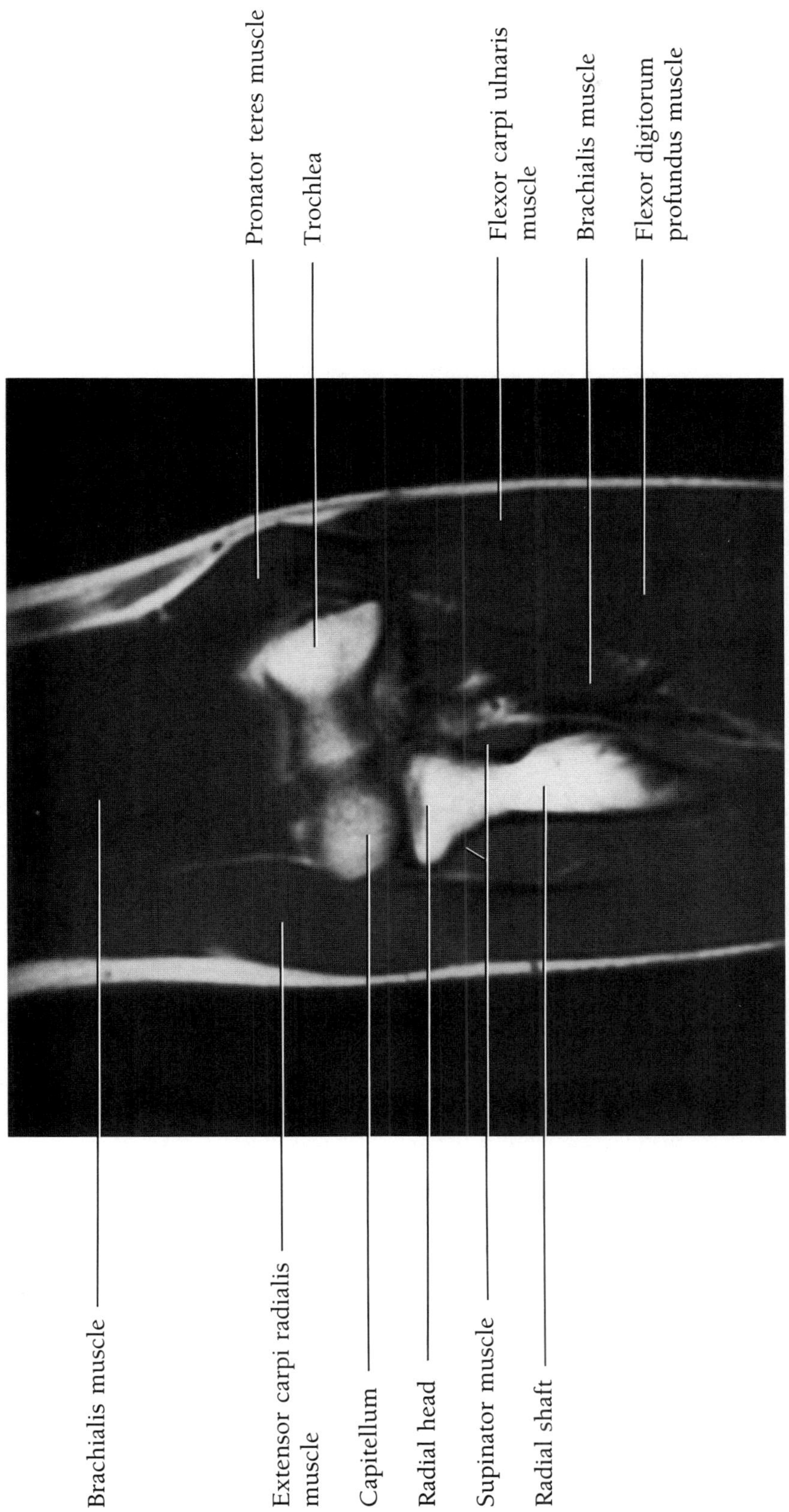

FIGURE 2–6(C)
Sagittal image of the normal elbow: Through the midportion of the elbow.

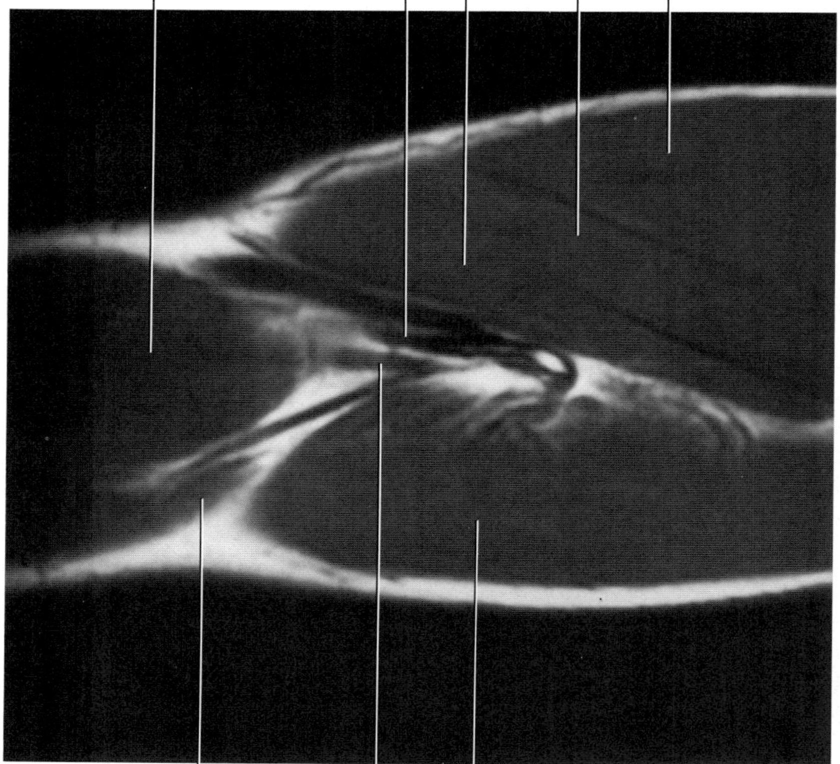

FIGURE 2–6(D)
Sagittal image of the normal elbow: Through the trochlea of the humerus and the trochlear notch of the ulna.

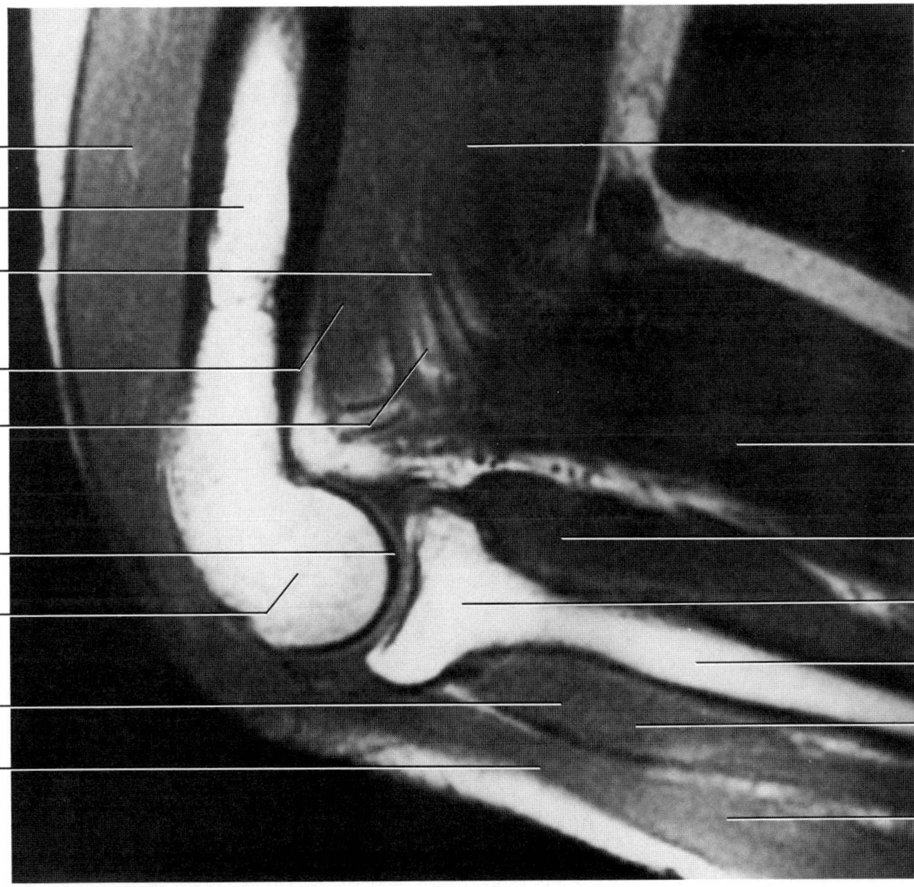

FIGURE 2–7(A)
Coronal image of the normal elbow: Through the posterior portion of the elbow.

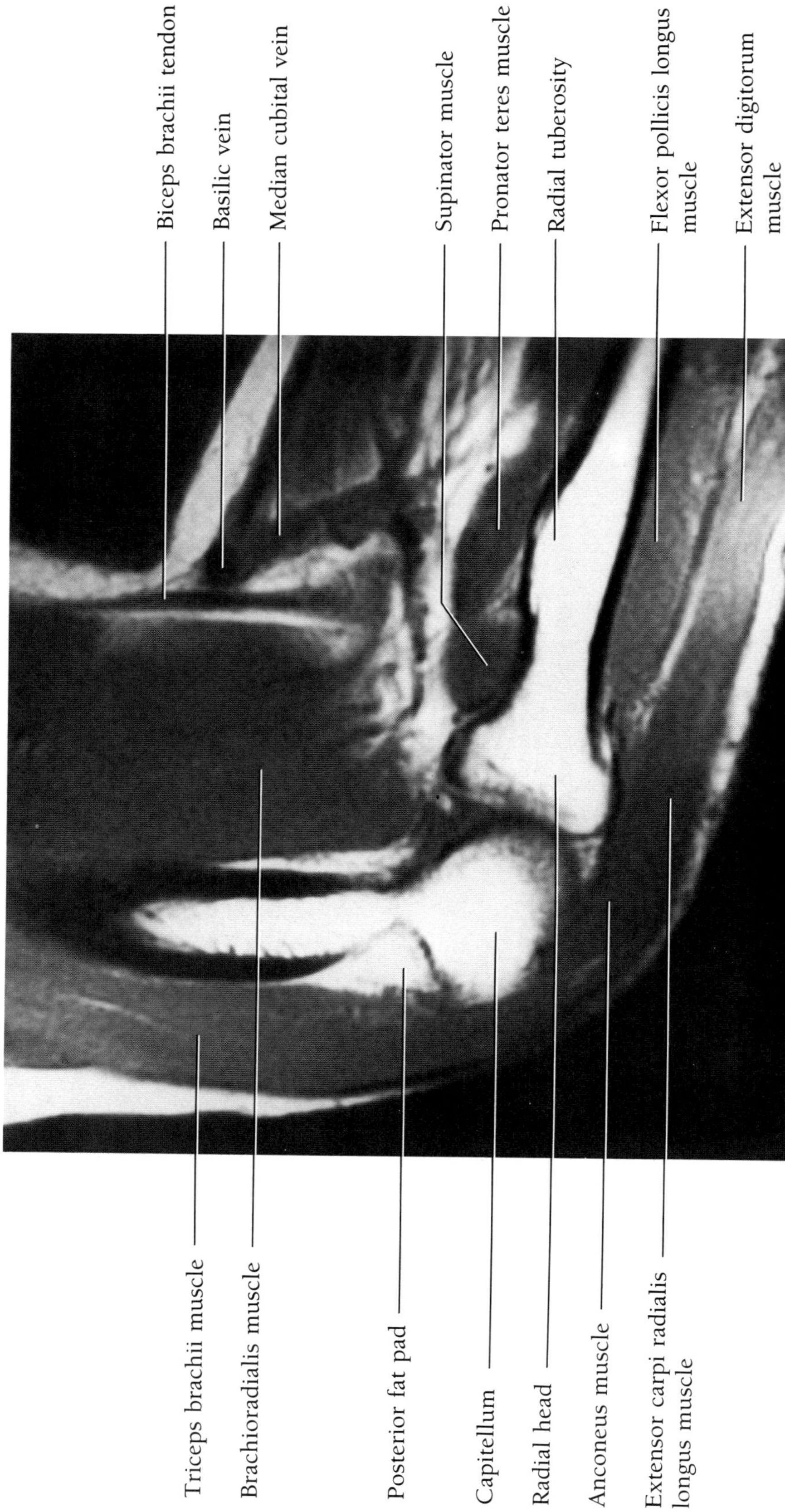

FIGURE 2–7(B)
Coronal image of the normal elbow: The midportion of the elbow.

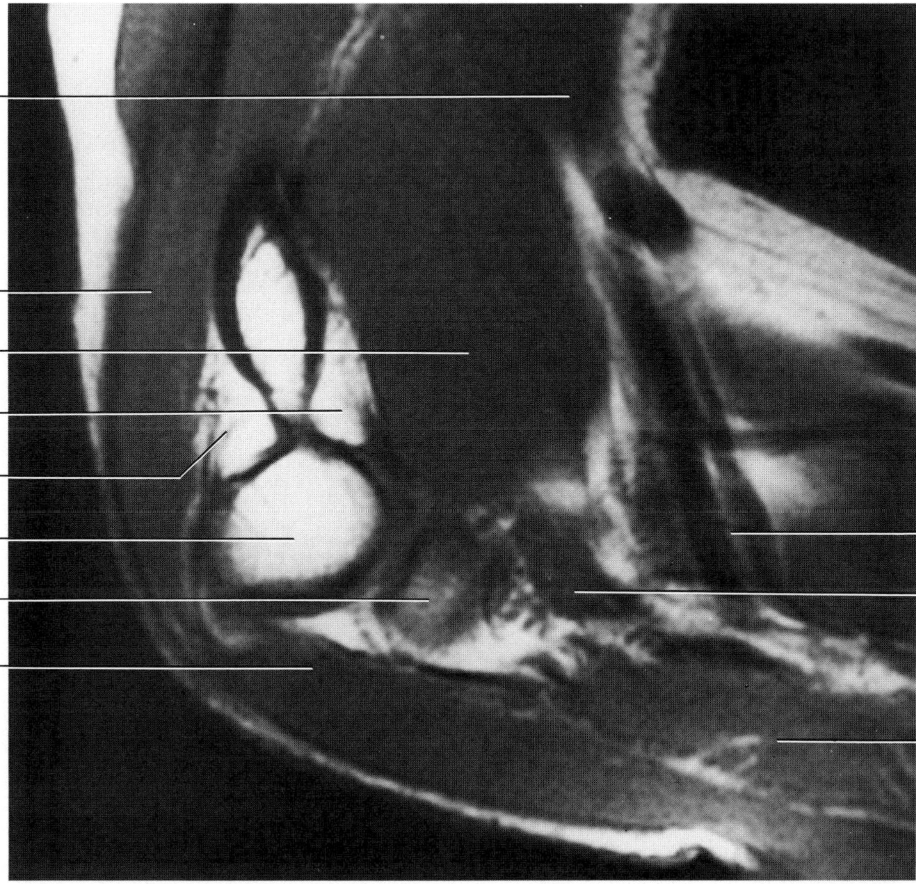

FIGURE 2–7(C)
Coronal image of the normal elbow: The anterior portion of the elbow.

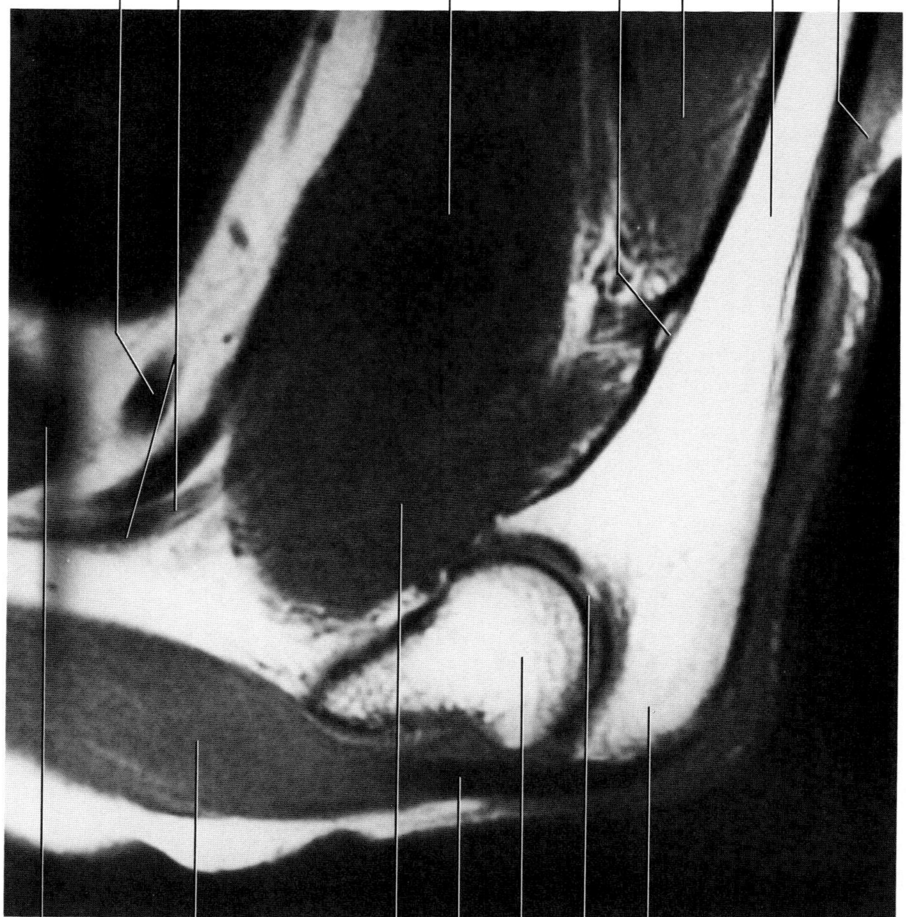

FIGURE 2–7(D)
Coronal image of the normal elbow: The antecubital fossa.

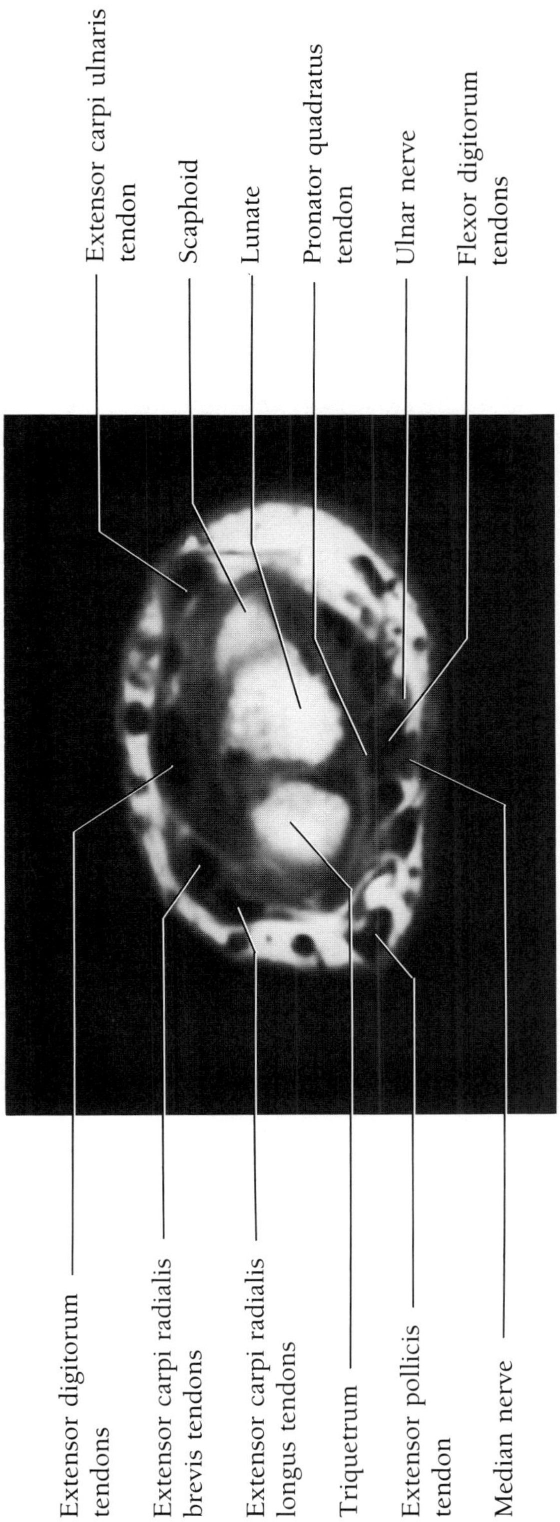

FIGURE 2–8(A)
Axial image of the wrist and hand: Through the distal radio-ulnar joint.

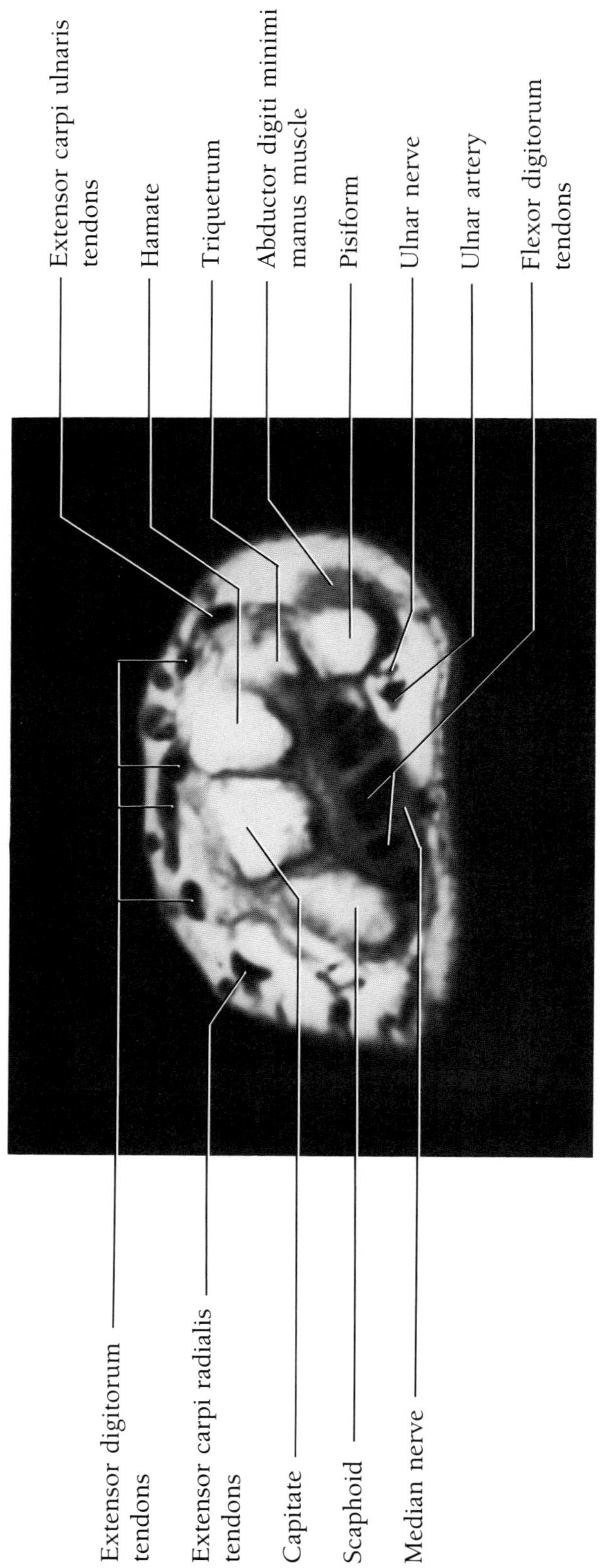

FIGURE 2–8(B)
Axial image of the wrist and hand: The proximal carpal tunnel.

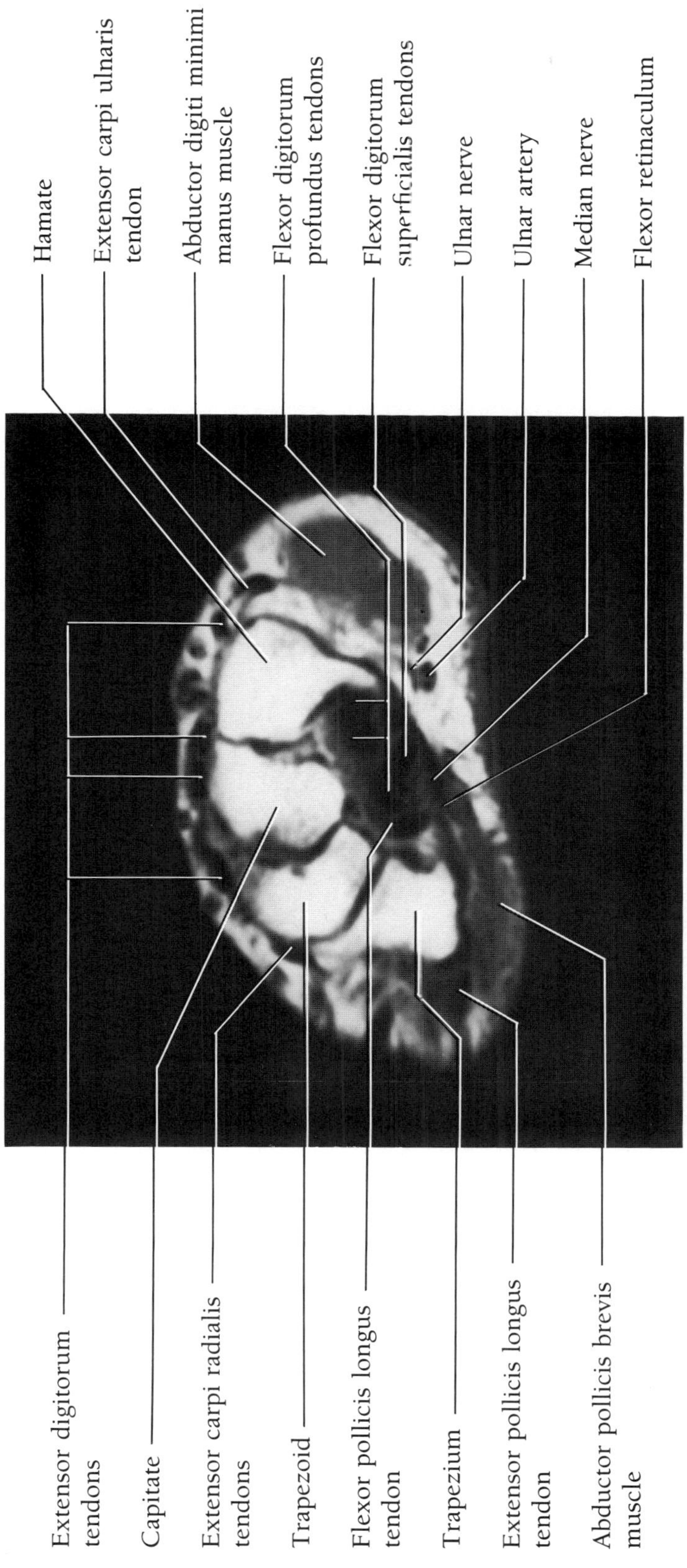

FIGURE 2–8(C)
Axial image of the wrist and hand: The midcarpal tunnel.

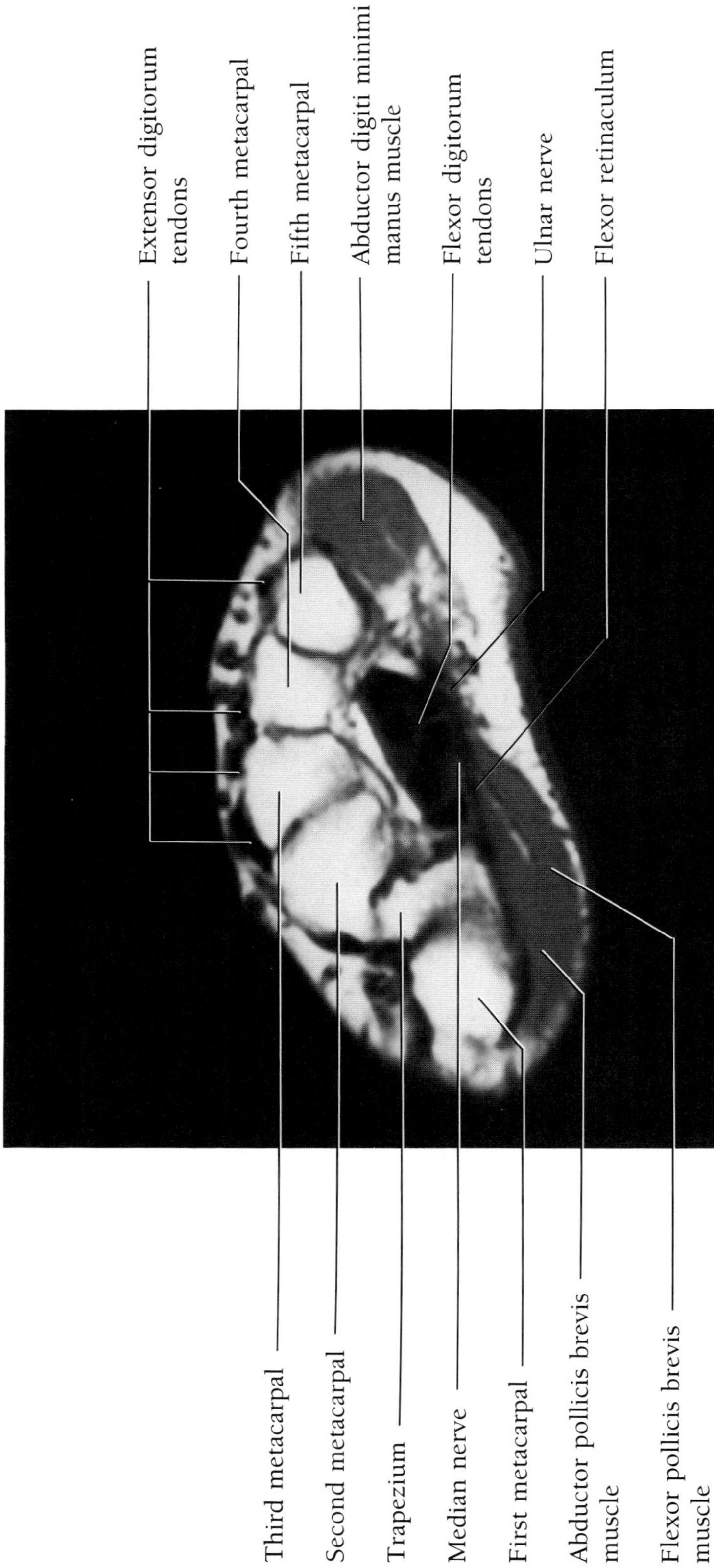

FIGURE 2–8(D)
Axial image of the wrist and hand: The distal carpal tunnel.

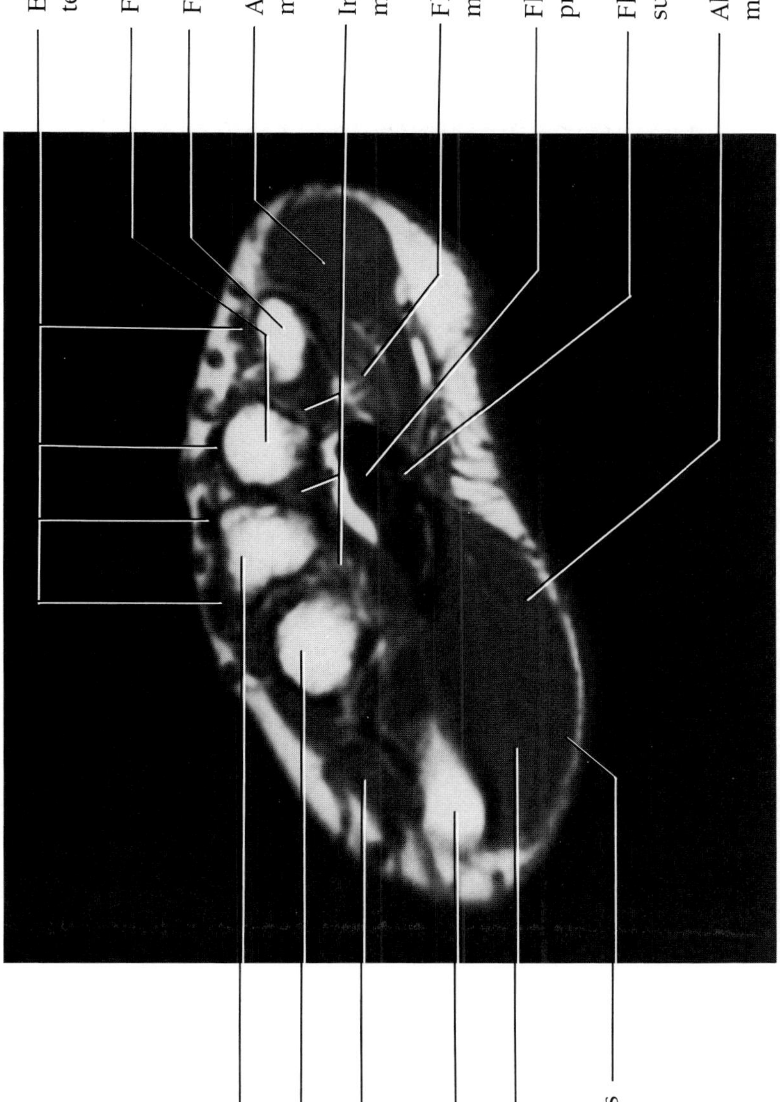

FIGURE 2–8(E)
Axial image of the wrist and hand: The bases of the metacarpals.

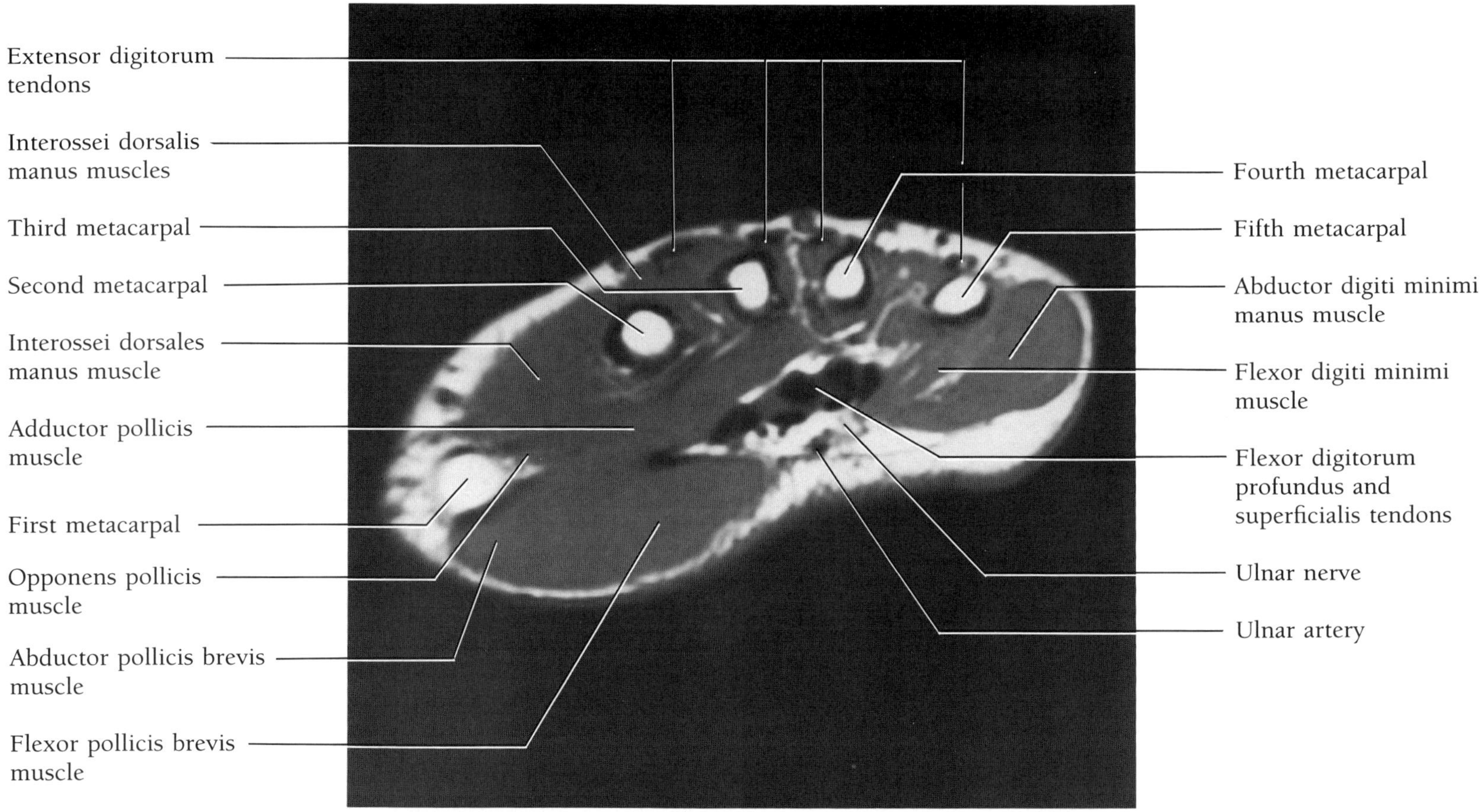

FIGURE 2–8(F)
Axial image of the wrist and hand: The midmetacarpal shafts.

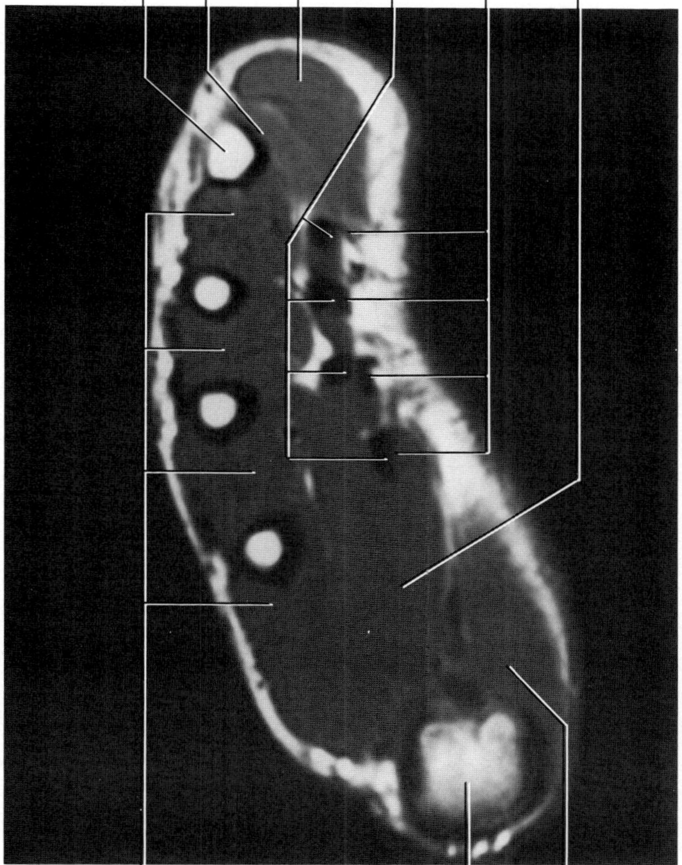

FIGURE 2–8(G)
Axial image of the wrist and hand: The distal metacarpal shafts and head of the first metacarpal.

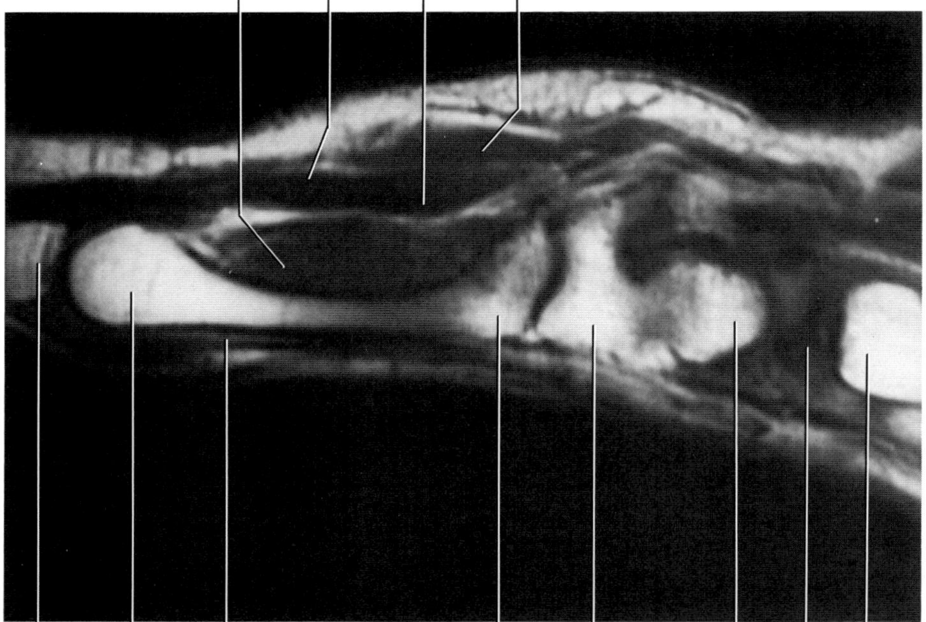

FIGURE 2–9(A)
Sagittal image of the wrist, hand, and fingers: Ulnar aspect.

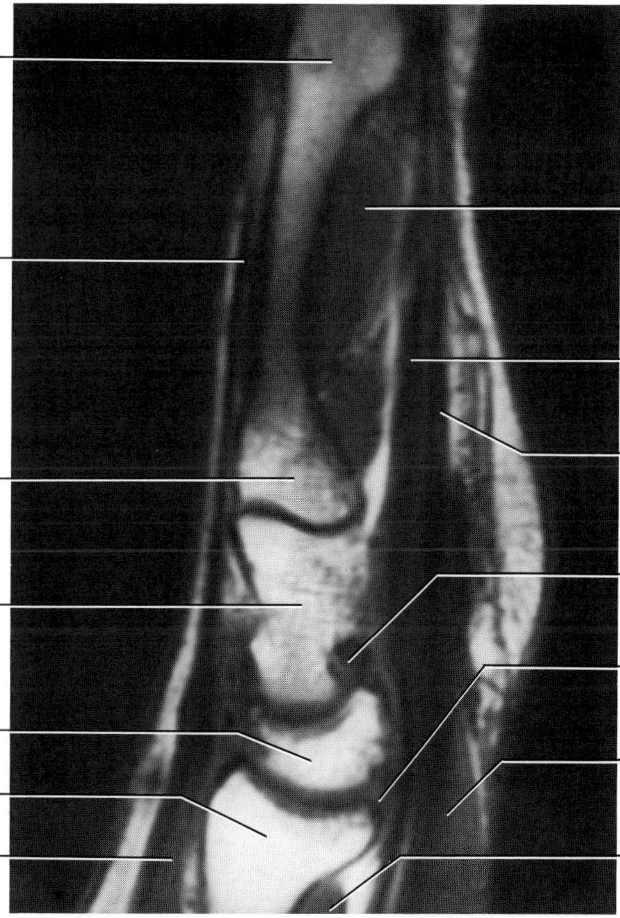

FIGURE 2–9(B)
Sagittal image of the wrist, hand, and fingers: Radial aspect.

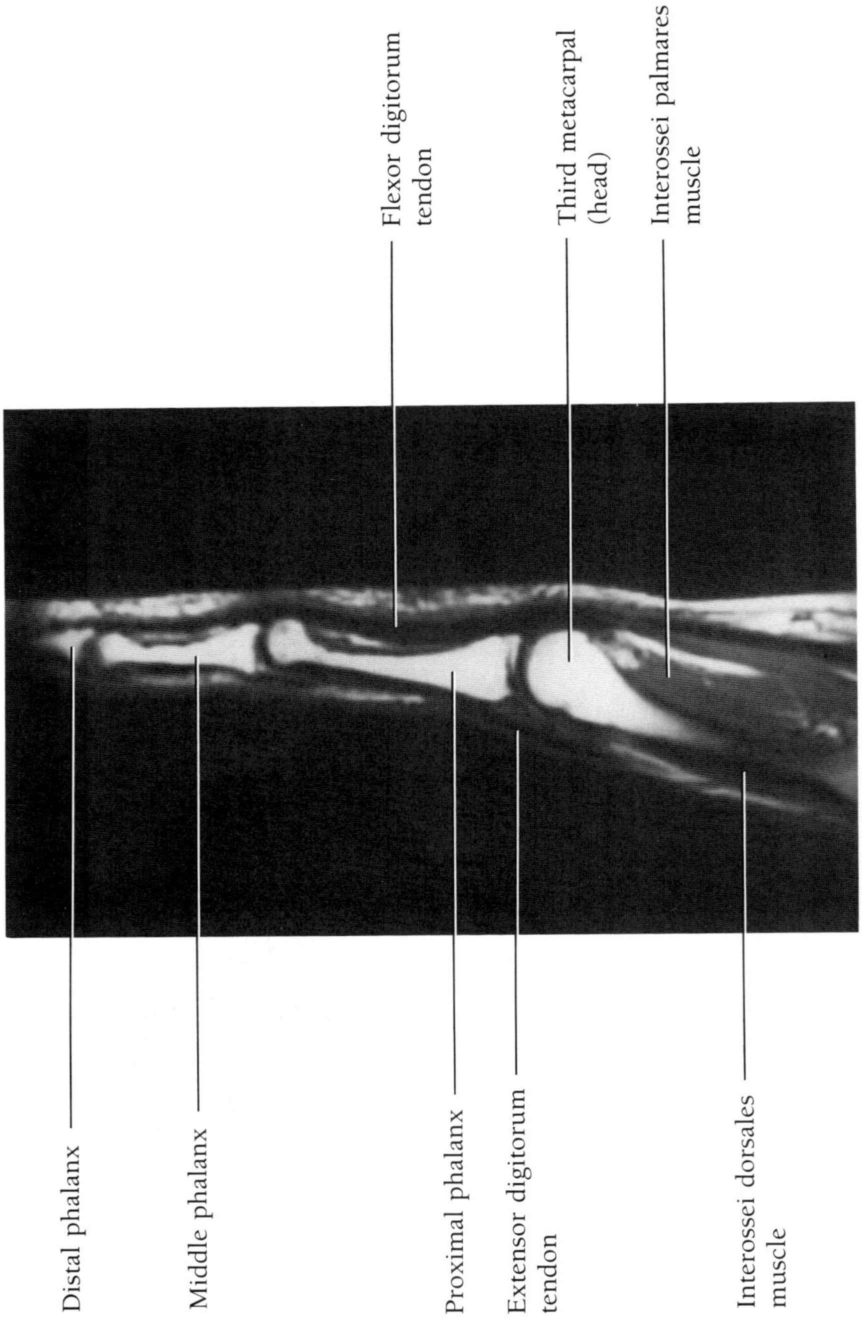

FIGURE 2–9(C)
Sagittal image of the wrist, hand, and fingers: Index finger.

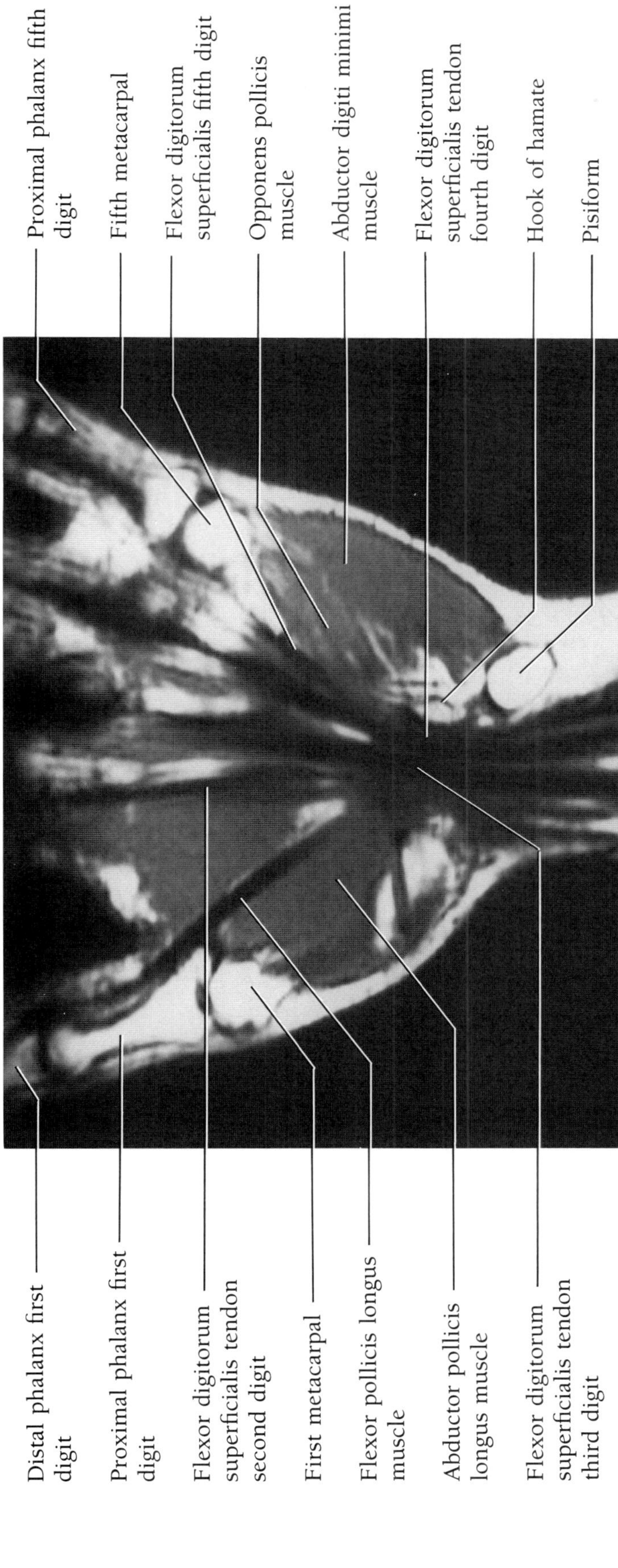

FIGURE 2–10(A)
Coronal image of the wrist and hand: Through the superficial tendons of the carpal tunnel.

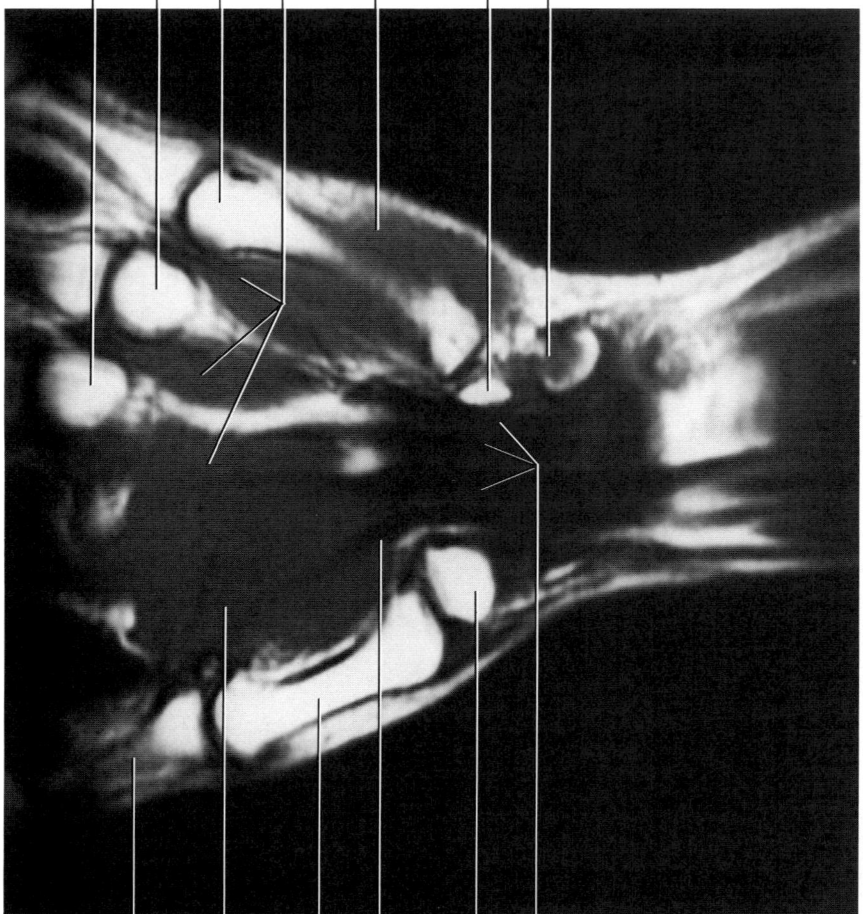

FIGURE 2–10(B)
Coronal image of the wrist and hand: Through the deep tendons of the carpal tunnel.

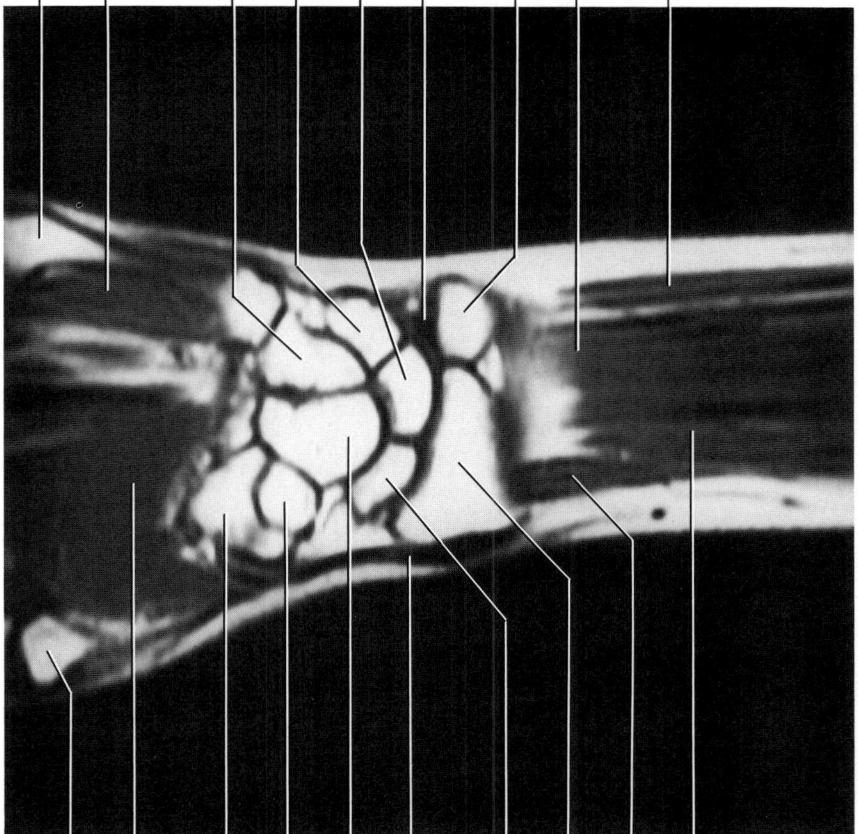

FIGURE 2–10(C)
Coronal image of the wrist and hand: Through the carpal bones.

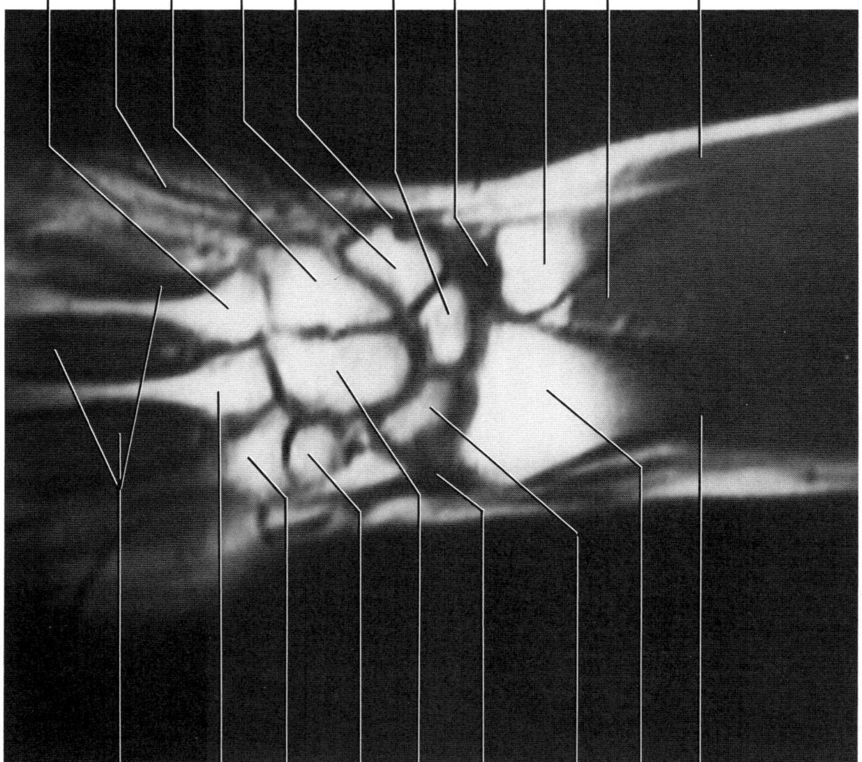

FIGURE 2–10(D)
Coronal image of the wrist and hand: Through the volar aspect of the wrist.

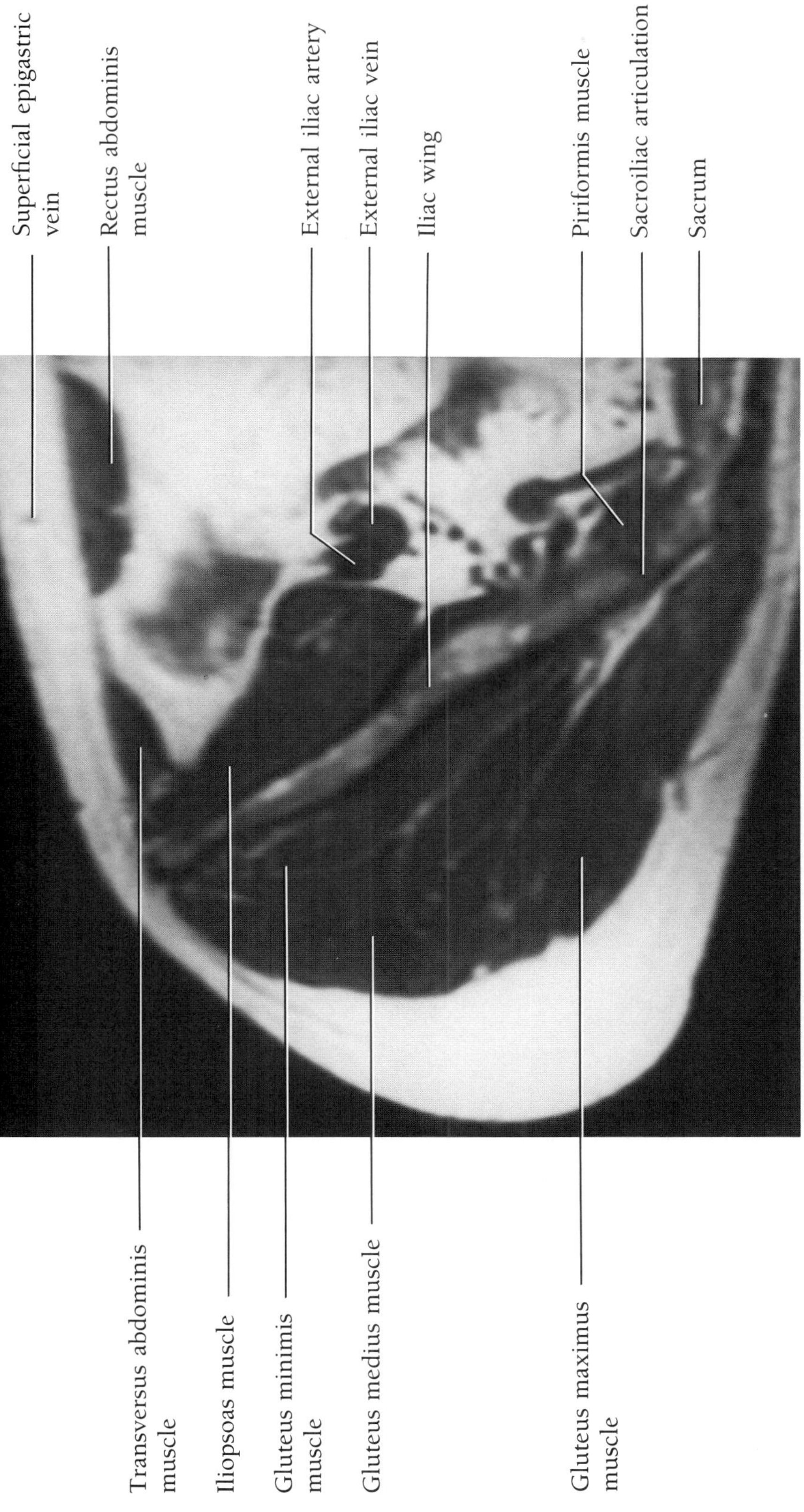

FIGURE 2–11(A)
Axial image through the hip: The level of the iliac wing.

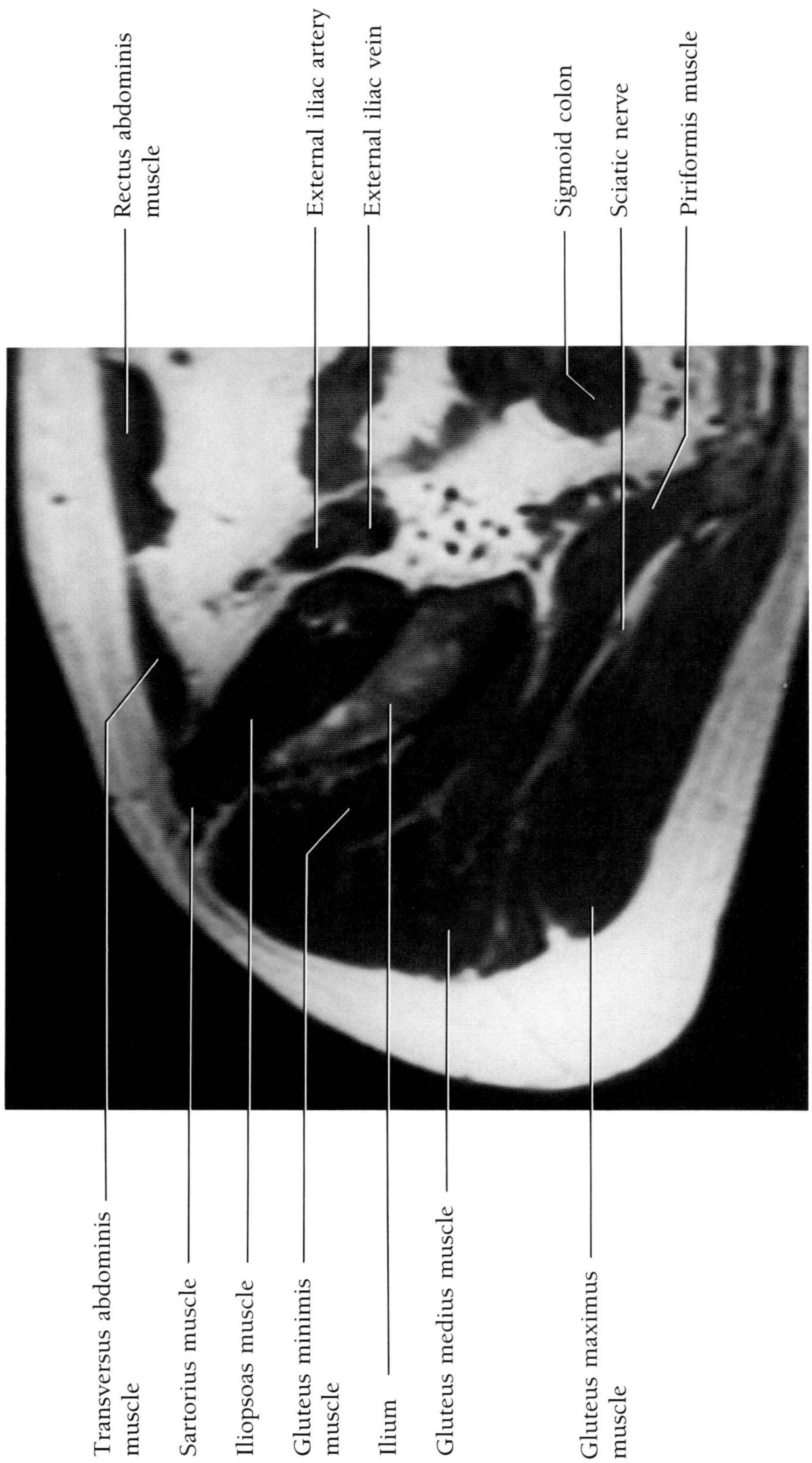

FIGURE 2–11(B)
Axial image through the hip: Superior acetabulum.

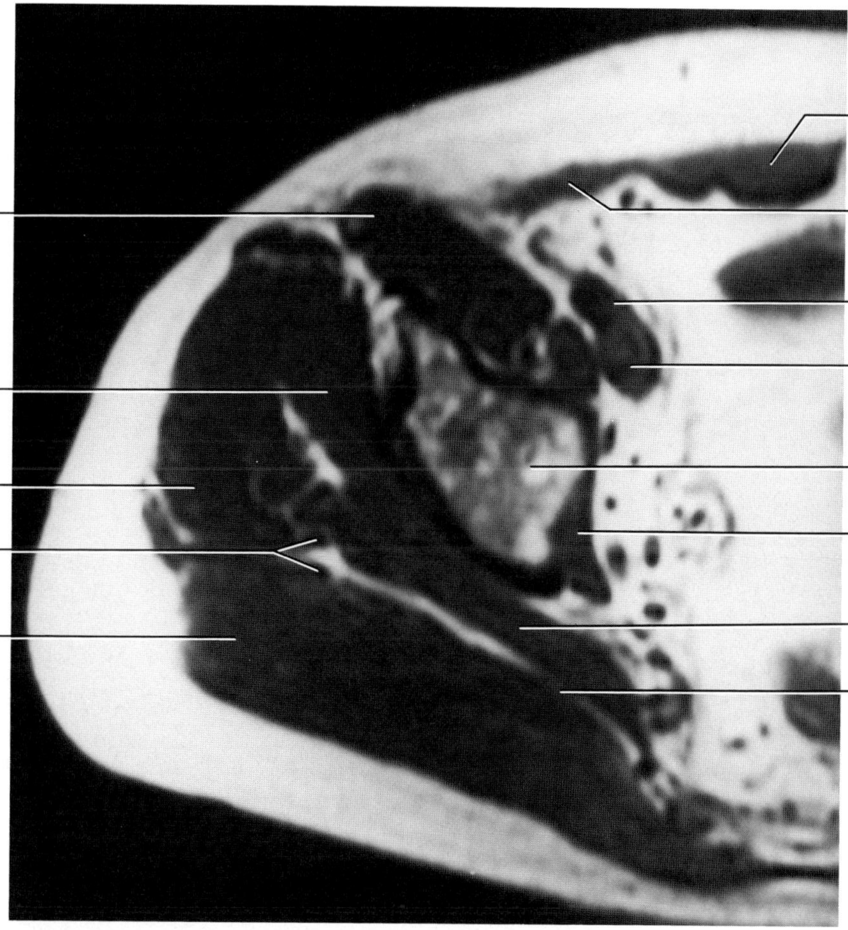

FIGURE 2–11(C)
Axial image through the hip: Acetabular roof.

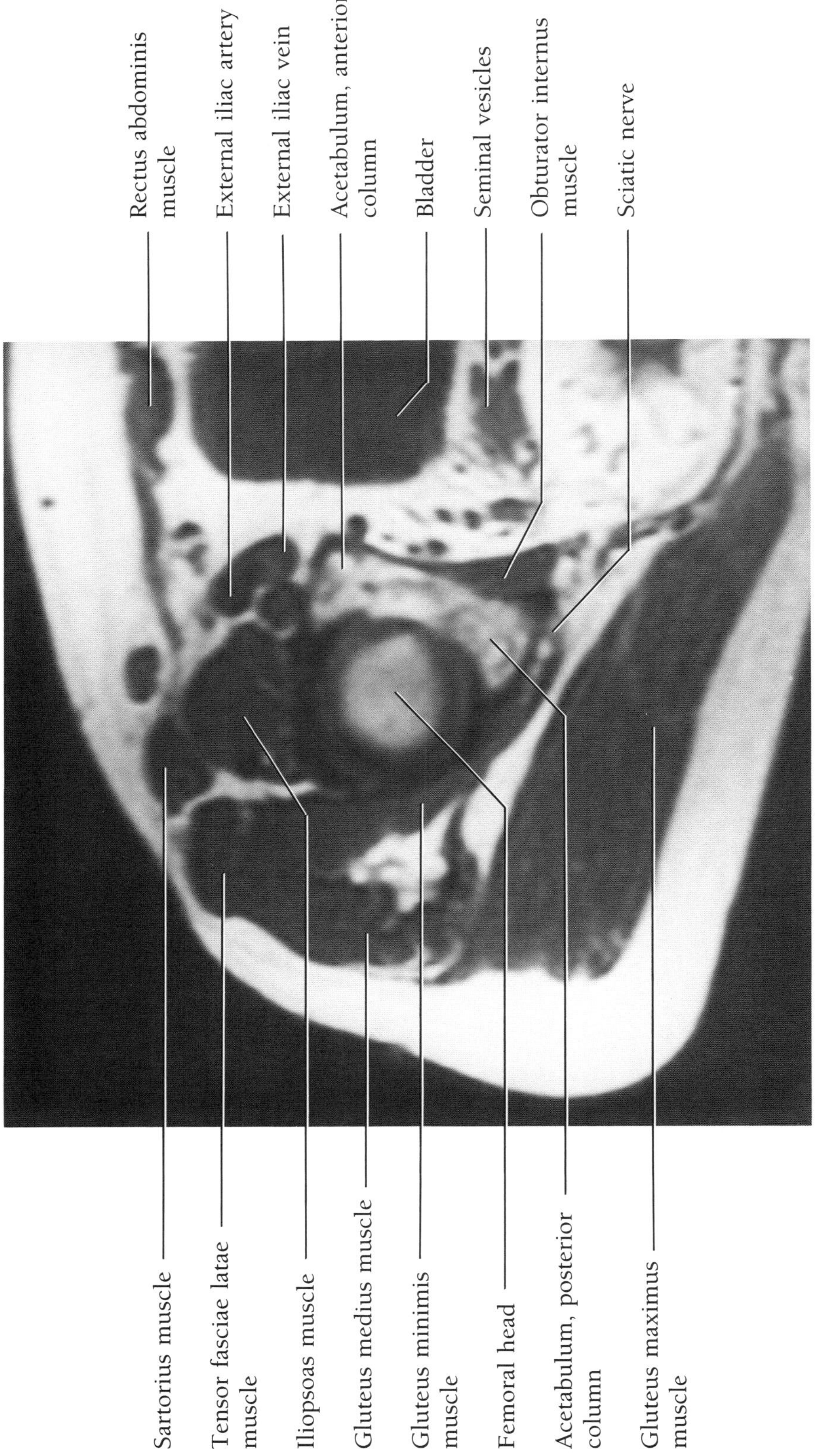

FIGURE 2–11(D)
Axial image through the hip: Superior portion of the hip joint.

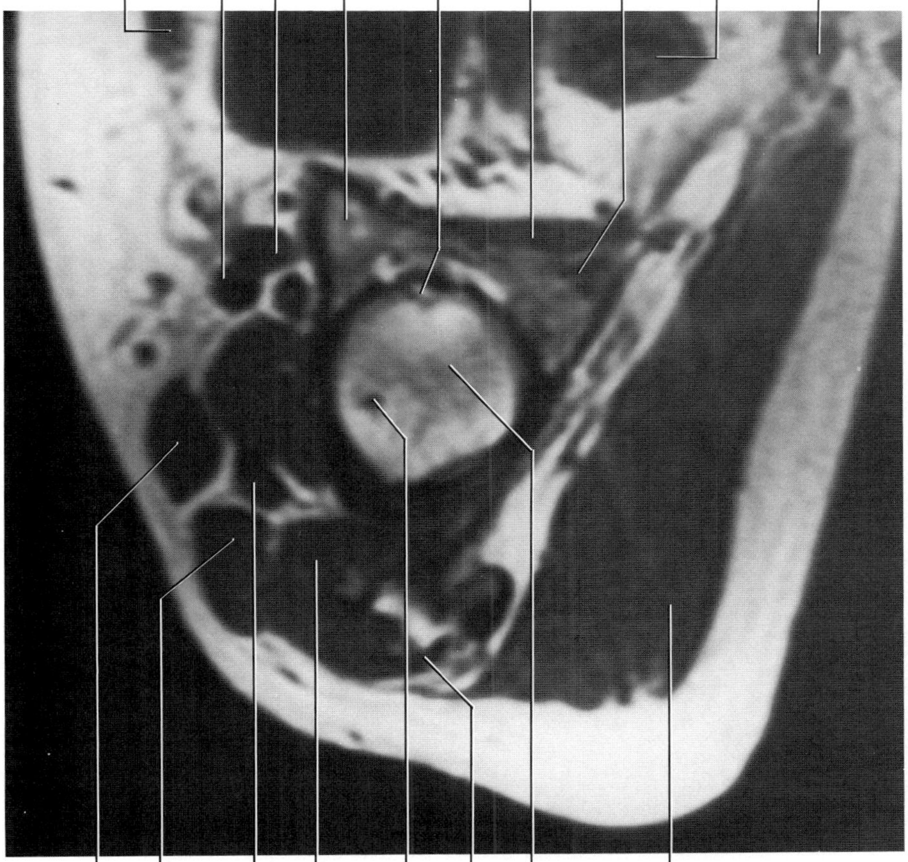

FIGURE 2–11(E)
Axial image through the hip: Midportion of the hip.

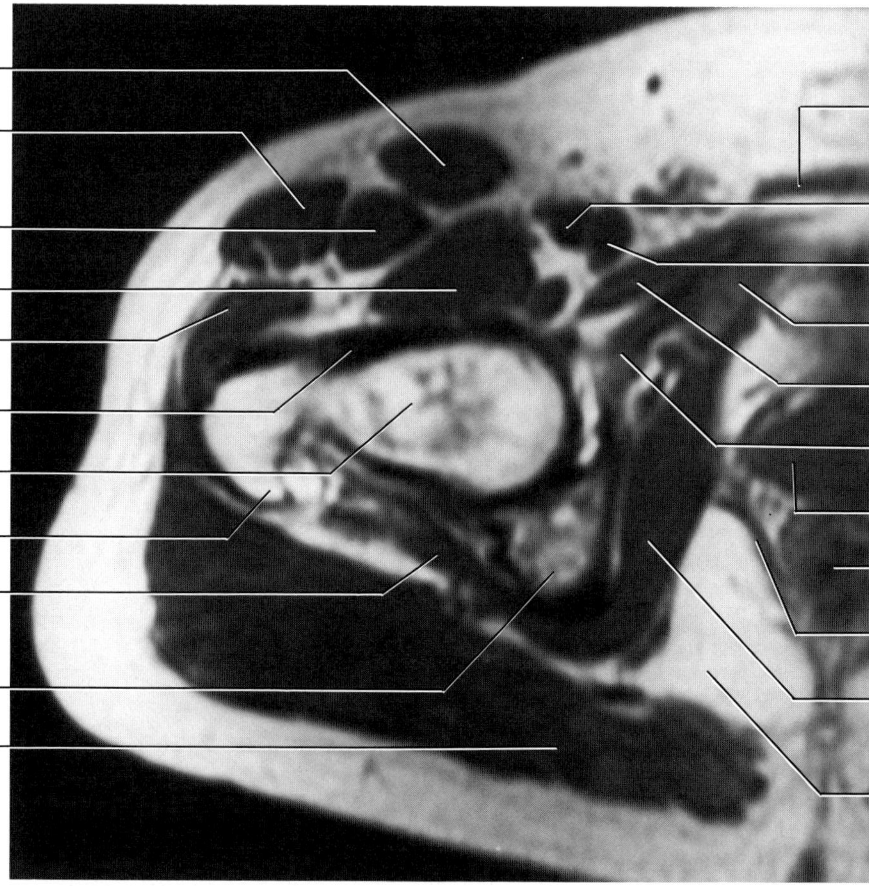

FIGURE 2–11(F)
Axial image of the hip: Through the femoral neck and greater trochanter.

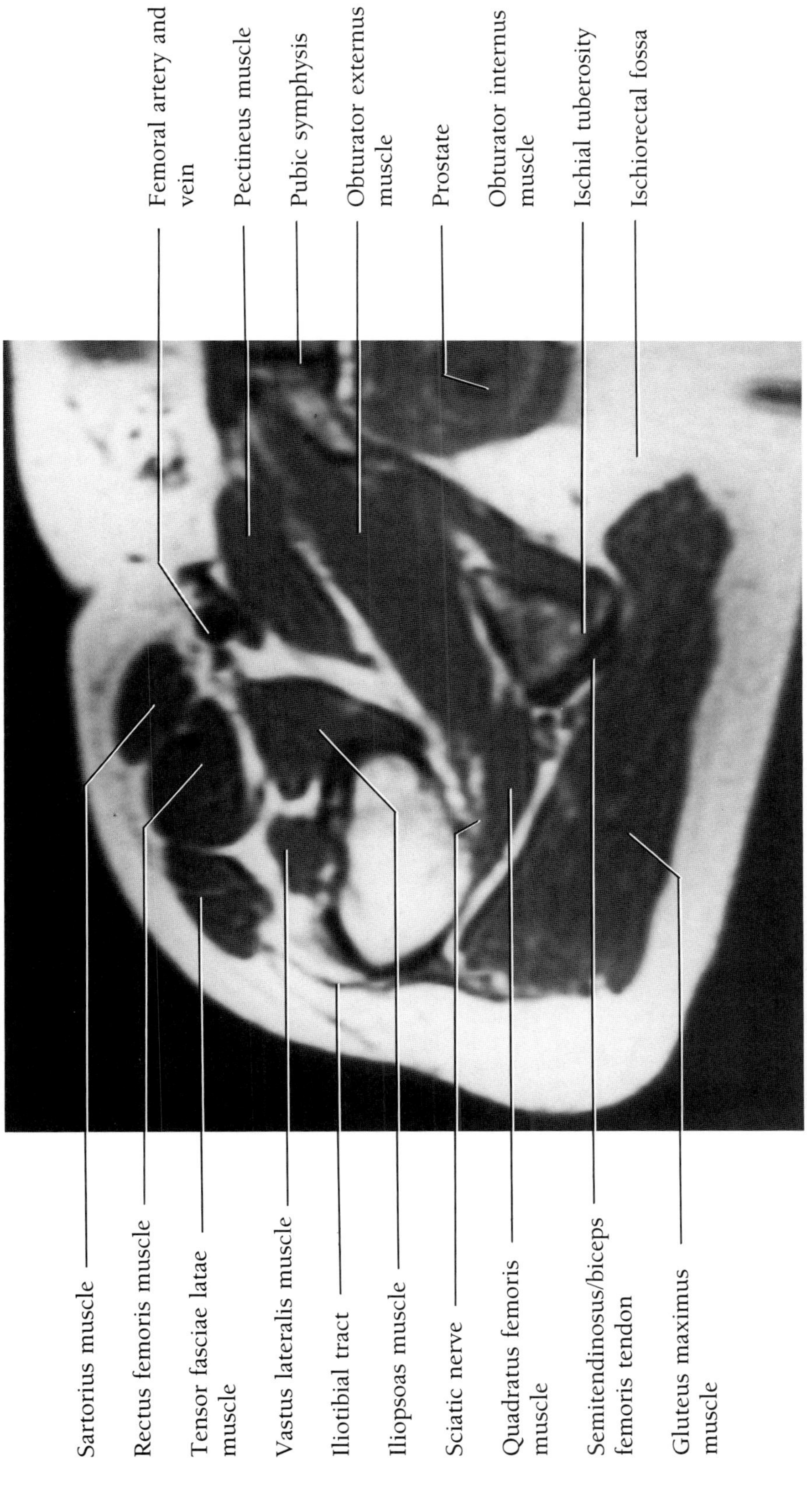

FIGURE 2–11(G)
Axial image of the hip: The proximal part of the femoral shaft.

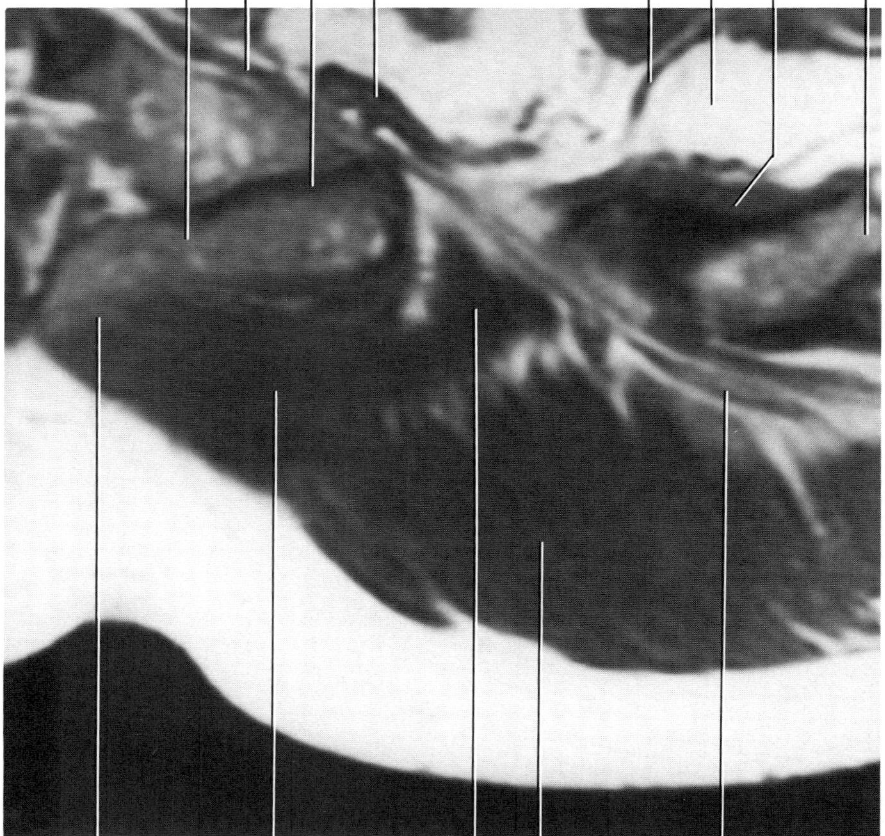

FIGURE 2–12(A)
Coronal image of the hip: The posterior part of the pelvis.

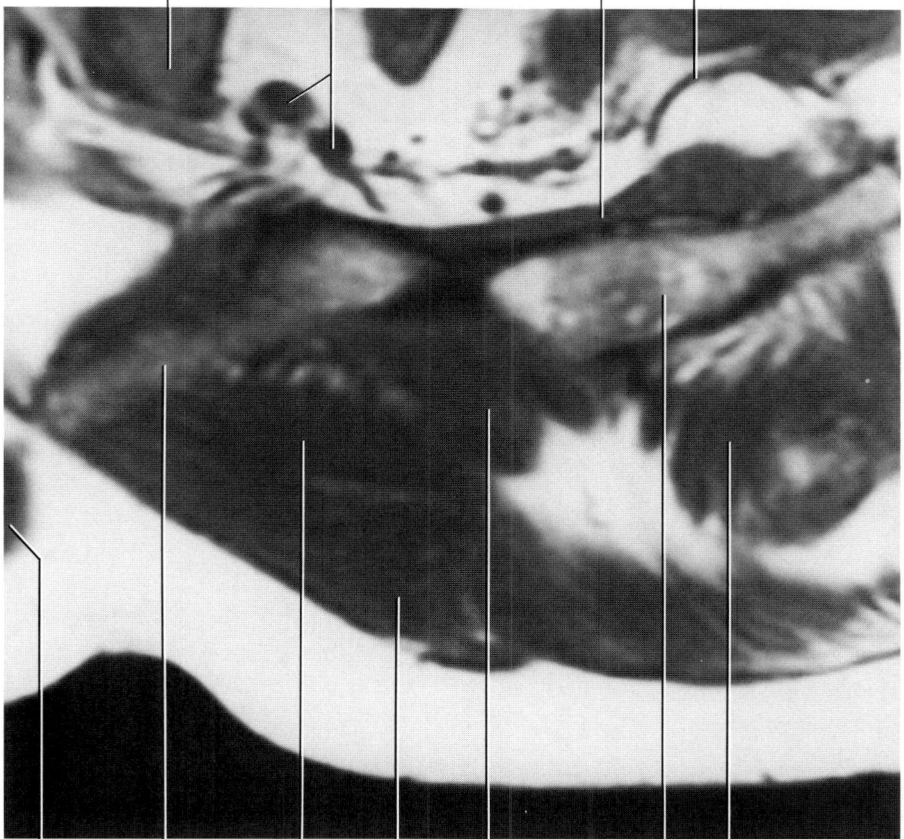

FIGURE 2–12(B)
Coronal image of the hip: The posterior aspect of the acetabulum.

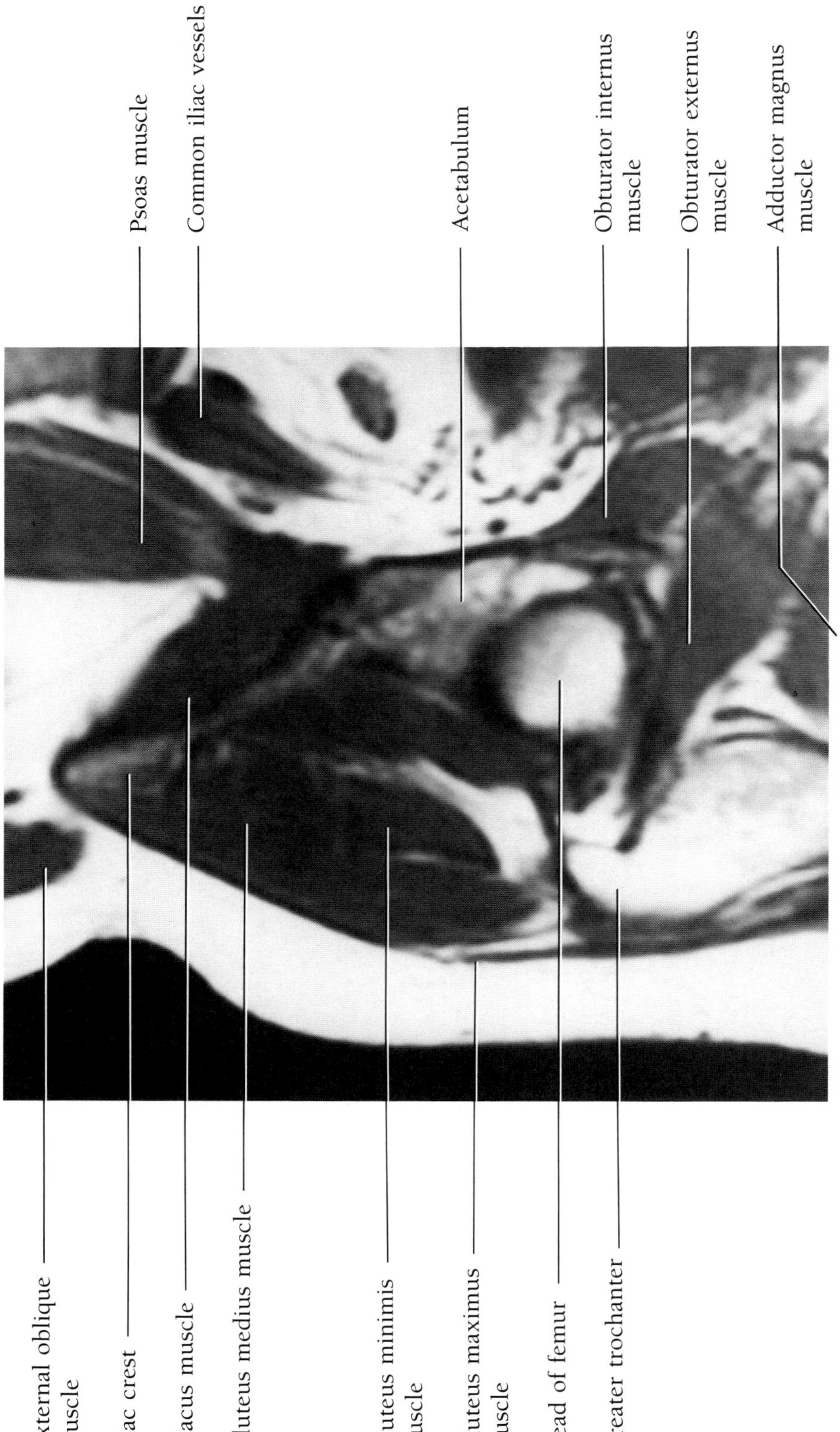

FIGURE 2–12(C)
Coronal image of the hip: The posterior portion of the hip joint.

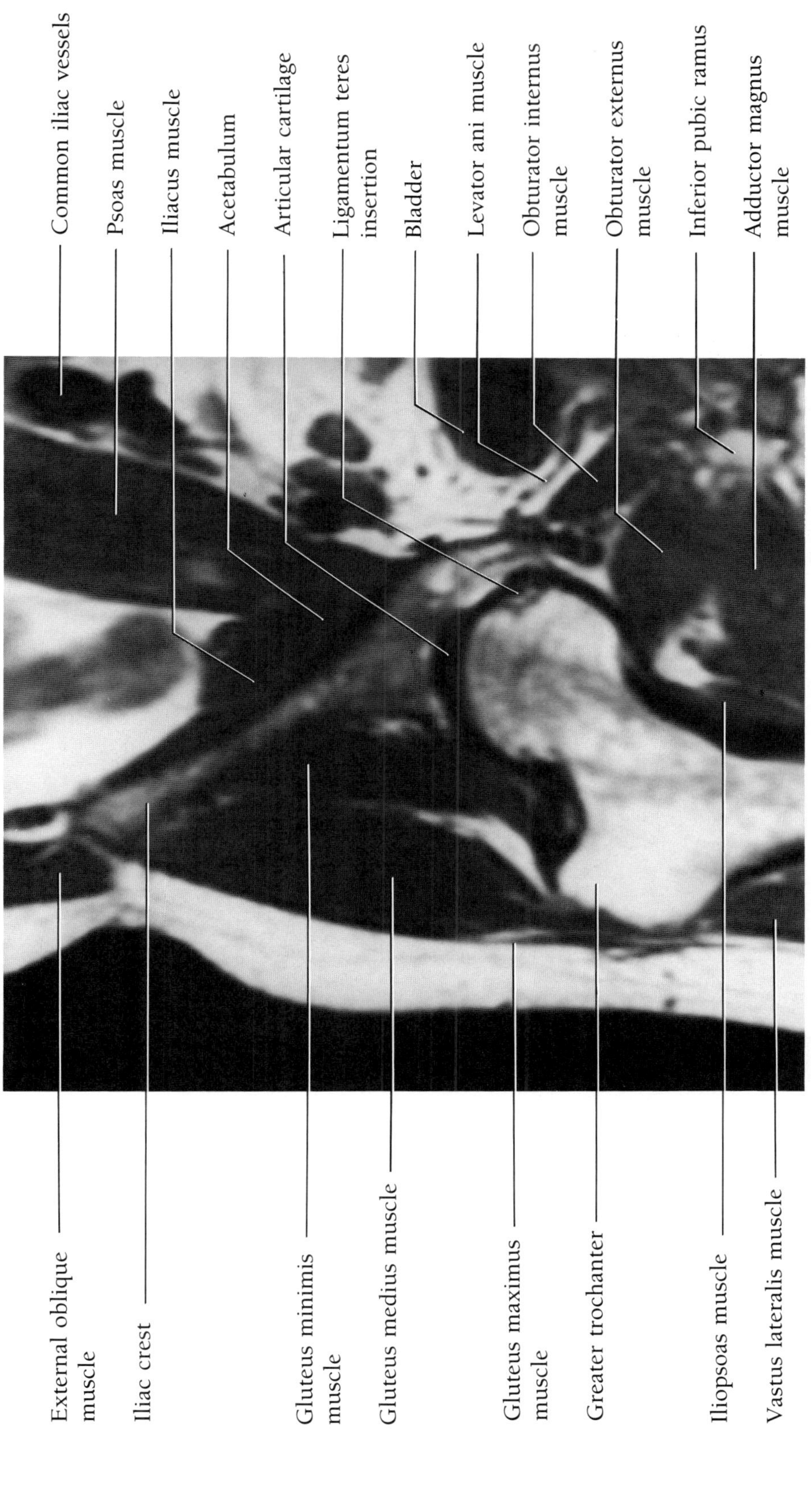

FIGURE 2–12(D)
Coronal image of the hip: Through the center of the hip.

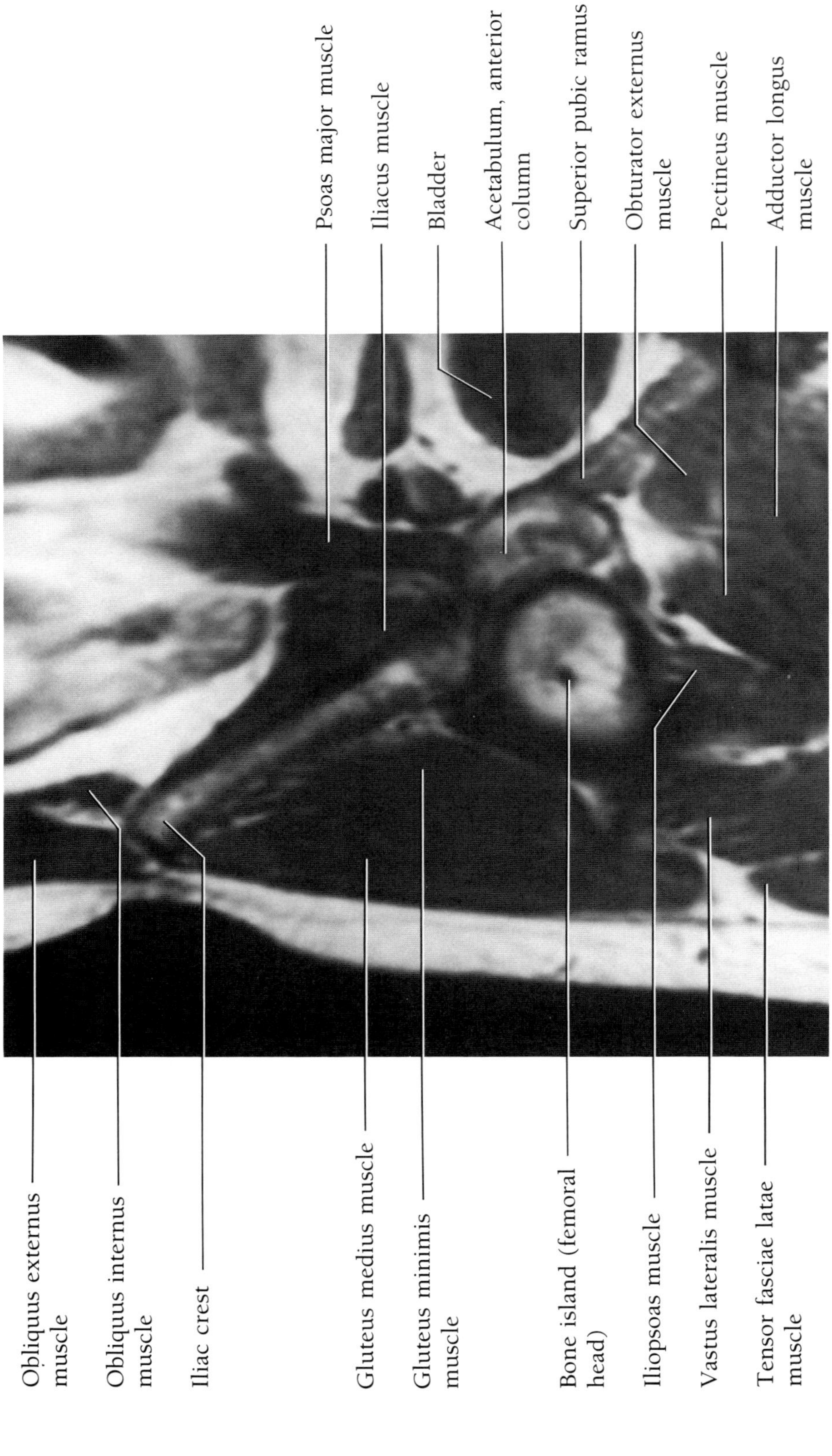

FIGURE 2–12(E)
Coronal image of the hip: The anterior portion of the hip joint.

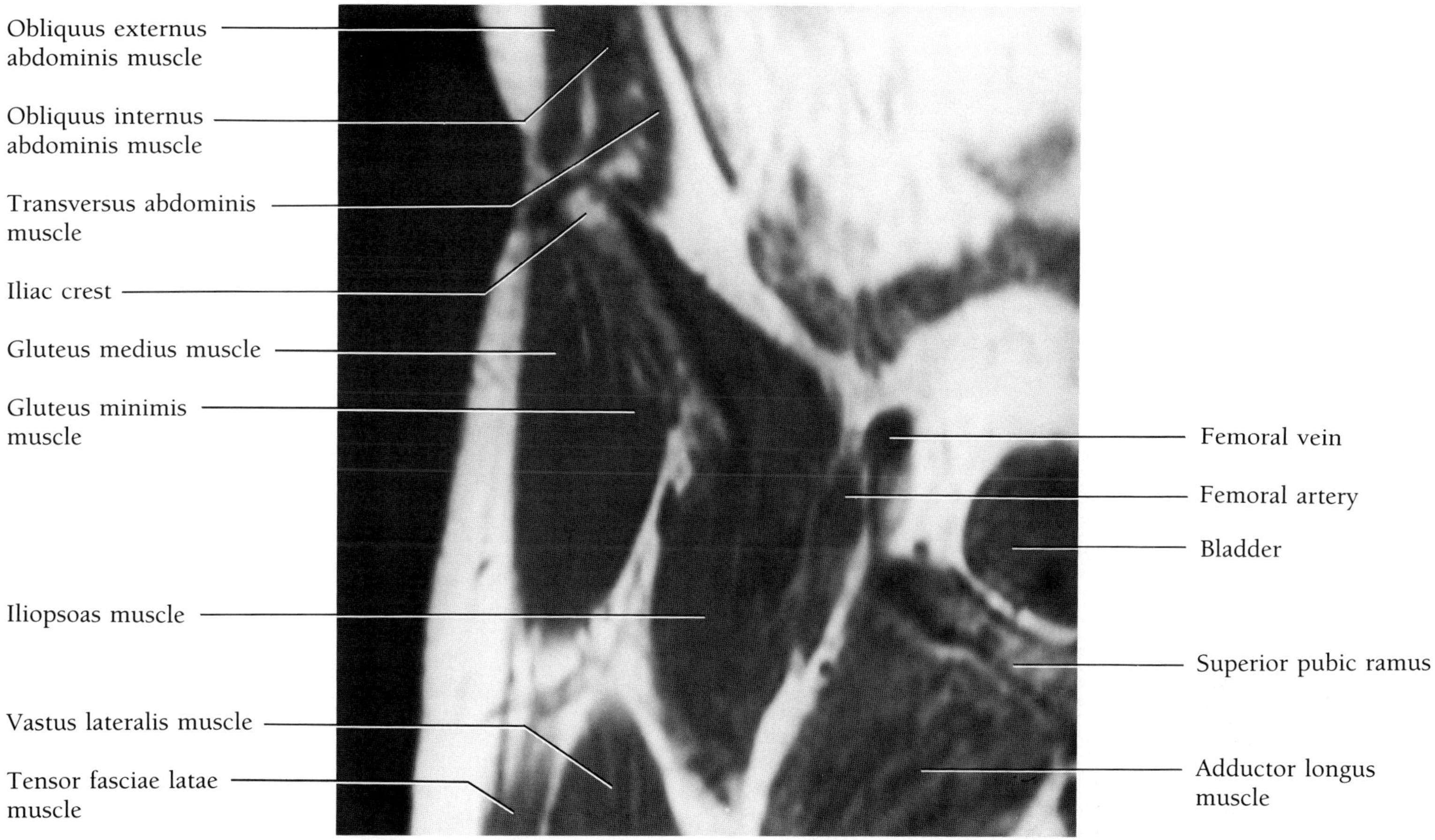

FIGURE 2–12(F)
Coronal image of the hip: Anterior to the hip joint.

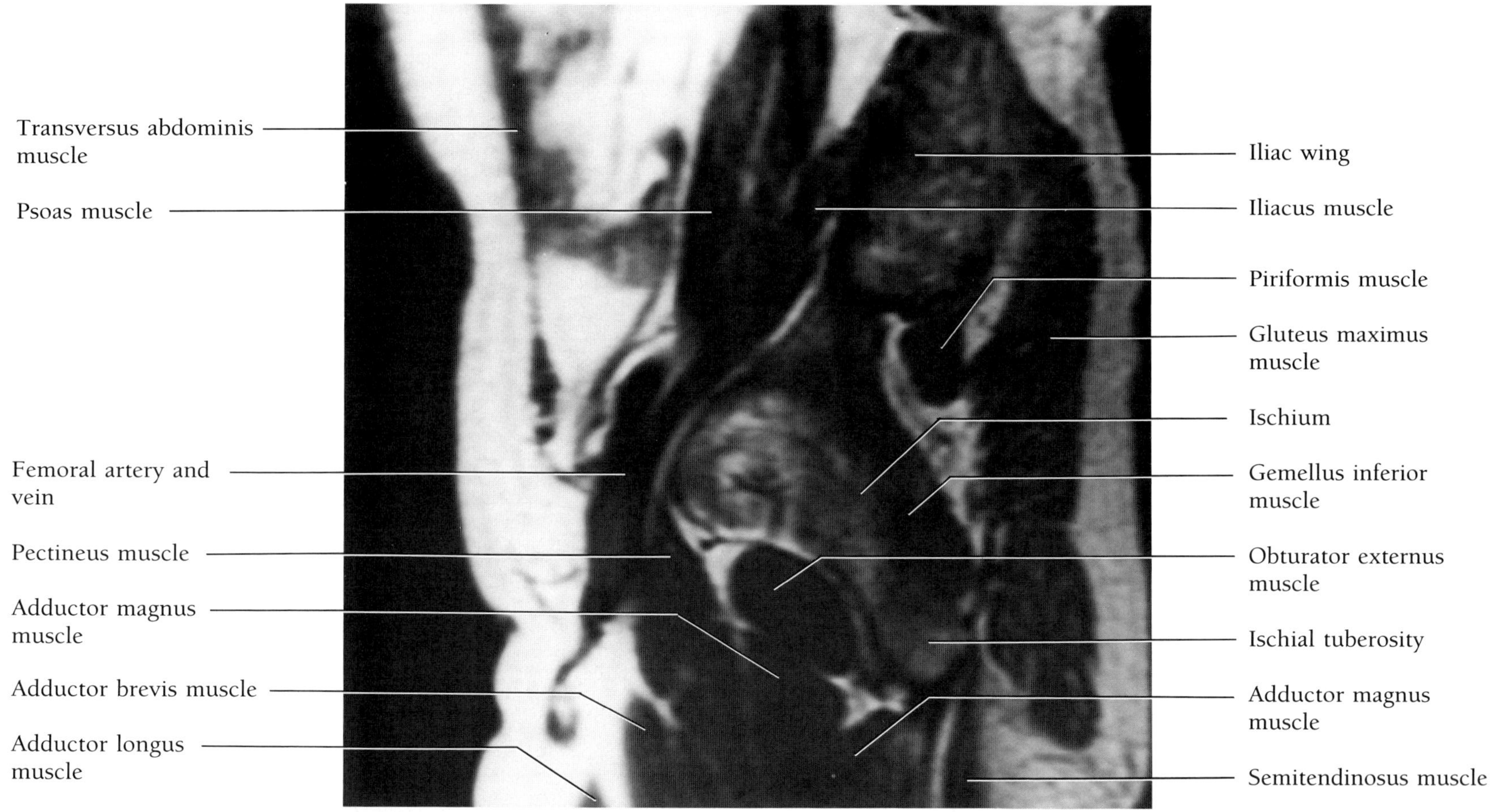

FIGURE 2–13(A)
Sagittal image of the hip: Through the acetabulum.

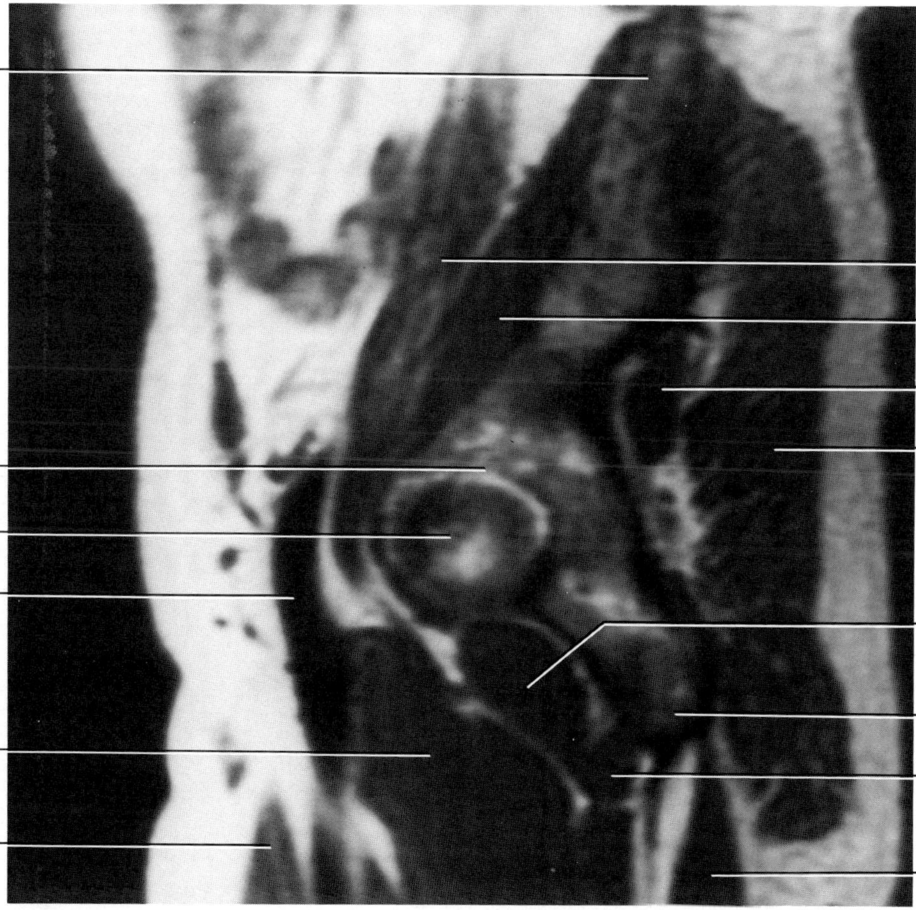

FIGURE 2–13(B)
Sagittal image of the hip: The medial portion of the hip joint.

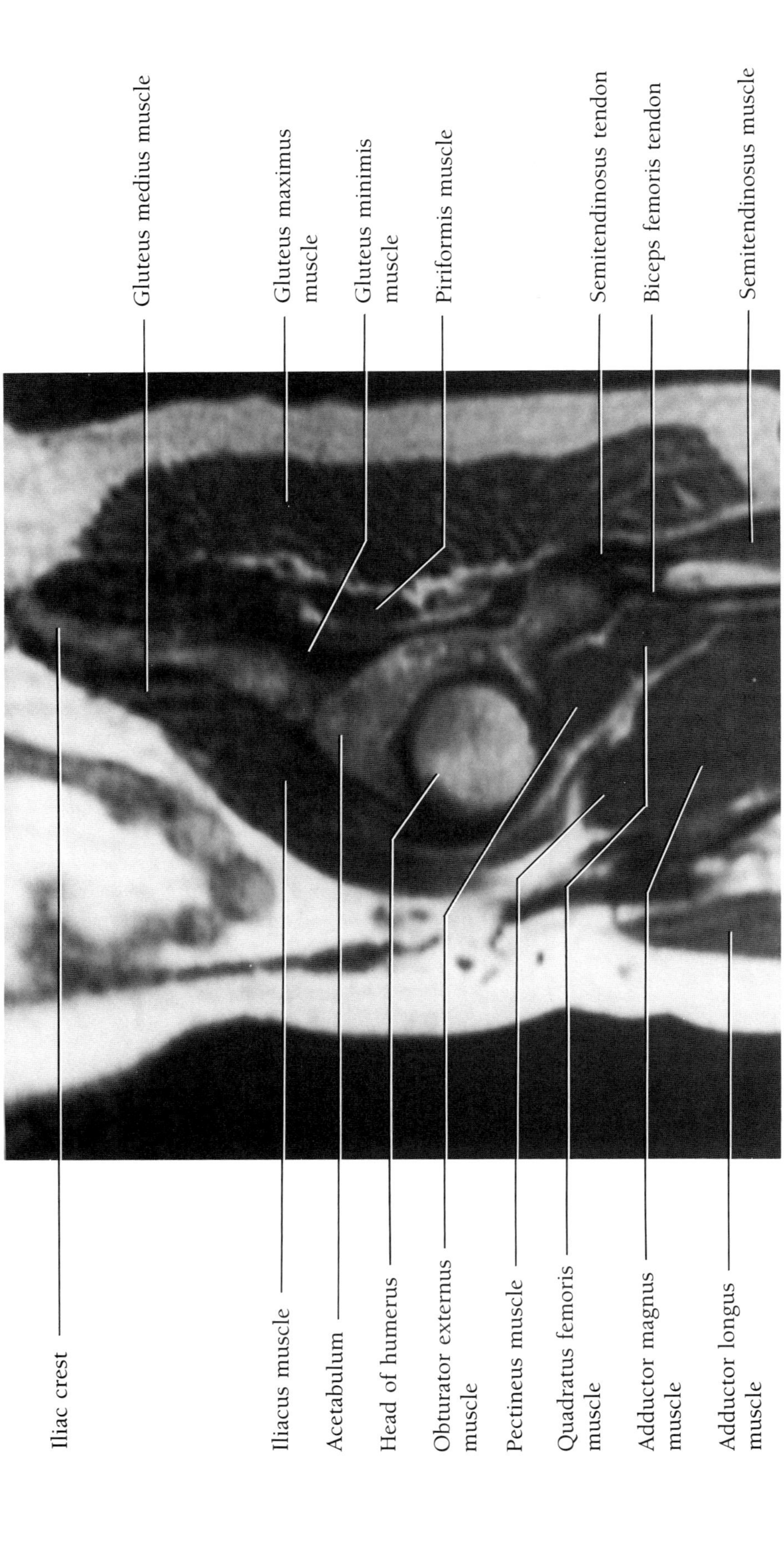

FIGURE 2–13(C)
Sagittal image of the hip: The central portion of the femoral head.

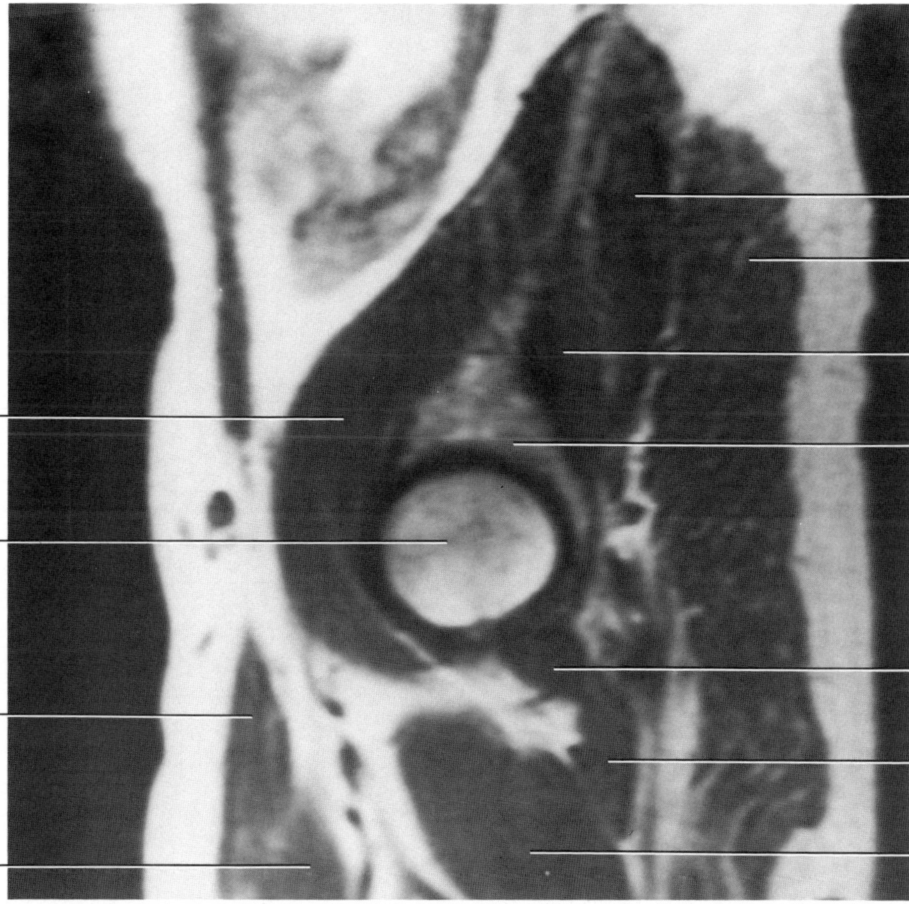

FIGURE 2–13(D)
Sagittal image of the hip: The lateral portion of the femoral head.

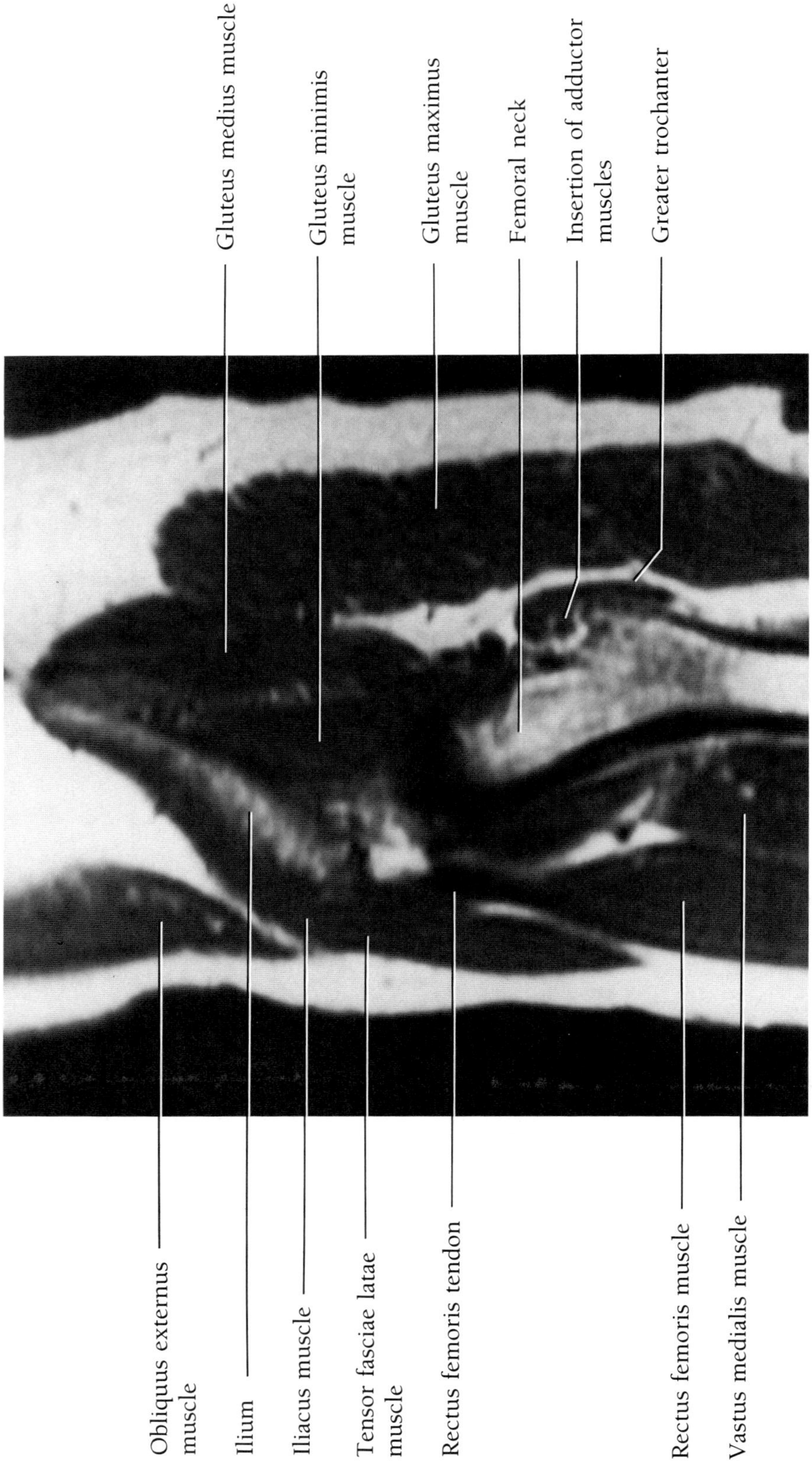

FIGURE 2–13(E)
Sagittal image of the hip: The femoral neck.

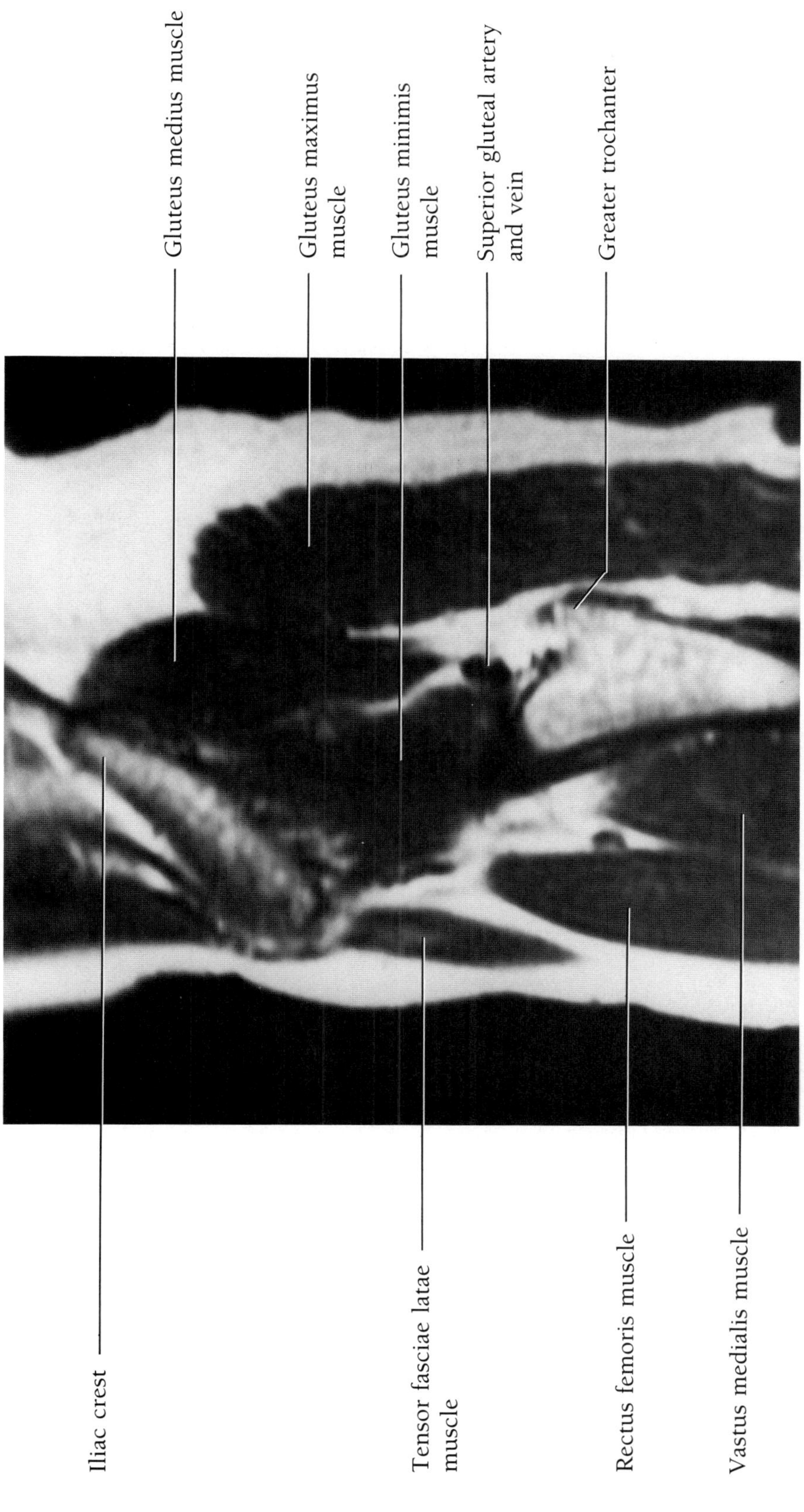

FIGURE 2–13(F)
Sagittal image of the hip: The greater trochanter.

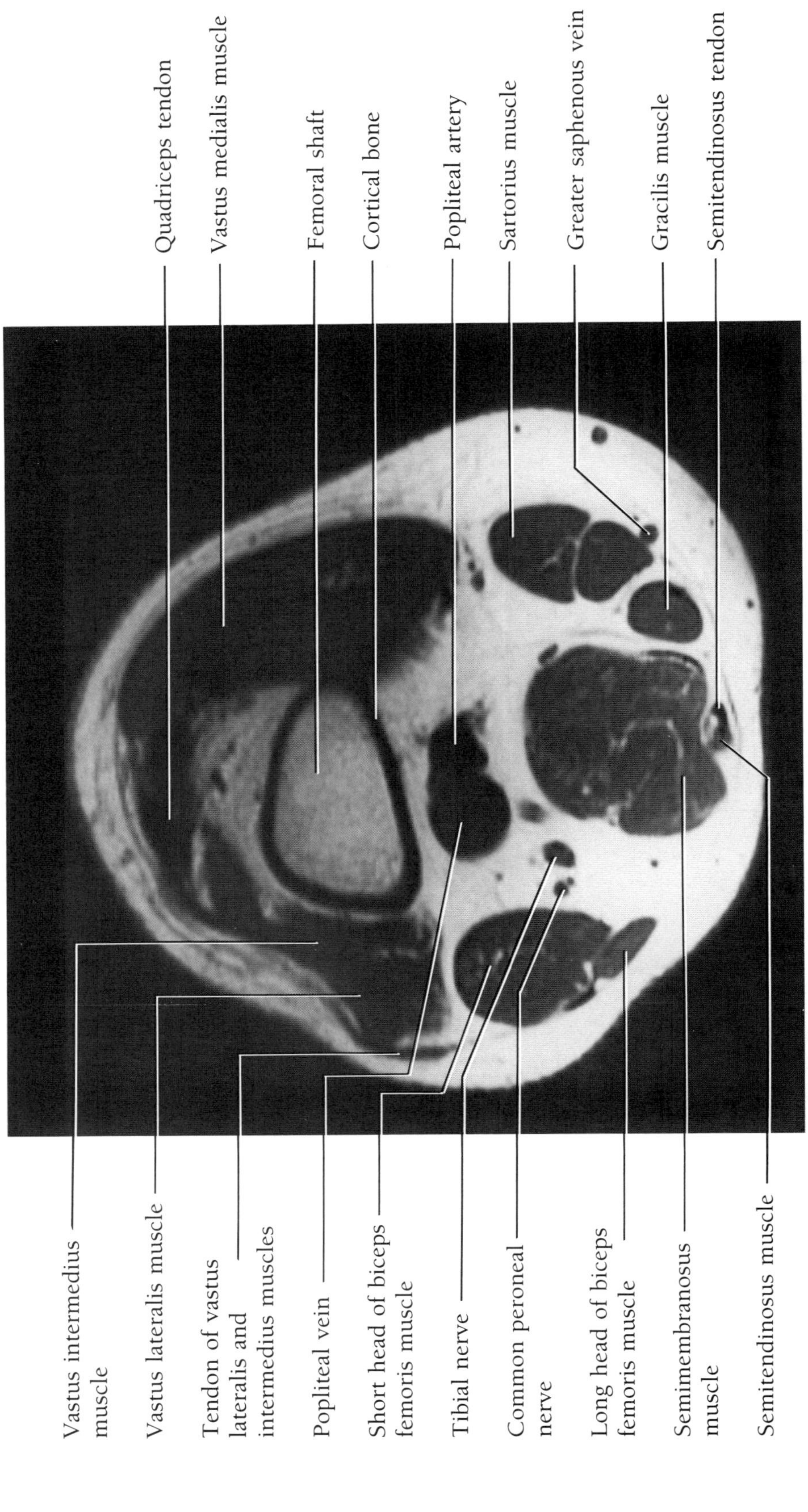

FIGURE 2–14(A)
Axial image of the knee: The level of the distal femur.

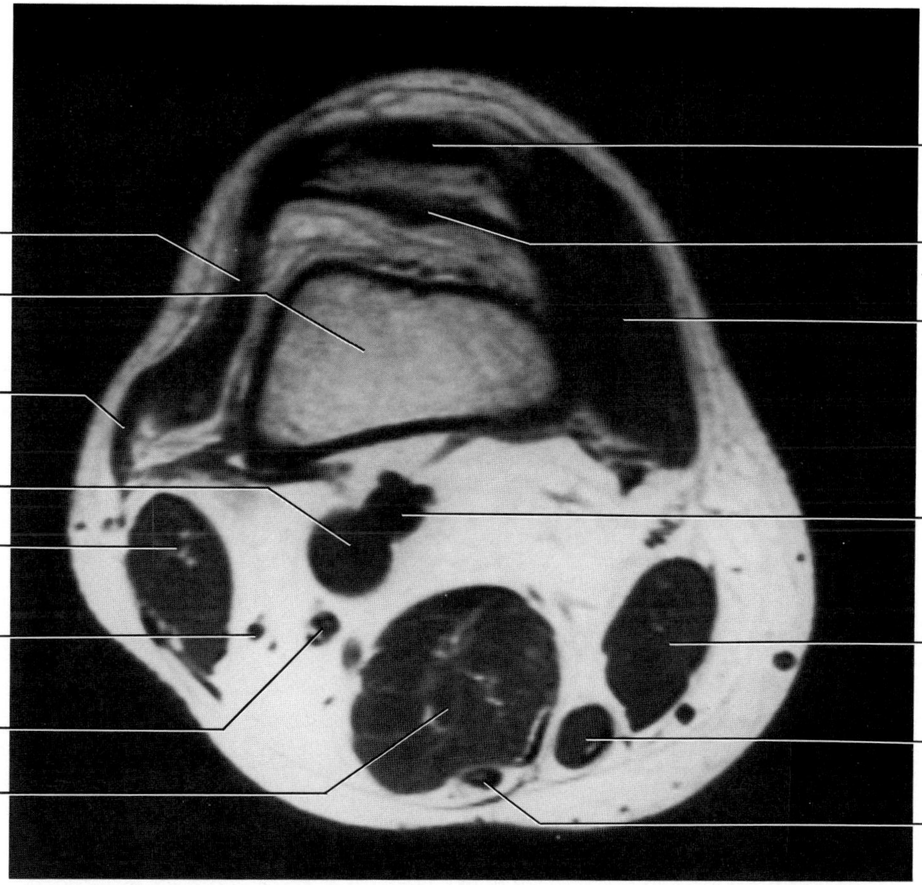

FIGURE 2–14(B)
Axial image of the knee: Just above the patella.

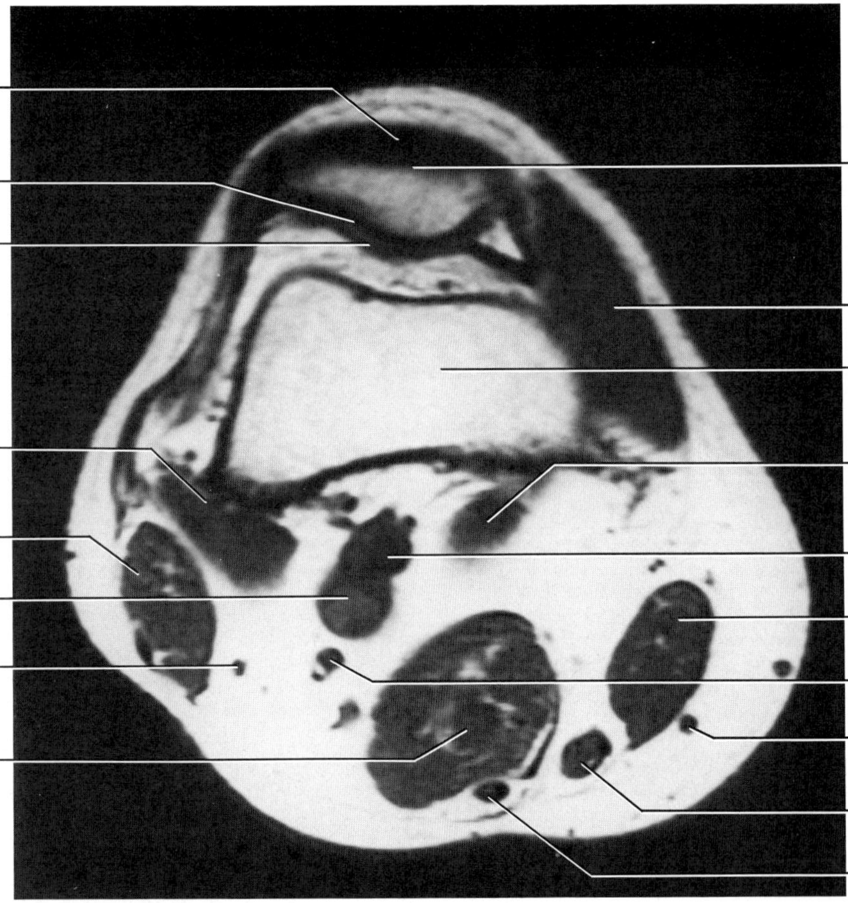

FIGURE 2–14(C)
Axial image of the knee: The superior portion of the patella.

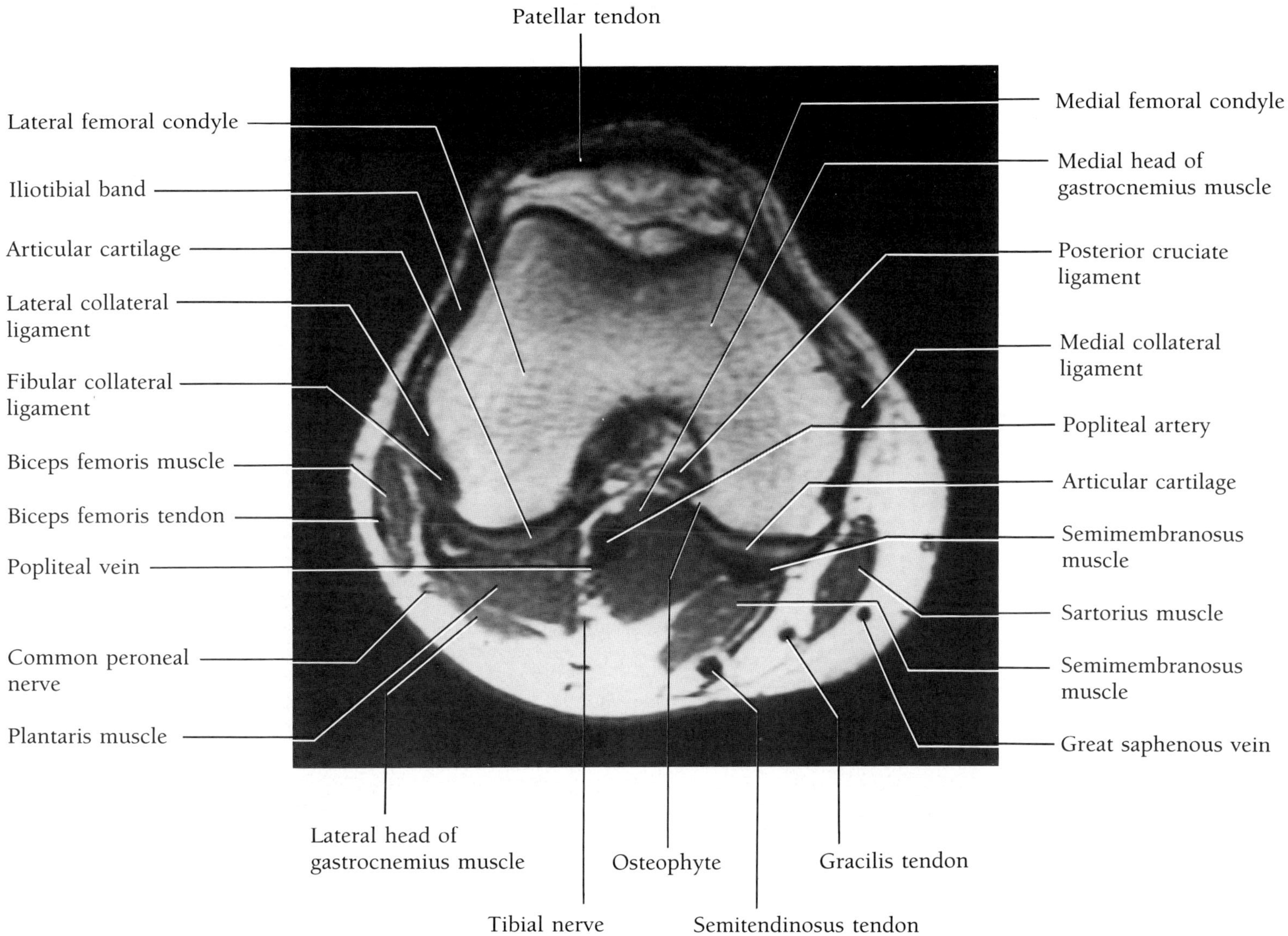

FIGURE 2–14(D)
Axial image of the knee: The femoral condyles.

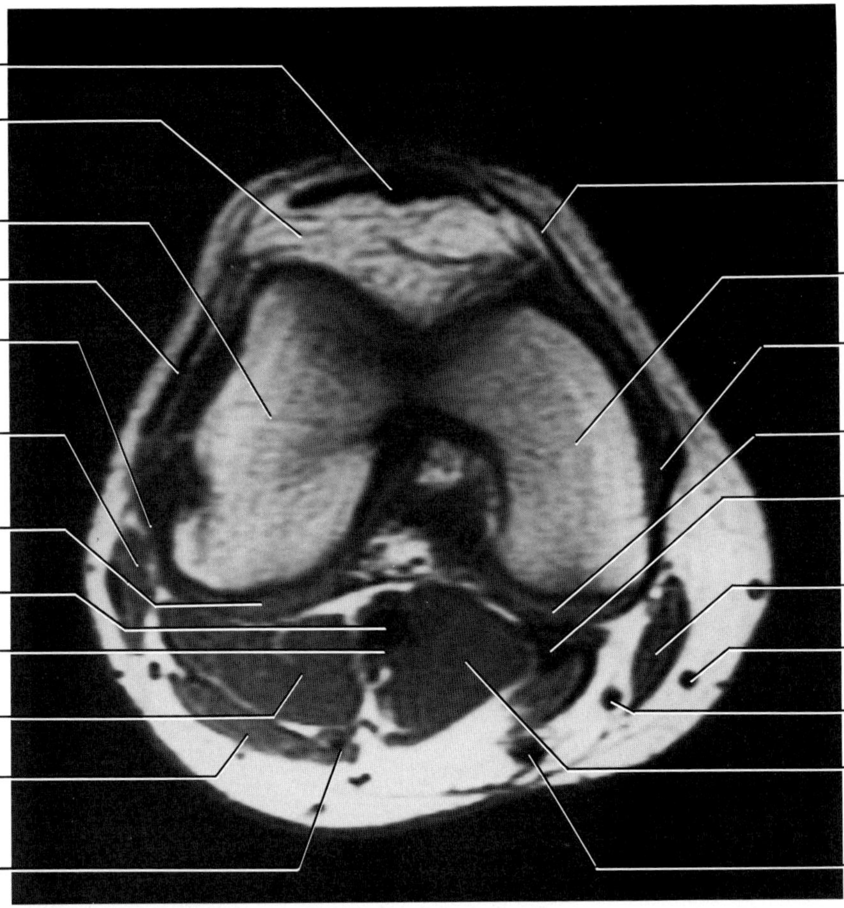

FIGURE 2–14(E)
Axial image of the knee: The distal femoral condyles.

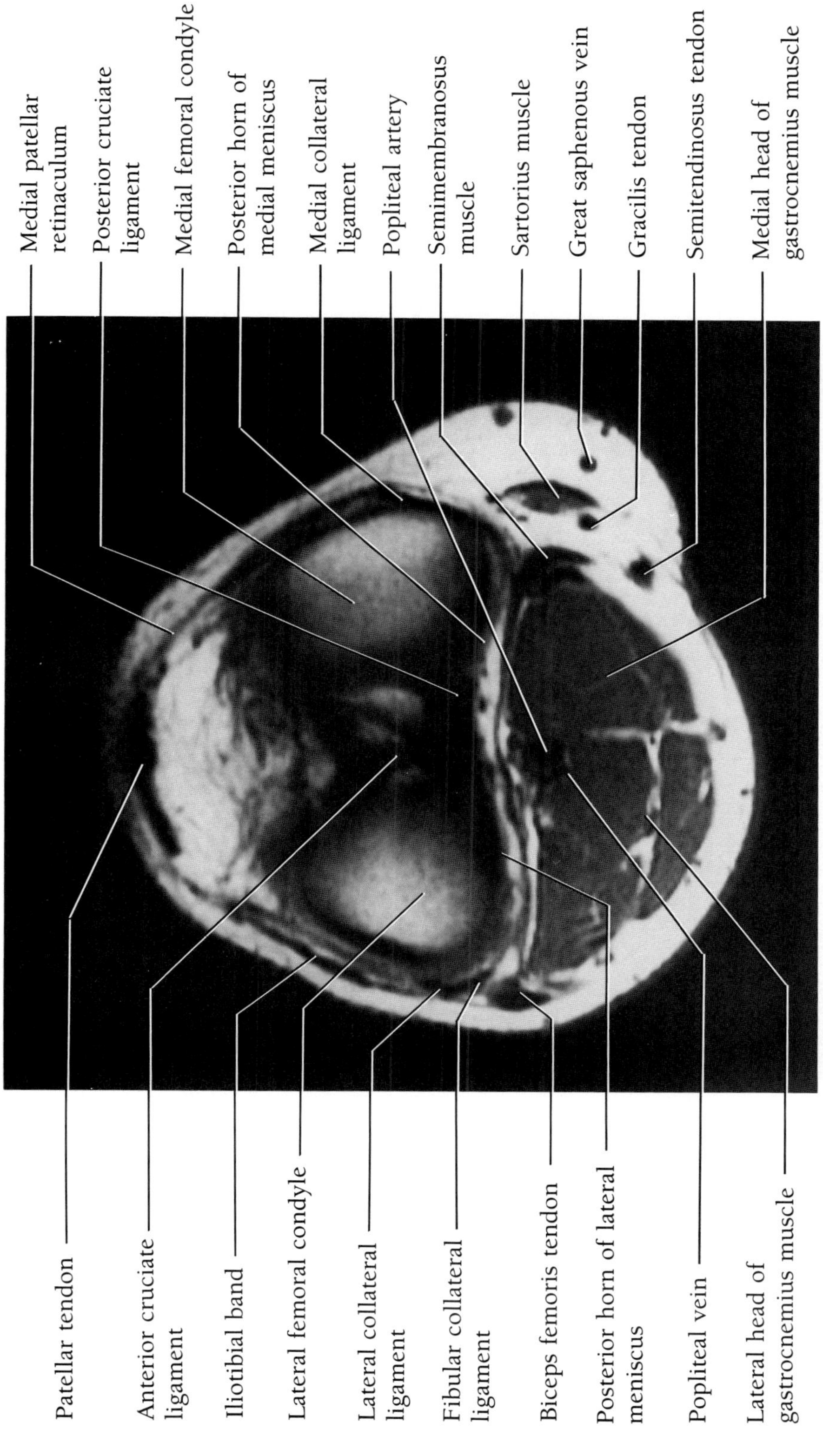

FIGURE 2–14(F)
Axial image of the knee: Through the menisci.

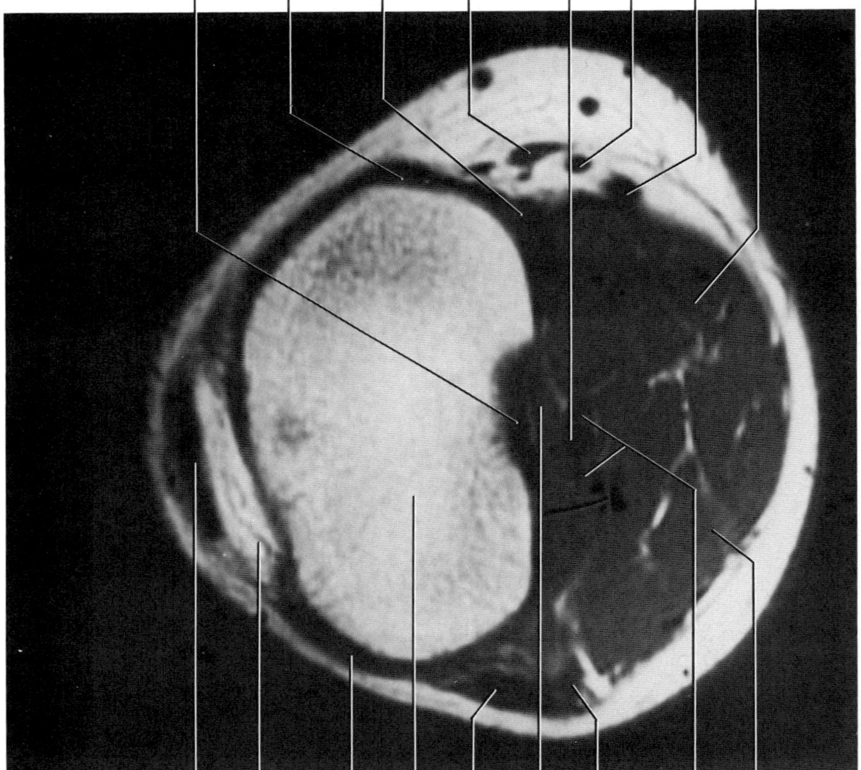

FIGURE 2–14(G)
Axial image of the knee: The tibial plateau.

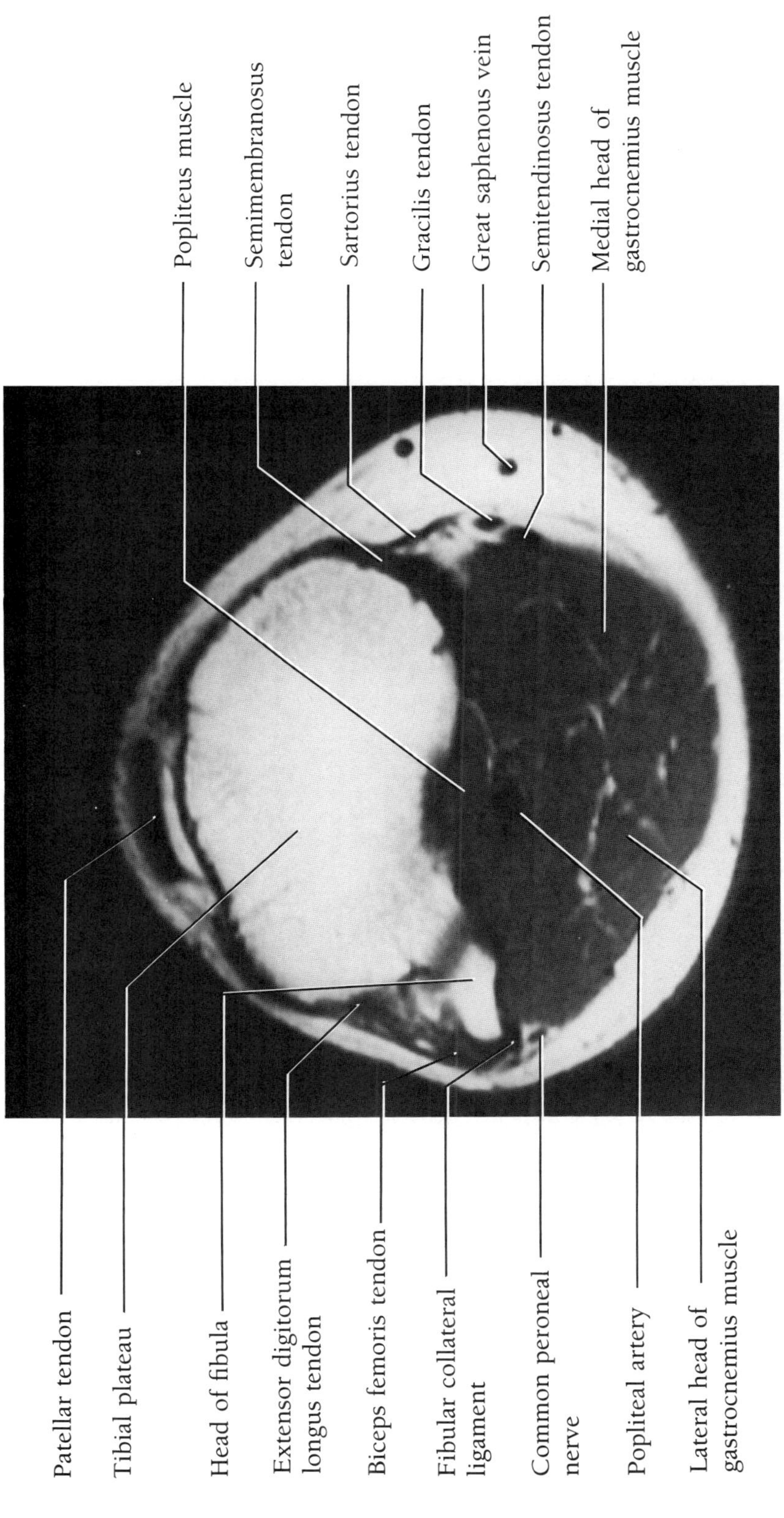

FIGURE 2–14(H)
Axial image of the knee: The fibulotibial joint.

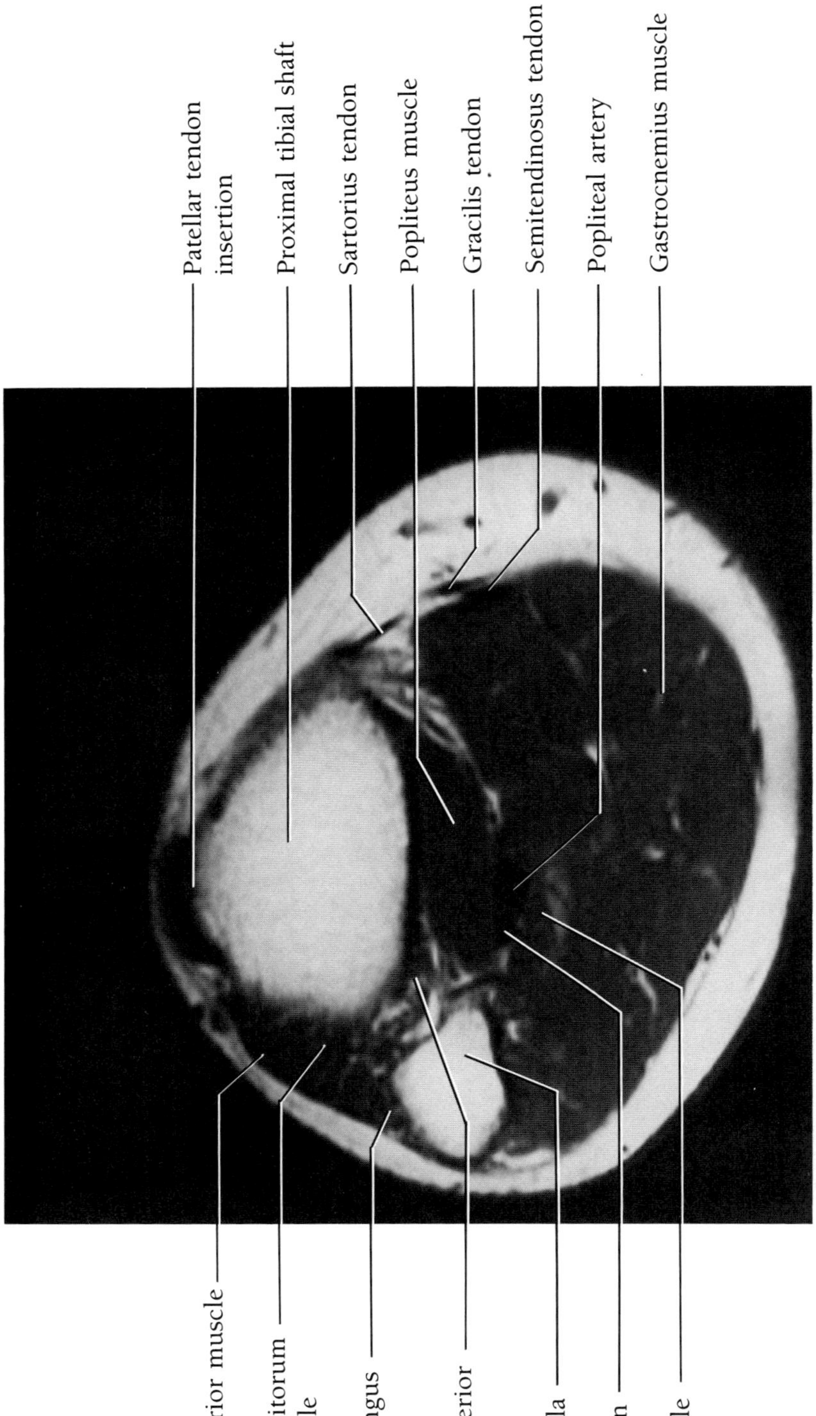

FIGURE 2–14(I)
Axial image of the knee: The proximal tibia and fibula.

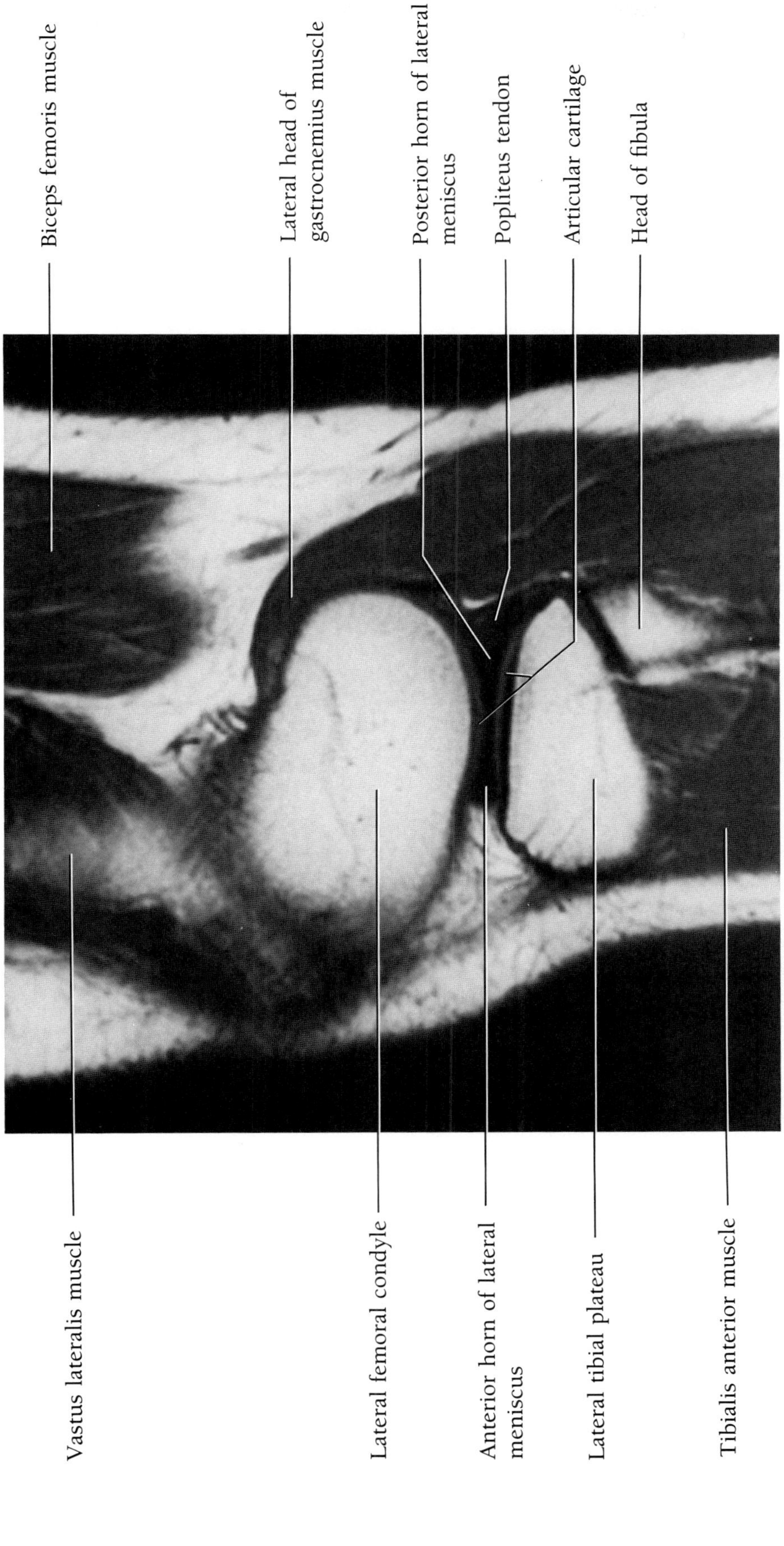

FIGURE 2–15(A)
Sagittal image of the knee: The lateral aspect of the joint.

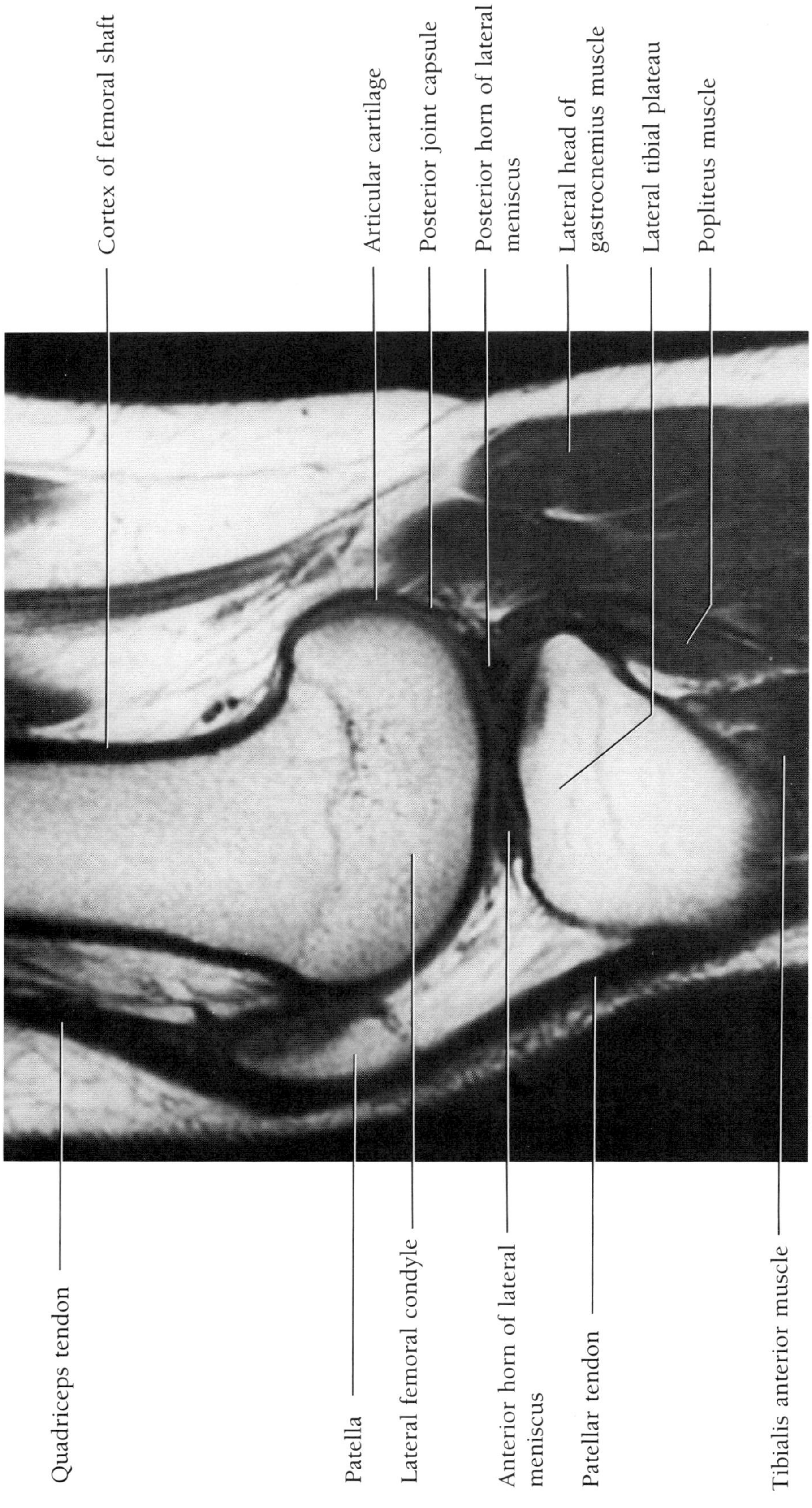

FIGURE 2–15(B)
Sagittal image of the knee: The lateral meniscus medial to the fibula.

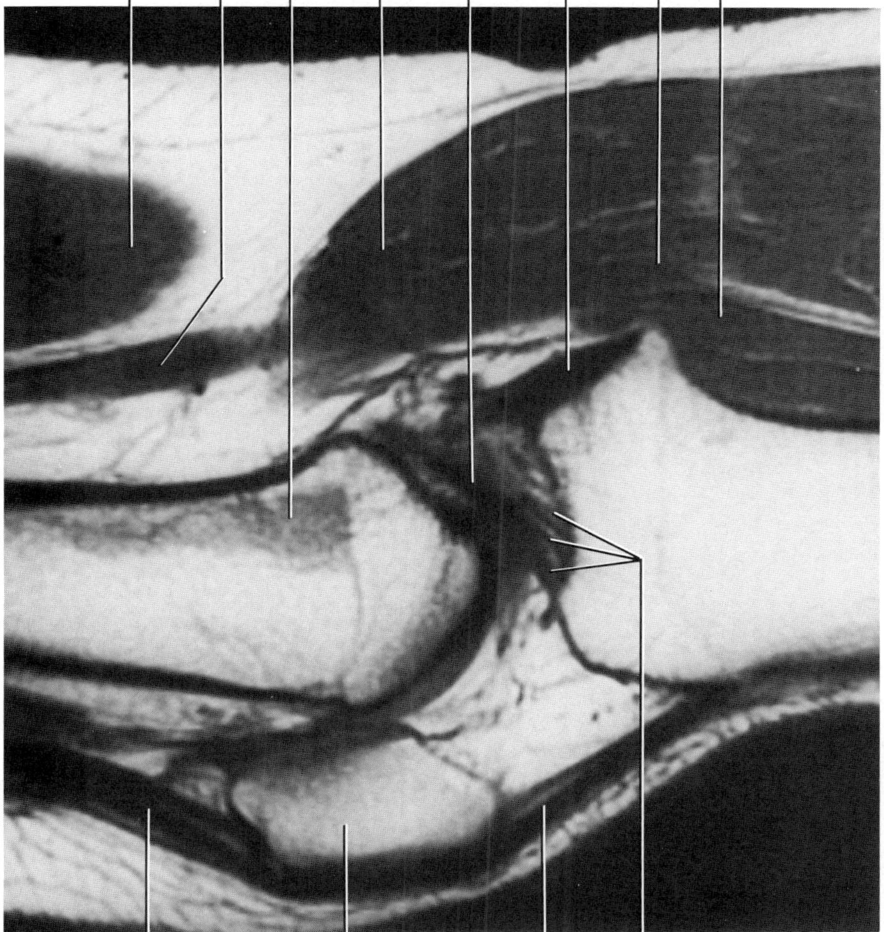

FIGURE 2–15(C)
Sagittal image of the knee: The anterior cruciate ligament.

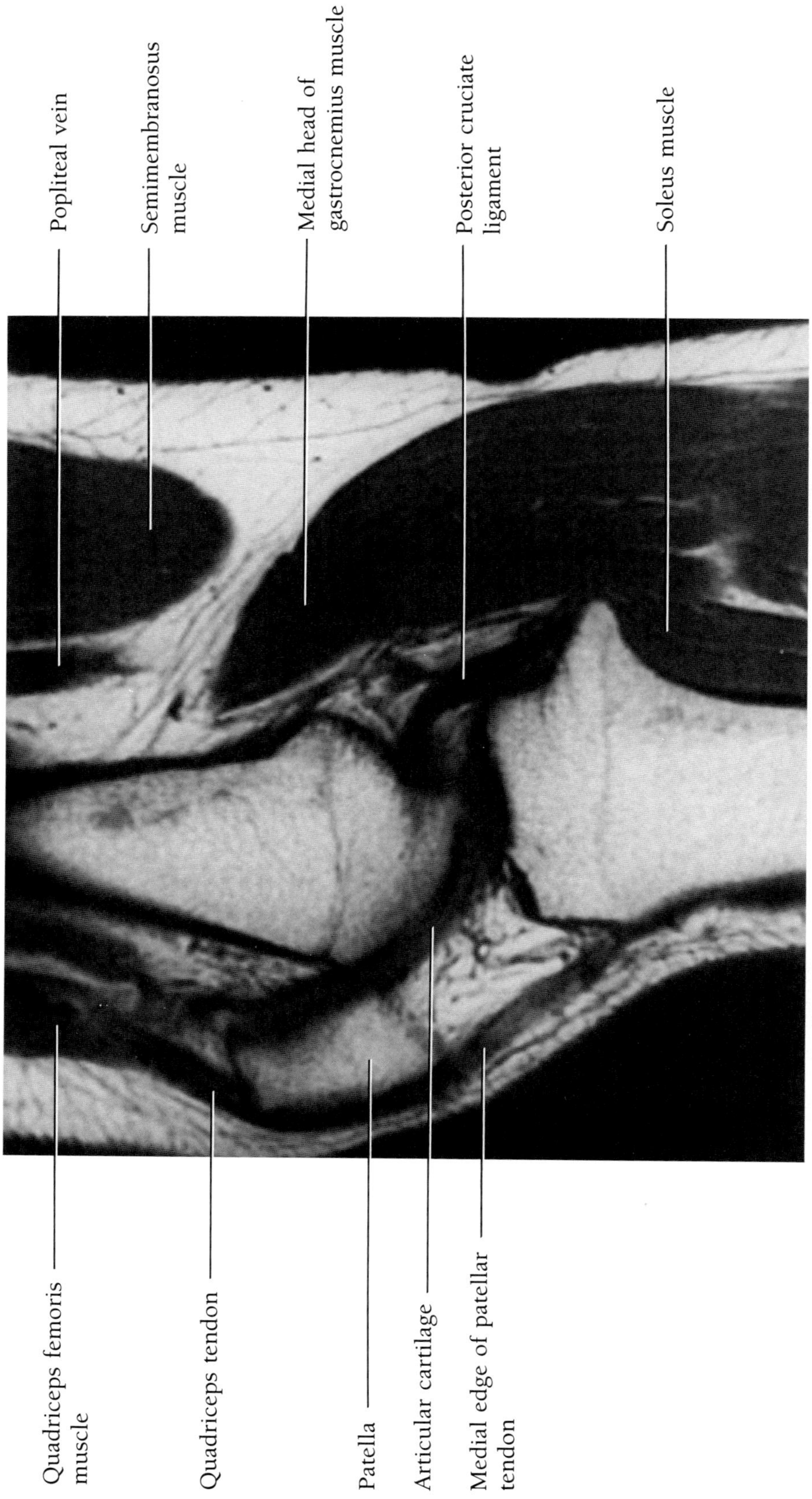

FIGURE 2–15(D)
Sagittal image of the knee: The posterior cruciate ligament.

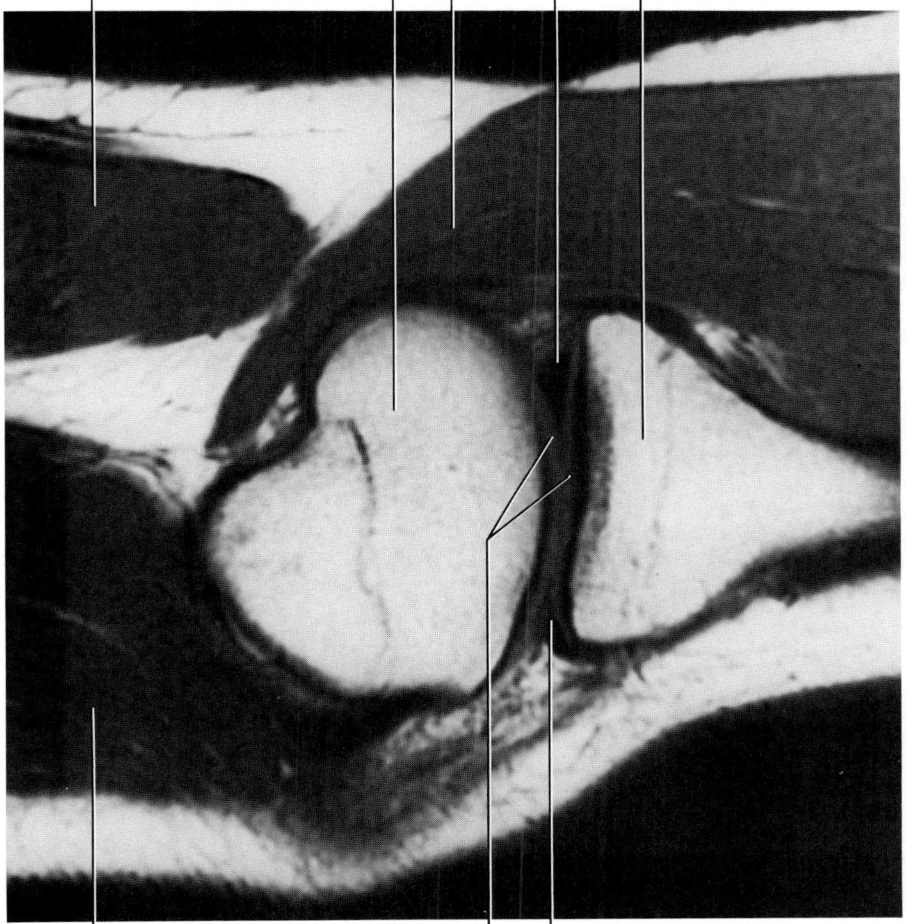

FIGURE 2–15(E)
Sagittal image of the knee: The central portion of the medial meniscus.

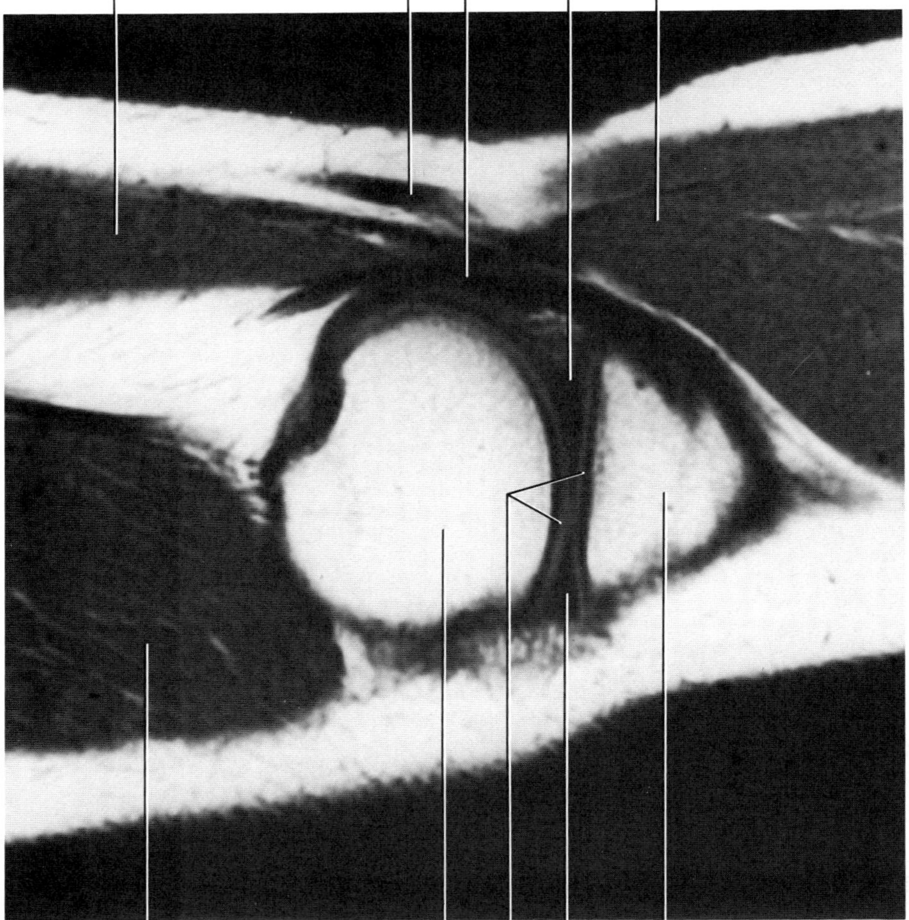

FIGURE 2–15(F)
Sagittal image of the knee: The medial portion of the medial meniscus.

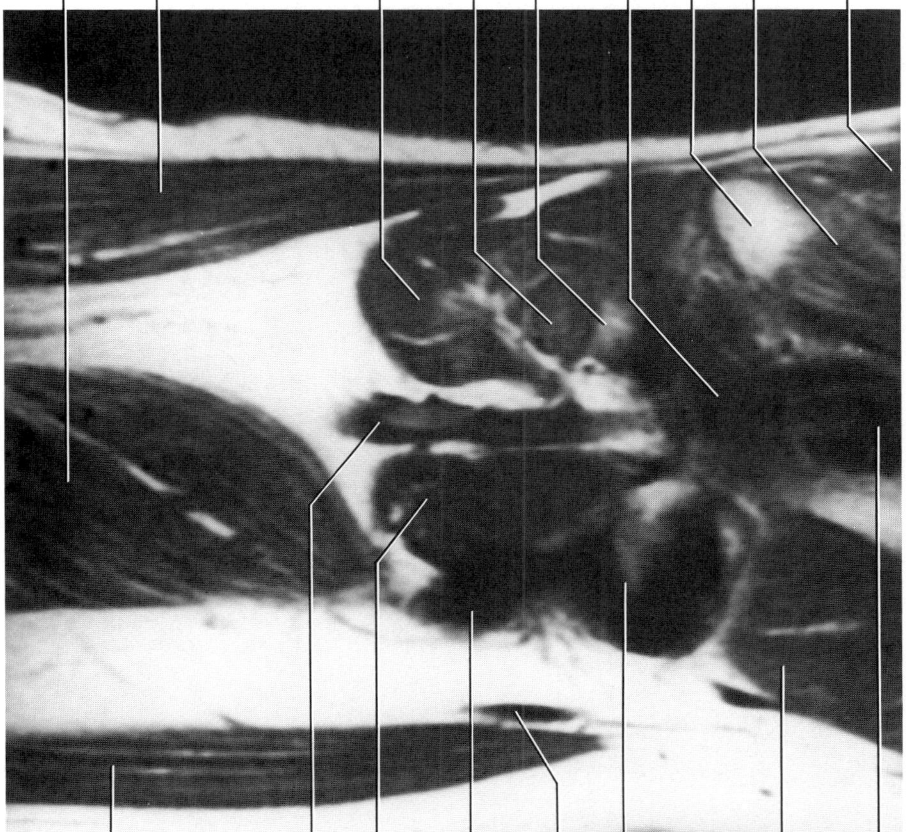

FIGURE 2–16(A)
Coronal image of the knee: Posterior to the condyles.

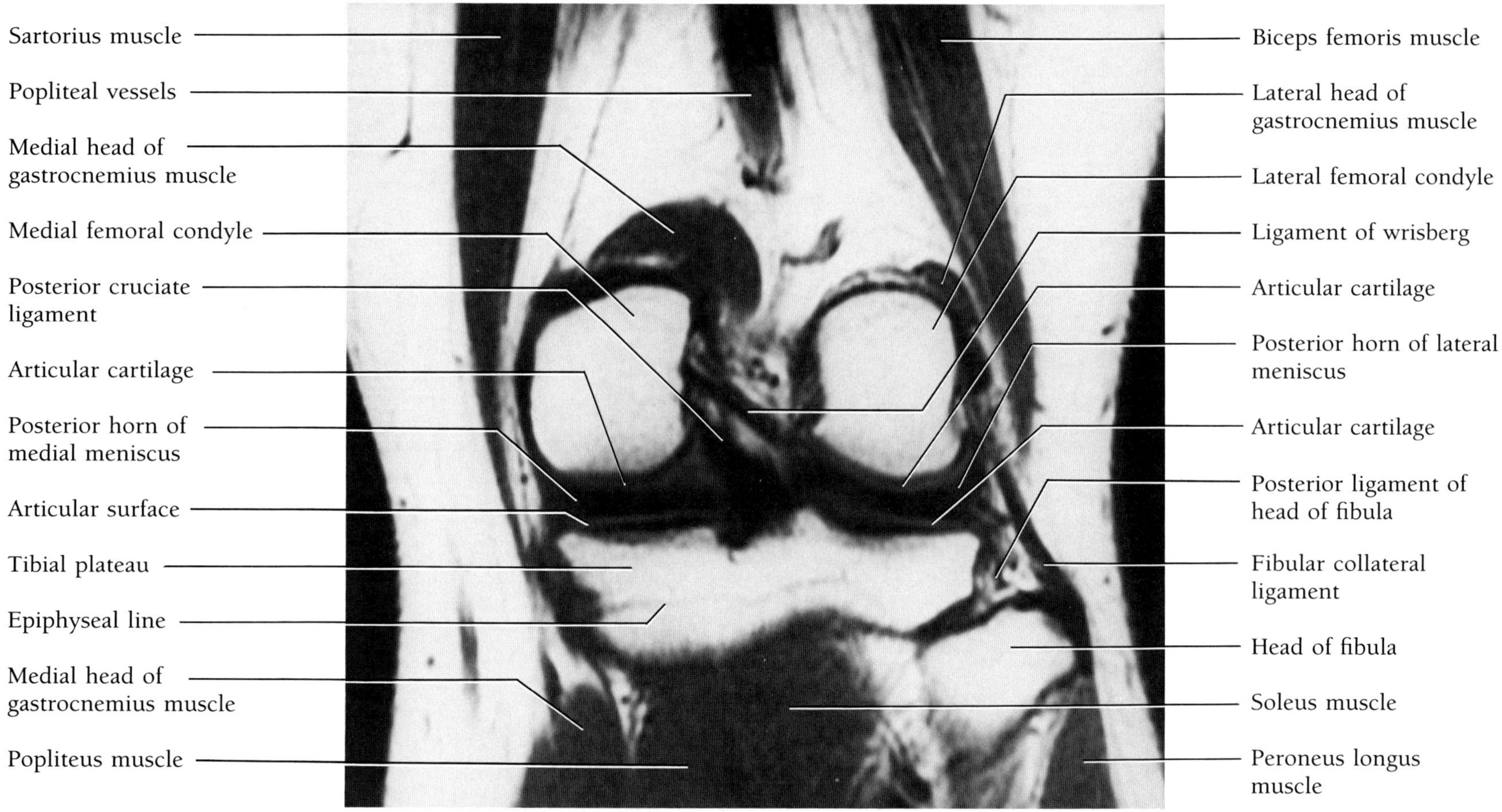

FIGURE 2–16(B)
Coronal image of the knee: The posterior portion of the condyles.

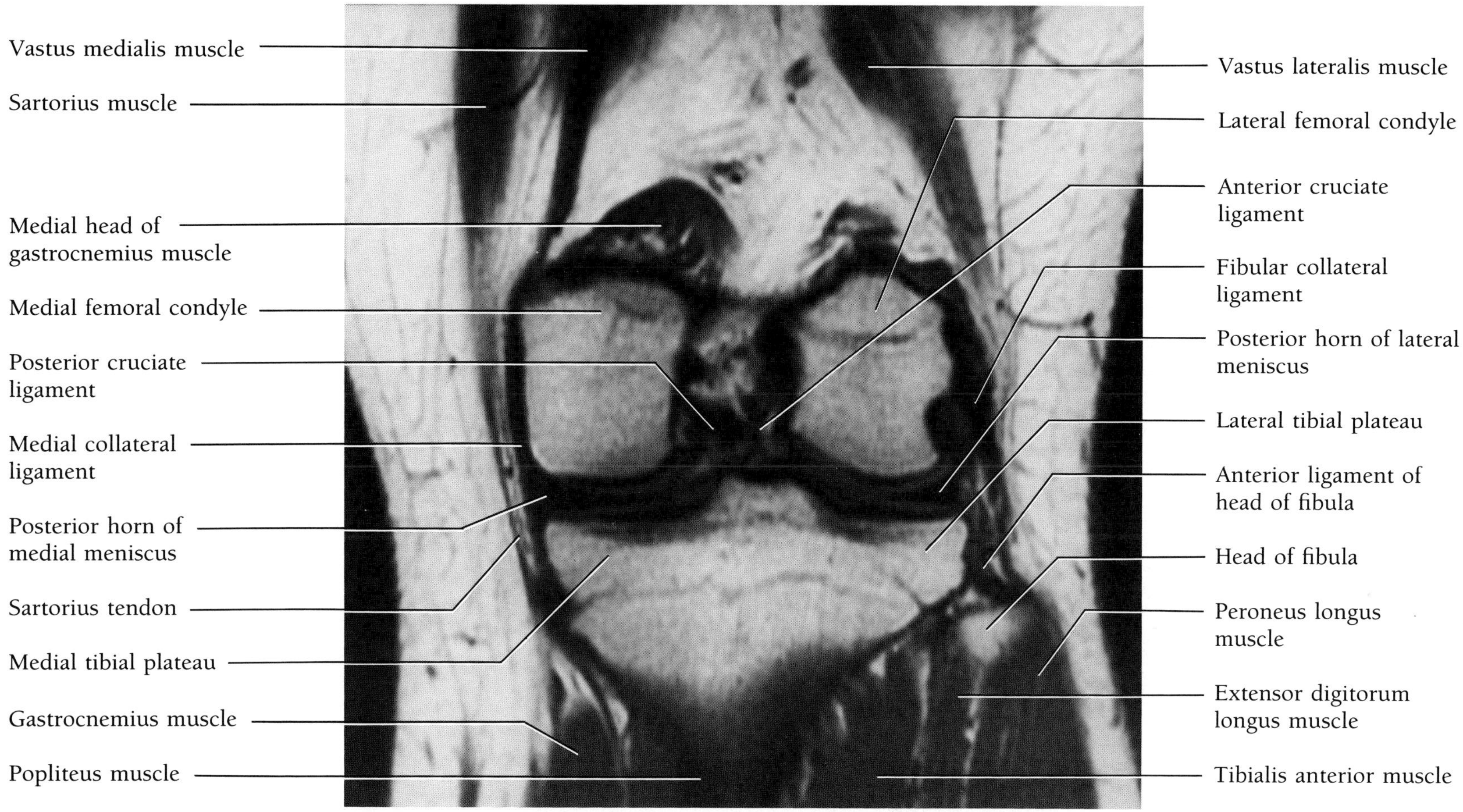

FIGURE 2–16(C)
Coronal image of the knee: The anterior portion of the fibula.

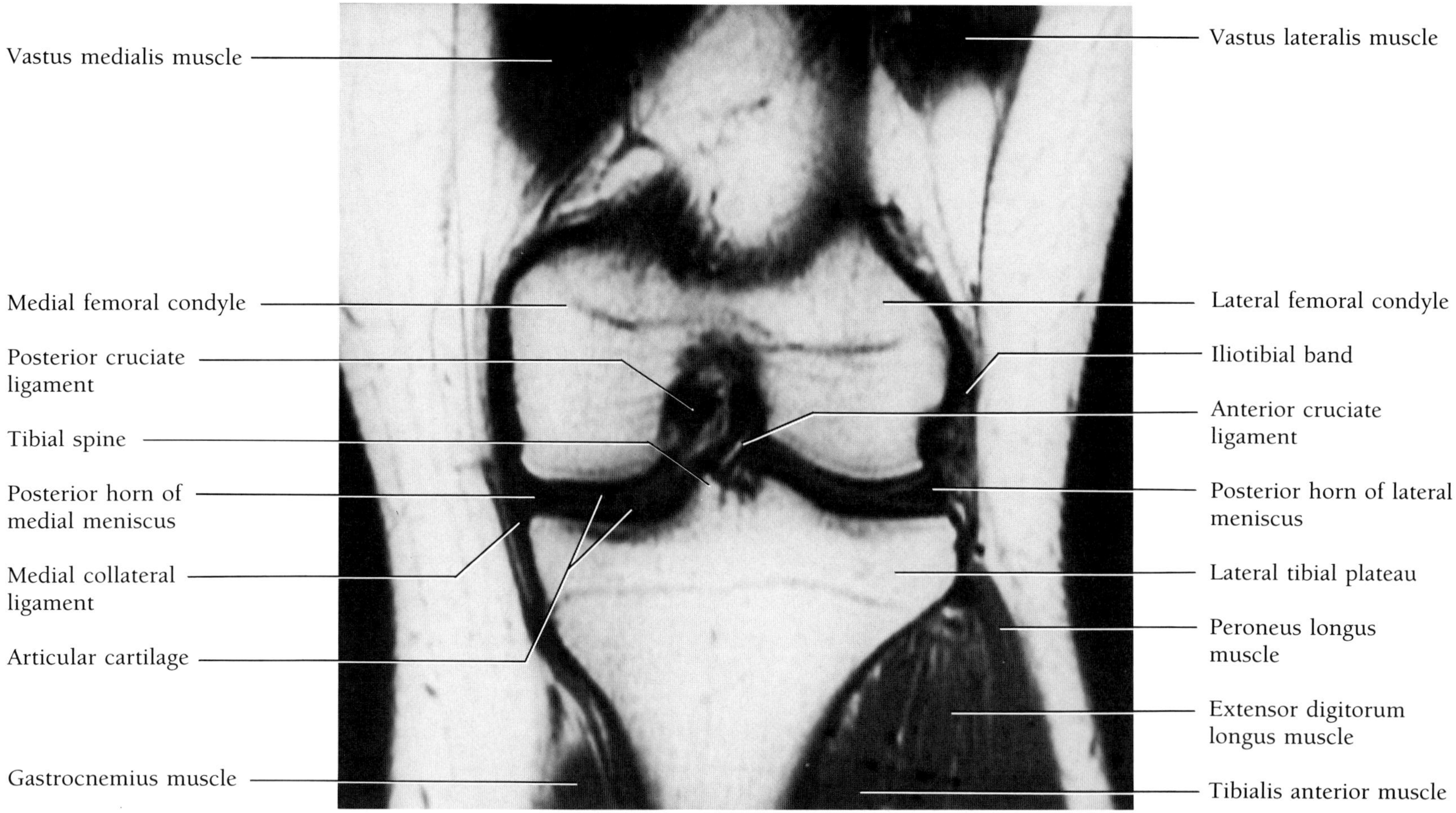

FIGURE 2–16(D)
Coronal image of the knee: The plane of the tibial spines.

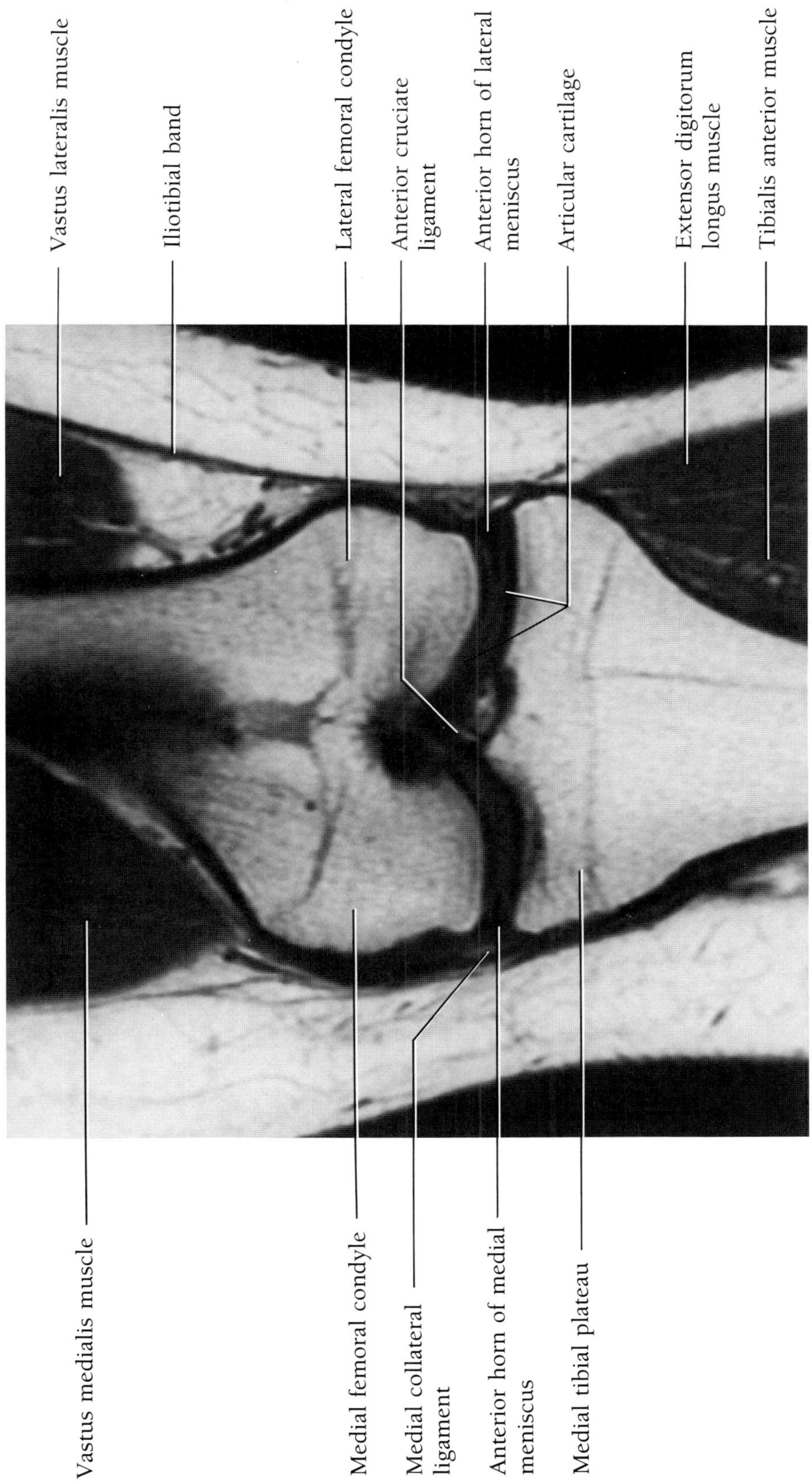

FIGURE 2–16(E)
Coronal image of the knee: Anterior to (D).

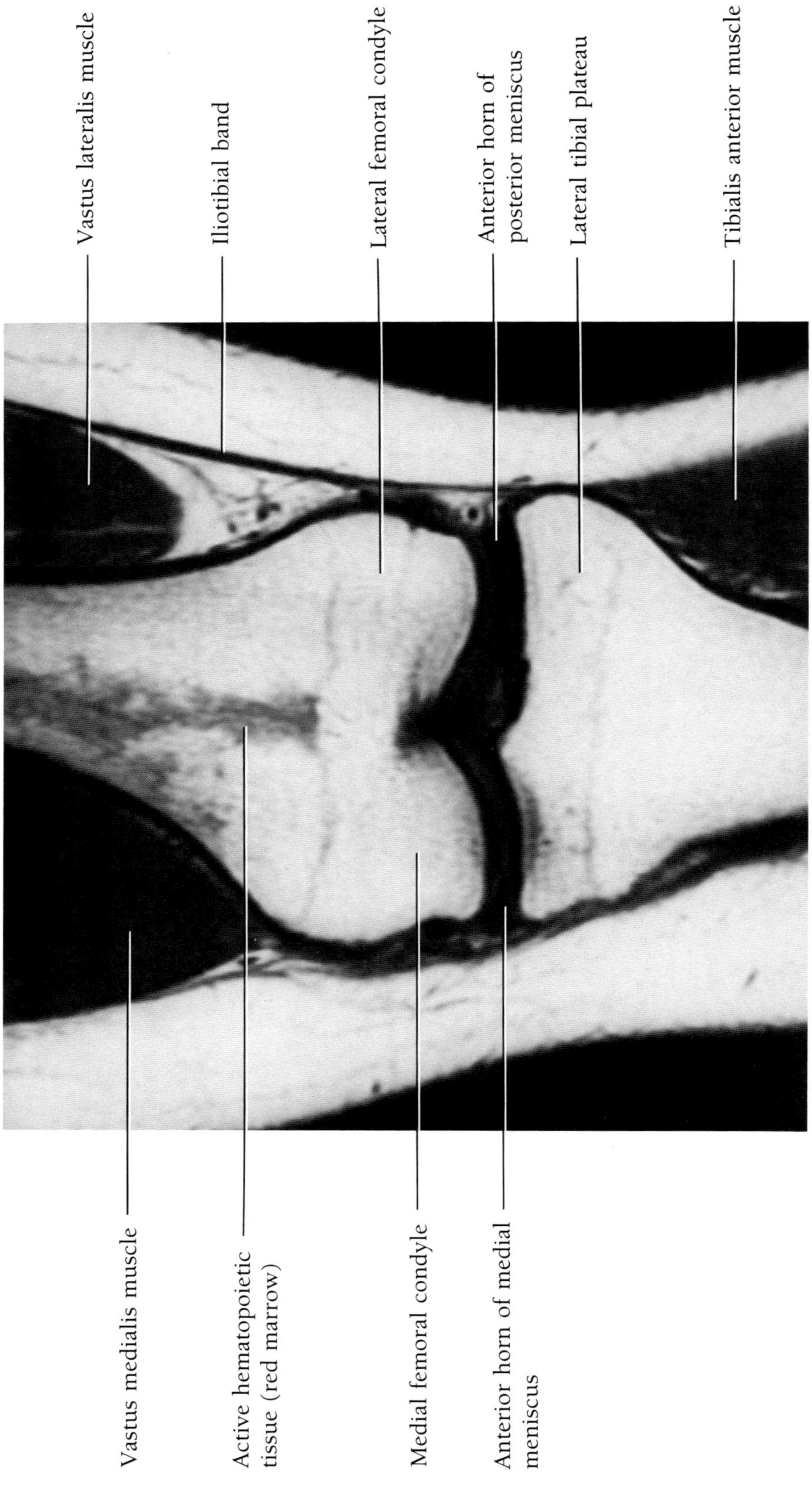

FIGURE 2–16(F)
Coronal image of the knee: Anterior to (E).

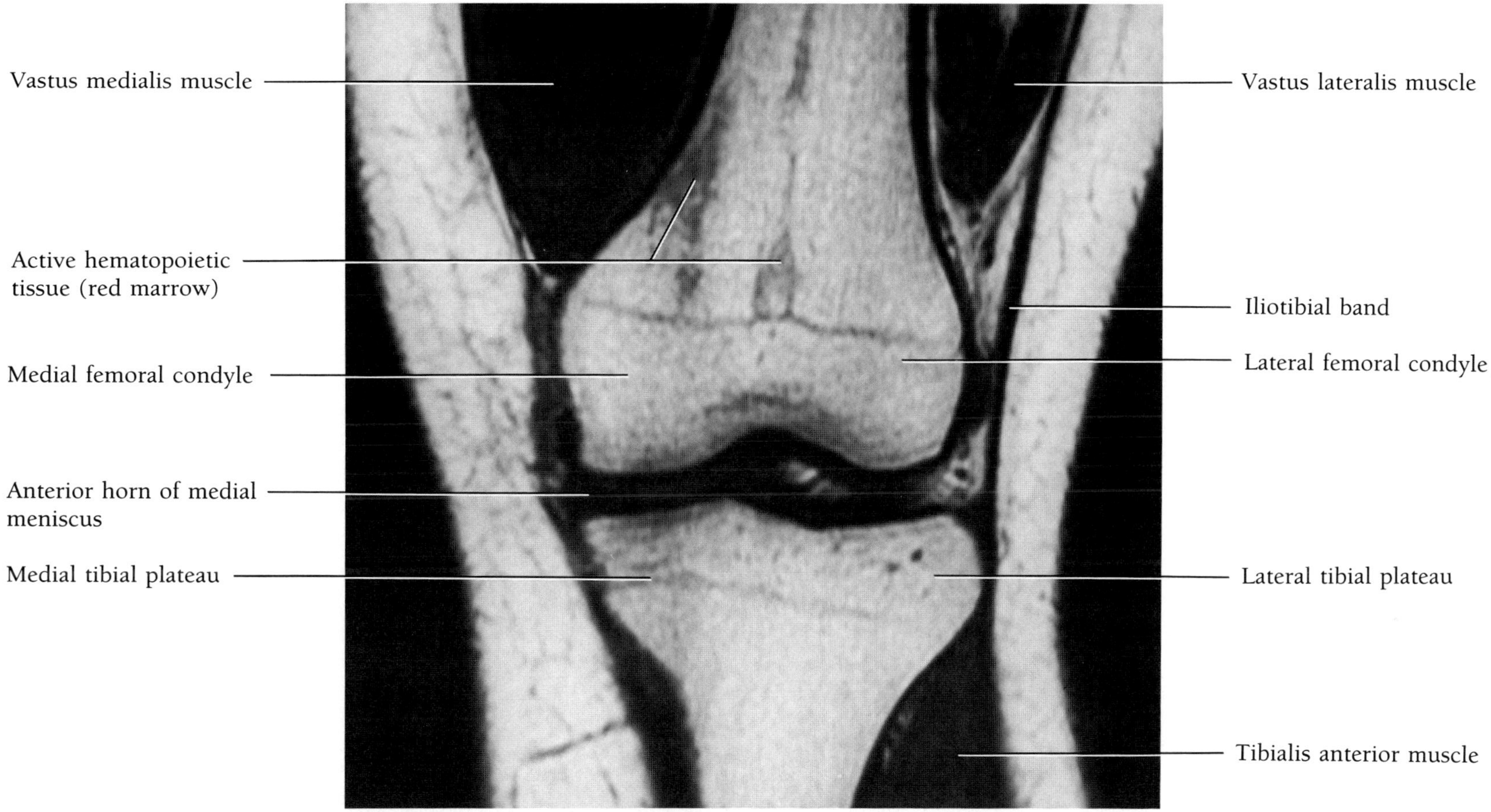

FIGURE 2–16(G)
Coronal image of the knee: Anterior to (F).

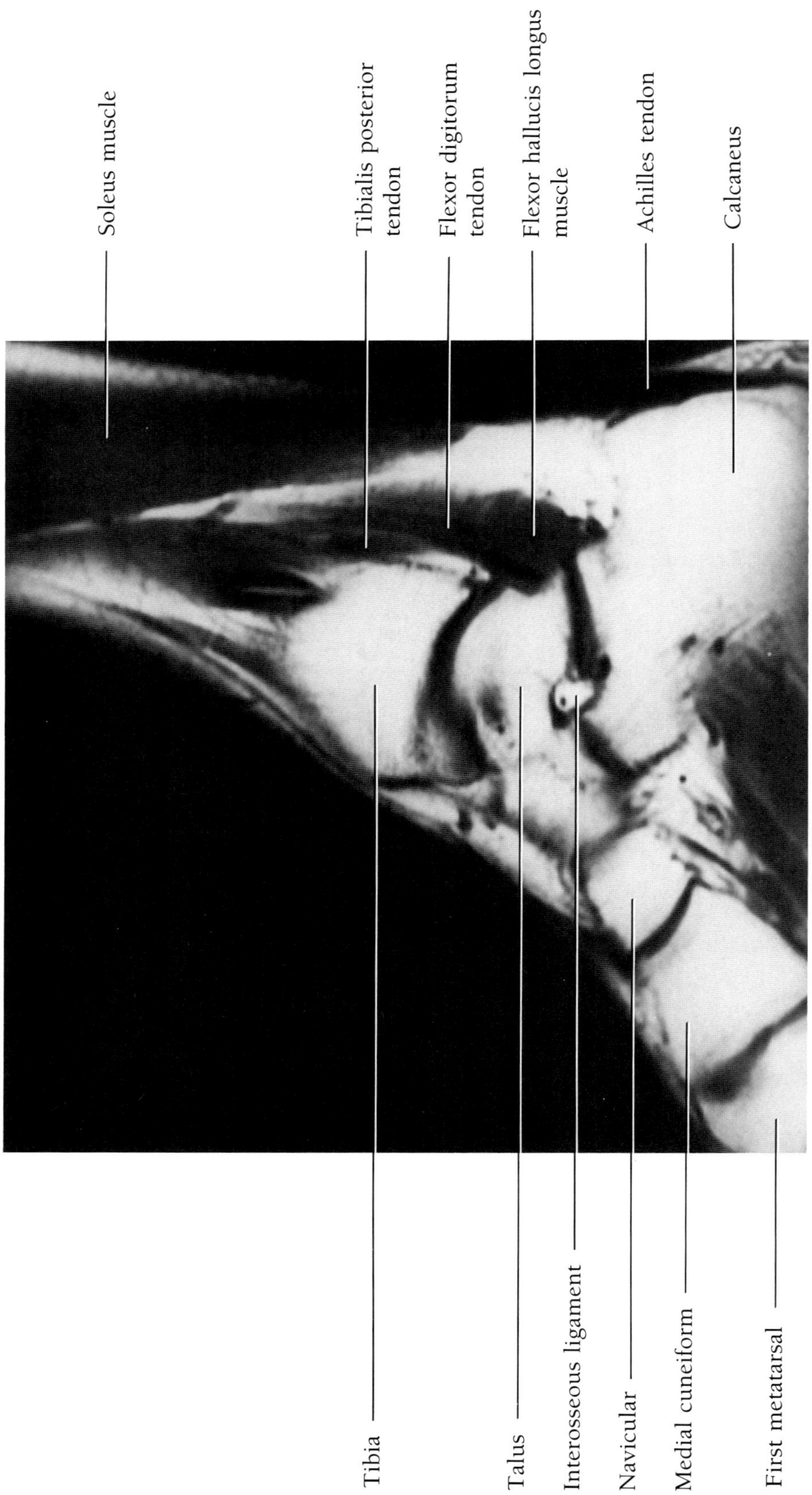

A

FIGURE 2–17
Sagittal images of the ankle and foot. A–E. *The medial aspect successively to the lateral aspect.*

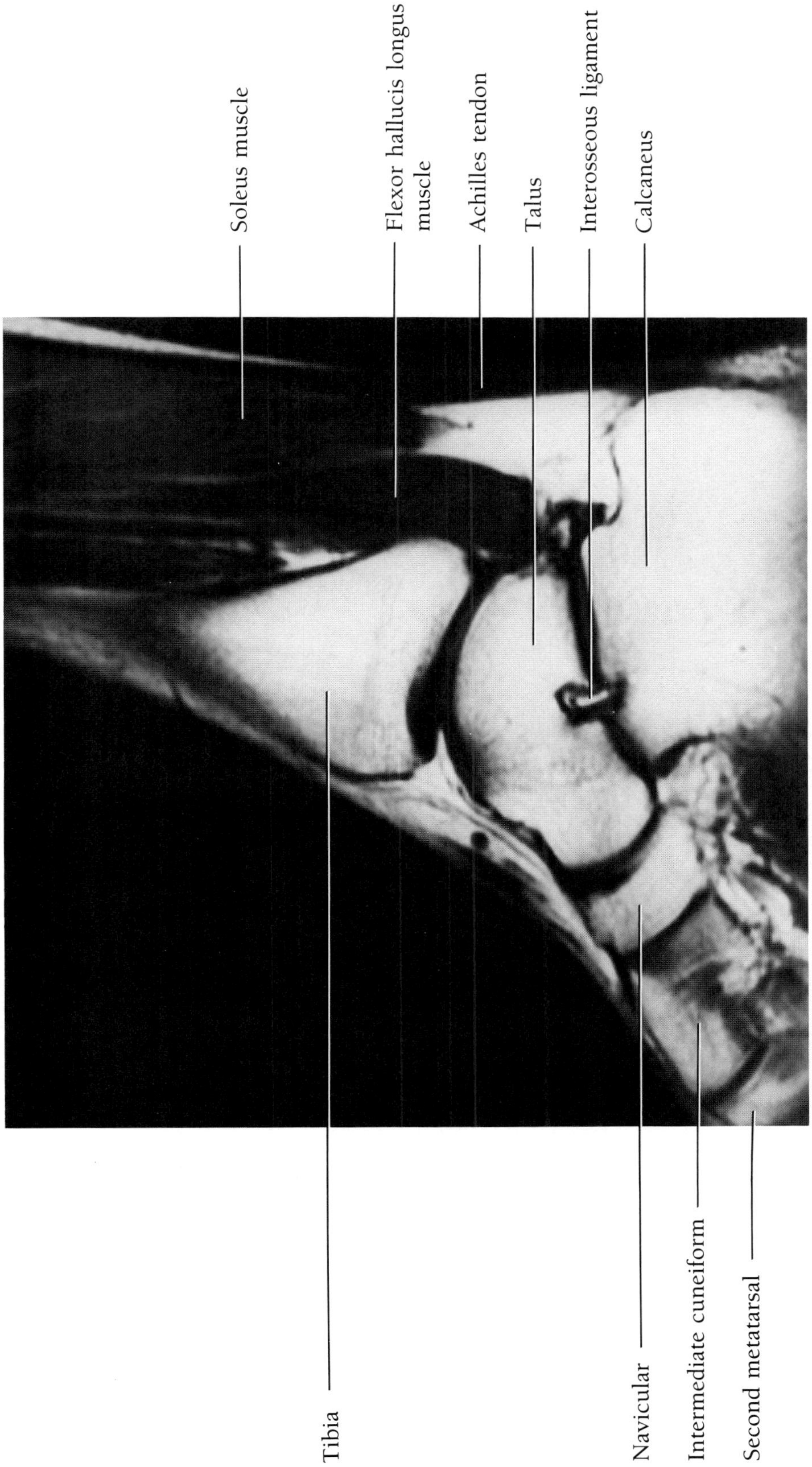
Soleus muscle
Flexor hallucis longus muscle
Achilles tendon
Talus
Interosseous ligament
Calcaneus
Tibia
Navicular
Intermediate cuneiform
Second metatarsal

FIGURE 2–17(B)

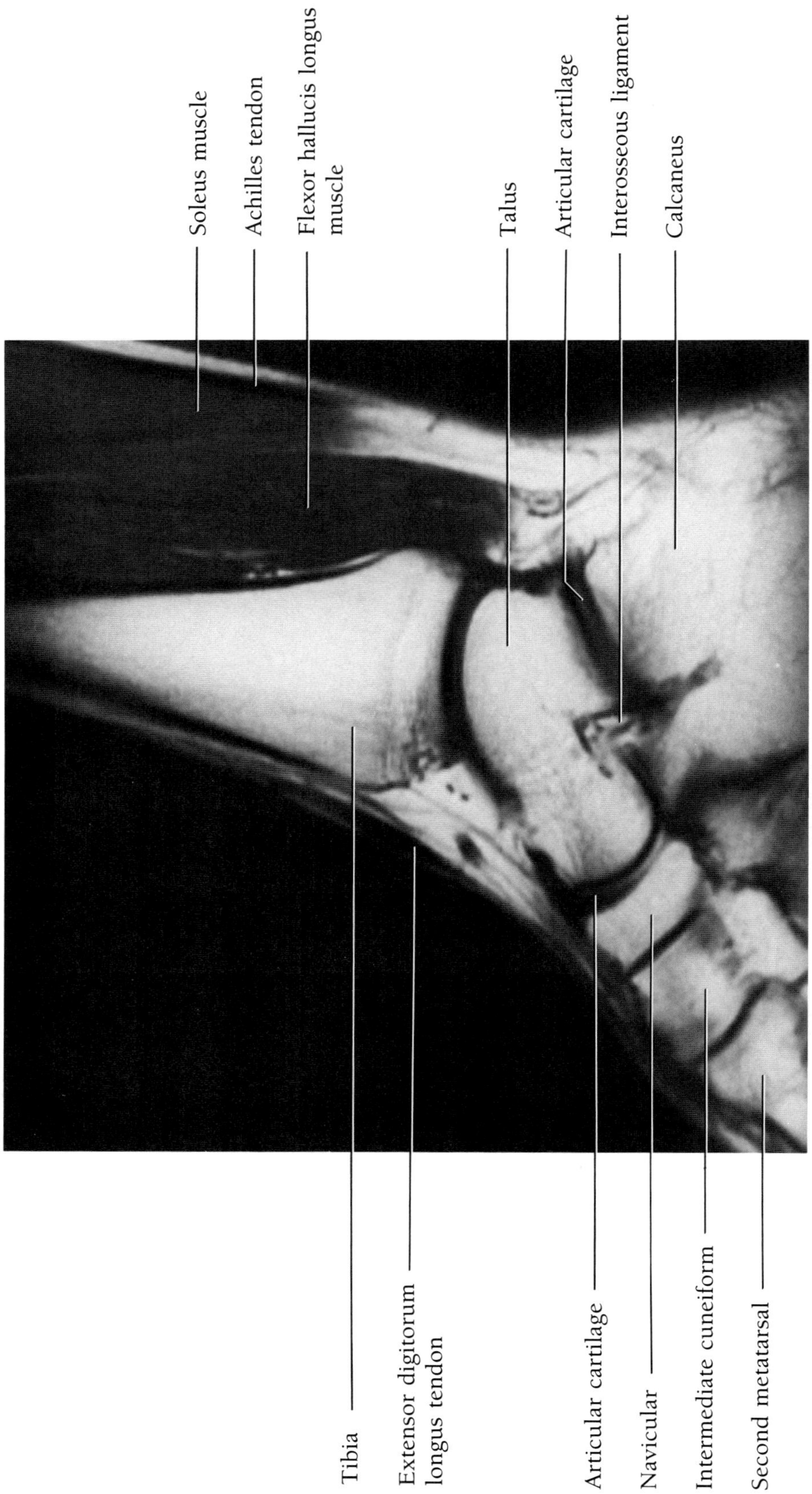

Soleus muscle
Achilles tendon
Flexor hallucis longus muscle
Talus
Articular cartilage
Interosseous ligament
Calcaneus
Tibia
Extensor digitorum longus tendon
Articular cartilage
Navicular
Intermediate cuneiform
Second metatarsal

FIGURE 2–17(C)

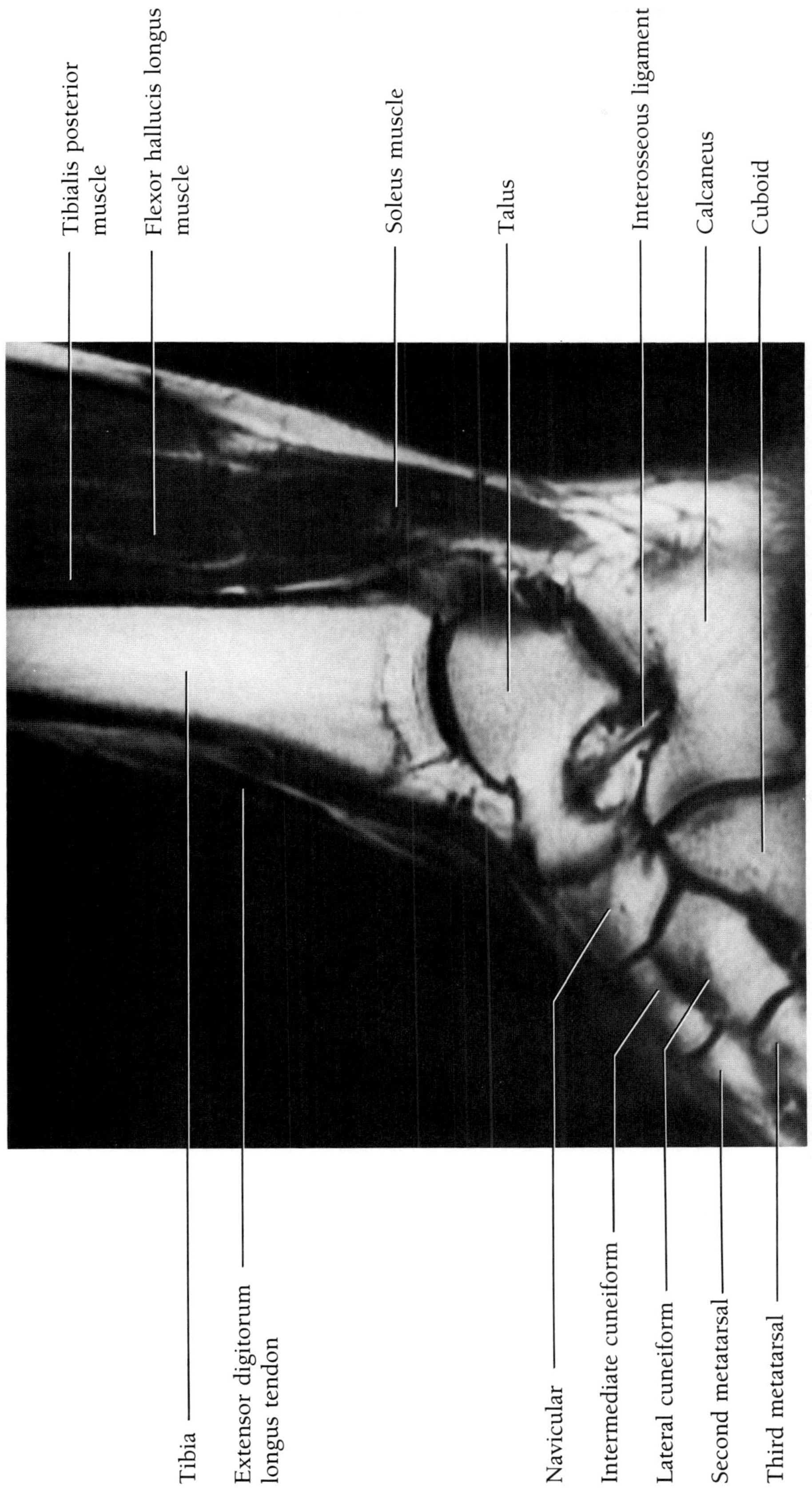
Tibialis posterior muscle
Flexor hallucis longus muscle
Soleus muscle
Talus
Interosseous ligament
Calcaneus
Cuboid
Tibia
Extensor digitorum longus tendon
Navicular
Intermediate cuneiform
Lateral cuneiform
Second metatarsal
Third metatarsal

FIGURE 2–17(D)

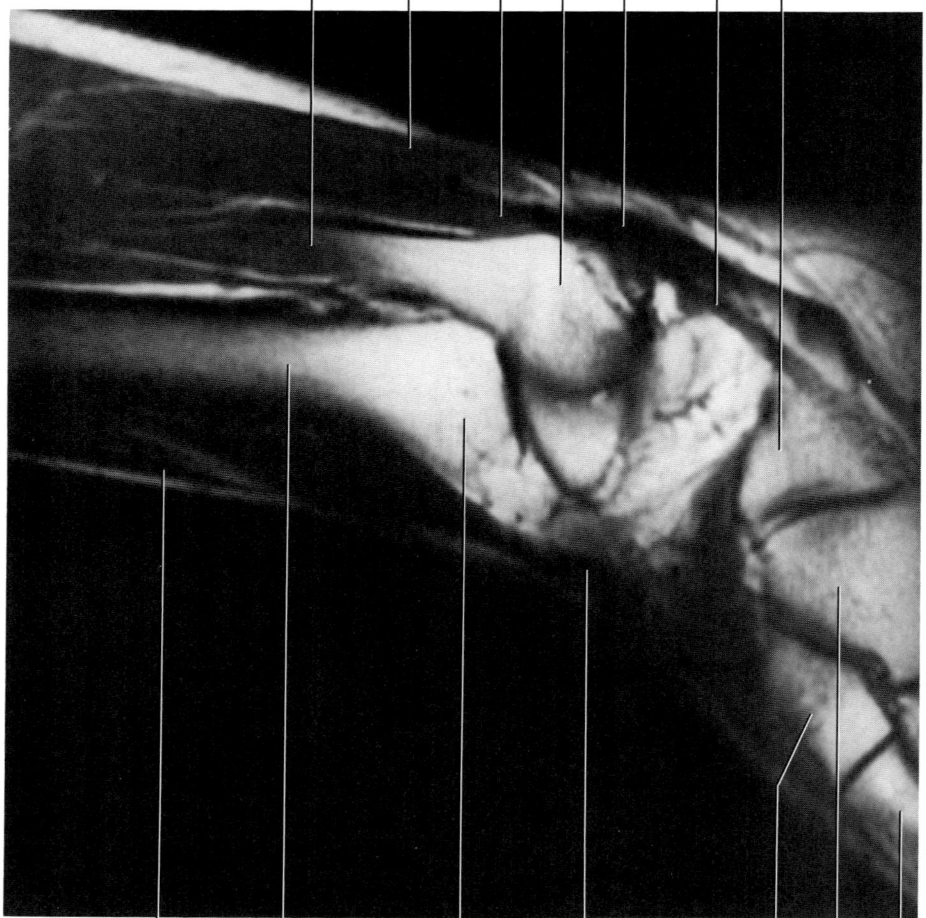

Flexor hallucis longus muscle
Peroneus longus muscle
Peroneus brevis muscle
Lateral malleolus
Peroneus longus tendon
Peroneus brevis tendon
Calcaneus
Extensor digitorum longus muscle and tendon
Tibial shaft
Distal tibia
Extensor hallucis longus tendon
Navicular
Cuboid
Intermediate cuneiform

FIGURE 2–17(E)

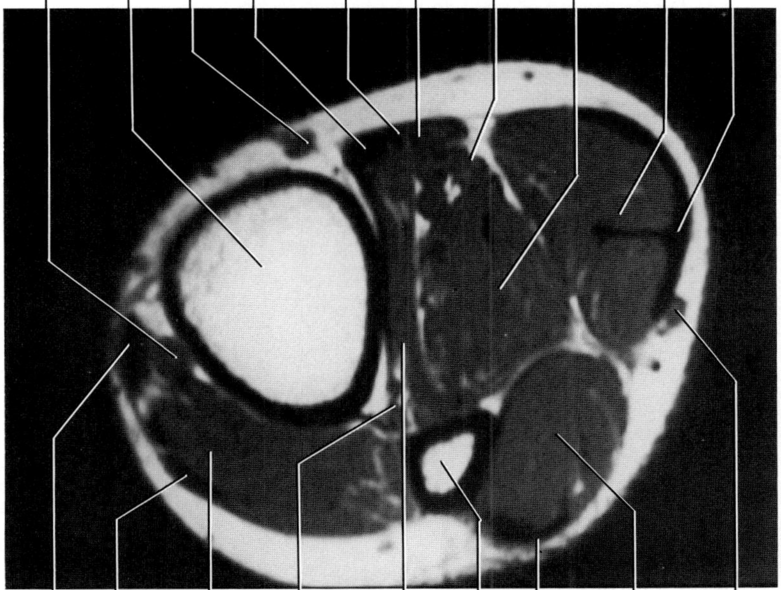

FIGURE 2–18(A)
Axial image of the foot and ankle: Through the distal part of the tibia and fibula.

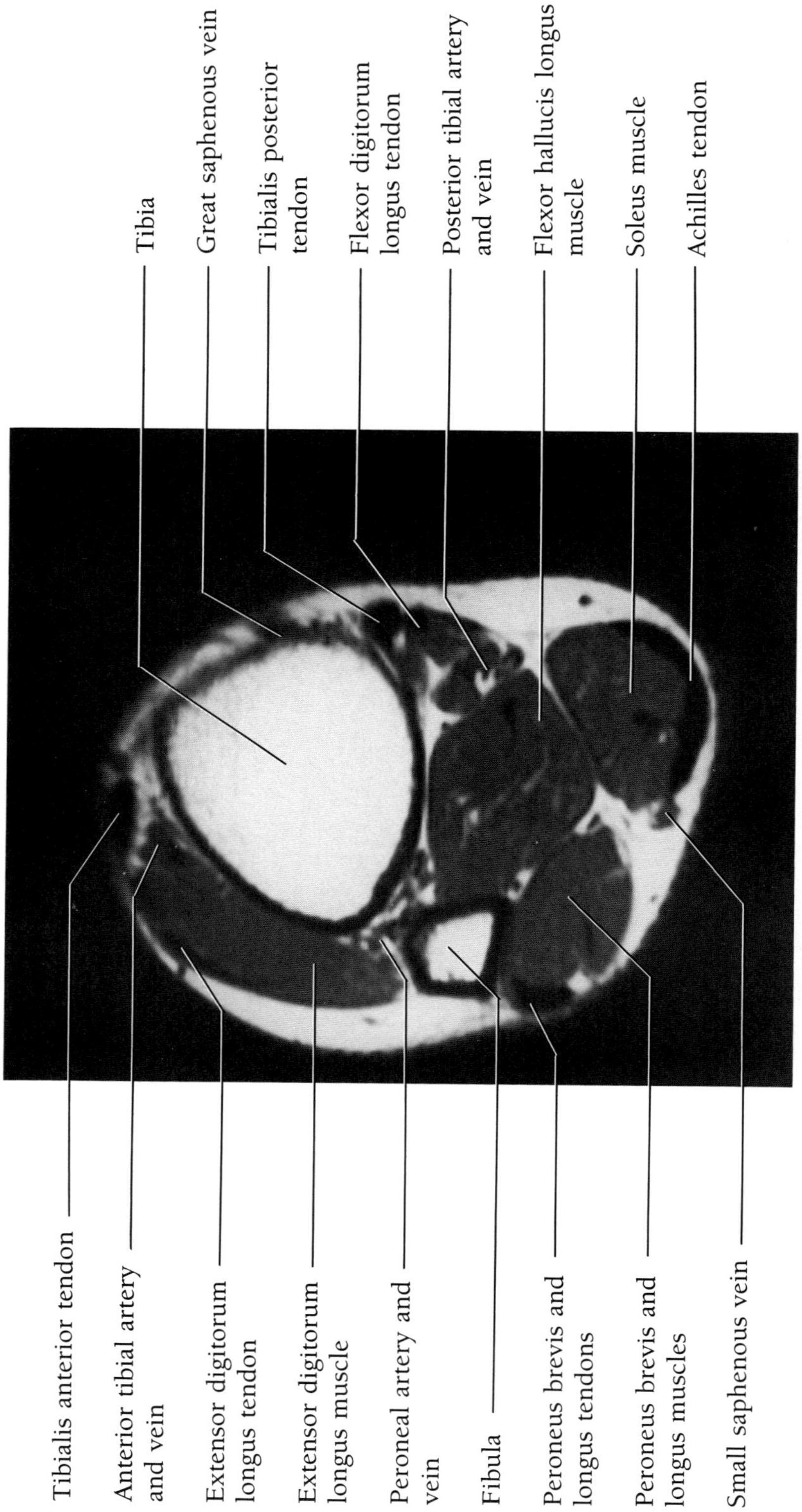

FIGURE 2–18(B)
Axial image of the foot and ankle: Just above the distal tibiofibular joint.

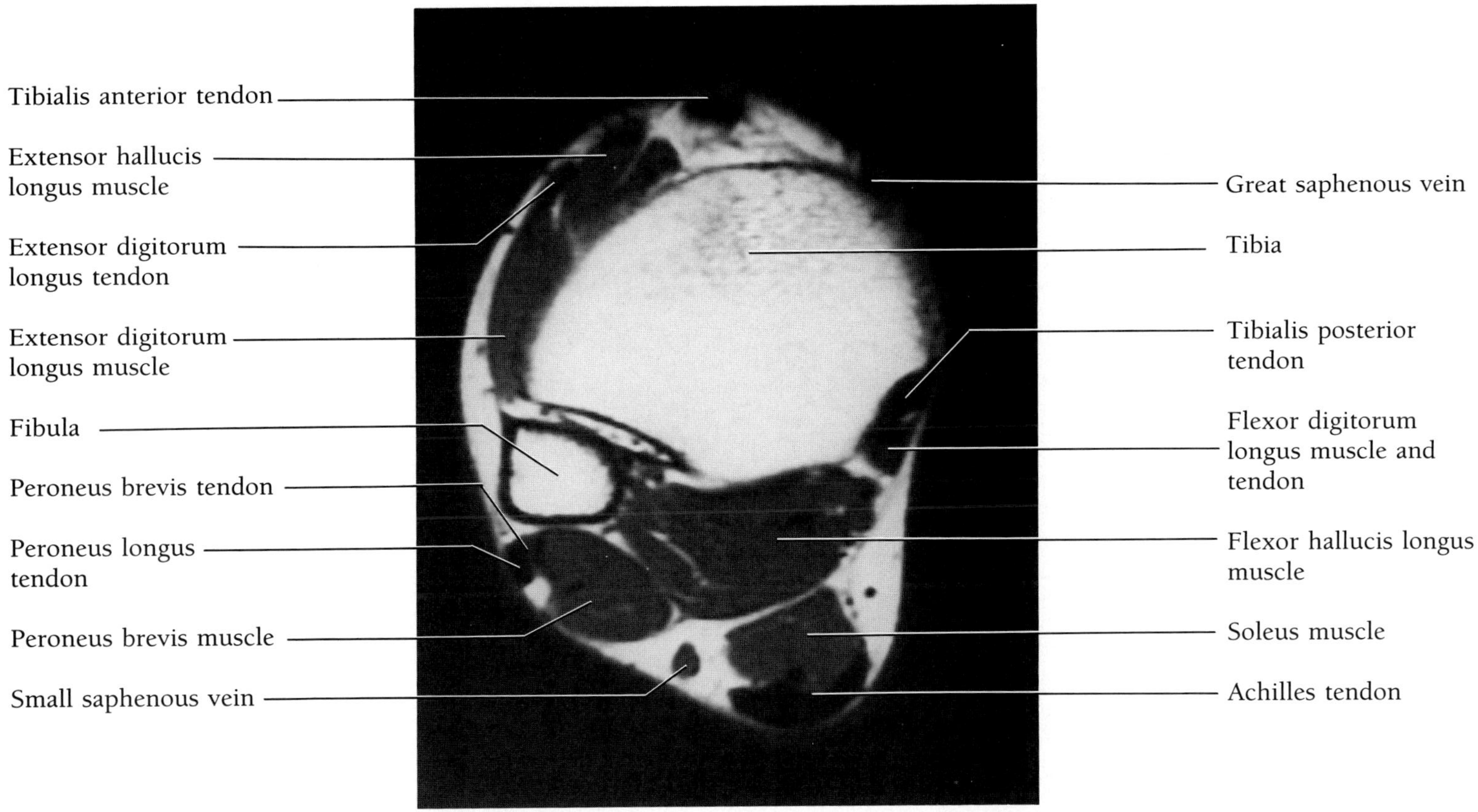

FIGURE 2–18(C)
Axial image of the foot and ankle: At the distal tibiofibular joint.

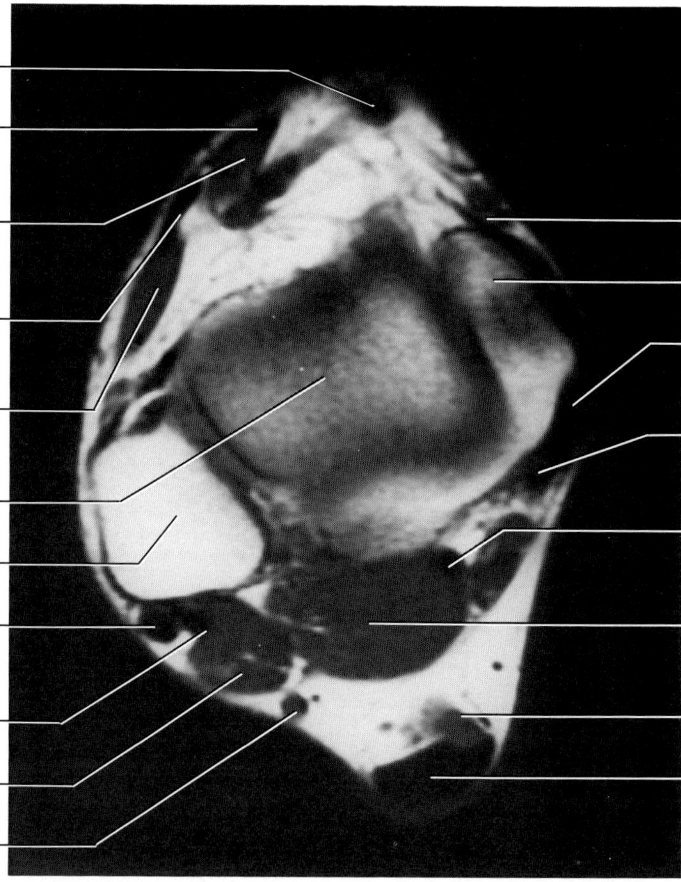

FIGURE 2–18(D)
Axial image of the foot and ankle: Through the tibiotalar joint space.

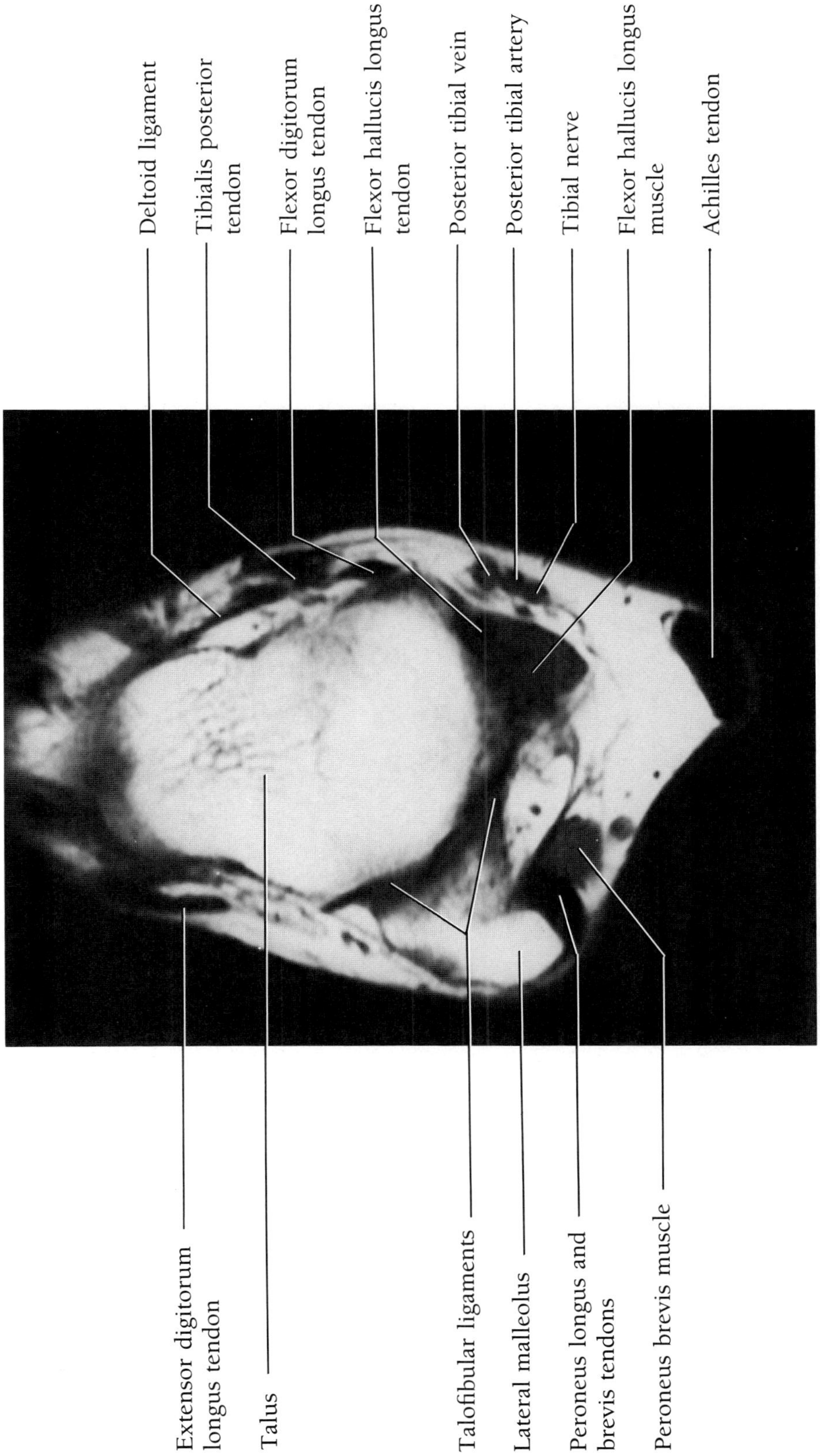

FIGURE 2–18(E)
Axial image of the foot and ankle: The distal lateral malleolus.

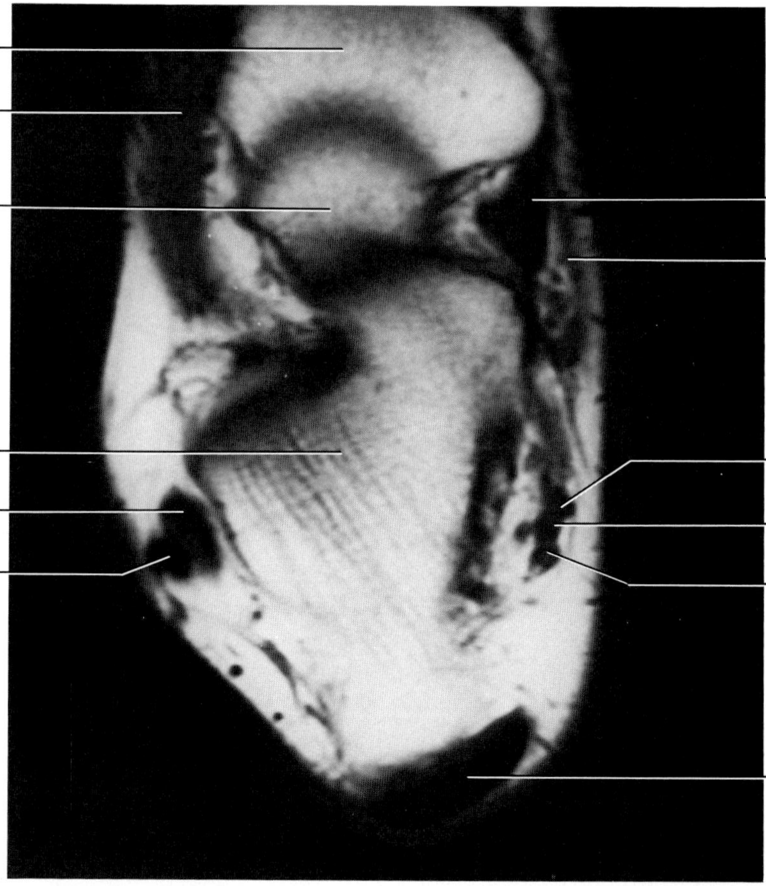

FIGURE 2–18(F)
Axial image of the foot and ankle: The midportion of the calcaneus.

3

Bone Scintigraphy in the Evaluation and Staging of Neoplasms

WILLIAM D. JOHNS

BARBARA J. McNEIL

The value of bone scintigraphy in the evaluation and staging of a variety of malignancies has been well documented in the past [1]. The use of bone-seeking radionuclides such as strontium 85 for detection of bone tumors was first suggested in the 1950s [2]. Further development of radiopharmaceuticals (technetium 99m [Tc-99m] phosphonates) improved camera resolution, and now emission CT allows detection of smaller and more subtle bony lesions than ever before.

Modern gamma cameras have an intrinsic resolution of less than 4-mm (full width at half maximum, FWHM) and provide a system resolution of approximately 9-mm FWHM using a Tc-99m source, a low-energy all-purpose collimator, and a distance of 10 cm from a source organ. Whole-body scanners that utilize moving tables result in a loss of several millimeters of resolution, primarily because of the increased organ-to-camera distance. For this reason, mul-

---

Supported in part by a grant from the National Cancer Institute (2 P01 CA41167-03A1).

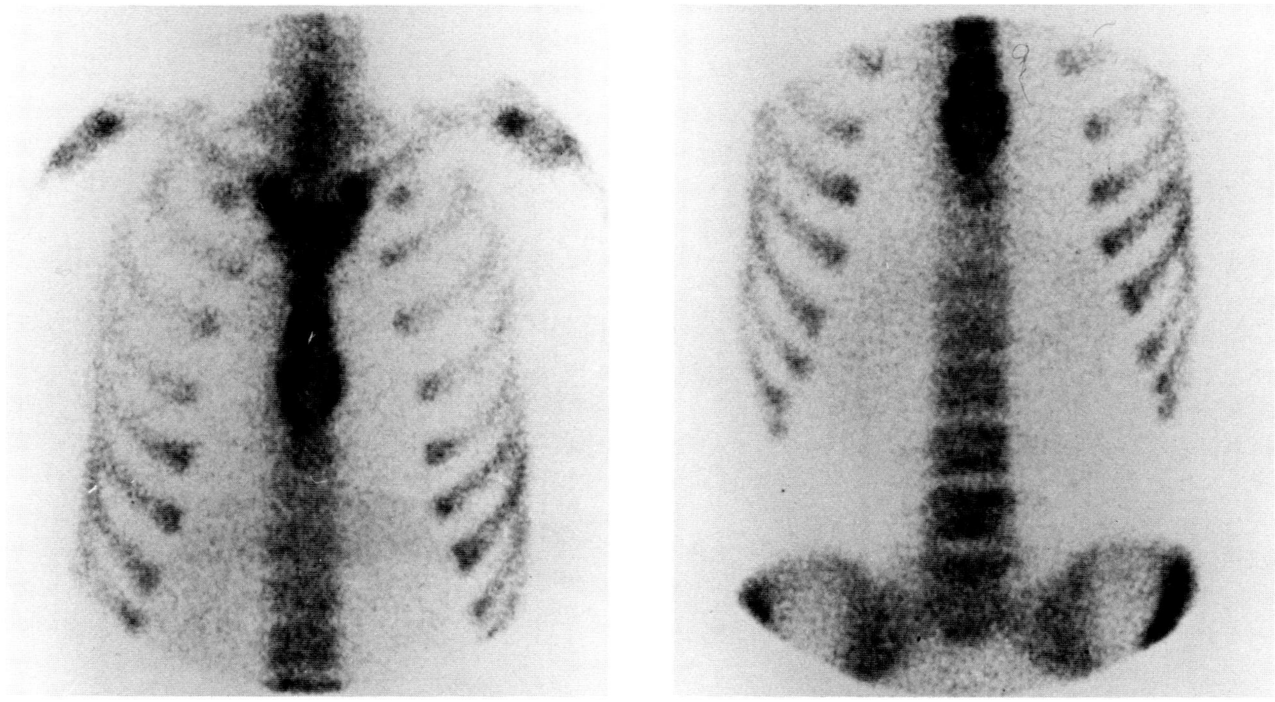

A B

FIGURE 3–1
Normal bone scan. A. *A normal anterior thorax.* B. *A normal anterior thorax and abdomen.*

tiple spot images obtained using a large field-of-view camera are preferred (Figure 3–1).

The most common indication for bone scanning in adults is for detection of metastatic malignancies and in some cases of primary bone tumors. Initial scans can be used for staging and in establishing a baseline for follow-up studies. There continues to be some controversy regarding the value and cost-effectiveness as well as the optimal frequency of follow-up bone scans necessary in various malignancies. This chapter reviews available data on the value of bone scanning in common tumors. Emphasis is placed on literature published since 1980.

Breast Carcinoma

BACKGROUND

Breast cancer is the fourth leading cause of death in women aged 35–54 years; 130,000 new cases are diagnosed annually in the United States, and one-third of patients eventually die of systemic manifestations of malignancy. Efforts over the past 30 years to improve therapy and to detect disease earlier have led to modest improvements in long-term survival [3]. It is hoped that the recent emphasis on the importance of frequent self-examinations and periodic screening mammograms will increase the proportions of cancers diagnosed early, and thus further improve survival.

Most radiographically detectable skeletal metastases from early untreated breast cancer are osteolytic, and at least 50% demineralization is required before they can be seen on conventional radiographs [4]. Bone scintigraphy may detect lesions months or years prior to radiographic changes and thus is the method of choice for searching for bony metastases in patients with breast cancer. For example, by 3 months 58% of patients with initially positive results on bone scans and negative results on radiographs will have positive findings on radiographs; by 6 months the percentage is 77% [5].

TABLE 3–1
Occurrence of bone metastases in patients with breast cancer

Authors	Stage	Patients with bone metastases (%)
Strender et al. (1981) [9]	Operable	15 by 3 yr
Roberts and Hayward (1983) [10]	Operable	20.8 by 5–8 yr
Pauwels et al. (1982) [12]	− Axillary nodes	12.3 by 4 yr
	+ Axillary nodes	21.5 by 4 yr
Chaudary et al. (1983) [13]	T1–T3a	10.4 by 2 yr
Pedrazzini et al. (1986) [11]	T1–T3a	6.9 by median of 26 mo

PRETREATMENT STUDIES

Combined data from a number of recent studies have indicated that bone scans performed at presentation show true-positive results in 2% of patients with stage I or II disease (see [1] for a summary of these studies). This percentage ranges from 1.8–3.5%, which is much less than the pre-1977 estimate of about 25%. In patients with stage III disease, the data indicate that true-positive results occur in just under 20% [1]. Virtually all clinicians use these data to justify pretreatment studies in patients with clinical stage III disease. There is considerably more controversy regarding their use in earlier stages, largely because of the potential for false-positive results. Nonetheless, many suggest that despite their low yield in patients with early stage malignancy, baseline scans improve the specificity of follow-up studies.

Positive findings on bone scans do not always indicate metastatic disease, even in patients with a known primary malignancy. In the above studies, this was particularly apparent in patients with early (stage I or II) disease in whom a unifocal bone metastasis was more common than was multifocal disease and thus may be confused with benign disease [6]. The false-positive yield in early disease is believed to be about 2% [1].

FOLLOW-UP PERIOD

As with other tumors, the extent of disease at the time of initial diagnosis influences the 5-year survival rate and the frequency of eventual metastases [7]. Overall, in patients with just one site of distant metastases, bone is the most common site; that is, 58% will have bony involvement [8]. This is important background information in discussing the yields from and the value of routine follow-up bone scans.

Earlier data on the rate of conversion from negative to positive results on scans were initially conflicting, but recent data suggest that detectable bone metastases seldom occur before 7 months and that 20% of patients will convert on bone scan within the first 3 years after diagnosis. For example, several years ago studies from Sweden and Great Britain [9,10] indicate that about 15–20% of patients with initially negative findings on bone scans had positive results within 3–8 years. A more recent study from the Ludwig group [11] indicated that among all patients with stages T1, T2, and T3a disease who had negative bone scans initially and who were followed for a median time of 26 months, 6.9% had their first recurrence in bone and that an additional 4.3% had bone disease as an additional later recurrence. Table 3–1 summarizes follow-up data from several studies [9–13].

Because of the incomplete data on rate of conversion from negative to positive findings on bone scans and the lack of consensus on the impact of early treatment of metastatic disease, there are no firm guidelines on the optimal frequency of follow-up bone scans. It may be argued that although the true-positive yield is low, because early adjuvant multichemotherapy results in prolonged disease-free survival and because morbidity due to pathologic fractures may be reduced with early radiation therapy, screening in the early stages is worthwhile.

Laboratory tests are usually not useful predictors of bone scan results. In the Ludwig group study [11], about 50% of patients having bone scans showing definite or possible evidence of

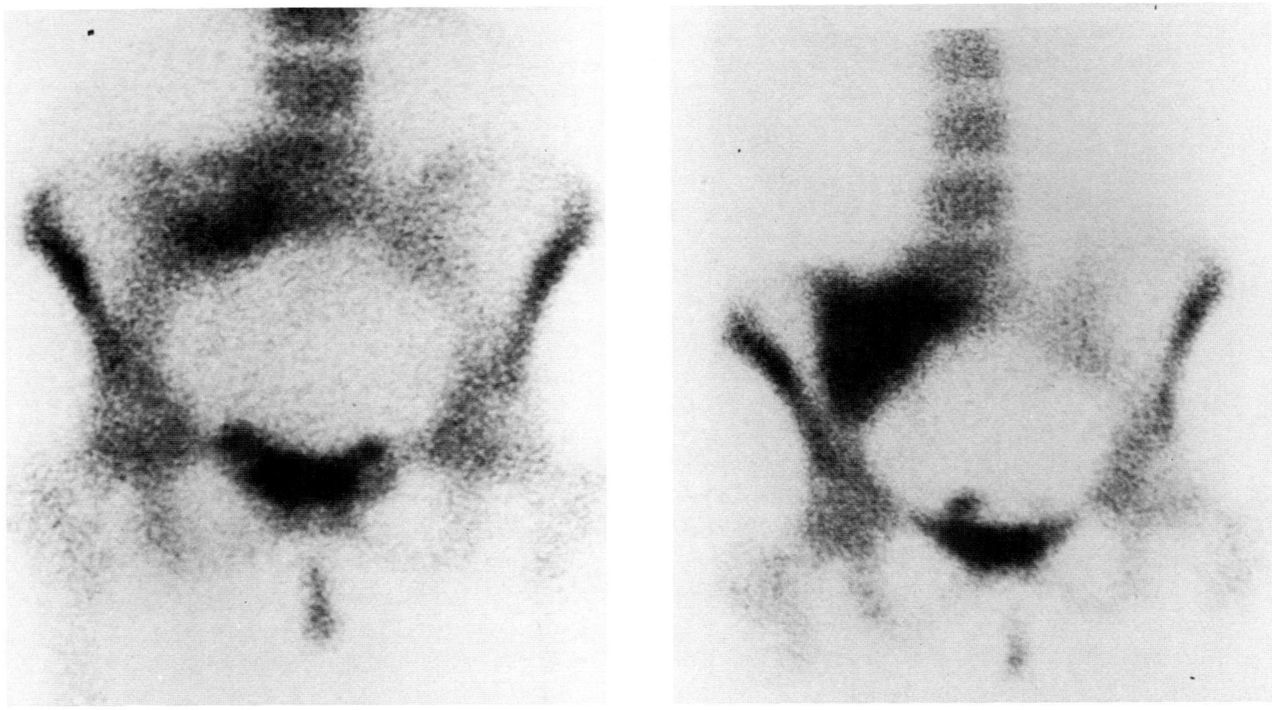

A B

FIGURE 3–2
Flare phenomenon in breast cancer. A. *An anterior view of the pelvis with increased uptake in the right sacroiliac region.* B. *The same view on a subsequent scan obtained 3 months after the patient began hormonal therapy. There is more intense uptake in the right sacroiliac region, with a correlative radiograph showing increased sclerosis.*

bone metastases had elevated alkaline phosphatase levels; in addition, only 30% of patients with abnormal alkaline phosphatase levels had normal bone scans. In the study of Pandya et al., only 12.1% of bone metastases were associated with abnormal results on chemistry tests [14].

A scintigraphic "flare phenomenon" is seen on follow-up scans in approximately 10–15% of breast cancer patients being treated with tamoxifen and/or chemotherapy [15–17] and usually occurs within 3 months of therapy initiation. This is to be distinguished from the symptomatic "flare" (i.e., increased bone pain) often seen shortly after therapy begins. Transient increased activity is seen in some bone lesions due to osteoblastic response and healing. Scans may occasionally show what appear to be "new" lesions; however, they are really sites of enhanced uptake in previously existing subtle lesions. This phenomenon must be considered in the evaluation of therapeutic response and should not be confused with disease progression. Radiographic correlation is helpful with evidence for healing (generally sclerosis) usually seen 8–12 weeks after initiation of therapy; however, these x-ray changes may not be seen until 5–6 months (Figure 3–2).

CONCLUSIONS

Baseline bone scans are useful in patients with stages I and II disease despite the low incidence of true-positive scans. Although the number of patients with true-positive scans is low in this group, the consequences may result in the initiation of therapy that has been shown to prolong disease-free survival and decrease morbidity. In addition, they improve the specificity of follow-up examinations. The higher incidence of positive findings on initial scans in patients with clinical stage III disease makes screening studies clearly indicated.

The frequency and length of time for which follow-up scans are necessary are not well defined; however, annual scans are suggested for 3–4 years after diagnosis, or more frequently if symptoms develop. Scans may also be useful in evaluating therapeutic response.

Lung Cancer

BACKGROUND

The increasing incidence of lung cancer and poor 5-year survival rate make this a major oncologic problem. In 1987 there were 150,000 new cases and between 1981 and 1987 lung cancer accounted for 20% of all cancer deaths [3]. It is the most common malignancy in men and is becoming increasingly common in women. Twenty-five percent of all lung cancers are of the small-cell variety. Of the remaining tumors, 33% are epidermoid, 25% are adenocarcinoma, and 16% are large cell [18].

PRETREATMENT STUDIES

Small-cell tumors commonly show bone metastases, with up to 40% of patients having skeletal lesions at presentation [19]. One study suggests that 18.9% of patients with small-cell lung cancer have bone metastases detected on scan (Table 3–2). About 17% will have evidence of bone marrow disease on a biopsy specimen [19]. There is not a one-to-one correlation between these groups, however; 22% have positive bone scan results with negative findings on examination of bone marrow biopsy specimens (Table 3–3) [19].

Patients with localized non–small-cell lung neoplasms (stage I or II) have a 35% five-year survival rate, while those with metastases rarely survive for 5 years regardless of the type of therapy. For asymptomatic patients without brain or liver metastases, 8% (stages I, II, and III) have bone metastases [20,21] and the small number of patients studied do not allow differentiation between early (stages I and II) and late (stage III) disease. For symptomatic patients, 32% (stages I and II) and 41% (stage III) have skeletal metastases [21]. These yields are higher than those for brain or hepatic metastases [21].

TABLE 3–2
Radiologic studies in patients with small-cell carcinoma and no obvious metastases (53 patients) [20]

Organ scanned	True-positive (%)
Bone	18.9
Liver	0.0
Brain (CT)	5.0

TABLE 3–3
Correlation between bone scan and bone marrow biopsy [19]

	Scan negative	Scan positive
Marrow negative	68 (61%)	25 (22%)
Marrow positive	9 (8%)	10 (9%)

FOLLOW-UP STUDIES

The frequency of required follow-up scans has not been well defined.

CONCLUSIONS

Patients with small-cell carcinoma should have bone scans on presentation. Although therapy is not altered by an abnormal scan in these patients, it may be used as a prognostic indicator.

For non–small-cell disease given the relatively high operative mortality rate of 5–10% [22], we believe that both asymptomatic and symptomatic patients with "operable" lung cancer should have routine preoperative bone scans.

Prostate Cancer

BACKGROUND

Prostate carcinoma is the third leading cause of death from cancer in American men, and at least 5% of all men over 50 years old die of prostate cancer. In 1988 there will be 99,000 new cases of prostate cancer and 28,000 deaths are expected [3]. Curability is closely related to stage and extent of disease at presentation [23]. Once the tumor is greater than 1 cm in diameter or 1 cc in volume, extracapsular extension is likely.

Detection of local extension and distant disease is important. Depending on the initial clin-

TABLE 3–4
Frequency of positive bone scans as a function of clinical stage in patients with prostate cancer

Authors	Clinical stage		
	A/T1	B/T2	C/T3
Biersack et al. (1980) [27]	1/14	12/62	20/51
Merrick et al. (1985) [28]	3/7	15/41	45/92
Shafer* (1976)	2/19	0/8	9/50
Total	6/40	27/111	74/193
	15%	24%	38%

*Unpublished data based on staging system used by Veterans' Administration (A,B,C,D). Biersack et al. and Merrick et al. used TNM system.

ical stage, 12–66% of patients will have disease in the periprostatic tissue [24,25]. In addition, most oncologists treat early disease (stages A and B) as "surgically operable" and stages C and D as more amenable to other modalities.

PRETREATMENT STUDIES

Currently both the Whitmore [26] and the tumor-node-metastases (TNM) systems are used for staging; in addition, Gleason [27] grades are sometimes superimposed. Thus, it is difficult to be sure that in the process of aggregating data across studies using different staging or grading systems, comparability exists among the patient populations. The data in Table 3–4 indicate results from three studies evaluating the usefulness of bone scans in patients with varying clinical stages at presentation. Averages show that about 15% of patients with T1 disease, 24% with T2, and 38% with T3 have bone metastases at presentation [28–30].

Neither skeletal surveys nor serum biochemistries (alkaline and acid phosphatases) are as sensitive for detecting metastatic disease as is bone scanning [31]. Only about 50% of patients with positive bone scan results will have elevated phosphatase values.

FOLLOW-UP STUDIES

There are few data on the frequency of conversion from negative to positive scan results in patients with prostatic cancer, in large part because of the fact that treatment is seldom instituted for asymptomatic bone metastases. The study by Merrick et al. [29] provide some data, but the exact interpretation is unclear. Among their patients with initial stage T0 or T1 disease and normal bone scans, 50% had bone metastases at 48 months and among those with stages T2–4, about 80% had bone metastases. Survival figures in this country at 5 years are considerably higher, particularly for early disease, and thus there may be some incomparability of disease definition from their data to a U.S. population.

The introduction of prostate surface antigen (PSA) will probably be useful in monitoring responses and recurrences after initial therapy for this disease. Currently, preliminary data indicate a good correlation between PSA levels and tumor volume [32]. Thus, additional studies on the complementarity of PSA assays and bone scans should be obtained over the next few years.

FLARE PHENOMENON

A scintigraphic "flare" phenomenon as seen in breast cancer patients receiving therapy may also be seen in prostate malignancies [33–35]. This is seen in 23% of patients receiving hormonal or chemotherapy. This is most commonly seen at approximately 3 months after initiation of therapy and should not be confused with disease progression. The same phenomenon has been observed in patients receiving hormone-like leuprolide therapy [36].

CONCLUSIONS
All patients with prostate cancer should have bone scans initially. Definitive statements about the value of bone scans in the follow-up period cannot be made. Their usefulness depends on the extent to which therapy will be changed as a result of the study and by the degree to which other modalities (e.g., PSA levels) can be more sensitive markers of disease progression.

Osteogenic Sarcoma

BACKGROUND
Osteosarcoma is a primary bone malignancy and is most often seen in children during the growth-spurt period. There are approximately 500 new cases diagnosed each year in the United States [37]. Lesions are usually found in the appendicular skeleton, with 40% in the femur, 15% in the tibia (Figure 3–3), and 15% in the humerus [37]. Current therapy involves surgical resection and chemotherapy [37].

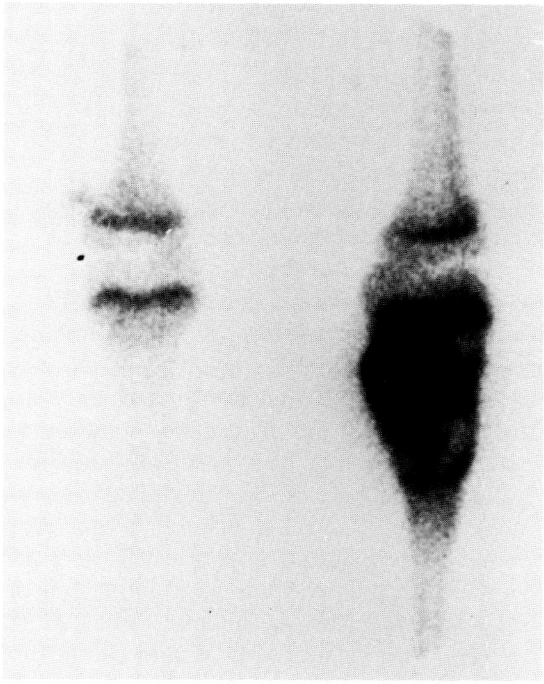

FIGURE 3–3
Osteosarcoma of the left tibia.

PRETREATMENT STUDIES
The bone scan is used primarily to evaluate distant disease since CT and MRI are superior modalities in evaluating local extension. At the primary site, uptake on the bone scan may extend beyond the true tumor margins, presumably as a result of marrow hyperemia, periosteal new bone formation, or medullary reactive bone. In addition, for tumors adjacent to the joint space, increased uptake in synovial joints may indicate tumor involvement requiring more extensive surgical resection [38]. When compared with bone scintigraphy, both CT and MRI have been shown to be better in accurately determining local extent of tumor [39,40].

In terms of distant disease approximately 3% of patients with osteosarcoma present with bone metastases. Table 3–5 summarizes the results of three studies [41–43].

FOLLOW-UP STUDIES
Bone scans have been used to detect tumor recurrence at the site of resection as well as distant

TABLE 3–5
Osteosarcoma: bone metastases at presentation

Authors	Total patients (no.)	Patients with metastases (no.)
Goldman et al. (1975) [40]	13	1
Goldstein et al. (1980) [41]	46	1
McKillop et al. (1981) [42]	48	1
Total	107	3 (3%)

spread after initial treatment. However, increased uptake at the tumor resection site due to recurrent tumor may be difficult to distinguish from residual activity due to surgical manipulation. Some increased uptake may be seen due to mechanical irritation after the patient has begun to use a limb prosthesis. Therefore, the possibility of false-positive interpretations exists.

During the follow-up period, there are only preliminary data on the percentage of patients developing bone metastases after treatment, and these studies suffer two limitations: (1) They may be based on data that do not reflect current treatment policies; and (2) all patients evaluated at presentation may not have been followed, thus biasing the data, that is, overestimating the percentage likely to develop disease. Recent estimates indicate that at 1 year nearly 75% of individuals placed on chemotherapy will be free of disease and at 2 years the number will be about 66% [44]. These percentages should be used in evaluating analogous percentages inferred from results of radiographic or scintigraphic studies.

One follow-up study at Children's Hospital in Boston showed a linear increase in bony metastases 5–29 months after initial diagnosis with a rate of increase of approximately 1% per month [45]. Most of these patients had previously developed lung metastases. However, bone lesions did develop before lung metastases in 19% of patients and bone lesions were the only manifestation of metastatic disease in 11% of patients [42].

In a study of solitary lesions outside the primary tumor, eight (89%) of nine proved to be metastatic [46]. The one false-positive solitary lesion occurred in the spine.

CONCLUSIONS

All patients with osteosarcoma should have baseline bone scans at the time of diagnosis. The frequency with which follow-up scans are done to detect the presence of metastatic disease or to follow responses to therapy in patients with metastases is variable. The current practice at Boston Children's Hospital is for patients to be scanned at approximately 3- to 6-month intervals for the first year, followed by yearly subsequent scans.

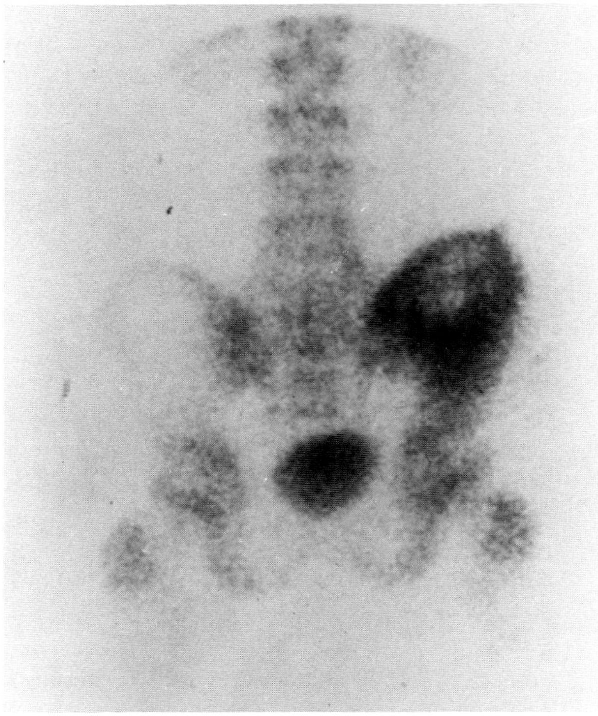

FIGURE 3–4
Ewing's sarcoma of the pelvis.

Ewing's Sarcoma

BACKGROUND

Ewing's sarcoma is a rare primary bone tumor occurring predominantly in the second decade of life. Males are affected more often than females. Lesions occur more commonly in the long bones, particularly the femur, but can also be found in the axial skeleton [37] (Figure 3–4). Current therapy may involve radiation or surgical ablation (depending on the location of the tumor) along with adjuvant chemotherapy [37].

PRETREATMENT STUDIES

Ewing's sarcoma usually shows intense homogeneous increased uptake with poorly defined margins but has been reported as a "cold" lesion on bone scan [47]. Initial bone scanning will detect bone metastases in 12% of patients [42,48]. In a recent study of 72 patients seen in India, metastases were most frequently seen in the axial skeleton [49].

In evaluating local extension of either Ewing's sarcoma or osteosarcoma into the soft tissue adjacent to bone or into the bone marrow, several studies have shown that MRI is superior to either CT or bone scanning [39,40].

FOLLOW-UP STUDIES

There have been few studies on the follow-up of patients of Ewing's sarcoma, and it is likely that in these studies not all patients evaluated at the time of presentation were followed up, thus leading to a skewing of the results. Nonetheless, these data suggest that about 39–45% of patients develop skeletal metastases during the follow-up period, at an average time of 14 months after presentation, with a median time of 11 months. One study demonstrated solitary lesions on bone scans of 43 (18%) of 244 patients with Ewing's sarcoma. Only 8 (19%) of these 43 solitary lesions proved to be metastases. Those abnormal areas that proved not to be metastatic were found most commonly in the arms and legs [49].

CONCLUSIONS

Every patient with Ewing's sarcoma should have an initial bone scan performed at the time of presentation. The frequency and length of time for which serial bone scans are carried out are variable. The usual pattern at Boston Children's Hospital is similar to that utilized for osteogenic sarcoma. Scans are obtained at 3- to 6-month intervals for the first year after diagnosis, and subsequently at yearly intervals or more frequently if symptoms develop.

Neuroblastoma

BACKGROUND

Neuroblastoma is the most common childhood tumor that develops skeletal metastases. Approximately 60% of patients have bone metastases detected on scans at some point in the disease course. Neuroblastoma accounts for 6–8% of all childhood malignancies [50]. Ninety percent of cases are seen in children less than 10 years old, with 50% occurring in those under 2 years old. Eighty-five percent of primary tumors occur in the abdomen [50]. Current therapy involves chemotherapy as the mainstay, with surgical resection and radiation therapy in some patients.

PRETREATMENT STUDIES

Most studies do not separate new from follow-up patients, and thus data on the value of bone scintigraphy are somewhat limited [51,52]. However, one study has shown that 11 (28%) of 40 have metastases on presentation and that radionuclide scans are superior to skeletal surveys in detecting metastases [53,54]. Most metastases are noted in the skull (22%) and long bones (femur, 20%; tibia, 12%). Symmetric increased uptake in metastases in the metaphyses of long bones is observed in 17–27% of patients. This must be considered when interpreting scans with this finding, which should be recognized as being abnormal. A small study in children with a known neuroblastoma demonstrated that 87% of solitary lesions were metastatic [46]. Extraosseous uptake may be seen in soft tissue metastases, the primary tumor, or ascites [55].

FOLLOW-UP STUDIES

The study by Baker et al. indicated that 17 (42%) of 40 children followed for an unspecified period of time developed new or recurrent skeletal lesions [53].

CONCLUSIONS

Initial staging bone scans should be performed on all patients with neuroblastoma. Current practice at the Boston Children's Hospital is to obtain follow-up studies at 3- to 6-month intervals to determine appropriate therapy and assess response to treatment.

Urinary Tract Neoplasms

Urinary tract malignancies occur mostly in adults and account for 5% of cancer-related deaths in males and 3% in females [3].

BLADDER CANCER

False-positive scans are noted in about 2% of the patients. Urothelial cell tumors tend to be locally invasive, with less than 30% for 5-year survival when the tumor extends beyond the bladder. At autopsy, skeletal metastases are observed in 25% of patients with bladder cancer [56]. At the time of presentation, 5% have skeletal metastases detected by scan, while 7.7% develop metastases during the course of follow-up [57].

The presence of bone metastases generally precludes radical therapy (total cystectomy and radical radiotherapy). Although some have suggested that the low yield of routine bone scintigraphy in patients with bladder cancer makes such studies cost-ineffective, decreases in the morbidity and mortality associated with not performing unnecessary surgery make bone scans seem worthwhile.

RENAL-CELL CARCINOMA

Skeletal metastases are found in 7.5% of patients with newly diagnosed renal-cell carcinoma [58]. A large portion of those with positive findings on scans also have positive findings on radiographs [59], and most patients with renal-cell carcinoma and skeletal metastases are symptomatic [59]. Approximately 10% of skeletal lesions are "cold" on scans [60].

Colon Cancer

The colon is the second most common site of malignancy in both males and females and accounts for approximately 20% of all cancer-related deaths. Peak incidence occurs in the fifth, sixth, and seventh decades. Primary therapy is surgical. The 5-year survival rate is 50% [3].

In a recent retrospective study, metastatic disease to bone occurred *at some time* in about 4% of patients with colon cancer and in 8.9% of patients with rectal cancer. Bone was the only site of metastases in 29% of these patients [61]. The vertebral column and pelvis were the most common sites of metastases. It is likely that these percentages are upper limits since the authors do not indicate whether all patients, or only those with symptoms, were continually screened for bone metastases. If the latter, then the above percentages are likely to be considerably lower.

Gynecologic Malignancies

Ovarian, endometrial, and cervical malignancies represent 14% of cancers diagnosed in women and account for 9% of all cancer-related deaths in women [3].

ENDOMETRIAL CANCER

Previous studies have indicated that fewer than 1% of patients with stage I or II endometrial cancer have bone metastases [62–64]. Approximately 15% of endometrial cancer patients have recurrent disease in bone, and in 90% this occurs in the first 3 years after diagnosis [65].

CERVICAL CANCER

The incidence of skeletal metastases in patients with cervical cancer varies, as might be expected, with disease stage: 0% in stage 0, 1% in stage I or II, 6% in stage III, and 10% in stage IV [62,66]. Another study demonstrated that 43% of patients with stage III or IV disease have findings suggestive of unilateral or bilateral renal obstruction on bone scan [67].

OVARIAN CANCER

Findings in patients with ovarian carcinoma are similar to those for the other gynecologic malignancies [68]. Essentially no bone metastases are found in stage I or II disease, with 8% found in stage III.

Multiple Myeloma

Multiple myeloma is a plasma-cell malignancy that may be associated with bony destruction. The incidence is currently approximately 3 in 100,000, and disease occurs at a median age of 60 years. It is slightly more common in males than females. The most common symptom noted is bone pain due to pathologic fracture.

Multiple studies have demonstrated that bone scintigraphy is less sensitive than a skeletal survey in the detection of myeloma lesions [69–71] (Table 3–6). Myeloma lesions are often "cold" on scans. On a site-by-site basis, bone scintigraphy is less sensitive than an x-ray survey. However, on a patient basis, nearly all patients with multiple myeloma will have at least one lesion detected by bone scan.

TABLE 3–6
Comparison of bone scans and plain radiographs on a site-by-site basis in patients with multiple myeloma [69,70]

	Sites (no.)	Abnormal site (%)
Bone scan only	81	23%
Radiograph only	146	42%
Both	122	35%

An area of potential clinical importance in patients with multiple myeloma is the extent to which an area of increased activity on scans corresponds to a pathologic fracture. In the series by Frank et al., 12 (75%) of 15 "hot" lesions were associated with a fracture, and most of these were in the dorsal and lumbar regions of the spine [70]. In the series by Woolfenden et al., 37 (36%) of 102 sites were abnormal on *both* scans and radiographs, and were associated with fractures [71].

Incidental Findings in Patients with Malignancies

Frequently, incidental abnormal findings are noted on bone scans in patients with neoplasms and are not necessarily related to primary or metastatic bone tumors. We list a few of them here. Uptake may be seen in a primary soft tissue tumor such as breast cancer (Figure 3–5) or neuroblastoma (Figure 3–6). Activity can also be seen in primary tumors or metastases in the lung or liver (Figure 3–7), in malignant ascites, and in pleural effusions.

Increased renal activity may be seen in patients with hypercalcemia (Figure 3–8) or occasionally with chemotherapy. A photopenic area in a kidney may be observed at the site of a primary renal tumor or metastasis (Figure 3–9) or indentation due to an adrenal metastasis (Figure 3–10). Renal obstruction or bladder indentation may be seen in patients with pelvic tumors such as ovarian cancer (Figure 3–11). Abnormal activity in bones in the appendicular skeleton

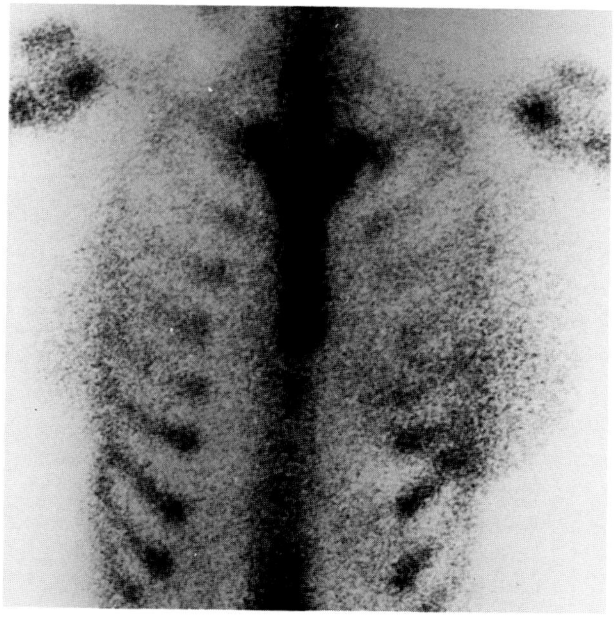

FIGURE 3–5
Soft tissue uptake in a primary tumor of the left breast.

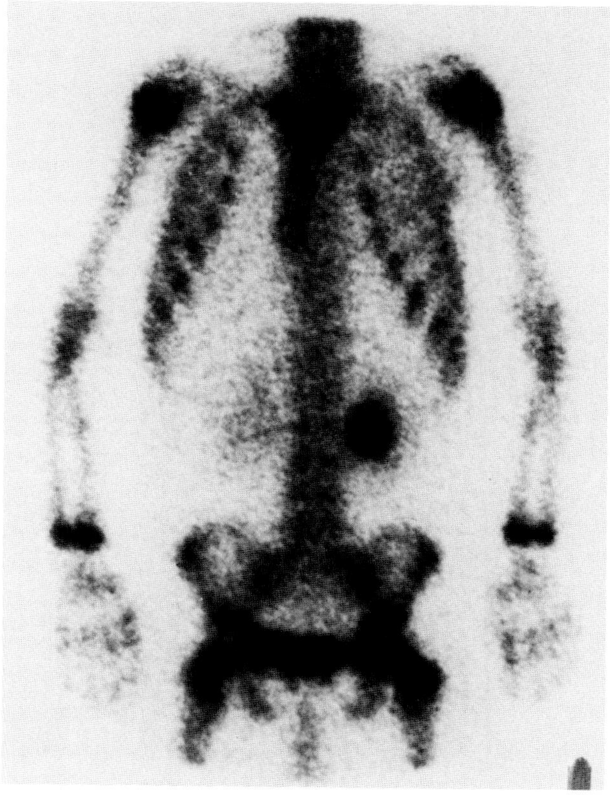

FIGURE 3–6
Uptake in a primary neuroblastoma of the left kidney.

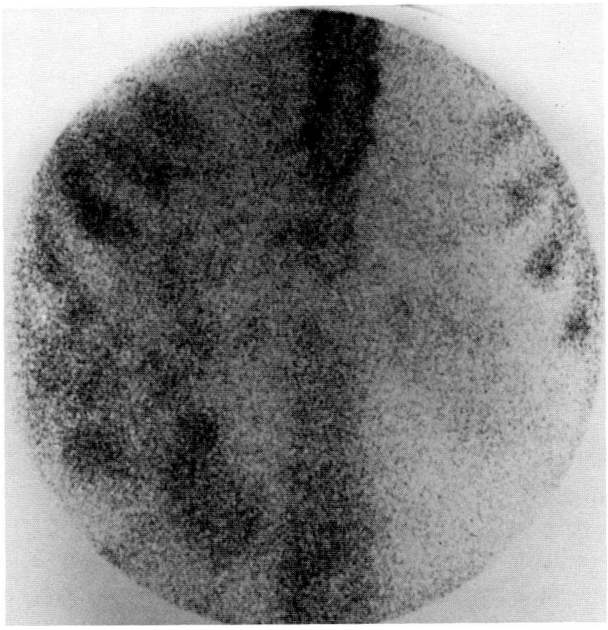

FIGURE 3–7
Hepatic uptake in a metastatic thyroid malignancy. A subsequent liver scan showed multiple focal defects consistent with metastases.

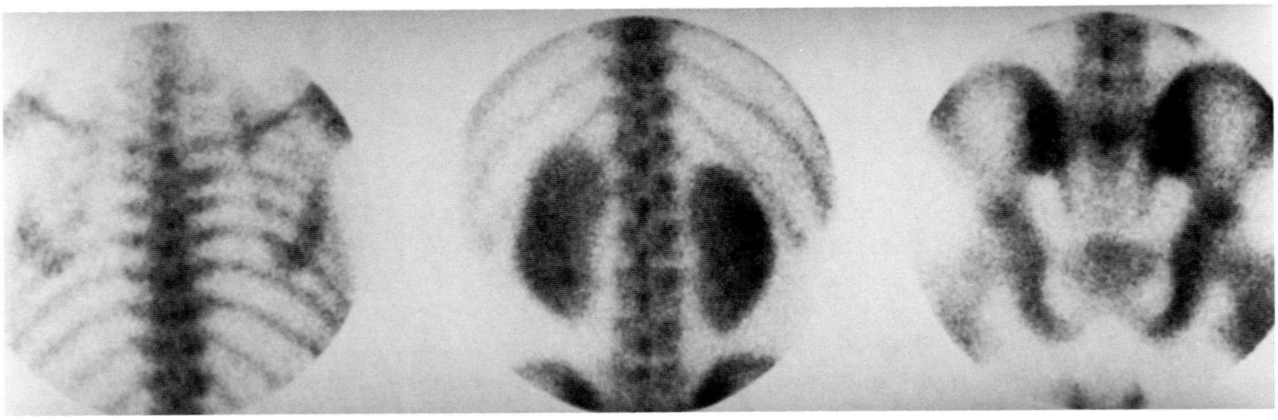

FIGURE 3–8
Persistent renal cortical activity 3 hours after a tracer was injected in a patient with metastatic cancer and hypercalcemia.

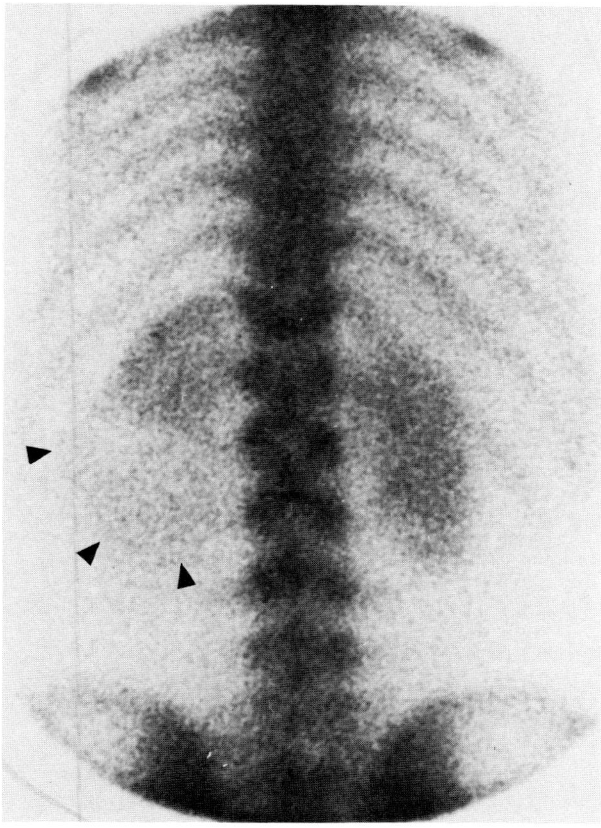

◀ *FIGURE 3–9*
Photopenic area in the inferior pole of the left kidney at the site of a primary renal-cell carcinoma.

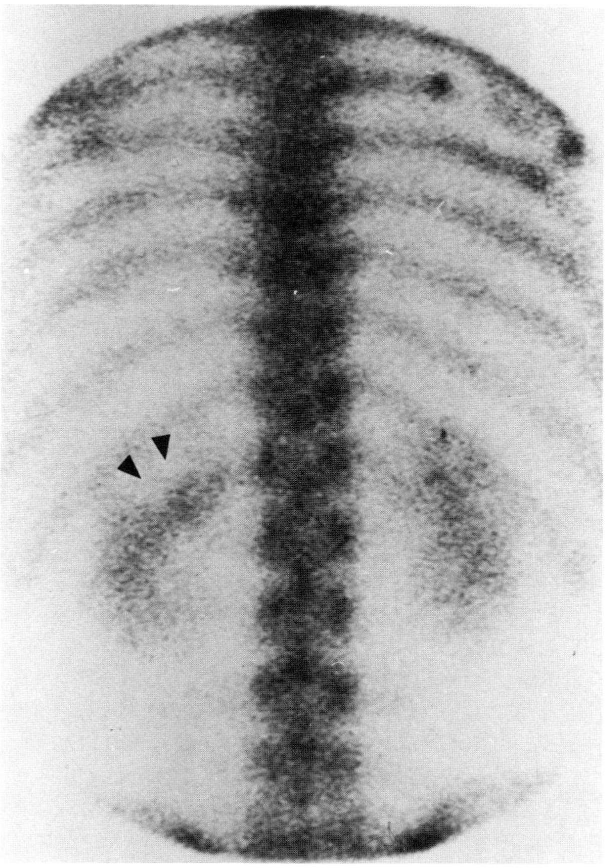

FIGURE 3–10
Indentation of the upper pole of the left kidney due to an adrenal metastasis from a primary lung malignancy.

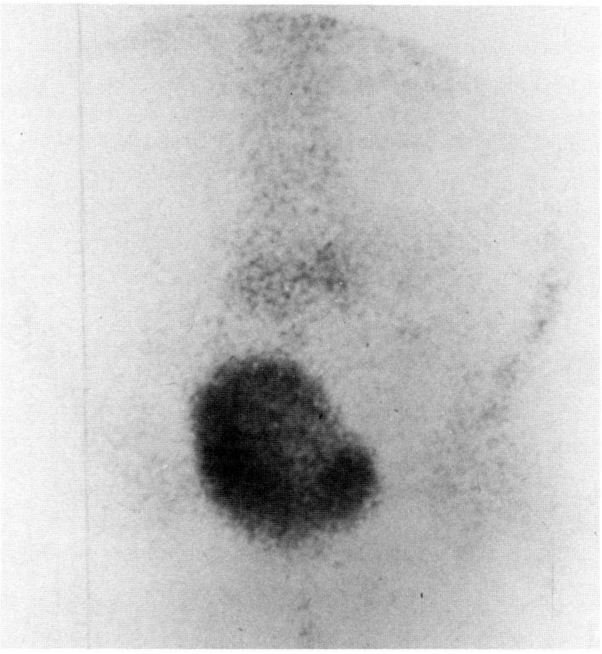

◀ *FIGURE 3–11*
Anterior view of the pelvis showing deviation of the bladder to the right due to a left ovarian tumor.

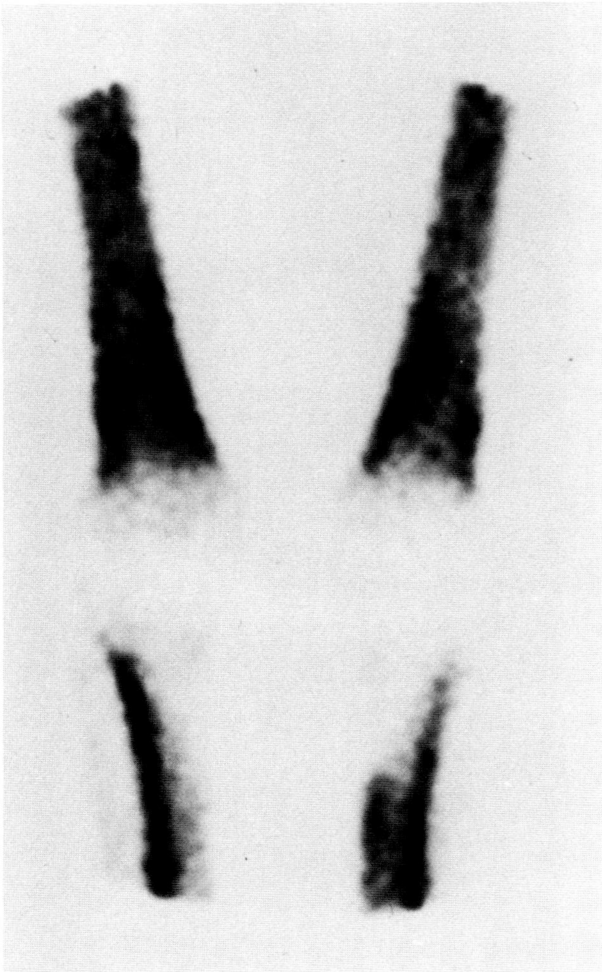

FIGURE 3–12
Hypertrophic osteoarthropathy of the lower extremities in a patient with a primary lung malignancy.

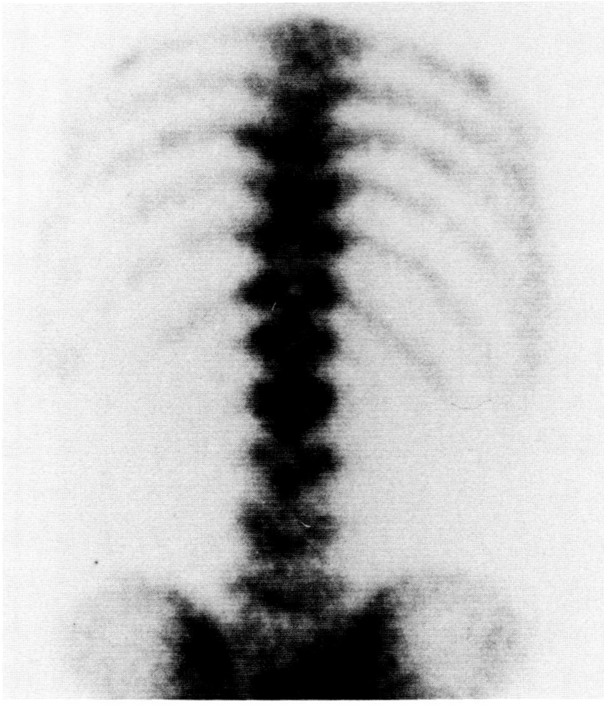

FIGURE 3–13
Diffuse increased uptake in the skeleton with absent renal activity in a patient with prostate cancer and diffuse bone metastases ("superscan").

may be seen in some patients with lung neoplasms with hypertrophic osteoarthropathy (Figure 3–12), which is unrelated to skeletal metastases. Absent renal uptake with intense activity in the bones (particularly the axial skeleton) is characteristic of a "superscan" (Figure 3–13). This may represent diffuse metastatic disease and is most often seen in patients with prostate or breast malignancies metastatic to bone.

REFERENCES

Breast cancer

1. McNeil BJ. Value of bone scanning in neoplastic disease. Semin Nucl Med 1984; 14:277–86.
2. Bauer GC, Wendeberg B. External counting of Ca-47 and Sr-85 in studies of localised skeletal lesions in man. J Bone Joint Surg [Br] 1959; 41:558–80.

3. Cancer statistics. 1988; 38(1):12.
4. Citrin DL, Tormey DC, Carbone PP. Implications of the Tc-99m diphosphonate bone scan on treatment of primary breast cancer. Cancer Treat Rep 1977; 61:1249–52.
5. Galasko CS. Skeletal metastases and mammary cancer. Ann R Coll Surg Engl 1972; 50:3–28.
6. Citrin DL. The role of the bone scan in the investigation and treatment of bone cancer. CRC Crit Rev Diagn Imaging 1980; 13:39–55.
7. Frazier TG, Copeland EM, Gallager HS, Paulus DD Jr, White EC. Prognosis and treatment in minimal breast cancer. Am J Surg 1977; 133:697–701.
8. Cutler SJ, Ardyce JA, Taylor SG III. Classification of patients with disseminated cancer of the breast. Cancer 1969; 24:861–69.
9. Strender LE, Lagergren C, Wallgren A, Liljevall S. Role of bone scans in the initial assessment of operable patients with breast carcinoma. Acta Radiol Oncol 1981; 20:187–91.
10. Roberts MM, Hayward JL. Bone scanning and early breast cancer: five year follow-up. Lancet 1983; 1:997–8.
11. Pedrazzini A, Gelber R, Isley M, Castiglione M, Goldhirsch A, for the Ludwig Study Group. First repeated bone scan in the observation of patients with operable breast cancer. J Clin Oncol 1986; 4:389–94.
12. Pauwels EK, Heslinga JM, Zwaveling A. Value of pre-treatment and follow-up skeletal scintigraphy in operable breast cancer. Clin Oncol 1982; 8:25–32.
13. Chaudary MA, Maisey MN, Shaw PJ, Rubens RD, Hayward JL. Sequential bone scans and chest radiographs in the postoperative management of early breast cancer. Br J Surg 1983; 70:517–8.
14. Pandya KJ, McFadden ET, Kalish LA, Tormey DC, Taylor SG, Falkson G. A retrospective study of earliest indicators of recurrence in patients on Eastern Cooperative Oncology Group adjuvant chemotherapy trials for breast cancer. A preliminary report. Cancer 1985; 55:202–5.
15. Bitran JD, Bekerman C, Desser RK. The predictive value of serial bone scans in assessing response to chemotherapy in advanced breast cancer. Cancer 1980; 45:1562–8.
16. Rossleigh MA, Lovegrove FT, Reynolds PM, Byrne MJ. Serial bone scans in the assessment of response to therapy in advanced breast cancer. Clin Nucl Med 1982; 7:397–402.
17. Rossleigh MA, Lovegrove FT, Reynolds PM, Byrne MJ, Whitney BP. The assessment of response to therapy of bone metastases in breast cancer. Aust NZ J Med 1984; 14:19–22.

Lung cancer

18. Minna JD, Higgins GA, Glatstein EJ. Cancer of the lung. In: Devita VT Jr, Hellman S, Rosenberg SA, eds. Cancer: principles and practice of oncology. Philadelphia: JB Lippincott, 1982: Chapter 14, pp. 287–328.
19. Levitan N, Byrne RE, Bromer RH, et al. The value of the bone scan and bone marrow biopsy in staging small cell lung cancer. Cancer 1985; 56:652–4.
20. Quinn DL, Ostrow LB, Porter DK, Shelton DK, Jackson DE Jr. Staging of non-small cell bronchogenic carcinoma. Relationship of the clinical evaluation to organ scans. Chest 1986; 89:270–5.
21. Hooper RG, Beechler CR, Johnson MC. Radioisotope scanning in initial staging of bronchogenic carcinoma. Am Rev Respir Dis 1978; 118:279–86.
22. Hardy JD. Hardy's textbook of surgery. Philadelphia: JB Lippincott, 1983:836.

Prostate cancer

23. Spirnak JP, Resnick MI. Clinical staging of prostatic cancer: new modalities. Urol Clin North Am 1984; 11:221–35.
24. Elder JS, Jewett HJ, Walsh PC. Radical perineal prostatectomy for clinical stage B2 carcinoma of the prostate. J Urol 1982; 127:704–6.
25. Jewett HJ, Eggleston JC, Yawn DH. Radical prostatectomy in the management of carcinoma of the prostate: probable causes of some therapeutic failures. J Urol 1972; 107:1034–40.
26. Whitmore WF Jr. Hormone therapy in prostatic cancer. Am J Med 1956; 21:697.
27. Gleason DF and the Veterans Administration Cooperative Urological Research Group. Histologic grading and clinical staging of prostatic carcinoma. In: Tannenbaum M, ed. Urologic pathology: The prostate. Philadelphia: Lea and Febiger, 1977:171–97.
28. Biersack HJ, Wegner G, Distelmaier W, Krause U. Bone metastases of prostatic carcinoma in relation to tumor size and grade of malignancy. Nuklearmedizin 1980; 19:29–32.
29. Merrick MV, Ding CL, Chisholm GD, Elton RA. Prognostic significance of alkaline and acid phos-

phatase and skeletal scintigraphy in carcinoma of the prostate. Br J Urol 1985; 57:715–20.

30. Shafer R. Unpublished results, 1985.
31. Schaffer DL, Pendergrass HP. Comparison of enzyme, clinical, radiographic, and radionuclide methods of detecting bone metastases from carcinoma of the prostate. Radiology 1976; 121:431–4.
32. Stamey TA, Yang N, Hay AR, McNeal JE, Freiha FS, Redwine E. Prostate-specific antigen as a serum marker for adenocarcinoma of the prostate. N Engl J Med 1987; 317:909–16.
33. Pollen JJ, Witztum KF, Ashburn WL. The flare phenomenon on radionuclide bone scan in metastatic prostate cancer. AJR 1984; 142:773–6.
34. Pollen JJ, Shlaer WJ. Osteoblastic response to successful treatment of metastatic cancer of the prostate. AJR 1979; 132:927–31.
35. Levenson RM, Sauerbrunn BJL, Bates HR, et al. Comparative value of bone scintigraphy and radiography in monitoring tumor response in systemically treated prostatic carcinoma. Radiology 1983; 146:513–8.
36. Johns WD, Kaplan WD. Leuprolide therapy for prostatic cancer. An association with scintigraphic "flare" on bone scan. Clin Nuc Med 1990; 15:485–7.

Osteogenic sarcoma

37. Welch KJ, Randolph JG, Ravitch MM, O'Neill JA Jr, Rowe MI. Pediatric surgery. 4th ed. Chicago: Year Book Medical, 1986:315–7.
38. Sandler MS, Heyman S, Watts H. Localization of 99mTc methylene diphosphonate within synovial fluid in osteosarcoma. AJR 1984; 143:349–50.
39. Boyko OB, Cory DA, Cohen MD, Provisor A, Mirkin D, DeRosa GP. MR imaging of osteogenic and Ewing's sarcoma. AJR 1987; 148:317–22.
40. Aisen AM, Martell W, Braunstein EM, McMillin KI, Phillips WA, Kling TF. MRI and CT evaluation of primary bone and soft-tissue tumors. AJR 1986; 146:749–56.
41. Goldman AB, Becker MH, Braustein P, Francis KC, Genieser NB, Firooznia H. Bone scanning-osteogenic sarcoma. Correlation with surgical pathology. Radiology 1976; 119:255.
42. Goldstein HA, McNeil BJ, Zufall E, Jaffe N, Treves S. Changing indications for bone scintigraphy in osteosarcoma. Radiology 1980; 135:177–80.
43. McKillop JH, Etcubanas E, Goris ML. The indications for and limitations of bone scintigraphy in osteogenic sarcoma: a review of 55 patients. Cancer 1981; 48:1133–8.
44. Link MR, Goorin AM, Miser AW, et al. The effect of adjuvant chemotherapy on relapse-free survival in patients with osteosarcoma of the extremity. N Engl J Med 1986; 314:1600–6.
45. McNeil BJ, Hanley J. Analysis of serial radionuclide bone images in osteosarcoma and breast carcinoma. Radiology 1980; 135:171–6.
46. Brown ML. Significance of the solitary lesion in pediatric bone scanning. Concise communication. J Nucl Med 1983; 24:114–5.

Ewing's sarcoma

47. Bushnell D, Shirazi P, Khedkar N, Blank J. Ewing's sarcoma seen as a "cold" lesion of bone scans. Clin Nucl Med 1983; 8:173–4.
48. Sty JR, Kun LE, Starshak RJ. Pediatric applications in nuclear oncology. Semin Nucl Med 1985; 15:171–200.
49. Nair N. Bone scanning in Ewing's sarcoma. J Nucl Med 1985; 26:349–52.

Neuroblastoma

50. Jaffe N. Malignant bone tumors. In: Lanzkowsky P, ed. Pediatric oncology. New York: McGraw-Hill, 1982: Chapter 10, pp. 293–309.
51. Howman-Giles RB, Gilday DL, Ash JM. Radionuclide skeletal survey in neuroblastoma. Radiology 1979; 131:497–502.
52. Podrasky AE, Stark DD, Hattner RS, Gooding CA, Moss AA. Radionuclide bone scanning in neuroblastoma: skeletal metastases and primary tumor localization of 99mTc MDP. AJR 1983; 141:469–72.
53. Baker M, Siddiqui AR, Provisor A, Cohen MD. Radiographic and scintigraphic skeletal imaging in patients with neuroblastoma. Concise communication. J Nucl Med 1983; 24:467–9.
54. Heisel MA, Miller JH, Reid BS, Siegel SE. Radionuclide bone scan in neuroblastoma. Pediatrics 1983; 71:206–9.
55. Mandell GA, Heyman S. Extraosseous uptake of technetium-99m MDP in secondary deposits of neuroblastoma. Clin Nucl Med 1986; 11:337–40.

Urinary tract neoplasms

56. Kishi K, Hirota T, Matsumota K, Kakizoe T, Murase T, Fujita J. Carcinoma of the bladder: a clin-

ical and pathological analysis of 87 autopsy cases. J Urol 1981; 125:36–9.

57. Davey P, Merrick MV, Duncan W, Redpath AT. Bladder cancer: the value of routine bone scintigraphy. Clin Radiol 1985; 36:77–9.
58. Rosen PR, Murphy KG. Bone scintigraphy in the initial staging of patients with renal-cell carcinoma. Concise communication. J Nucl Med 1984; 25:289–91.
59. Clyne CA, Frank JW, Jenkins JD, Smart CJ. The place of the 99mTc-polyphosphonate bone scan in renal carcinoma. Br J Urol 1983; 55:174–5.
60. Kim EE, Bledin AG, Gutierrez C, Haynie TP. Comparison of radionuclide images and radiographs for skeletal metastases from renal cell carcinoma. Oncology 1983; 40:284–6.

Colon cancer

61. Bonnheim DC, Petrelli NJ, Herrera L, Walsh D, Mittelman A. Osseous metastases from colorectal carcinoma. Am J Surg 1986; 151:457–9.

Gynecologic malignancies

62. Mettler FA Jr, Christie JH, Garcia JF, Baldwin MH, Wicks JD, Bartow SA. Radionuclide liver and bone scanning in the evaluation of patients with endometrial carcinoma. Radiology 1981; 141:777–80.
63. Kamath CR, Maruyama Y, DeLand FH, Van Nagell JR. Value of bone scanning in detecting occult skeletal metastases from adenocarcinoma of the endometrium. Diagn Gynecol Obstet 1982; 4:155–8.
64. Harbert JC, Rocha L, Smith FP, Delgado G. The efficacy of radionuclide liver and bone scans in the evaluation of gynecologic cancers. Cancer 1982; 49:1040–2.
65. Salazar OM, Feldstein ML, DePapp EW, et al. Endometrial carcinoma: analysis of failures with special emphasis on the use of initial preoperative external pelvic radiation. Int J Radiat Oncol Biol Phys 1977; 2:1101–7.
66. Kamath CR, Maruyama Y, DeLand FH, Van Nagell JR. Role of bone scanning for evaluation of carcinoma of the cervix. Gynecol Oncol 1983; 15:171–85.
67. Katz RD, Alderson PO, Rosenshein NB, Bowerman JW, Wagner HN Jr. Utility of bone scanning in detecting occult skeletal metastases from cervical carcinoma. Radiology 1979; 133:469–72.
68. Mettler FA Jr, Christie JH, Crow NE Jr, et al. Radionuclide bone scan, radiographic bone survey, and alkaline phosphatase: studies of limited value in asymptomatic patients with ovarian carcinoma. Cancer 1982; 50:1483–5.

Multiple myeloma

69. Ludwig H, Kumpan W, Sinzinger H. Radiography and bone scintigraphy in multiple myeloma: a comparative analysis. Br J Radiol 1982; 55:173–81.
70. Frank JW, Le Besque S, Buchanan RB. The value of bone imaging in multiple myeloma. Eur J Nucl Med 1982; 7:502–5.
71. Woolfenden JM, Pitt MJ, Durie BGM, et al. Comparison of bone scintigraphy and radiography in multiple myeloma. Radiology 1980; 134:723–8.

4

Radionuclide Imaging of Bone Disease in Children

DAVID A. SUMMERVILLE

S. TED TREVES

Whereas the general indications for bone scintigraphy in childhood are similar to those encountered in the adult population and include infection, tumors, low back pain, trauma, and avascular necrosis, the frequency with which bone scans are performed for the various general indications differ between the two populations [1]. The indications for performing skeletal scintigraphy at Boston Children's Hospital in 1986 are summarized in Table 4–1. From this table, one can see that bone scintigraphy used for the detection of osseous metastases arising from soft tissue tumors accounted for only 13% of studies performed and bone scans for the detection of all malignant conditions accounted for only 20% of studies performed. This is in marked contrast to the experience of most nuclear medicine departments that deal primarily with adults in which bone scans used for the evaluation of malignant neoplasms account for a majority of studies.

In addition to the general indications for skeletal scintigraphy differing for the two populations, the specific disease entities for which bone scans are ordered in the pediatric popula-

TABLE 4–1
Indications for skeletal scintigraphy performed at The Children's Hospital, Boston, in 1986

Indication	Percent of studies
Low back pain	25
Infection	21
Osteomyelitis	16
Septic joint	5
Neoplasms	27
Primary bone tumors	
Benign	7
Malignant	7
Metastases from soft tissue tumors	13
Stress fractures	10
Trauma	5
Avascular necrosis	5
Arthritis	4
Other	3

tion are usually quite dissimilar from those for which bone scans are ordered in adults. This occurs because the developing skeleton is subject to many disorders not encountered in adults. Even when the disease entities appear to be the same, such as in osteomyelitis, the pathophysiology and therapeutic considerations often differ significantly in the separate populations. Thus, the physician interpreting pediatric bone scans must be familiar with a knowledge of bone pathology that is not required by those whose practice is limited to adults. In addition, the physician practicing pediatric nuclear medicine must be aware of how the normal appearance of a bone scan varies with the age of the individual.

The purpose of this chapter is to discuss the applications of bone scintigraphy in the diagnosis of bone disorders in children. In doing so, we concentrate on disorders that are unique to the developing skeleton.

Methods and Dosimetry

The doses administered in pediatric nuclear medicine are designed to be the lowest necessary to allow a satisfactory examination. Although this policy increases the time necessary to obtain high-resolution images, it reduces the radiation exposure to the patient to a minimum. A number of formulas have been developed for determining the dose administered to children. Those that scale the adult dose to the body surface area or weight of the child are generally adequate for children other than infants. In infants there is a minimum administered dose below which the study will be inadequate. This minimum dose is determined empirically and varies with the type of study. For static bone scans, in which an intravenous injection of technetium 99m methylene diphosphonate (^{99m}Tc-MDP) is given 2–4 hours prior to imaging, the usual administered dose adjusted for body weight is 200 μCi/kg. A minimum dose of 500 μCi, the smallest dose we have yet found to yield consistently good studies, is employed in babies whose body weight is less than 2.5 kg.

When clinically indicated, a so-called three-phase bone scan is performed. The three-phase bone scan includes the 2–4 hour delayed images, a radionuclide angiogram of the anatomic region of interest (recorded on a computer at 1 frame/sec for 180 seconds), and a static image (the so-called "blood-pool" image) which is obtained immediately after the radionuclide angiogram.

The delayed images obtained 2–4 hours after injection of ^{99m}Tc-MDP include routine views of the entire skeleton and are obtained on a large-field-of-view camera employing a high resolution, parallel-hole collimator. An initial image of 500,000 counts is taken of the anterior part of the chest. Images of the remainder of the skeleton are then obtained for the amount of time required for the initial 500,000-count image. Additional images of 500,000 counts are obtained of suspicious areas of bone and may consist of images obtained in a different projection than the standard images or may consist of views that magnify the region of interest. Magnification images are often obtained with a pinhole collimator, the value of which cannot be overemphasized in the pediatric population. Such pinhole collimator images are of particular benefit when scanning the hips for an area of decreased radiopharmaceutical uptake resulting from compromised circulation to the femoral head, or when looking for abnormal foci of activity abutting a growth plate. Not infrequently, the magnified images disclose abnormalities not readily appreciated from the images obtained with the parallel-hole collimator.

SPECT is becoming an increasingly important tool in identifying and localizing bony lesions in the pediatric population. SPECT has the same relationship to planar imaging that CT has to plain radiography; that is to say, with SPECT a slice image, or more precisely, a series of slice images is obtained by the mathematic reconstruction of a data set of planar images. This technique is performed with a rotating gamma camera and requires the collection of data on a computer equipped with the appropriate software. The resulting images, which can be displayed with the same degree of resolution in transaxial, coronal, and sagittal planes, provide both improved image contrast and more complete spatial information than can be obtained with planar imaging alone.

Acquisition of the tomographic images requires that a sequence of planar images be obtained by the camera making a trajectory of 180 or 360 degrees around the patient. Since these studies require that the patient lie still for approximately 20 minutes, they are generally easier to perform in older pediatric patients. In the younger population, sedation often is necessary.

The nature of the tomographic process is such that far greater attention to the technical aspects of image acquisition is required than is necessary with planar imaging. A discussion of the various procedural details required for obtaining quality SPECT images is beyond the scope of this work; several excellent books [2,3] and reviews are available [4,5].

Two procedural difficulties encountered in performing bone scans in children are: (1) Many younger children are unable to remain sufficiently still for the time required to obtain an image; and (2) in younger children, who are incapable of voiding on request, the presence of a bladder full of radioactivity may obscure pathologic features of the hips and pelvis. The first of these problems can often be circumvented by technologists who are well trained in dealing with children and by performing the study in a quiet, dimly lit room free of extraneous noise and traffic. While such gentle handling often results in a sleeping child, there are times when wrapping the child in a sheet or sedating the child is the only realistic alternative. When a child is unable to void during the study and good images of the pelvis are required for the study, we often have the child return to the department several hours later. In rare instances, emptying the bladder via catheter may be unavoidable.

The absorbed radiation doses for ^{99m}Tc-MDP for children of average body habitus are listed by age group in Table 4–2. In interpreting such data, it must be emphasized that accurate data on absorbed radiation doses in the pediatric population are generally not available due to a lack of detailed information on the biodistribution and kinetics of radiopharmaceuticals in children, particularly in premature babies and infants. Most biodistribution data used for the calculation of absorbed radiation doses by children are extrapolated from studies of either healthy adult humans or animals. Since it is well known that the biokinetics of radiopharmaceuticals in children, particularly in infants, often differs from that of adults, absorbed dose calculations for children must be viewed as estimates.

GALLIUM 67 CITRATE

Gallium scintigraphy is occasionally used as an adjunct to bone scanning in the diagnosis of osteomyelitis. When scintigraphy is performed in a patient with bone infection, a usual dose of

TABLE 4–2
Estimated absorbed doses from intravenous injection of ^{99m}Tc-MDP

	Age (yr)		
	Newborn	5	15
Body weight (kg)	3.5	19.1	54.5
Administered dose (mCi)	0.7	3.8	10.9
Absorbed doses (rad)			
Bladder wall	1.5	1.5	2.0
Bone surfaces	—	2.1	2.2
Kidneys	0.09	0.14	0.19
Red marrow	—	0.33	0.36
Ovaries	0.05	0.12	0.15
Testes	0.04	0.10	0.12
Total body	0.05	0.08	0.10

TABLE 4–3
Estimated absorbed doses from intravenous injection of ^{67}Ga citrate

	Age (yr)		
	Newborn	5	15
Body weight (kg)	3.5	19.1	54.5
Administered dose (mCi)	0.25	0.76	2.2
Absorbed dose (rad)			
Small intestine	0.6	0.8	0.9
Wall of large intestine	3.0	2.2	2.2
Kidneys	1.1	1.1	1.3
Liver	1.3	1.2	1.6
Red marrow	—	2.2	2.4
Bone surface	—	15.0	16.0
Spleen	2.1	1.6	1.9
Ovaries	0.6	0.7	0.9
Testes	0.5	0.4	0.5
Total body	0.7	0.6	0.7

40 μCi/kg, with a minimum dose of 0.25 mCi and a maximum of 3.0 mCi, is employed. Estimates of absorbed radiation doses are given in Table 4–3. Although imaging for acute infection can be performed at 4–6 hours after injection, a better lesion-to-background ratio is obtained at 24 or 48 hours.

INDIUM-LABELED WHITE BLOOD CELLS

First introduced in 1976, the use of indium 111 (^{111}In)–labeled white blood cells for the detection of inflammatory processes has gained wide acceptance and application [6–9]. Labeling the white blood cells requires a minimum of 20 ml of the patient's blood, extracting the platelet-free white blood cell fraction, and incubating the white blood cells with a complex of ^{111}In, usually ^{111}In-oxine. The labeling process takes 2–4 hours, and requires a laminar flow hood. Imaging is performed approximately 24 hours after intravenous administration of the labeled cells. The estimated absorbed radiation doses for an administered dose of 12 μCi/kg have been calculated for a 5-year-old child weighing 20 kg, and are, in rads: whole body, 0.12–0.17; liver, 0.67–0.77; spleen, 4.1–5.6; and marrow, 0.15–1.1 [8].

Infection of Bone and Joints

OSTEOMYELITIS

Acute osteomyelitis is primarily a disease of childhood; over 85% of cases reported in several large series have occurred in children less than 16 years old [10,11]. Although osteomyelitis may arise by inoculation of bone from a contiguous focus of infection, acute osteomyelitis of childhood occurs primarily by hematogenous seeding. The slow blood flow in the metaphyseal sinusoids adjacent to the growth plate with subsequent bacterial lodgement has been postulated as the reason for the predominant metaphyseal location of osteomyelitis in children [12]. Although up to 30% of cases of hematogenous osteomyelitis of childhood occur at sites other than the metaphyses of tubular bones, these cases usually occur adjacent to cartilage where the vascular arrangement is similar to that of the metaphysis, such as the ischiopubic synchondroses [13]. Osteomyelitis involving the epiphysis is primarily a disorder of infants and is believed to occur by spread of the infection from the metaphysis through vascular channels that cross the growth plate in children less than 1 year old [11,14]. These vascular channels generally disappear by the age of 15 months, leaving an avascular growth plate that acts as a barrier to the spread of infection in older individuals [14]. Epiphyseal osteomyelitis is occasionally encountered in children older than 1 year, however, and can arise either by extension from a metaphyseal focus, perhaps due to the presence of persistent transphyseal vascular channels, or may occur by direct hematogenous seeding in the epiphysis.

Diagnosis and treatment Prompt diagnosis and treatment of osteomyelitis in children are crucial. Delay in treatment may result in damage to the growth plate with subsequent growth arrest, or may spread from the metaphyseal location to the joint space and, if treatment is delayed, result in articular damage and a future of orthopedic problems. Although such extension to the joint space may occur by transphyseal extension, the infection most commonly occurs in the hip and shoulder where the joint capsules extend onto the metaphysis and seeding of the joint arises from the direct transcortical extension of the metaphyseal infection.

Early diagnosis in children is often difficult because clinical signs and symptoms as well as laboratory and radiographic abnormalities may be absent, subtle, or nonspecific. Early changes of osteomyelitis may not become apparent on plain radiographs for several days to a week after the onset of symptoms. In contrast, the scintigram usually demonstrates increased uptake of the radiopharmaceutical 24–48 hours after the onset of symptoms. Scintigraphic changes indicative of osteomyelitis are often apparent at the time of presentation and may predate radiographic changes by several days to a week [15]. As such, bone scintigraphy should be considered part of the early evaluation of children with signs and symptoms of osteomyelitis, particularly when radiographs are considered normal or nondiagnostic. When scintigraphy is used in the evaluation of a patient suspected of having osteomyelitis, it is important that the results of the scan be correlated with the clinical, laboratory, and radiographic results.

In contrast to the situation in adults, in whom the precise role of skeletal scintigraphy in the diagnosis of osteomyelitis remains a matter of controversy, the role of bone scintigraphy in the diagnosis of pediatric bone infection is well defined. This is due to the fact that in adults, in whom suspected hematogenous bone infection usually arises in areas of preexisting bone pathology, it is often difficult to determine whether increased activity on the 2–4-hour delayed bone scan merely reflects the underlying osseous lesion or whether the abnormal activity indicates secondary infection. In children, however, osteomyelitis usually involves a region of bone not affected by other osseous pathologic factors. Thus, in the appropriate clinical setting, abnormal activity on the bone scan is a sensitive and specific finding for infection when the abnormal activity on the scan involves a radiographically normal region of bone or when radiographs show changes compatible with the diagnosis of osteomyelitis.

Although skeletal scintigraphy represents the most powerful diagnostic modality when osteomyelitis is suspected in the pediatric population, the successful detection of lesions requires careful attention to technical detail. Since hematogenous osteomyelitis frequently involves the metaphysis and abuts the growth plate, difficulty may arise in resolving abnormal activity from normal intense growth plate activity. High-resolution images with careful positioning of the patient to obtain equivalent views of the contralateral unaffected side are required to detect such activity. Magnification views with pinhole collimation are often necessary to demonstrate the abnormal activity, particularly in infants. The importance of pinhole collimation in the detection of childhood epiphyseal osteomyelitis is well established [16]. When images are obtained with the pinhole collimator, it is important that an image of the unaffected side also be obtained for comparison (Figure 4–1). Careful positioning is particularly important when osteomyelitis of the proximal humeral metaphysis is suspected since the apparent width of the growth plate varies with the position of the humerus. Improper positioning can result in either a false-positive or a false-negative study.

Although the diagnosis of osteomyelitis can usually be made from the 2–4-hour delayed scintigraphic images, there are two circumstances in which the diagnosis may prove difficult. As discussed previously, detection of osteomyelitis may prove difficult at a site where there is preexisting bone pathology. Difficulties in interpretation can also arise when osteomyelitis is suspected at a site where there is an overlying soft tissue infection. In such cases, increased activity in the bone may occur as a result of increased blood flow to the region secondary to the soft tissue infection and may not represent bone infection. In both cases, the three-phase bone scan or gallium scintigraphy often prove helpful.

The addition of the radionuclide angiogram and immediate postinjection blood-pool image to the standard 2–4-hour delayed images increases the accuracy of the bone scan in the diagnosis of osteomyelitis in the presence of cellulitis or other bone abnormality [17]. On the three-phase study, osteomyelitis is characterized by increased blood flow to the area of the lesion seen on the

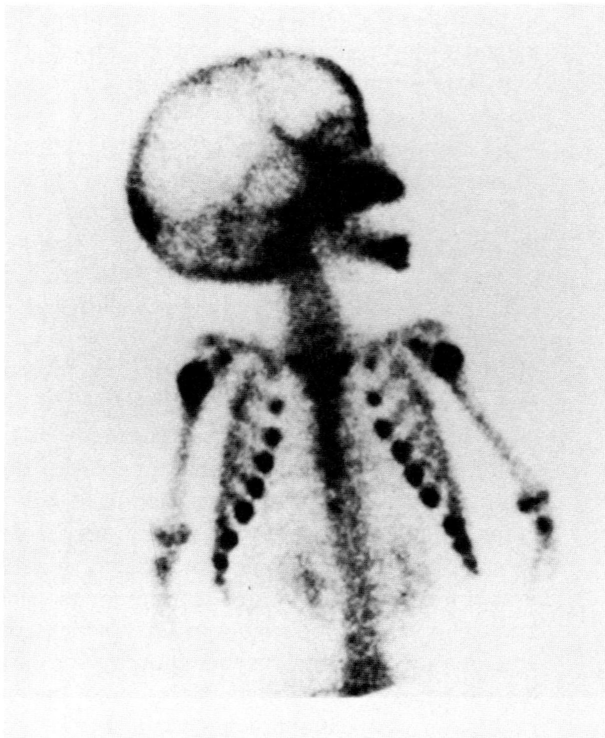

A

FIGURE 4–1
A. Anterior image obtained with a parallel-hole collimator from an 8-month-old girl with suspected osteomyelitis of the right proximal humeral metaphysis. Radiographic findings in the region were normal. B, C. Magnification images obtained in the posterior projection of the right (B) and left (C) shoulders. Increased uptake involving the right metaphysis proved to be osteomyelitis.

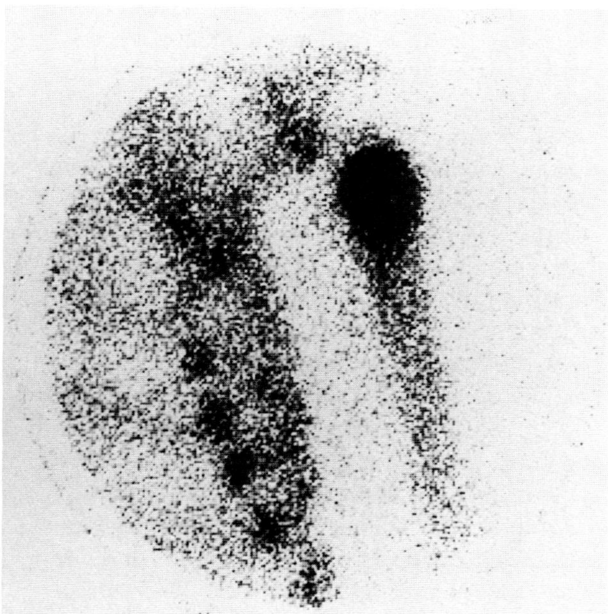

B

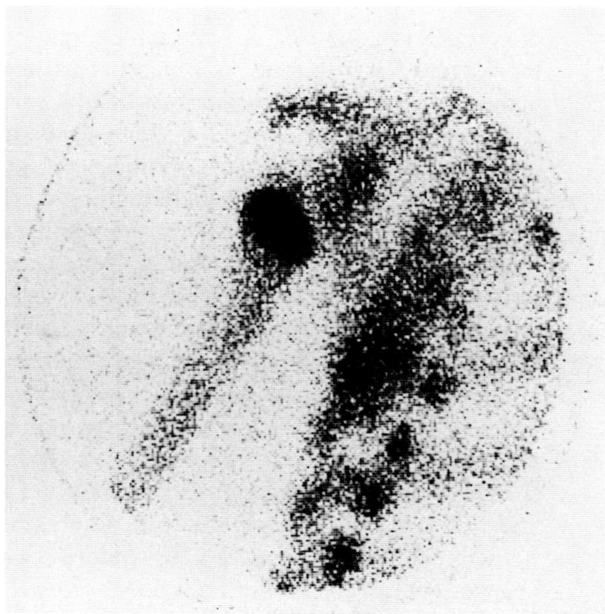

C

radionuclide angiogram, hyperemia ascribable to bone on the blood-pool image, and focal intense uptake in bone on the delayed images. Noninfected osseous lesions generally show either normal or mildly increased blood flow and hyperemia. In one study, the use of the three-phase bone scan markedly changed the ability to distinguish between noninfectious skeletal disease and osteomyelitis, with the sensitivity to detect noninfectious lesions improving from 66 to 97% [17]. In addition, the same study demonstrated the enhanced ability of the three-phase bone scan in distinguishing between isolated soft tissue infection and soft tissue infection superimposed on osteomyelitis. Three-phase scintigraphy of bone showed that in 98 patients, the false-positive rate for osteomyelitis was decreased from 25 to 6% without altering the 92% sensitivity (Figures 4–2 and 4–3).

The usefulness of a 24-hour delayed image to distinguish osteomyelitis from isolated soft tissue infection has recently been investigated [18,19]. In cases of osteomyelitis, the ratio of lesion to background activity is observed to increase between 4 and 24 hours after administration of the bone-seeking agent, whereas the ratio is reported to decrease or remain the same in cases of soft tissue infection.

Gallium scanning may be particularly useful in the evaluation of infection at a site with a preexisting osseous abnormality [20–22]. In a study of 163 adult patients suspected of having osteomyelitis at a site of previous infection or fracture, a pattern of uptake on the gallium scan, which showed a larger area of relatively more uptake by the lesion on the gallium scan than on the bone scan, proved to be active osteomyelitis in the 12 patients exhibiting such a pattern [22]. When the relative intensities of the suspected lesions were similar on the two scans, the likelihood of active disease was increased, but false-positive findings were encountered. When the uptake on the gallium scan was normal or less than that on the bone scan, the likelihood of osteomyelitis was significantly diminished; however, false-negative results did occur. In another study involving children suspected of acute osteomyelitis at a site of preexisting bone pathology, the pattern of focally increased gallium activity proportionally equal to or greater than the increased activity seen on bone scan increased the specificity from 0.25 to 0.63 while maintaining a sensitivity of 1.0 in 11 patients [20]. At times, gallium 67 (^{67}Ga) concentrates in areas of inflammation even when the ^{99m}Tc-MDP scan is normal [23–25]. Gallium imaging, at times, can also provide earlier detection of osteomyelitis than is possible with ^{99m}Tc-MDP imaging in areas adjacent to growth plates.

It must be noted that although an abnormal finding on scintiscan obtained in the presence of signs and symptoms of bone infection is specific for osteomyelitis, the absence of osteomyelitis cannot be entirely excluded by a normal scan. Some patients have shown normal or decreased uptake in areas of subsequently proved osteomyelitis [26–31]. This may reflect a combination of bacterial proliferation and host immune response resulting in an occlusion of local blood vessels. This explanation is reinforced by a report that high-pressure abscesses will cause an absence of radiotracer uptake on the scan [23]. Although some workers have reported poor detectability of osteomyelitis with the bone scan, particularly in the neonatal population [31], other workers have found bone scintigraphy to be a sensitive tool in the diagnosis of osteomyelitis [32]. These workers stress the necessity of high-resolution and magnification views in order to visualize the increased juxtametaphyseal activity in neonatal osteomyelitis. Whatever the cause, the occasional normal finding on scintiscan in patients with osteomyelitis means that a normal result is no guarantee that a patient does not have osteomyelitis. When the clinical impression of osteomyelitis persists, a repeat bone scan several days later, or a ^{67}Ga citrate study, may resolve the issue.

One modality that has been gaining increasing interest over the last few years is the use of ^{111}In-labeled white blood cells for the detection of osteomyelitis. The theoretic advantage of the labeled white blood cells is that they accumulate only in regions of infection and, unlike ^{99m}Tc-MDP and ^{67}Ga, show no affinity for bone. Thus, any focal accumulation of labeled white blood cells in bone is considered osteomyelitis. A recent study by Merkel et al. in adults suggests that

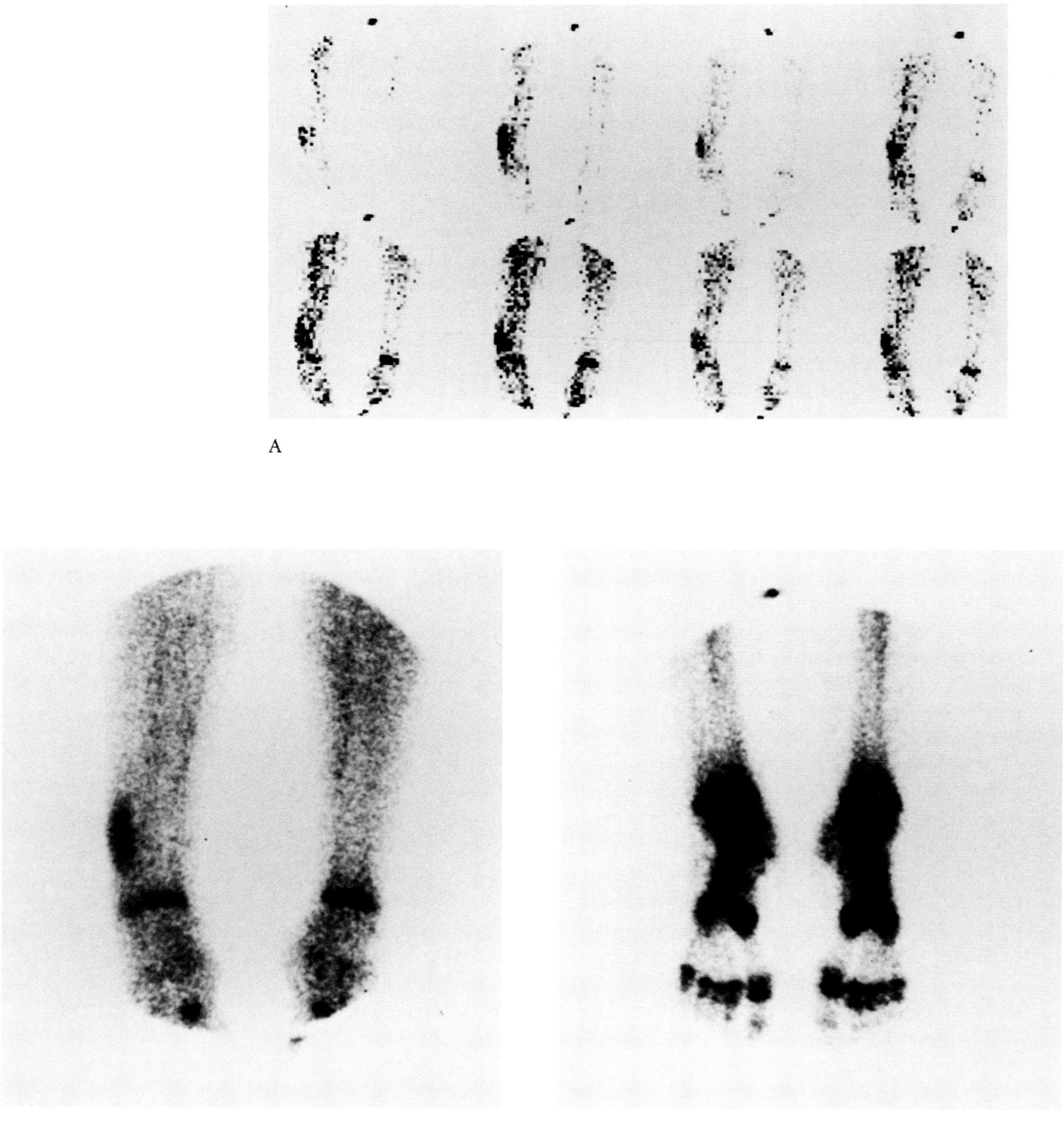

A

B

C

FIGURE 4–2

A 15-year-old boy had swelling and inflammation at the distal aspect of the right lower extremity. Both the flow and tissue phase images show increased activity corresponding to the soft tissue swelling seen on physical examination; no abnormal activity ascribable to bone is seen. Delayed images have a normal appearance, indicating a soft tissue infection without associated osteomyelitis. A. Radionuclide angiogram. B. Tissue phase image. C. Four-hour delayed image.

^{111}In-labeled white blood cells are more sensitive and specific for diagnosing osteomyelitis than is ^{67}Ga scintigraphy [33]. Unfortunately, only limited data are available regarding the usefulness of the labeled white blood cells for the detection of osteomyelitis in children [8]. Whether labeled white blood cell studies will replace ^{67}Ga scintigraphy for the detection of osteomyelitis when bone scan findings are equivocal awaits further study.

One unusual condition that must be mentioned is the recently described entity of chronic relapsing multifocal osteomyelitis (CRMO). This condition, first described in 1972 by Giedion et al. [34], is characterized by intermittent episodes of bone pain, fever, and an increased erythrocyte sedimentation rate (ESR) [35,36]. Although the pain may be limited to only one site, suggesting an acute osteomyelitis, skeletal surveys or bone scintigraphy usually identify multiple sites of involvement. The lesions, which characteristically occur adjacent to the growth plates of both long and short tubular bones [37], show lytic changes consistent with a hematogenous osteomyelitis. Although the histologic findings are generally consistent with such a diagnosis, evaluation of special histologically stained sections and cultures fails to detect an organism.

The scintigraphic findings are variable and reflect the episodic nature of the disorder. Uptake of the bone scanning agent is usually intense during acute exacerbations. Skeletal scintigraphy can detect lesions of CRMO before they become apparent radiographically or clinically. During periods of clinical improvement, uptake of the radiotracer of the lesions gradually decreases and eventually becomes normal. Skeletal scintigraphy has proved to be a useful tool both in the diagnosis of this disorder and in following its unpredictable course.

SEPTIC ARTHRITIS

Suspected joint infection represents a common indication for skeletal scintigraphy in children. Typically the patient presents with local pain, swelling, and fever; a rapid ESR and elevated white blood cell count may be present. The physical examination usually reveals decreased mobility of the joint and resistance to passive mo-

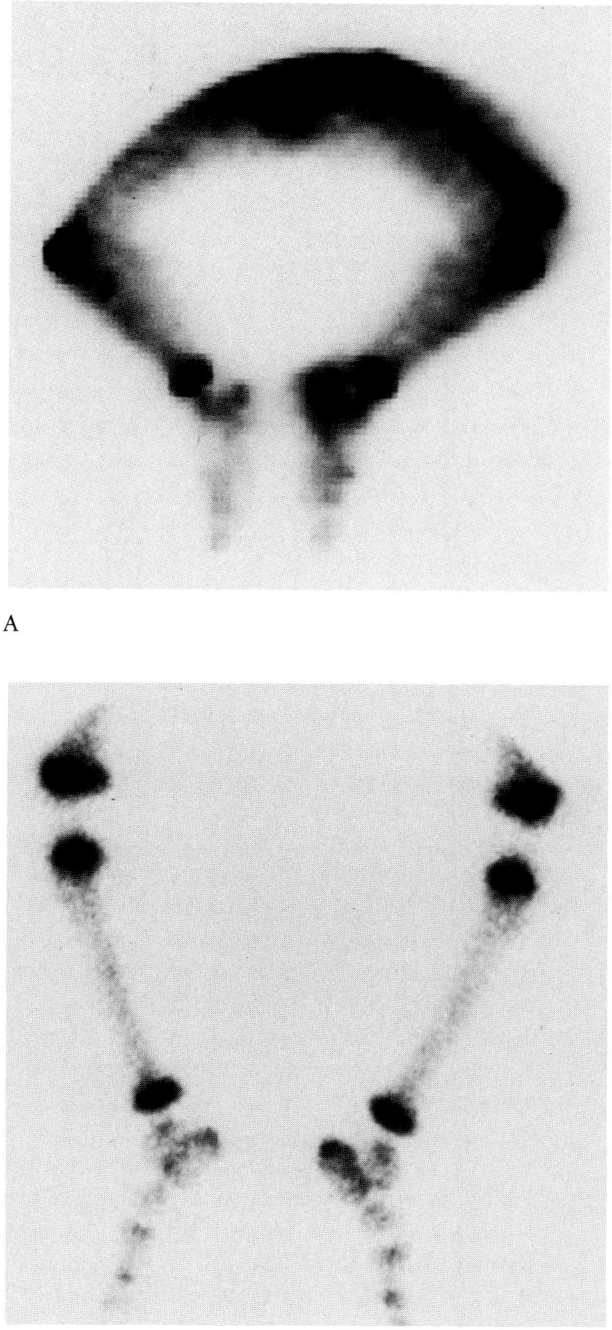

A

B

FIGURE 4–3

A. *Increased activity involving the heel of an infant on the tissue phase image of a three-phase bone scan. The patient had marked inflammation of the heel.* B. *Focally increased activity in the calcaneus proved to be osteomyelitis.*

tion. Although these signs and symptoms are helpful when present, at times the only indication of a septic joint may be refusal to move the affected limb.

Although potential routes for infection of the joint include direct inoculation secondary to trauma or medical intervention, or spread from a contiguous focus of infection, septic arthritis in children usually results from the hematogenous spread of bacteria from a distal, primary infection. Although any joint may be involved, hematogenous seeding of the appendicular skeleton, especially the hips and knees, is commonly observed in infants and children. In older children, infections involving the axial skeleton are observed with increasing frequency.

The primary role of skeletal scintigraphy in the evaluation of the child with a suspected joint infection is to rule out other bone or joint pathologic entities that may account for the patient's symptoms, and to evaluate the bone adjacent to the suspect joint for evidence of osteomyelitis. The scintigraphic findings observed with a localized joint infection are often absent or subtle, and may be observed with other joint abnormalities including toxic (transient) synovitis or a variety of rheumatic disorders [38]. Thus, the scintigraphic findings of abnormal periarticular uptake of the bone-seeking agent on the delayed images, although consistent with septic arthritis and implicating a joint abnormality as the likely source of the patient's pain, are not specific. Further, a normal bone scan does not exclude septic arthritis. In the presence of a normal bone scan, the decision to proceed with arthrocentesis should depend entirely on the clinical suspicion.

When the diagnosis of septic arthritis appears inescapable on the basis of the clinical presentation and radiographic findings, the bone scan still serves a valuable role in the evaluation of the adjacent bone for coexistent osteomyelitis. The presence of coexistent osteomyelitis often sufficiently alters the management of the patient that a bone scan should be performed in virtually every patient with a presumptive diagnosis of septic arthritis.

For those children who present with pain and difficulty in moving a limb, but in whom the signs and symptoms of septic arthritis are either not present or compelling, the bone scan can be of great value both in evaluating the joint for conditions other than infection, and in evaluating the remainder of the extremity for unsuspected pathology. The bone scan is of particular benefit in infants and toddlers in whom the only evidence of a problem may be the inability to move, or bear weight on, an extremity. It is not unusual for us to perform a bone scan on a toddler who is referred with normal radiographs of the hips and an indication of refusal to bear weight on a lower extremity, and to find osseous pathology elsewhere in the leg or foot (see Figure 4–2).

Of all the joints that may be infected, particular attention must be paid to infections of the hip. Pyarthrosis of the hip is one of the true emergencies of infancy and childhood; immediate drainage is necessary to prevent epiphyseal destruction. Because the joint capsule of the hip encloses the growth plate, direct inoculation of this joint can occur from an osteomyelitis of the femoral neck metaphysis. In the presence of septic arthritis of the hip, scanning may reveal decreased or even absent radiopharmaceutical uptake in the femoral head. This is due to ischemia resulting from increased synovial fluid pressure in the articular capsule. If pressure in the capsule is insufficient to inhibit blood flow, the scan may be normal or may show only faintly increased uptake in bones surrounding the involved joint [17]. As indicated previously, this finding is not, however, diagnostic of septic arthritis and has been reported in cases of toxic (transient) synovitis [38].

Neoplasia

PRIMARY MALIGNANT BONE TUMORS

Osteosarcoma Osteosarcoma, the most common primary malignant bone tumor in young people, occurs primarily in the 10–20-year age group, with approximately two-thirds of all cases being diagnosed before the age of 20, and only 10% of all cases being detected in children less than 10 years old. Although virtually any bone may be involved, two-thirds of these tumors arise in the long bones of the lower extremities and generally involve the metaphysis near the growth plate. Of the flat bones, the ilium is most often involved, but the sacrum, sternum, ribs, and skull (including the jaw) are all vulnerable [39].

Osteosarcoma is a rapidly growing neoplasm that generally arises in the medullary canal and spreads through the haversian canals, eventually destroying the cortex. Extensive soft tissue involvement is usually observed at presentation. Two rarer and more indolent forms of osteosarcoma, the parosteal and periosteal types, arise from the outer cortex or periosteum and carry a better prognosis. One other form of osteosarcoma must be mentioned and that is the rare multifocal osteosarcoma in which multiple sites of involvement are observed at the time of presentation. Whether the multiple lesions develop synchronously or represent the rapid, atypical metastatic spread of the tumor has not been determined.

Skeletal scintigraphy is useful both in localizing the primary lesion and in detecting skeletal metastases. Uptake of the bone agent by the primary tumor is particularly intense in osteosarcoma and usually extends slightly beyond the limit of tumor, probably as a result of hyperemia and reactive new bone formation [40]. Review of a small number of patients at our institution demonstrated that the bone scan and radiographic studies correlated well for 12 of 13 patients, and that on pathologic examination there was no extension of the tumor beyond the limit of surgical resection. In the single exception, radiotracer uptake was slightly increased throughout the shaft while the tumor was confined to the distal metaphysis. Chew and Hudson [40] found an extended uptake pattern in 11 of 18 preoperative scans when they compared the scintigraphic findings with the surgical specimens.

Skeletal scintigraphy appears most useful in the early detection of local recurrence and metastases in osteosarcoma. Although bone scintigraphy can detect soft tissue (particularly lung) metastases, scintigraphy is less sensitive than plain chest radiographs, conventional chest tomography, and CT in the diagnosis of pulmonary metastases; the sensitivity for the detection of pulmonary metastases is less than 20% [41]. We have encountered cases where lung metastases, not previously identified by other radiographic modalities, were detected on the bone scan. When a lesion in the lung fields is identified on bone scan, it is important to localize its position in relation to the lung, bone, or chest wall. This can usually be readily performed by obtaining oblique views, by use of the slant-hole collimator, or by SPECT.

The significance of skeletal scintigraphy in the detection of osseous metastases has changed in the chemotherapeutic era. Prior to the routine use of chemotherapy in the treatment of osteosarcoma, only a small percentage of children were observed to have osseous metastases at presentation and the development of osseous metastases almost never occurred before the appearance of pulmonary metastases. The use of chemotherapeutic agents has altered this pattern; in a combined series of 64 patients who had undergone chemotherapy, bone metastases without soft tissue metastases occurred in 7 patients (11%), and bone metastases preceded lung metastases in 12 (19%) [41–43].

In interpreting the bone scan in patients who have undergone amputation of a lower limb and who are being evaluated for recurrence or metastases, one must be aware of those benign bone scan abnormalities that are associated with the presence of a prosthetic device. These abnormalities include increased activity (1) at the end of the stump, (2) in the region of the ipsilateral hip joint, and (3) in the soft tissues surrounding the prosthesis. In addition, the contralateral hip and lower extremity may show stress-related changes resulting from an altered gait.

Ewing's sarcoma Ewing's sarcoma, which accounts for 10–15% of all primary bone tumors, is characterized histologically by uniform, densely packed small cells with round nuclei but without distinct cytoplasmic borders or prominent nucleoli. The vast majority occur in the first two decades of life with the peak incidence between the ages of 5 and 9 years in girls and 10–14 years in boys [44].

The clinical and radiographic presentation of Ewing's sarcoma may mimic osteomyelitis, with the patient presenting with fever, a leukocytosis, and an increased ESR. Pain and swelling

FIGURE 4–4

A. Increased activity involving the proximal part of the femur. B, C. Increased activity involving the femoral midshaft. Final diagnoses were: A. Ewing's sarcoma. B. Eosinophilic granuloma. C. Osteomyelitis.

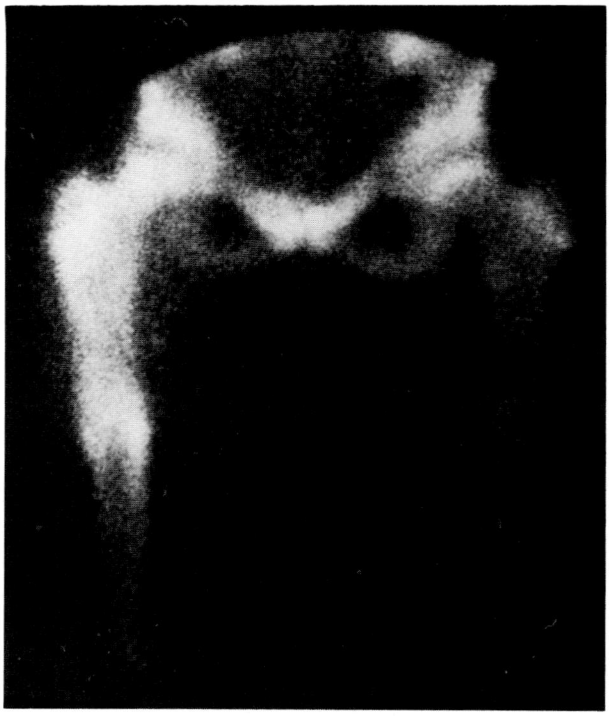

A

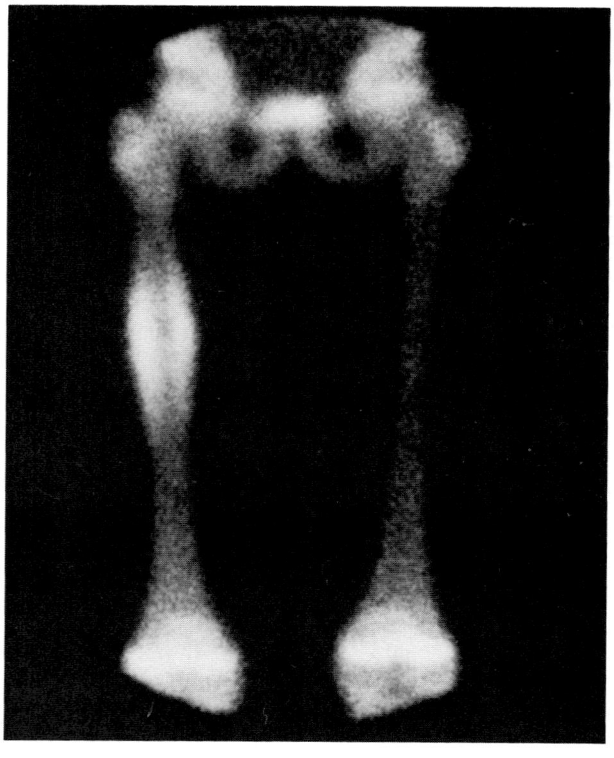

B

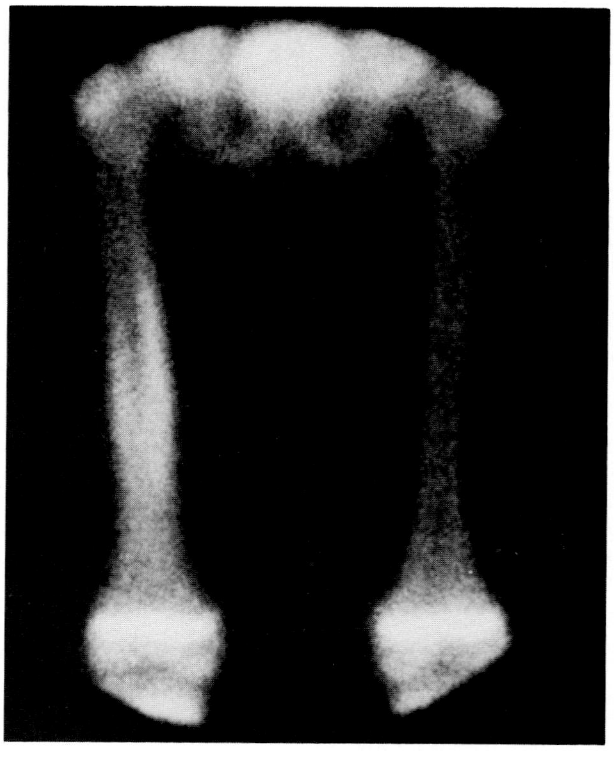

C

with a local increase in temperature are common. Both conditions may involve the metaphysis, although Ewing's sarcoma more commonly involves the midshaft. In one review, the most common site of the primary tumor was the pelvis (25%) followed in order of frequency by ribs, femora, spine, tibiae, fibulae, and scapulae [45]. Another series of 167 cases showed that the long tubular bones—femora (27%), tibiae and fibulae (17%), and the humeri (9%)—were the most often affected. Among the flat bones, the pelvis was involved in 18% of patients [46].

Because of the importance of metastatic disease in the choice of primary and follow-up therapy for this tumor, bone scintigraphy should be done at presentation and at frequent intervals thereafter. The scintigraphic appearance of Ewing's sarcoma consists of intense uptake of the bone-seeking agent in the lesion [47,48]. When the lesion occurs in the midshaft, the differential diagnosis must include eosinophilic granuloma in addition to Ewing's sarcoma and osteomyelitis (Figure 4–4). Unlike osteosarcoma, in which 2% of the patients have bone metastases at presentation, 11% of patients with Ewing's sarcoma present with bone metastases and another 45% develop bone metastases during the first 2 years after presentation.

Histiocytosis X Histiocytosis X is a collective term that includes three entities: Letterer-Siwe disease, Hand-Schüller-Christian disease, and eosinophilic granuloma. They are all characterized by a granulomatous collection of histiocytic cells, often referred to as Langerhan's cells, which are associated with variable numbers of granulocytes, lymphocytes, and prominent collections of eosinophils. The spectrum of disease varies from benign involvement of a single bone (in unifocal eosinophilic granuloma) to widely disseminated disease involving bone and soft tissues that undergoes a rapidly fatal course.

The osseous disease associated with the histiocytoses arises in hematopoietically active medullary sites and commonly involves the skull, femora, ribs, vertebrae, and pelvis. Long bone lesions are encountered as commonly in the diaphysis as in the metaphysis. The scintigraphic appearance of the lesions associated with the histiocytoses is quite variable, and reflects the stage

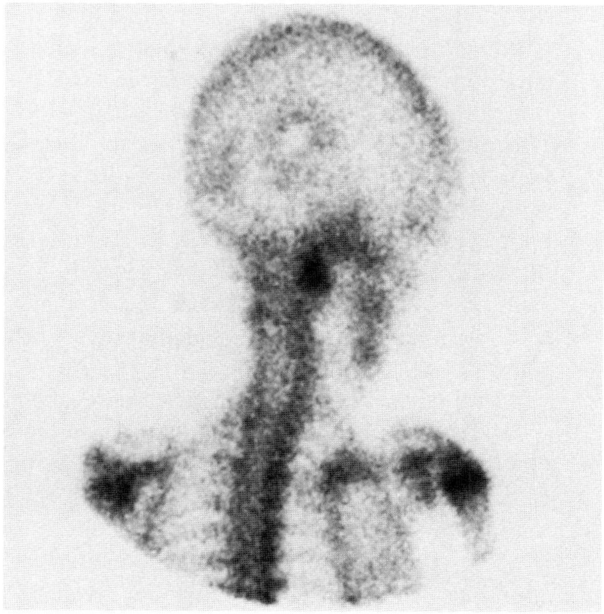

FIGURE 4–5
Typical appearance of eosinophilic granuloma involving the skull of a 6-year-old boy, demonstrating a rim of increased activity surrounding a central photopenic region.

of the lesion. Early lesions, which are confined to the medullary cavity without cortical involvement, may appear normal on bone scan. Marked increased uptake by the lesions is observed when active destruction of the cortex occurs (Figure 4–5) [49]. When active bone destruction lessens and the morphology of the lesion becomes characterized by fibrosis with bony edges exhibiting only mild sclerosis, the lesion may appear either normal on bone scan or as a photopenic defect that might be encircled by a rim of increased activity (see Figure 4–4).

It is the ability of the mature histiocytic lesions to appear normal on bone scan that results in the well-documented discrepancy between findings associated with the histiocytoses on bone scan and those observed on plain radiographs. Although various series have shown some variability, in general the sensitivity of the bone scan in detecting bony lesions observed on radiographs is only about 60% [50–53].

Skeletal scintigraphy will occasionally detect lesions not seen on skeletal survey; presumably these represent younger lesions in which bone destruction is insufficient to be radiographically observable. In view of these findings, the role of skeletal scintigraphy should be considered adjunctive to the radiographic bone survey, which remains the examination of choice to identify bone lesions in this disease complex [50,51,53]. A negative finding on radiographic study may occasion a radionuclide study to rule out early bone involvement. A recent report suggests that skeletal scans may be somewhat more reliable than radiographs for the detection of recurrence [54].

Despite the infiltrate of inflammatory cells observed with the histiocytoses, ^{67}Ga scintigraphy has no role in the evaluation of bony disease. In one study of five patients with 20 osseous lesions, only 2 lesions were detected by ^{67}Ga imaging, and both of these were apparent on bone scan [55]. ^{67}Ga scintigraphy may be of occasional use, however, in evaluating nonosseous disease.

BENIGN BONE TUMORS AND TUMOR-LIKE CONDITIONS

Although bone scintigraphy is a sensitive technique for the detection of many benign tumors and tumor-like conditions, the diagnosis is usually made radiographically and the majority of these patients do not undergo skeletal scintigraphy. Skeletal scintigraphy may at times be useful in helping to distinguish benign from malignant lesions and in evaluating the skeleton for multiple lesions. Because these lesions may be discovered when evaluating a patient with bone pain, or may be detected as incidental findings, it is important that the physician interpreting the scans be aware of the scintigraphic patterns associated with these lesions.

The scintigraphic appearance of the benign tumors and tumor-like conditions can generally be separated into those that demonstrate avid uptake of the bone-seeking agent, and appear on the bone scan as regions of increased activity, and those that demonstrate little avidity for the agent. Depending on the size of these nonavid lesions and the degree to which they take up the bone agent, the lesions can appear as focal areas of photopenia on the scan or have an activity similar to the surrounding bone and not be readily apparent. The lesions that generally show an uptake greater than that of the surrounding bone include giant-cell tumors [56], osteoblastomas [57,58], and chondroblastoma [59]. Alternatively, nonossifying fibromas [60] and solitary bone cysts generally are not avid; however, increased uptake can at times be observed at the osseous borders of these lesions. This rim of increased activity at times may obscure the decreased uptake of the lesion itself when images are obtained using a parallel-hole collimator; magnified images obtained with a pinhole collimator will often demonstrate the true nature of the lesion. The aneurysmal bone cyst demonstrates variable avidity for the bone agent [61,62]. Uptake of the bone agent by osteochondromas (exostoses) is variable. Marked uptake of the radiopharmaceutical in an osteochondroma, particularly one that had previously not shown much avidity for the bone agent, is worrisome for sarcomatous change (Figures 4–6 and 4–7).

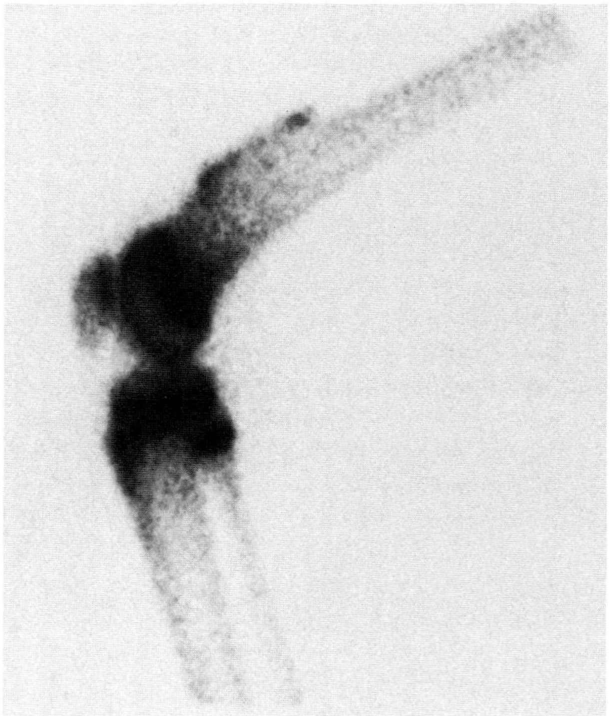

FIGURE 4–6
Lateral image obtained from the leg of a 14-year-old boy, demonstrating mild increased uptake of the bone scan agent by a well-documented benign osteochondroma arising from the distal femoral metaphysis.

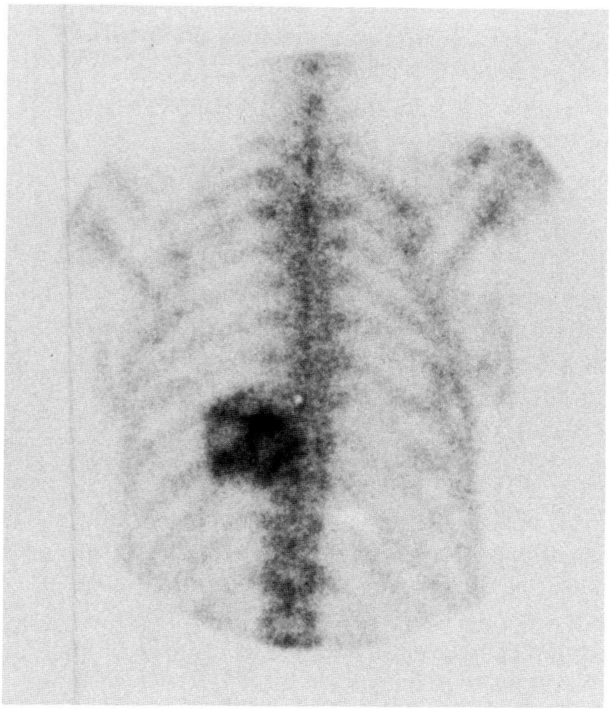

FIGURE 4–7
Markedly increased uptake of the bone-seeking agent in a midthoracic lesion in a 13-year-old boy with hereditary multiple exostoses. Despite the intense uptake, the lesion was benign.

The one condition in which scintigraphy plays an invaluable role is the osteoid osteoma. Osteoid osteoma is a common tumor-like condition of adolescents and young adults, usually occurring between the ages of 10 and 25. The lesion is characterized by pain, usually starting as minor discomfort and progressing to significant pain. Anatomically it consists of a well-demarcated core (nidus) of osteoid in a stroma of loose vascular connective tissue that is surrounded by a zone of dense, sclerotic bone. The radiographic appearance consists of a radiolucent sphere surrounded by a radiodense zone [38]; these features may be more apparent on CT than on plain radiographs [63]. The absence of a radiologic abnormality does not preclude the diagnosis of osteoid osteoma since symptoms may antedate radiographic changes by 6 months to 1 year.

Bone scintigraphy is invaluable in patients with a clinical suspicion of osteoid osteoma in whom radiographs are either normal or in whom radiographs fail to demonstrate the small radiolucent nidus. The scintigraphic appearance is that of a well-localized area of intensely increased activity on the delayed images, which may be surrounded by a halo of mild-to-moderate increased activity [58,63,64]. A false-negative study is rare; the sensitivity of skeletal scintigraphy for the detection of osteoid osteoma is virtually 100% [65,66]. Osteoid osteomas are generally also visualized on the blood-pool images of three-phase bone scans. This characteristic has been reported as being of some diagnostic significance, particularly in distinguishing osteoid osteoma of the neural arch from spondylolysis [67,68].

Intraoperative localization of osteoid osteomas can present a major problem, since visual and radiologic localization of the lesion often proves difficult and complete excision of the nidus is necessary to prevent recurrence. Even the experienced surgeon guided by intraoperative radiographs will occasionally miss these lesions, requiring that the patient undergo a second procedure (Figure 4–8). The intraoperative use of skeletal scintigraphy permits accurate localization of the lesion within the open operative field and can ensure removal of the lesion before the wound is closed.

Some centers perform intraoperative localization of osteoid osteomas with a radiosensitive probe [69] and obtain images of the excised bone outside the operating room [70]. Such images are useful when the lesion has been removed en bloc, and demonstrate the increased uptake of the nidus with less intense activity in the surrounding bone, but may be difficult or impossible to interpret when an en bloc resection has not been performed.

Our approach is to perform intraoperative imaging with a portable gamma camera equipped with a pinhole magnification collimator. Covering the camera head with a sterile plastic C-arm cover allows imaging of the operative field. By using a sterile radioactive probe in concert with the gamma camera, the surgeon can accurately localize the lesion. In addition, images of the operative field can be obtained in the operating suite after excision to document removal of the nidus [71]. We have not encountered one patient

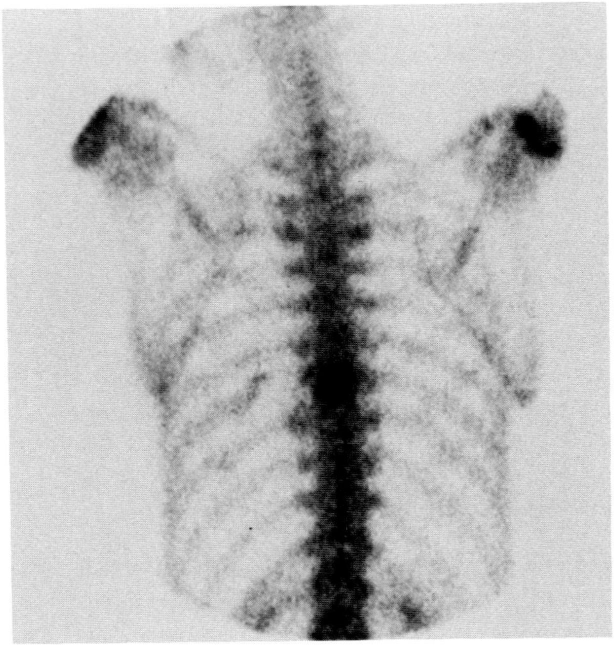

FIGURE 4–8
Posterior image showing a discrete focus of increased activity, typical for osteoid osteoma, at T-8 obtained from a 14-year-old girl who had undergone attempted resection of an osteoid osteoma at this site 4 months previously. This image was identical to a previous bone scan performed at another institution prior to surgery. The attempted resection was performed without the benefit of intraoperative skeletal scintigraphy.

in whom reoperation was necessary after the osteoid osteoma was excised in conjunction with this technique.

METASTASES FROM SOFT TISSUE MALIGNANCIES

With the exception of those neoplasms, such as neuroblastoma, in which diffuse bone marrow involvement may occur, evaluation of the bone scan performed for the detection of osseous metastases is the same for both childhood and adult malignancies. Any focal abnormality that cannot be ascribed to a benign cause by correlation with the clinical history or with radiographs must be considered highly suspicious for metastatic disease. This is particularly true in children who are too young to have developed the arthritic changes commonly seen in adults. Although metastatic disease can arise from a wide variety of malignant processes in children, we limit this discussion to those neoplasms we encounter most frequently.

Neuroblastoma Except for tumors of the central nervous system, neuroblastoma is the most common solid tumor of infancy and childhood. The majority of neuroblastomas manifest clinically before the age of 5, and at least half are detected in children less than 2 years old. Congenital neuroblastomas have been described [72,73].

Neuroblastomas (and its related childhood neoplasms, ganglioneuroblastoma and ganglioneuroma) are believed to arise from primitive neural crest tissue at any site along the craniospinal axis. Common sites include the adrenal medulla, the retroperitoneum, the thorax, the cervical ganglia, and the head and neck. The first clinical manifestations result from metastatic involvement or the finding of a large intra-abdominal mass. Neuroblastoma spreads by local invasion and hematogenous metastatic spread to the liver, bone marrow, skeleton, and intracranial dura. Since 50% or more of these children have disseminated disease at presentation, skeletal scintigraphy is routinely included in the early evaluation and staging of the disease.

The primary prognostic factors of neuroblastoma are the age of the patient at the time of diagnosis, the site of the primary tumor, the stage of the tumor at the time of diagnosis, and the degree of differentiation of the tumor toward ganglioneuroblastoma or ganglioneuroma. Although several staging systems have been proposed, the one most commonly employed is that of Evans et al. [74]. In this system, stage I represents disease localized to the organ of origin, stage II represents tumor extending in continuity with the organ of origin but not crossing the midline, stage III indicates tumor in continuity with the organ of origin but extending beyond the midline, and stage IV represents disseminated disease. This staging method also distinguishes a special category of metastatic disease, stage IV-S. Patients included in this stage are those who would otherwise be stage I or II but who also have remote disease confined to one or more of the following sites: liver, skin, or bone marrow, but not bone. This stage (which is not included in all staging schemes) is seen almost exclusively in patients less than 1 year old and has a much better prognosis than does stage IV disease in children of similar age. Skeletal scintigraphy, by defining bone involvement, thus plays a critical role in the staging of neuroblastoma.

Once the tumor is diagnosed as being a neuroblastoma, evaluation proceeds toward establishing the extent of the tumor. This workup variously includes intravenous urography, radiography of the chest, abdominal ultrasound or CT, and evaluation of the skeletal system. Although the evaluation of the skeletal system has traditionally been performed radiographically with a skeletal survey, more recent practice utilizes bone scintigraphy instead of the skeletal survey.

Skeletal scintigraphy in patients with neuroblastoma is quite characteristic [75]. Often the bone-seeking radiopharmaceutical will concentrate in the primary tumor, with varied degrees of intensity. The intensity of such uptake has not been shown to relate to the degree of malignancy or prognosis [76]. Even primary lesions that show no radiographic evidence of calcification may reveal increased uptake of the radiopharmaceutical. This property of ^{99m}Tc-MDP to concentrate in the primary tumor aids in defining the extent of the tumor and in detecting residual tumor.

Skeletal scintigraphy is a sensitive detector of neuroblastoma metastatic to bone and the findings will be abnormal weeks before radiographic changes are apparent [77]. In one recent series the false-negative and false-positive rates for the detection of skeletal involvement were 4.8 and 9.5%, respectively [78]. A typical pattern of increased uptake can be seen in the proximal part of the humeri, distal aspect of the femora, and proximal part of the tibiae. These lesions are often symmetric, and in some instances the scintiscans may be wrongly interpreted as normal.

Retinoblastoma Retinoblastoma is a tumor of childhood, the cells of which are derived from the fetal neuronal retina. It usually develops in the posterior portion of the retina and is bilateral in some 30% of cases. It is one of the few tumors that has been associated with a specific chromosomal abnormality. Because they arise in the eye, retinoblastomas are often diagnosed early when the prognosis for cure is excellent. Because they can metastasize by hematogenous dissemination to the bone or bone marrow, bone scans are of value in detecting disseminated disease.

The genetic abnormality associated with retinoblastoma also carries an increased incidence of other malignancies; about 1% of patients surviving the hereditary form of retinoblastoma will develop osteosarcoma, usually in late childhood or adolescence. This may arise in the area irradiated for treatment of the primary retinoblastoma or may occur at a site distant from the radiation field.

Rhabdomyosarcoma Rhabdomyosarcoma of childhood presents as a mass usually of the head and neck or genitourinary system, although it has been reported in almost every area of the body. Three separate histologic patterns are commonly recognized: embryonal, alveolar, and pleomorphic (with a fourth type, botryoid, being considered a form of embryonal rhabdomyosarcoma), each type being more typically associated with a specific age group. The prognosis, which prior to the development of contemporary chemotherapy had a 5-year survival of less than 20%, depends mainly on site and stage and is relatively independent of histologic type. Rhabdomyosarcomas tend to undergo extensive local invasion as well as to metastasize to lymph nodes, lungs, liver, and both bone and bone marrow. Although radiographic procedures usually define the extent of the tumor, skeletal scintigraphy has proved useful in defining the extent of bone involvement both in the initial lesions and in recurrences, as well as in detecting metastatic involvement.

Trauma

Although the vast majority of fractures will be detected radiographically and do not require skeletal scintigraphy, bone scanning at times can be of great assistance and may provide the only physical evidence of fracture. Direct trauma to bone or even the periosteum alone produces an immediate physiologic response that results in increased uptake of the radiotracer [79,80], which can be detected on the bone scan within 24–48 hours of injury and before changes are radiographically visible. This makes skeletal scintigraphy an extremely sensitive modality for the diagnosis of trauma. The pattern of abnormal uptake can be focal, well defined, or diffuse, depending on the nature of the injury. When fractures are not visible radiologically or are in areas difficult to evaluate on routine radiographs, such as the spine, ribs, carpals, tarsals, or scapula, radionuclide scanning is particularly helpful. In addition, in infants who have difficulty ambulating and in whom the abnormality cannot be easily localized by physical examination, the bone scan will often pinpoint the abnormality (Figure 4–9).

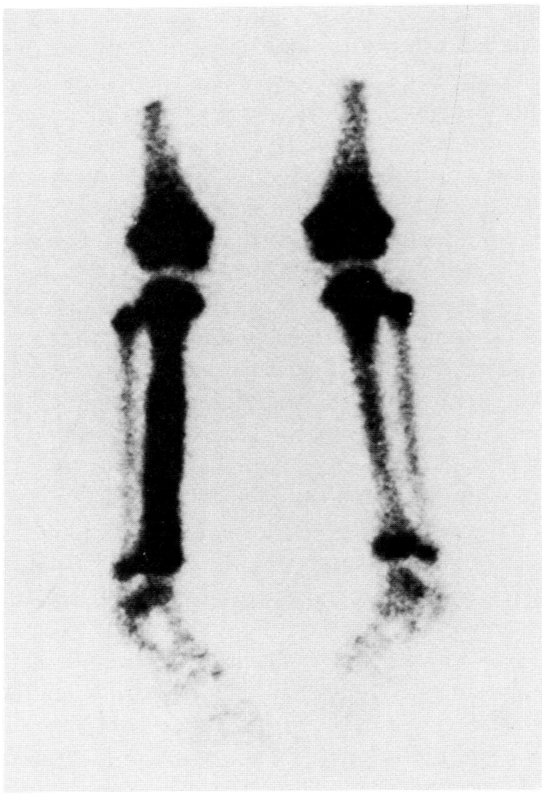

FIGURE 4–9
Anterior images obtained from an 8-month-old girl who refused to bear weight on her right lower extremity. A hip abnormality was suspected. Marked increased uptake throughout the tibial shaft resulted in a presumptive diagnosis of fracture. Characteristic radiographic findings appeared 1 week later.

Increased uptake at the site of a fracture may be seen for as long as several years after the injury. In adults, at least 60% of fracture sites regain a normal bone scan appearance within 1 year and approximately 90% return to normal within 2 years [67]. Since the radiographic changes associated with the healing of fractures generally occur earlier in children than in adults, it seems likely that the time at which the bone scan returns to normal after fracture should be shorter in children, but this remains to be documented. Thus, the bone scan can be useful in detecting remote injuries that may not be apparent on plain radiographs. This characteristic makes the bone scan an important tool in the evaluation of child abuse and neglect (Figure 4–10). Bone scanning may provide evidence of old bone injury when current radiographs detect only recent fracture or may provide the only evidence of skeletal injury when child abuse is suspected on other grounds. In one study that compared radiographic and scintigraphic skeletal surveys of 261 children who were suspected victims of child abuse, radiography demonstrated positive results in 105 children and produced false-negative results in 32, whereas scintigraphy demonstrated positive results in 120 and produced false-negative results in only 2 [81]. Despite such studies, controversy surrounding the role of skeletal scintigraphy in the diagnosis of child abuse or neglect still exists [82–87]. As has been eloquently discussed by Kleinman [88], much of this controversy appears to arise from the technical difficulties inherent in obtaining the near-perfect studies often needed to detect abnormalities in this population. Without compulsive attention to the many technical factors required to obtain such a study, such telling abnormalities as the metaphyseal corner fracture, a virtually pathognomonic indicator of child abuse, can easily be missed.

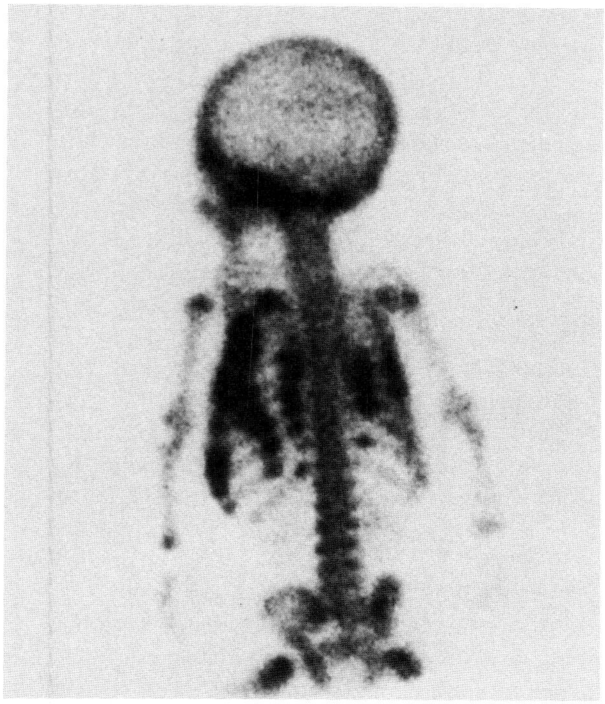

FIGURE 4–10
Posterior bone scan obtained from a 2-month-old girl. The multiple rib fractures, particularly those involving the costovertebral junction, are virtually diagnostic of child abuse.

Benign Orthopedic Disorders

Skeletal scintigraphy is useful in the diagnosis and evaluation of a wide variety of benign orthopedic conditions of childhood and adolescence. Not only is the technique useful in evaluating disorders associated with the developing skeleton, such as idiopathic necrosis of the femoral head, but also as more children are entering rigorous, year-round athletic training programs, it is becoming increasingly important in the evaluation of sports-related bone injuries.

ASEPTIC NECROSIS

Osteonecrosis of bone occurs in both adult and pediatric populations, and is associated with a variety of predisposing factors including steroid use. It has long been recognized, however, that osteonecrosis of specific bones of the developing skeleton occurs without an identifiable cause. Each of these disorders, collectively referred to as the osteochondroses, is associated with a specific age group.

Idiopathic aseptic necrosis occurs in a variety of sites including the femoral head (Legg-Calvé-Perthes disease), the tarsal navicular (Köhler's disease), the capitellum of the humerus (Panner's disease), the carpal lunate (Kienböck's disease), the second metatarsal head (Freiberg's infraction), and the phalanges of the hand (Thiemann's disease), among others [89]. Skeletal scintigraphy plays a major role in the diagnosis of early avascular necrosis since bone scan changes are often demonstrable prior to the development of radiographic abnormalities.

Legg-Calvé-Perthes disease The most common of these disorders is Legg-Calvé-Perthes disease in which the femoral head, including bone and marrow, undergoes necrosis. Both the articular cartilage, which derives its nutrients from the joint fluid, and the epiphyseal growth plate are spared. This condition occurs in children between the ages of 4 and 8 years and has a distinct male predominance. Either hip can be affected; bilateral disease occurs in 10% of cases [90].

The fundamental pathologic abnormality in Legg-Calvé-Perthes disease is osteonecrosis involving the bone and marrow, which, during the eventual process of bone resorption and healing, can lead to structural failure and collapse. The extent of damage and degree of subsequent bone collapse are variable [89]. Early diagnosis depends on being able to detect changes during the early avascular period before structural abnormalities have occurred. It is during this early period of avascularity that the bone scan is of particular benefit.

The most striking abnormality visible by bone scintigraphy is that of an avascular femoral head. Early in the disease there is absence of radiotracer uptake in all or a portion of the femoral head [91]. Radionuclide bone scanning has a reported sensitivity and specificity of greater than 90% in this phase of the disease [92–97], and attempts have been made to quantitate the degree of femoral head uptake and prognosis [98,99]. SPECT has been shown to increase the sensitivity of skeletal scintigraphy in diagnosing avascular necrosis of the femoral head in adults [100]. Whether it can increase the sensitivity of detection in children compared to pinhole collimator images remains to be proved.

Radiologically, there may be no detectable change during the first few weeks of the disease. Later changes include a slight widening of the joint space. A linear, subchondral lucency of the epiphysis then appears and is accompanied by increased bone density. The scintiscan at this stage may still reveal no uptake in the femoral head and sometimes may be accompanied by reactive, increased uptake involving the acetabulum and proximal part of the femur [99]. After a variable period, this is followed by reossification and gradual reformation of the femoral head. Revascularization of the femoral head is accompanied by increasing uptake on the bone scan [93]. Uptake by the revascularized femoral head typically increases to the point where uptake in the affected femoral head is greater than on the normal, contralateral side. Over a variable period of time, this abnormal uptake returns to normal if there has been no significant structural damage.

Freiberg's disease Freiberg's disease, a disorder characterized by pain and swelling about the second metatarsal of the foot accompanied by radiologic changes involving the metatarsal head, is typically seen in adolescent girls between the ages of 13 and 18 [101]. Although it typically involves the second metatarsal, other metatarsal bones may be affected. The disorder is generally believed to result from an ischemic event or events, perhaps secondary to trauma, resulting in avascular necrosis. The usual scintigraphic appearance at presentation is one of increased uptake in the region of the second metatarsal head, corresponding to the flattening and sclerosis observed radiographically [102]. That these changes are the result of an earlier episode of avascular necrosis is suggested by a recent report demonstrating a region of relative photopenia that involved the second metatarsal head in a patient with foot pain who later developed radiographic abnormalities consistent with Freiberg's infraction [103].

Köhler's disease Köhler's disease is a self-limited disorder of the tarsal navicular bone, causing pain and swelling in the foot (Figure 4–11). It is most commonly observed between the ages of 3 and 7 years. Although the abnormalities associated with this disorder have been associated with avascular necrosis, confirmation is lacking. The typical bone scan findings of absent uptake by the navicular bone are consistent, however, with this hypothesis [104]. The bone scan can be particularly useful in making the diagnosis of this disorder, since the radiographic abnormalities associated with Köhler's disease can be similar to normal patterns of ossification.

OSTEOCHONDRITIS DISSECANS

Osteochondritis dissecans represents a rather common orthopedic condition characterized by ischemic necrosis and subsequent collapse of a fragment of the peripheral osseous epiphysis with attached articular cartilage. This fragment can then undergo three fates: It can remain in situ, it can undergo slight displacement, or it can undergo separation (dissection) from the epiphysis and become "loose" within the joint cavity or attached to the synovial lining. Although this disorder is known in adults, it is most common in the 15–25-year age group. The precipitating cause of this condition remains a matter of debate; however, a history of trauma can often be elicited.

The most typical location of osteochondritis dissecans is the distal condylar surfaces of the femur, with the medial condyle being affected in 85% of cases. In about 10% of cases the defects are bilateral. The classic defect involves the inner (lateral) aspect of the medial condyle, although any aspect of the condyle may be involved (Figure 4–12). Osteochondritis dissecans may also involve the talar dome, patella, elbow, and wrist.

It has been shown that the known sensitivity of skeletal scintigraphy in following the course of healing fractures is also applicable to the osteochondral fractures associated with osteochondritis dissecans [105,106]. During the period of active fragmentation and subsequent healing,

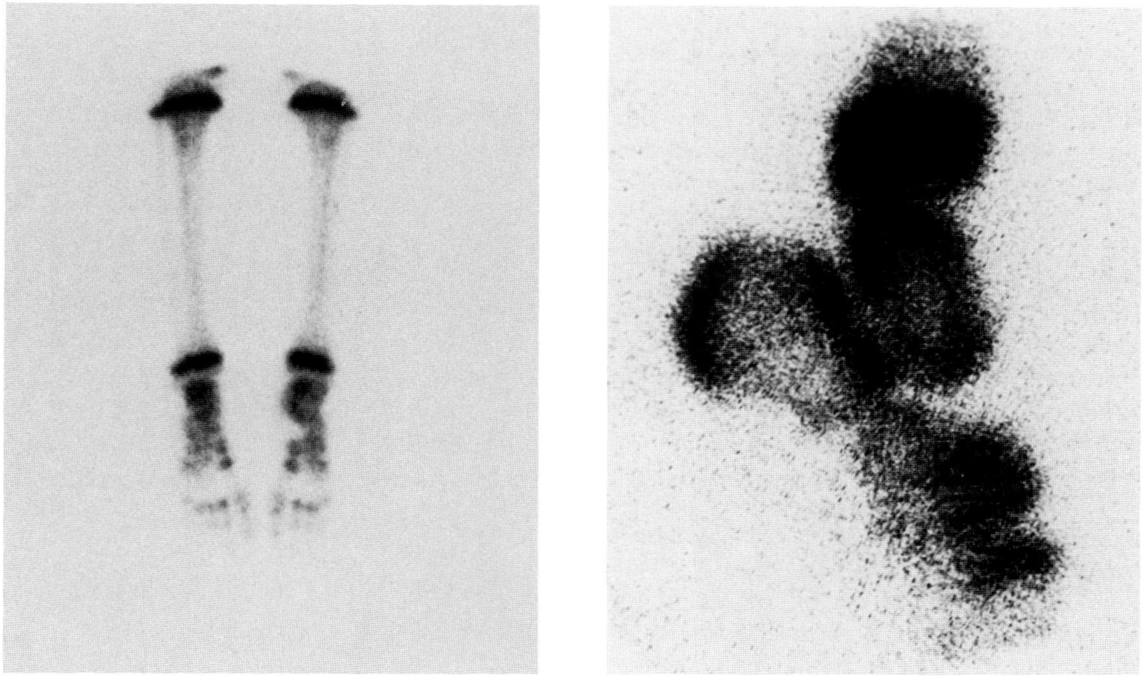

A B

FIGURE 4–11
Bone scan images of the feet obtained from a 5-year-old boy with unexplained foot pain. A. A photopenic region corresponding in position to the nonperfused right tarsal navicular bone is seen on standard views obtained with a parallel-hole collimator. B. The same is seen on lateral magnified views obtained with a pinhole collimator.

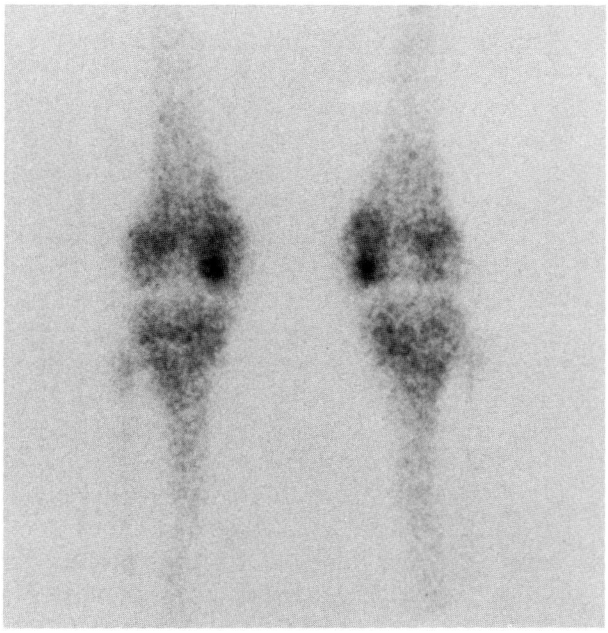

FIGURE 4–12
Typical appearance of bilateral osteochondritis dissecans.

there is an area of increased uptake on the bone scan, with the level of uptake indicating the extent of healing or progression. Using magnification scintigraphy, it is occasionally possible to observe the photon-deficient area, corresponding to the region of avascular necrosis, surrounded by an area of increased activity.

STRESS FRACTURES

Stress fractures represent the focal microscopic disruption of bone that occurs when bone is subjected to forces in excess of its mechanical capability to withstand. Stress fractures can result from a single event in which the applied force on the bone exceeds its mechanical capability, resulting in a fatigue fracture, or may result from the seemingly paradoxic physiologic weakening of bone that occurs when it is subjected to multiple episodes of unusual stress. This weakening results from osteoclastic resorption of bone, which, after new osteon formation, will result in an increased mechanical capability of the bone to withstand the new stresses [107,108]. If the stress that initiates the osteoclastic activity continues and is of sufficient strength, microfractures of the weakened cortex occur, with the result being an overt stress fracture. The former type of stress fracture, which results from a single event, has been described in military recruits who have metatarsal stress fractures after a sustained march [109,110]; the latter is typified by the casual jogger who increases his or her training regimen in preparation for a marathon.

Stress fractures have recently become associated with a growing list of athletic endeavors including running, football, tennis, gymnastics, hockey, basketball, swimming, and baseball [111–113]. We have seen a large number in women who regularly perform aerobics, and stress fractures seem to be an occupational hazard to professional ballet dancers (Figures 4–13 and 4–14) [114]. Not surprisingly, stress fractures of specific locations are associated with specific activities and result both from the manner in which the force of gravity is transmitted along the bones and through the actions of muscle contraction and ligamentous pull. The excessive force of muscular contraction has been postu-

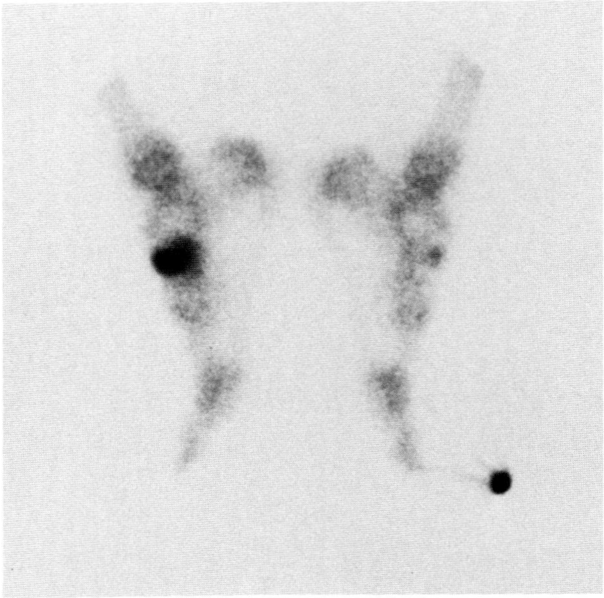

FIGURE 4–13
Stress fracture involving the right navicular bone in a ballet dancer.

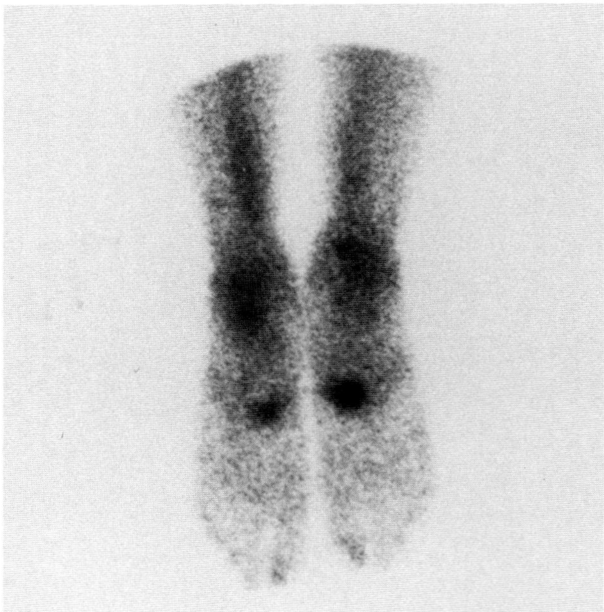

FIGURE 4–14
Bilateral stress fractures of the feet in a ballet dancer.

lated as the cause of stress fractures involving the fibula [115,116], and appears to be the mechanism by which stress fractures of the scapula and pubic ramus are sustained. Although almost any bone may be affected, stress fractures are most commonly observed in the long bones of the lower extremity. These are particularly common in runners, with the typical location being the posterior tibial cortex at the junction of the middle and distal third [117].

Suspicion that a patient has a stress fracture is usually generated from the characteristic history and physical findings. A history of the insidious onset of bone pain over several weeks or months associated with increasing physical activity, and a physical examination showing swelling and palpable bony tenderness associated with the area of pain, are typical of stress fractures [118,119]. Although a strong suspicion of a stress fracture can arise from the history and physical findings, diagnosis requires observing typical abnormalities on radiographs and bone scans. The importance of obtaining the plain film is twofold: It may establish the diagnosis of stress fracture, and it can exclude the presence of other less common disorders in the differential diagnosis such as osteosarcoma, osteomyelitis, and osteoid osteoma. Often, however, no abnormal findings are seen on the plain radiograph, and at this point the key to making the diagnosis of stress fracture is skeletal scintigraphy.

Skeletal scintigraphy demonstrates the stress fracture 1–3 weeks before x-ray changes are apparent [120]. In one study, Prather et al. noted an average duration of 26 days from the onset of symptoms to radiographic confirmation, while the scan findings were positive an average of approximately 15 days from the onset of symptoms [121]. In this study, four patients with clear signs and symptoms of stress fracture and typical changes on bone scan never developed radiographic abnormalities, suggesting that early treatment may prevent progressive bone changes. Another study found the bone scan results to be positive 1–3 weeks before x-ray changes occurred [120]. When a bone scan is done for a suspected stress fracture, other, often unsuspected stress-related changes are commonly encountered. In our experience, more than 60% of athletes with tibial abnormalities have exhibited bilateral changes. In addition, nonspecific changes are frequently encountered in the knees, feet, and contralateral extremity.

The sensitivity of bone scanning for stress fractures is excellent, approaching 100%. The specificity is lower; in one series the specificity for the diagnosis of stress fracture was 16 of 21, or 76% [121]. Differentiation of stress fracture from other causes (such as infection or tumor) cannot, therefore, be made unambiguously from the bone scan alone, but must rely on the clinical picture, other radiologic procedures, and perhaps biopsy. Since typical stress fractures commonly occur in the diaphysis of long bones, other disorders in the differential diagnosis must include Ewing's sarcoma, osteomyelitis, and eosinophilic granuloma.

Stress fractures of long bones generally have a distinctive appearance on bone scan and characteristically appear as longitudinal fusiform foci of intensely increased activity that appear confined to the cortical surface of the involved bone. When the stress factor is recent (approximately 2–4 weeks old), changes will be apparent on all three phases of the three-phase scan. With healing, first the radionuclide angiogram and then the blood-pool image will return to normal. The intensity of the stress fracture on the delayed images returns slowly to normal and may be abnormal through 8–10 months [117].

Not all stress-related lesions have this characteristic fusiform appearance, and stress-related changes vary from small, ill-defined cortical areas of increased activity to large transcortical areas of intensely increased activity. A four-stage grading system for stress fractures has recently been proposed; the time required for resolution of the injury has been shown to be proportional to the grade [122]. Whether those lesions showing only small, ill-defined cortical areas of uptake should be considered stress fractures is debatable. We prefer to call small foci of ill-defined activity stress changes, reserving the designation stress fracture for those lesions having the characteristic fusiform appearance. This distinction has proved useful in guiding clinicians in their choice of treatment.

One type of sports-related injury deserves

special mention, and this is the so-called "shin splint" lesion. This widely used term has been applied to pain located along the anterior portion of the tibia that characteristically occurs in runners. In contrast to the focal accumulation of the bone-seeking agent that is observed with developing or overt stress fractures, shin splints typically appear scintigraphically as diffuse nonfocal periosteal uptake of tracer confined primarily to the posterior tibial cortex [117,123]. It has been suggested that these scintigraphic findings represent disruptions of Sharpey's fibers connecting periosteum to bone and result from the abnormal excursion of various muscle and tendon complexes [123]. If so, shin splints represent a different pathologic entity than stress fractures.

LOW BACK PAIN

Low back pain in children and adolescents usually represents organic disease and therefore clinical and radiologic evaluation is required [124]. The causes are multiple and include mechanical back pain arising from mechanical strain on the ligaments, muscles, and facet joints; fracture; infection; and tumors. Distinguishing low back pain arising from organic bone disease from that arising from strain to the muscles and ligaments is required for the proper choice of treatment. Among the imaging methods, bone scintigraphy has proved to be of great benefit in the detection of bone disorders of the lower back, often detecting abnormalities when radiographs are normal.

Bone scintigraphy has proved particularly useful in the diagnosis of low back pain arising from spondylolysis. Spondylolysis, a cause of lower back pain of increasing incidence in young athletes, represents a fracture of the pars interarticularis [125,126]. Spondylolysis was once thought to be a congenital lesion; however, recent evidence suggests forced hyperextension of the lower spine as the main cause of the defect [127]. Treatment is generally conservative, involving specific exercises and an antilordotic brace, with spinal fusion being reserved for those who continue to be symptomatic or develop significant spondylolisthesis.

The role of imaging in the evaluation of a suspected pars interarticularis defect is twofold: to confirm the diagnosis of spondylolysis; and to exclude other anatomic causes of lower back pain such as neoplasia, infection, discogenic pain, or apophyseal fracture of a vertebral body. Imaging in the diagnosis of spondylolysis at our institution involves both radiographic procedures and bone scintigraphy.

Findings consistent with spondylolysis on skeletal scintigraphy are focal increased uptake of the bone-seeking agent in the region of the pars interarticularis [128,129]. These defects may only be apparent on oblique views of the spine, which should be performed on every patient with a complaint of low back pain. Although sites of active bone pathology involving the vertebral column may be readily apparent on planar bone scintigraphy, difficulties in detecting subtle lesions are commonly encountered. Overlap of osseous elements, particularly involving the posterior structures of the neural arch, may obscure the presence of a bony abnormality particularly when imaged in a single projection. SPECT, by synthesizing the data from multiple projections, produces a precise image of the distribution of radioactivity in a single plane, and thus achieves a resolution unattainable by planar imaging. SPECT has been reported to increase the sensitivity of detecting abnormal pars activity when compared to planar imaging [129]. Using this technique, we have discovered osseous abnormalities involving the vertebral column in one-fourth of patients with low back pain in whom planar scintigraphy demonstrated normal findings.

Findings on bone scintigraphy and radiographs may be either concordant or discordant; radiographs can reveal a pars defect when scintigraphic findings are normal, and normal radiographic findings can be associated with focal areas of abnormal pars uptake on the bone scan. Positive bone scans when radiographs are normal have been associated with recent pars fractures, whereas normal scintigraphic studies in the presence of radiographic abnormalities have been associated with old injuries. Thus, an asymptomatic pars defect may represent an incidental radiographic finding in a search for the cause of low back pain, with the pain actually resulting from an unrelated abnormality.

KNEE PAIN

Chronic knee pain can arise from a number of conditions and often presents a diagnostic challenge to the clinician. Bone scintigraphy has proved valuable in the diagnosis of osteochondritis dissecans, reflex sympathetic dystrophy, osteonecrosis of the femoral condyle, and fractures of the patella when clinical examination and planar radiographs were nondiagnostic, and may detect other osseous pathologic features as a cause of referred knee pain. One recent study compared SPECT images with conventional radiography and planar bone scintigraphy in 27 patients with chronic knee pain and showed SPECT to be highly sensitive in detecting knee disorders [130]. Although its role in the diagnosis of disorders of the knee needs to be more fully evaluated, bone scintigraphy can at times prove a valuable adjunct, particularly when other imaging modalities have proved unrewarding.

FOOT PAIN

Although foot pain in childhood and adolescence is a common complaint, the value of skeletal scintigraphy in the evaluation of the painful foot has received limited attention. Yet, skeletal scintigraphy has the ability to detect a number of disorders that may not be apparent radiographically, including metatarsal stress fractures, sesamoiditis or fracture, plantar fasciitis, tarsal navicular stress fracture and avascular necrosis, retrocalcaneal bursitis, tarsal coalitions, and, in a recent report, ligamentous and capsular injuries [131]. Several of the more common disorders are discussed below.

Tarsal coalitions The term *tarsal coalition* refers to the presence of a congenital abnormal fusion of two or more tarsal bones, usually occurring between the calcaneus and navicular or between the calcaneus and talus. These fusions, which can be fibrous, cartilaginous, or osseous, often present during childhood and adolescence with pain and stiffness of the foot that is aggravated by walking. The findings of pes planus, peroneal tendon spasm, and limited inversion include this disorder in the complex known as peroneal spastic flatfoot, which can also be seen with infection, tuberculosis, trauma, and rheumatoid arthritis. Because the clinical findings associated with tarsal coalitions are nonspecific, diagnosis depends on the detection of the fusion by radiographic methods or bone scintigraphy.

Although plain radiographs demonstrate many tarsal coalitions, the complex anatomy of the talocalcaneal articulation makes coalitions of the sustentaculum tali, the medially located middle talocalcaneal joint, difficult to distinguish. Radionuclide bone scans have proved to be of value in the diagnosis of these coalitions, as the coalition shows focally increased uptake [132,133] (Figure 4–15), and may be accompanied by a second focus of abnormal activity involving the superior aspect of the talus or talonavicular joint [132]. This second focus of activity has been attributed to abnormal mechanical forces arising secondary to the coalition. This uptake is often far better appreciated on pinhole views. When plain radiographs do not reveal a cause for the foot pain, scintigraphy may provide the only evidence of the painful coalition.

Sesamoiditis/fracture Pain beneath the first metatarsal head can arise from a variety of conditions including bruising of the sesamoid(s), fracture, or problems in articulation between the sesamoid and metatarsal head [134]. Differentiation between sesamoid fracture and a bipartite sesamoid is often difficult on plain radiographs [135]. Increased uptake of the bone-seeking agent on skeletal scintigraphy has been used to differentiate acute fracture from the anatomic variant bipartite sesamoid [131,134].

Retrocalcaneal bursitis With vigorous overuse, an inflammatory response may develop in the retrocalcaneal bursa, which lies between the Achilles tendon and the posterior-superior aspect of the calcaneus. Plain radiographs may be normal or may demonstrate a heel spur anterior to the Achilles tendon. The diagnosis may be made scintigraphically by the finding of increased uptake of the tracer involving the superior-posterior aspect of the calcaneus, in the region of the insertion of the Achilles tendon.

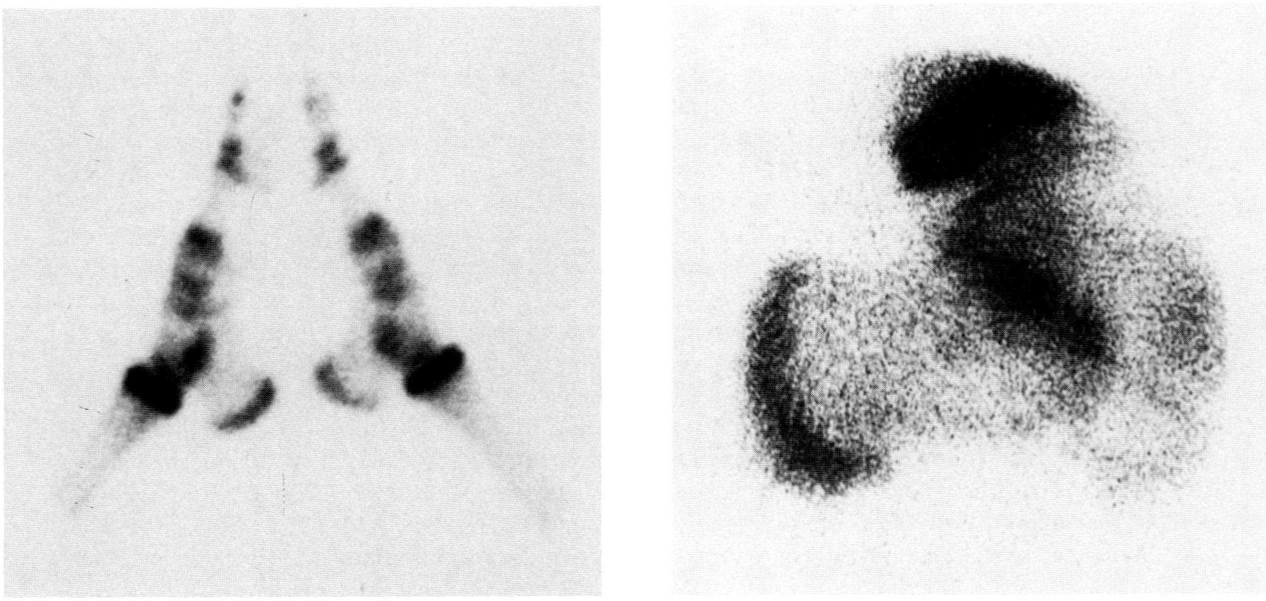

A B

FIGURE 4–15
Images of the feet of an 11-year-old girl with restricted subtalar motion and pain but negative findings on plain radiographs. A. Images obtained with a parallel-hole collimator show bilateral foci of abnormal uptake in the region of the subtalar joints. B. This abnormal activity is better seen using a pinhole collimator. Bilateral subtalar coalitions were confirmed by CT.

Plantar fasciitis Pain and tenderness beneath the anterior portion of the heel are most frequently caused by strain or tear of the plantar fascia in the region of its attachment to the calcaneus. Plain radiographs are usually normal; occasionally a spur is observed. Delayed bone scan images demonstrate increased activity of the anterior-inferior portion of the calcaneus, in the region of insertion of the plantar fascia. This activity may be best appreciated on the blood-pool images of the three-phase bone scan [134].

RHEUMATIC JOINT DISORDERS

Although skeletal scintigraphy is considered a useful diagnostic tool in the evaluation of rheumatic disease in adults [136], its utility in children has been less well established. In one study, skeletal scintigraphy was performed in 26 children with either definite or probable juvenile rheumatoid arthritis; all 26 patients had abnormal findings on a bone scan. In the experience of the authors, false-negative or false-positive examinations were rare in juvenile rheumatoid arthritis.

REFLEX SYMPATHETIC DYSTROPHY

Reflex sympathetic dystrophy (RSD) is a syndrome of uncertain etiology characterized by pain, tenderness, vasomotor instability, swelling, and dystrophic skin changes, usually involving an extremity [137]. Although this syndrome is more common in the adult population, it is observed in the pediatric population. The diagnosis in children can be quite difficult since all of the characteristic findings may not be observed (the dystrophic skin changes are uncommonly observed in children), and the radiographic findings are often normal.

Bone scan findings characteristic of RSD in the adult population are well described and con-

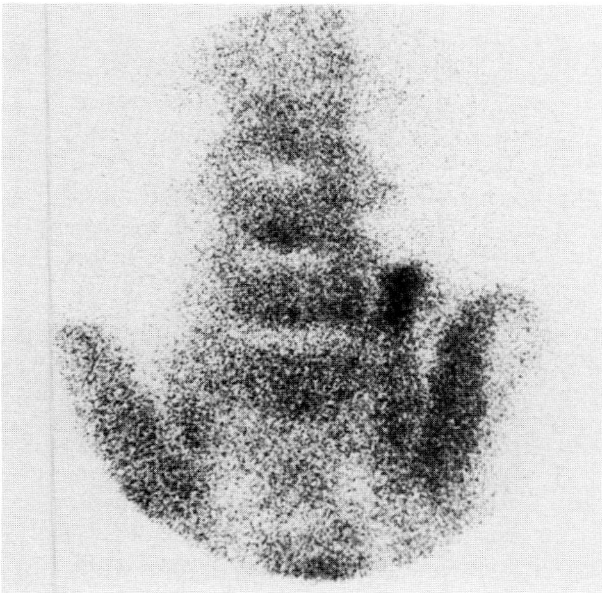

FIGURE 4–16
Magnified images, using a pinhole collimator, from a 5-year-old girl with a month-long history of left-sided low back pain, showing increased paravertebral activity. Surgery demonstrated the lesion to be myositis ossificans.

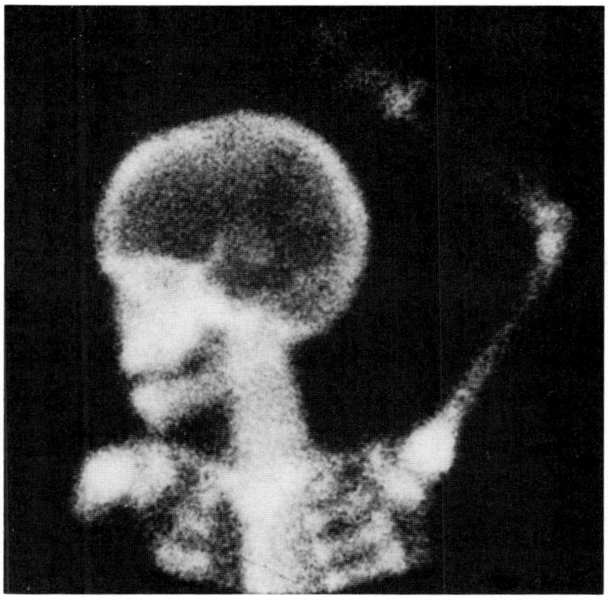

FIGURE 4–17
Left lateral bone scan of the skull in a patient with bilateral retinoblastoma and a morphologically similar calcified lesion involving the pineal gland (trilateral retinoblastoma). Uptake of the bone-seeking agent in the pineal lesion is seen.

sist of diffusely increased activity of the involved extremity on the delayed images. This increased uptake is particularly apparent in the periarticular regions [138]. This finding has been shown to be both a sensitive and a specific marker for RSD in adults. Changes on the radionuclide angiogram and blood-pool (tissue plane) images associated with RSD are more variable [139,140].

Recent reports have suggested that the syndrome in children may have a different bone scan appearance than that observed in adults. These reports suggest that decreased perfusion and bone uptake on skeletal scintigraphy may be at least as common as increased uptake in RSD in children [141]. In addition, children with RSD may show normal bone scans. The differences in scan findings between the pediatric and adult populations remain to be explained.

Soft Tissue Activity

Soft tissue activity on ^{99m}Tc-MDP bone scans has been reported secondary to a wide variety of causes. In general, this soft tissue uptake reflects the deposition of calcium salts in the soft tissues. This calcification may accompany tissue degradation or necrosis (dystrophic calcification) or may occur whenever the solubility product of the calcium salt is exceeded (metastatic calcification). In addition, a number of malignant soft tissue tumors will demonstrate increased uptake of the bone agent. Such calcifications are commonly encountered in the pediatric age group in neuroblastoma and in the soft tissue metastases of osteosarcoma. Benign conditions in which soft tissue uptake of the bone agent is commonly observed include recent intramuscular injections and myositis ossificans (Figures 4–16 and 4–17).

REFERENCES

1. Treves ST, Kirkpatrick JA. Bone. In: Treves ST, ed. Pediatric nuclear medicine. New York: Springer, 1985:1.
2. Croft BY. Single photon emission computed tomography. Chicago: Year Book Medical, 1986.
3. Sorensen JA (ed). Single photon emission computed tomography. New York: Society of Nuclear Medicine, 1980.
4. Heller SL, Goodwin PN. SPECT instrumentation: performance, lesion detection, and recent innovations. Semin Nucl Med 1987; 17:184–99.
5. Collier BD Jr, Hellman RS, Krasnow AZ. Bone SPECT. Semin Nucl Med 1987; 17:247–66.
6. McAfee JG, Thakur ML. Survey of radioactive agents for in vitro labeling of phagocytic leukocytes. 1. Soluble agents. J Nucl Med 1976; 17:480–7.
7. Thakur ML, Coleman RE, Mayhill CG, et al. Preparation and evaluation of 111-In-labeled leukocytes as an abscess imaging agent in dogs. Radiology 1976; 119:731–6.
8. Gainey MA, McDougall IR. Diagnosis of acute inflammatory conditions in children and adolescents using In-111 oxine white blood cells. Clin Nucl Med 1984; 9:71–4.
9. Sfakianakis GN, Sheikh W, Heal A, et al. Comparisons of scintigraphy with In-111 leukocytes and Ga-67 in the diagnosis of occult sepsis. J Nucl Med 1982; 23:618–26.
10. Winters JL, Cahen I. Acute hematogenous osteomyelitis: a review of sixty-six cases. J Bone Joint Surg [Br] 1960; 42:691–704.
11. Trueta J. The three types of acute hematogenous osteomyelitis: a clinical and vascular study. J Bone Joint Surg [Br] 1959; 41:671–80.
12. Hobo T. Zur Pathogenese der akuten haematogenen Osteomyelitis, mit Berucksichtigung der Vitalfarbungslehre. Acta Sch Med Univ Imp Kioto 1921–2; 4:1–29.
13. Nixon GW. Hematogenous osteomyelitis of metaphyseal-equivalent locations. AJR 1978; 130:123–9.
14. Ogden JA. Changing patterns of proximal femoral vascularity. J Bone Joint Surg [Am] 1974; 56:941–50.
15. Treves ST, Khettry J, Broker FH, et al. Osteomyelitis: early scintigraphic detection in children. Pediatrics 1976; 57:173–86.
16. Rosenbaum DM, Blumhagen JD. Acute epiphyseal osteomyelitis in children. Radiology 1985; 156:89–92.
17. Maurer AH, Chen CDP, Camargo EE, et al. Utility of three phase skeletal scintigraphy in suspected osteomyelitis. Concise communication. J Nucl Med 1981; 22:941–9.
18. Alazraki N, Dries D, Datz F, Lawrence P, Greenberg E, Taylor A Jr. Value of a 24-hour image (four phase bone scan) in assessing osteomyelitis in patients with peripheral vascular disease. J Nucl Med 1985; 7:711–7.
19. Israel O, Gips S, Jerushalmi J, Frenkel A, Front D. Osteomyelitis and soft-tissue infection: differential diagnosis with 24 hour/4 hour ratio of Tc-99m MDP uptake. Radiology 1987; 163: 725–6.
20. Lewin JS, Rosenfield NS, Hoffer PN, et al. Active osteomyelitis in children: combined Tc-99m and Ga-67 imaging. Radiology 1986; 158:795–804.
21. Schauwecker DS, Park H-M, Mock BH, et al. Evaluation of complicatory osteomyelitis with Tc-99m MDP, In-111 granulocytes, and Ga-67 citrate. J Nucl Med 1984; 25:849–53.
22. Tumeh SS, Ariabadi P, Weissman BN, et al. Chronic osteomyelitis: bone and gallium scan patterns associated with active disease. Radiology 1986; 158:685–8.
23. Erasmie U, Hirsch G. Acute haematogenous osteomyelitis in children. The reliability of skeletal scintigraphy. Z Kinderchir 1981; 32:360–6.
24. Handmaker H, Giammona ST. Improved early diagnosis of acute inflammatory skeletal-articular disease in children: a two-radiopharmaceutical approach. Pediatrics 1984; 73:661–9.
25. Pope TL, Teague WG, Kossack R, et al. Pseudomonas sacroiliac osteomyelitis: diagnosis by gallium citrate Ga-67 scan. Am J Dis Child 1982; 136:649–50.
26. Berkowitz ID, Wenzel W. "Normal" technetium bone scans in patients with acute osteomyelitis. Am J Dis Child 1980; 134:828–30.
27. Jones DC, Cady RB. "Cold" bone scans in acute osteomyelitis. J Bone Joint Surg [Br] 1981; 63:376–8.
28. Nelson HT, Taylor A. Bone scanning in the diagnosis of acute osteomyelitis. Eur J Nucl Med 1980; 5:267–9.
29. Sullivan DC, Rosenfield NS, Ogden J, et al. Problems in the scintigraphic detection of osteomyelitis in children. Radiology 1980; 135:731–6.

30. Wald ER, Mirro R, Gartner JC. Pitfalls in the diagnosis of acute osteomyelitis by bone scan. Clin Pediatr 1980; 19:597–600.

31. Ash JM, Gilday DL. The futility of bone scanning in neonatal osteomyelitis: concise communication. J Nucl Med 1980; 21:417–20.

32. Bressler EL, Conway JJ, Weiss SC. Neonatal osteomyelitis examined by bone scintigraphy. Radiology 1984; 152:685–8.

33. Merkel KD, Brown ML, Dewanjee MK, Fitzgerald RH Jr. Comparison of indium-labeled-leukocyte imaging with sequential technetium-gallium scanning in the diagnosis of low-grade musculoskeletal sepsis. J Bone Joint Surg [Am] 1985; 67:465–76.

34. Giedion A, Holthusen W, Masel LF, Vischer D. Subacute and chronic "symmetrical" osteomyelitis. Ann Radiol 1971; 15:329–42.

35. Bjorksten B, Boquist L. Histopathological aspects of chronic recurrent multifocal osteomyelitis. J Bone Joint Surg [Br] 1980; 62:376–80.

36. Solheim LF, Paus B, Liverud K, Stoen E. Chronic recurrent multifocal osteomyelitis. Acta Orthop Scand 1980; 51:37–41.

37. Probst FP, Bjorksten B, Gustavson KH. Radiological aspect of chronic recurrent multifocal osteomyelitis. Ann Radiol 1978; 21:115–25.

38. Bower GD, Sprague P, Geijsel H, et al. Isotope bone scans in the assessment of children with hip pain or limp. Pediatr Radiol 1985; 15:319–23.

39. Aegerter E, Kirkpatrick JA. Orthopedic diseases. Physiology, pathology, radiology. 4th ed. Philadelphia: WB Saunders, 1975.

40. Chew FS, Hudson TM. Radionuclide bone scanning of osteosarcoma: falsely extended uptake patterns. AJR 1982; 139:49–54.

41. Goldstein H, McNeil BJ, Zufall E, et al. Changing indications for bone scintigraphy in patients with osteosarcoma. Radiology 1980; 135:177–80.

42. McNeil BJ. Value of bone scanning in neoplastic disease. Semin Nucl Med 1984; 14:277–86.

43. McKillop JH, Elcubanas E, Goris ML. The indications for and limitations of bone scintigraphy in osteosarcoma. Cancer 1981; 48:1133–8.

44. Glass AG, Fraumeni JF. Epidemiology of bone cancer in children. J Natl Cancer Inst 1970; 44:187–99.

45. Goldstein H, McNeil BJ, Zufall E, et al. Is there still a place for bone scanning in Ewing's sarcoma? J Nucl Med 1980; 21:10–2.

46. Huvos A. Bone tumors. Diagnosis, treatment and prognosis. Philadelphia: WB Saunders, 1979.

47. Jaffe N, Traggis D, Sallan S, et al. Improved outlook for Ewing's sarcoma with combination chemotherapy (vincristine, actinomycin D and cyclophosphamide) and radiation therapy. Cancer 1976; 38:1925–30.

48. Rosen C, Caparros B, Mosende C, et al. Curability of Ewing's sarcoma and considerations for future therapeutic trials. Cancer 1978; 41:888–9.

49. Bell EG, McAfee JG, Mahon DF. Bone scanning. In: Schneider PB, Treves S, eds. Nuclear medicine in clinical practice. Amsterdam: Elsevier, 1978, pp. 61–83.

50. Parker BR, Pinckney L, Etcubanas E. Relative efficiency of radiographic and radionuclide bone surveys in the detection of skeletal lesions of histiocytosis X. Radiology 1980; 134:377–80.

51. Kumar R, Balachandrau S. Relative roles of radionuclide scanning and radiographic imaging in eosinophilic granuloma. Clin Nucl Med 1980; 5:538–46.

52. Westra SJ, van Woerden H, Postma A. Radionuclide bone scintigraphy in patients with histiocytosis X. Eur J Nucl Med 1983; 8:303–6.

53. Siddiqui AR, Jashjian JH, Lazarus K, et al. Nuclear medicine studies in evaluation of skeletal lesions in children with histiocytosis X. Radiology 1981; 6:270–1.

54. Arne M, Vazebrock W, Brassow F. A comparison of radiographic and bone scan findings in histiocytosis X. Skeletal Radiol 1983; 9:170–3.

55. Makhija MC, Davis G. Gallium-67 imaging in pulmonary eosinophilic granuloma. Clin Nucl Med 1986; 9:139–42.

56. Levine E, DeSmet AA, Neff JR, et al. Scintigraphic evaluation of giant cell tumors of bone. AJR 1984; 143:343–8.

57. Makhija MC, Stein IH. Bone imaging in osteoblastoma. Clin Nucl Med 1983; 8:141.

58. Sty J, Simons G. Intraoperative 99m technetium bone imaging in the treatment of benign osteoblastic tumors. Clin Orthop 1982; 165:223–7.

59. Humphry A, Gilday DL, Brown RG. Bone scintigraphy in chondroblastoma. Radiology 1980; 137:497–9.

60. Greyson ND, Pang S. The variable bone scan appearance of nonosteogenic fibroma of bone. Clin Nucl Med 1981; 6:242–5.

61. Makhija MC. Bone scanning in aneurysmal bone cyst. Clin Nucl Med 1981; 6:500–1.
62. Hudson TM. Scintigraphy of aneurysmal bone cysts. AJR 1984; 142:761–5.
63. Helms CA, Hattner RS, Vogler JB. Osteoid osteoma: radionuclide diagnosis. Radiology 1984; 151:779–84.
64. Smith FW, Gilday DL. Scintigraphic appearance of osteoid osteoma. Radiology 1980; 137: 191–5.
65. Sty JR, Hernandez R, Starshak RJ. Body imaging in pediatrics. Orlando: Grune & Stratton, 1984:87–9.
66. Fehring TK, Green NE. Negative radionuclide scan in osteoid osteoma. A case report. Clin Orthop 1984; 185:245–9.
67. Matin P. The appearance of bone scans following fractures, including immediate and long-term studies. J Nucl Med 1979; 20:1227–31.
68. Blom J, Pauwels EK, Pisa LN, et al. Misleading ^{67}Ga uptake and serial bone scintigraphy in osteoid osteoma. Diagn Imaging 1983; 52:276–9.
69. Colton CL, Hardy JG. Evaluation of a sterilizable radiation probe as an acid to the surgical treatment of osteoid osteoma. Technical note. J Bone Joint Surg [AM] 1983; 65:1019–22.
70. Ghelman B, Vigorita VJ. Postoperative radionuclide evaluation of osteoid osteomas. Radiology 1983; 146:509–12.
71. Taylor GA, Shea N, O'Brien T, et al. Osteoid osteoma. Localization by intraoperative magnification scintigraphy. Pediatr Radiol 1986; 16:313–16.
72. Bachmann KD. Das neuroblastoma sympathicum. Klinik ude Prognose von 1030 Fallen. Kinderheilk 1962; 86:710–24.
73. Peterson DR, Bill AH Jr, Kirkland IS. Neuroblastoma trends in time. J Pediatr Surg 1969; 4:244–9.
74. Evans AE, D'Angio GJ, Randolph J. A proposed staging for children with neuroblastoma. Cancer 1971; 27:374–8.
75. Levinson ED, Spencer RP. Radionuclide and ultrasound images as potential prognostic indicators in neuroblastoma. Invest Radiol 1981; 16:373–8.
76. Rosenfield N, Treves S. Osseous and extraosseous uptake of fluorine-18 and technetium-99m polyphosphate in children with neuroblastoma. Radiology 1974; 111:127–33.
77. Howman-Giles RB, Gilday DL, Ash JM. Radionuclide skeletal survey in neuroblastoma. Radiology 1979; 131:497–502.
78. Podresky AE, Stark DD, Hattner RS, et al. Radionuclide bone scanning in neuroblastoma: skeletal metastases and primary tumor localization of ^{99m}Tc-MDP. AJR 1983; 141:469–72.
79. Fordham EW, Ramachandran PC. Radionuclide imaging of osseous trauma. Semin Nucl Med 1974; 4:411–29.
80. Harcke HT. Bone imaging in infants and children: a review. J Nucl Med 1978; 19:324–9.
81. Sty JR, Starshak RJ. The role of bone scintigraphy in the evaluation of the suspected abused child. Radiology 1983; 146:369–75.
82. Diament MJ. Should the radionuclide skeletal survey be used as a screening procedure in suspected child abuse victims? Radiology 1983; 148:573. letter to editor.
83. Sty JR, Starshak RJ. Should the radionuclide skeletal survey be used as a screening procedure in suspected child abuse victims? Radiology 1983; 148:573–4. letter to editor.
84. Merten DF, Radkowski MA, Leonidas JC. Should the radionuclide skeletal survey be used as a screening procedure in suspected child abuse victims? Radiology 1983; 148:574. letter to editor.
85. Conway JJ. Should the radionuclide skeletal survey be used as a screening procedure in suspected child abuse victims? Radiology 1983; 148:574–5. letter to editor.
86. Merten DF, Radkowski MA, Leonidas JC. Should the radionuclide skeletal survey be used as a screening procedure in suspected child abuse victims? Radiology 1983; 148:575–6. letter to editor.
87. Berdon WE. Should the radionuclide skeletal survey be used as a screening procedure in suspected child abuse victims? Radiology 1983; 148:576. letter to editor.
88. Kleinman PK, ed. Diagnostic imaging of child abuse. Baltimore: Williams & Wilkins, 1987:5–28.
89. Kenzora JE, Glimcher MJ. Pathogenesis of idiopathic osteonecrosis: the ubiquitous crescent sign. Orthop Clin North Am 1985; 16:681–96.
90. Sutro CJ, Pomeranz MM. Perthes disease. Arch Surg 1936; 34:360–5.
91. Bonnavens F, Hernandez A, D'Ambrosia R. Bone

scintigraphic changes in osteomas of the femoral head. Orthop Clin North Am 1985; 16:697–701.

92. Calver R, Venugopal V, Dorgan J, et al. Radionuclide scanning in early diagnosis of Perthes' disease. J Bone Joint Surg [Br] 1981; 63:379–82.
93. Cavailloles F, Bok B, Bensahel H. Bone scintigraphy in the diagnosis and followup of Perthes' disease. Eur J Nucl Med 1982; 7:327–30.
94. Bensahel H, Bok B, Cavailloles F, Csukonyi Z. Bone scintigraphy in Perthes disease. J Pediatr Orthop 1983; 3:302–10.
95. Fasting OJ, Bjerkreim T, Langeland N, et al. Scintigraphic evaluation of the severity of Perthes' disease in the initial stage. Acta Orthop Scand 1980; 51:655–60.
96. Lucie RS, Fuller S, Burdick DC, et al. Early prediction of avascular necrosis of the femoral head following femoral neck fractures. Clin Orthop 1981; 161:207–14.
97. Sutherland AD, Savage JP, Paterson DC, et al. The nuclide bone scan in the diagnosis and management of Perthes' disease. J Bone Joint Surg [Br] 1980; 62:300–6.
98. Fisher RL, Roderique JW, Brown DC, et al. The relationship of isotopic bone imaging findings to prognosis in Legg-Perthes disease. Clin Orthop 1980; 150:23–9.
99. LaMont RL, Muz J, Heilbronner D, et al. Quantitative assessment of femoral head involvement in Legg-Calve-Perthes disease. J Bone Joint Surg [Am] 1981; 63:746–52.
100. Collier BD, Carrera GF, Johnson RP, et al. Detection of femoral head avascular necrosis in adults by SPECT. J Nucl Med 1985; 26:979–87.
101. Freiberg AH. The so called infraction of the second metatarsal bone. J Bone Joint Surg 1926; 8:257–61.
102. Silberstein ED. Bone scintigraphy. Mount Kisco, NY: Futura, 1984:168.
103. Mandell GA, Harcke HT. Scintigraphic manifestations of infraction of the second metatarsal (Freiberg's disease). J Nucl Med 1987; 28:249–51.
104. McCauley RGK, Kahn PC. Osteochondritis of the tarsal navicular. Radioisotopic appearance. Radiology 1977; 123:705–6.
105. Cahill BR, Gerg BC. 99m-Technetium phosphate compound joint scintigraphy in the management of juvenile osteochondritis dissecans of the femoral condyles. Am J Sports Med 1983; 8:393–4.
106. Yeung DW. Radionuclide imaging in osteochondritis dissecans. Clin Nucl Med 1981; 6:122.
107. Wilson E. Stress fractures. Radiology 1969; 92:481–6.
108. Johnson LC, Stradford HT, Geis RW, et al. Histogenesis of stress fractures. J Bone Joint Surg [Am] 1963; 45:1542–7.
109. Griffis CG, Protzmann RR. Stress fractures in men and women undergoing military training. J Bone Joint Surg [Am] 1977; 59:825–9.
110. Hopson CN, Perry DR. Stress fracture of the calcaneus in women marine recruits. Clin Orthop 1977; 128:159–62.
111. Ferris WD, Krissoff WB. Runner's injuries. Sports Med 1979; 12:54–64.
112. Morris JM, Provost RA. Fatigue fracture of the femoral shaft. J Bone Joint Surg [Am] 1969; 51:487–98.
113. Eugher W. Stress fracture of the medial tibial plateau. J Bone Joint Surg [Am] 1977; 59:767–9.
114. Burrows HJ. Fatigue infraction of the middle of the tibia in ballet dancers. J Bone Joint Surg [Br] 1956; 38:893–4.
115. Devas MB, Sweetnam R. Stress fractures of the fibula—a review of 50 cases in athletes. J Bone Joint Surg [Br] 1956; 38:818–29.
116. Symeonides PP. High stress fractures of the fibula. J Bone Joint Surg [Br] 1980; 62:192–3.
117. Rupani HD, Holder LE, Espinola DA, et al. Three-phase radionuclide bone imaging in sports medicine. Radiology 1985; 156:187–96.
118. Marty R, Denny J, McKamey MR, et al. Bone trauma and related benign disease: assessment by bone scanning. Semin Nucl Med 1976; 6:107–20.
119. Geslieu JE, Thrall JH, Espinosa JE, et al. Early detection of stress fractures using Tc-99m polyphosphate. Radiology 1976; 121:638–47.
120. Wilcox JR, Moniot AC, Green JP. Bone scanning in the evaluation of exercise related stress injuries. Radiology 1977; 123:699–703.
121. Prather JL, Nusynowitz ML, Snowdy HA, et al. Scintigraphic findings in stress fractures. J Bone Joint Surg [Am] 1977; 59:869–74.
122. Zwas ST, Elkanovitch R, Frank G. Interpretation and classification of bone scintigraphic findings in stress fractures. J Nucl Med 1987; 28:452–7.

123. Holder LE, Michael R. The specific scintigraphic pattern of "shin splints in the lower leg." Concise communication. J Nucl Med 1984; 25: 865–9.

124. Micheli LJ. Low back pain in the adolescent: differential diagnosis. Am J Sports Med 1979; 7:362–4.

125. Troup JDG. The etiology of spondylolysis. Orthop Clin North Am 1977; 8:57–64.

126. Cyron M, Hutton WC, Troup JDG. Spondylolytic fractures. J Bone Joint Surg [Br] 1976; 58:462–6.

127. Teitz CC. Sports medicine concourse in dance and gymnastics. Pediatr Clin North Am 1982; 29:1399–1421.

128. Papanicolaou N, Wilkinson RH, Emans JB, et al. Bone scintigraphy and radiography in young athletes with low back pain. AJR 1985; 145:1039–44.

129. Collier BD, Johnson RP, Carrera GF, et al. Painful spondylolysis or spondylolisthesis studied by radiography and single-photon emission computed tomography. Radiology 1985; 154:207–11.

130. Collier BD, Johnson RP, Isitman AT, et al. Chronic knee pain assessed by SPECT: comparison with other modalities. Radiology 1985; 157:795–802.

131. Maurice HD, Newman JH, Watt I. Bone scanning of the foot for unexplained pain. J Bone Joint Surg [Br] 1987; 69:448–52.

132. Glodman AB, Pavlov H, Schneider R. Radionuclide bone scanning in subtalar coalitions: differential considerations. AJR 1982; 138:427–38.

133. Deutsch AL, Resnick D, Campbell G. Computed tomography and bone scintigraphy in the evaluation of tarsal coalition. Radiology 1982; 144:137–40.

134. Holder LE, Matthews LS. The nuclear physician and sports medicine. In: Freeman IM, Weissman HS, eds. Nuclear medicine annual 1984. New York: Raven Press, 1984:81–140.

135. Zinman H, Keret D, Reis ND. Fracture of the medial sesamoid bone of the hallux. J Trauma 1982; 21:581–2.

136. Namey TC. Nuclear medicine and special radiologic imaging and technique in the diagnosis of rheumatic diseases. In: Kelley WN, Harris ED, Ruddy S, Sledge CB, eds. Textbook of rheumatology. Philadelphia: WB Saunders, 1985: 608–40.

137. Carlson T, Jacobs AM. Reflux sympathetic dystrophy syndrome. J Foot Surg 1986; 25: 149–53.

138. Kozin F, Soin J, Ryan L, et al. Bone scintigraphy in the reflux sympathetic dystrophy syndrome. Radiology 1984; 38:437–43.

139. Holder LE, Mackinnon SE. Reflex sympathetic dystrophy in the hands: clinical and scintigraphic criteria. Radiology 1984; 152:517–22.

140. Greyson ND, Tepperman PS. Three-phase bone studies in hemiplegia with reflux sympathetic dystrophy and the effect of disease. J Nucl Med 1984; 25:423–9.

141. Laxer RM, Allen RC, Malleson PN, et al. Technetium 99m-methylene diphosphonate bone scans in children with reflux neurovascular dystrophy. J Pediatr 1985; 106:437–40.

5

Radionuclide Bone Scanning for Diagnosis of Joint Pain

ROBERT SCHNEIDER

Complaints of joint pain may be due not only to arthritis, but also to multiple other conditions, including osteonecrosis, stress fractures, neoplastic disease, and infection. When the diagnosis is not apparent from clinical and radiographic findings, radionuclide bone scanning may be helpful [1–3]. Bone scanning is a sensitive method of detecting abnormalities in the bones and joints often before radiographic abnormalities can be seen. The specificity of bone scanning can be increased by careful attention to the pattern of abnormal uptake in an individual joint and by abnormalities found in the remainder of the skeleton. An example of this is seen in Figure 5–1. A 62-year-old woman was referred to a rheumatologist because of complaints of knee and ankle pain. Radiographs showed a normal appearance. A bone scan showed increased uptake in both knees, affecting the right and left medial femoral condyles and left medial tibial plateau. There was also increased uptake in both calcanei and in multiple ribs. This pattern is consistent with insufficiency fractures and suggests the presence of

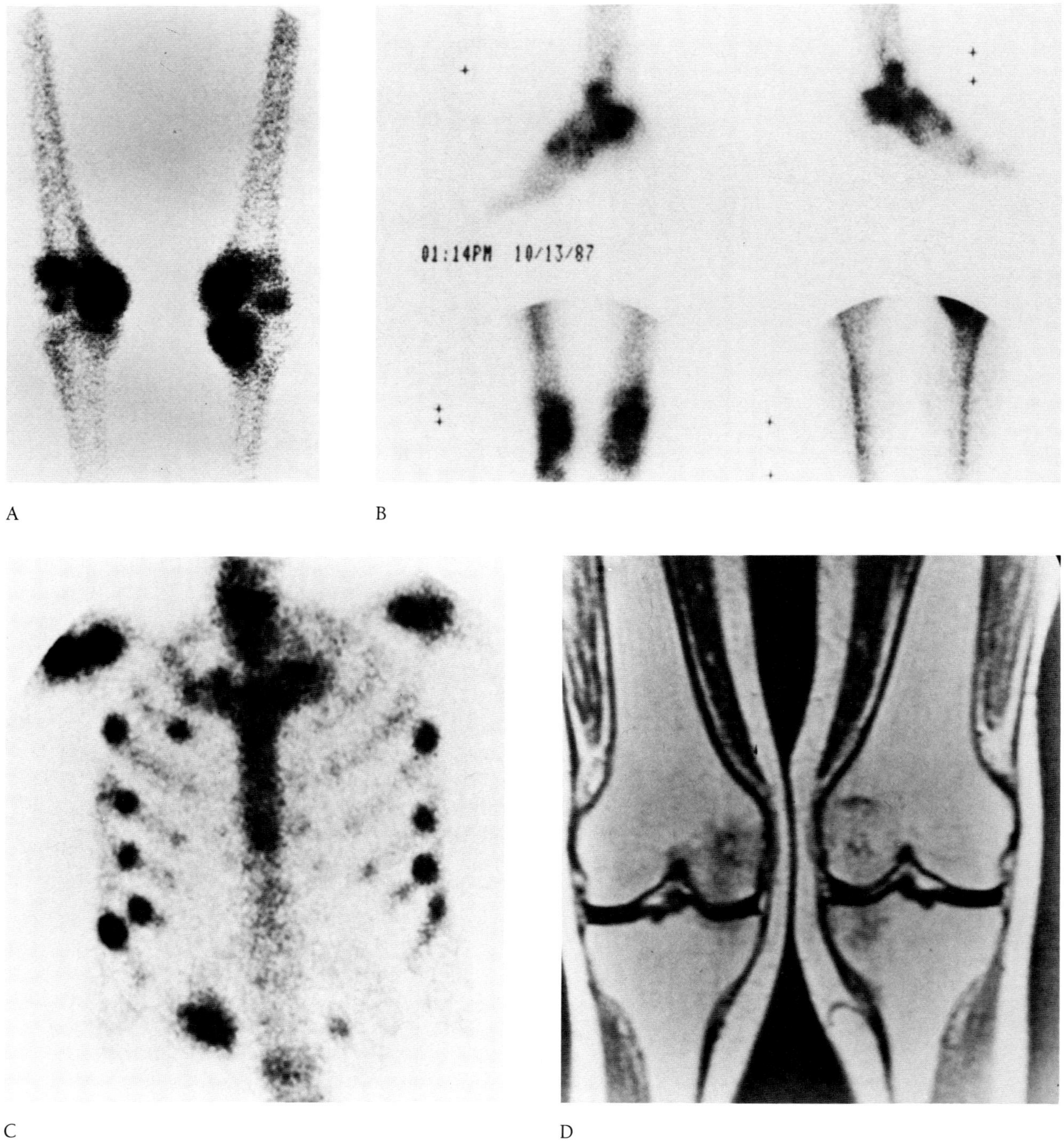

FIGURE 5–1
Osteomalacia with multiple insufficiency fractures. A. *Increased uptake in the right and left medial femoral condyles and left medial tibial plateau.* B. *Increased uptake in both calcanei.* C. *Focal areas of increased uptake in multiple ribs in a pattern typical of fractures.* D. *T1-weighted MRI scan of the knees, showing decreased signal intensity in the right and left medial femoral condyles and left medial tibial plateau.*

osteomalacia. This was later confirmed clinically. An MRI scan of the knees in this same patient showed areas of decreased signal intensity in both knees, corresponding to the sites of high uptake on the bone scan, but was misinterpreted as being due to osteonecrosis. Despite the advent of new diagnostic imaging methods, such as CT and MRI scanning, radionuclide bone scanning has a role to play in the evaluation of joint pain. Compared to MRI scanning, it is relatively inexpensive and has the ability to scan the entire skeleton.

Three-phase bone scanning is suggested for the evaluation of pain around individual joints. In the first phase, the dynamic flow study, in which scans are obtained every 3–5 seconds for 1 minute, shows the radionuclide in the blood vessels. The second phase, the blood-pool scan, consists of a static image done immediately after the dynamic flow study, and shows radionuclide in the extravascular space. In the third phase, delayed images are done 2 or more hours after injection and show uptake in bone. The dynamic flow and blood-pool scans are called the early phase and are helpful in demonstrating synovitis, inflammation, or other causes of hyperemia around the joint, and may be helpful in demonstrating soft tissue lesions that show minimal or no bone uptake on delayed static images (Figure 5–2). Magnification or pinhole collimator views may provide better detail of the joints. This is most valuable in children where high uptake in the growth plate may obscure the abnormality (Figure 5–3). SPECT may be used to supplement routine planar images to provide better detail. This has been of most value in the spine, especially in detecting stress fractures of the pars interarticularis that can cause back pain and may not be visible on planar images because of the uptake in the overlying vertebral body (Figure 5–4) [4]. SPECT has also been used to evaluate painful hips and knees [5,6]. In the hips, radionuclide in the bladder may cause artifacts, making interpretation of the scan difficult. Subtraction of bladder activity by data-processing techniques has been reported to be helpful in alleviating this problem [7].

In evaluating a painful joint, comparison with the opposite side is helpful, as variations occur in the relative intensity of uptake in joints compared with that in the surrounding bone and soft tissue. In conditions with high metabolic turnover of the bone, such as in some cases of renal osteodystrophy and hyperparathyroidism, there is often high uptake in the joints due to increased bone turnover, but no joint disease (Figure 5–5) [8]. It may be difficult to determine whether symmetric high uptake in joints is normal or abnormal. High uptake in both patellae has been found to be a common finding that may be normal or associated with a wide variety of disorders, and is not of diagnostic value [9].

Technetium 99m (^{99m}Tc) phosphate complexes are the preferred scanning agents for evaluation of joint pain. Although these phosphate complexes do not localize directly in the synovium, synovitis can be detected by the presence of diffuse increased vascularity corresponding to the joint capsule on the dynamic flow and blood-pool scans, and by diffuse increased uptake on the delayed static images. A variety of radionuclides, including ^{99m}Tc pertechnetate, are taken up in the synovium. Synovial scanning, however, cannot detect the bony abnormalities that may be the cause of joint pain, and is less sensitive than bone scanning with technetium phosphate complexes for the detection of arthritis [10]. Therefore, synovial scanning is rarely used clinically and has been replaced by bone scanning with technetium phosphate complexes for the evaluation of joint pain. Gallium 67 (^{67}Ga) citrate and indium 111 (^{111}In)–labeled white blood cell scanning may be used to supplement bone scanning for the evaluation of infection.

Arthritis

Bone scanning is usually not needed for the diagnosis of generalized inflammatory arthritis, such as rheumatoid arthritis and gout, as the diagnosis of these conditions is made by clinical and laboratory findings. The course of these diseases can be followed by physical examination of the involved joints without the need for bone scanning [11]. Bone scanning, however, may be used when clinical findings are uncertain, and can determine whether or not joint inflammation is actually present.

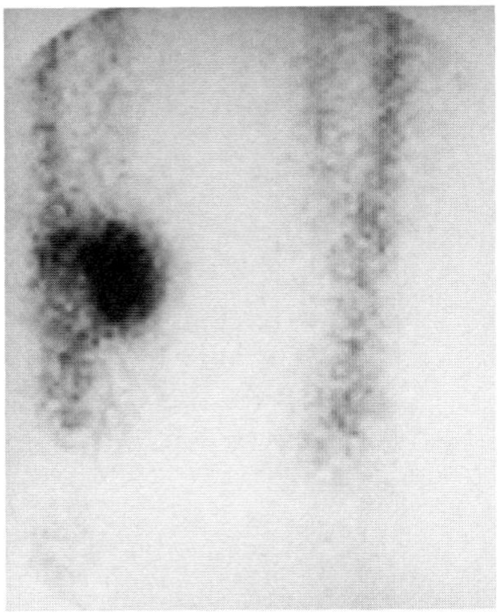

B

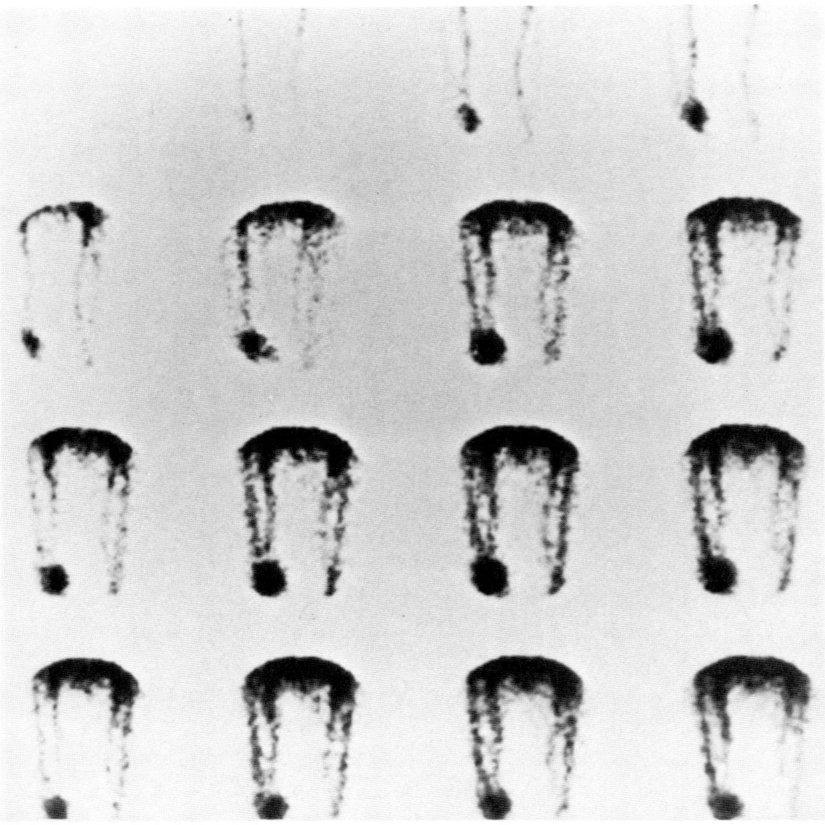

A

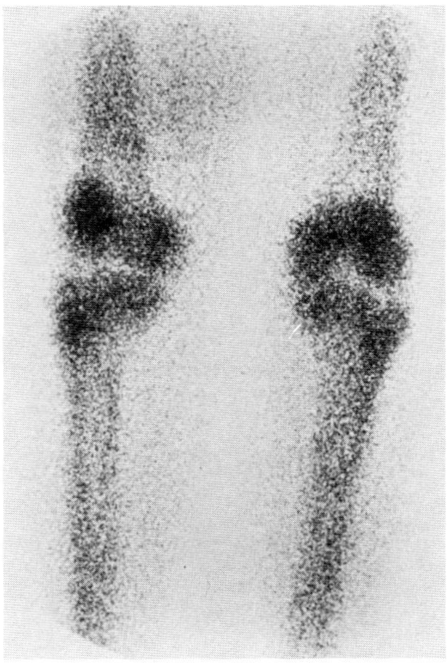

C

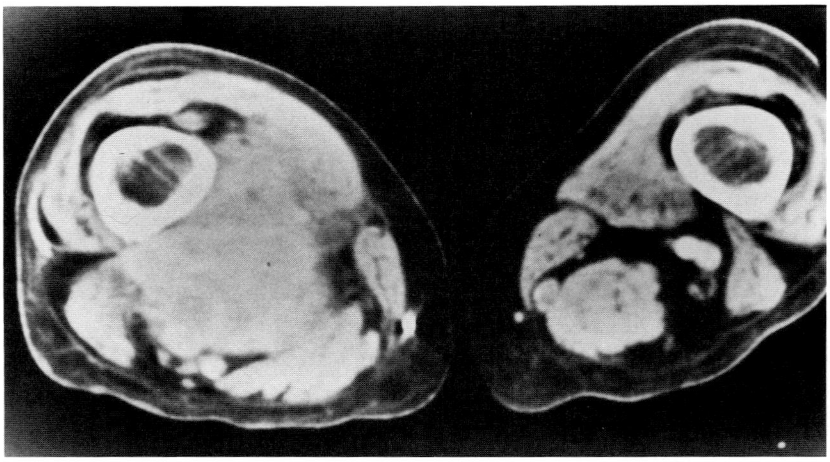

D

FIGURE 5–2
Metastatic disease to soft tissues above the knee. A. Dynamic flow study. B. Blood-pool scan. Both studies show marked increased vascularity in the distal part of the thigh, above the knee. C. Delayed static images. Only very mild increased uptake is present in the soft tissues of the thigh just above the knee, with no abnormal increased uptake in the bone or joint. D. CT scan. A soft tissue mass is present in the posterior distal aspect of the thigh.

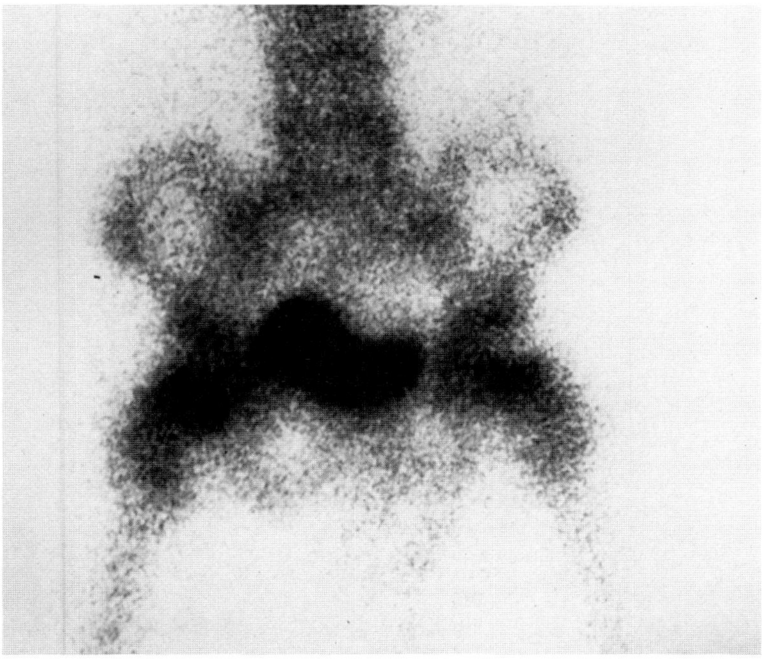

A

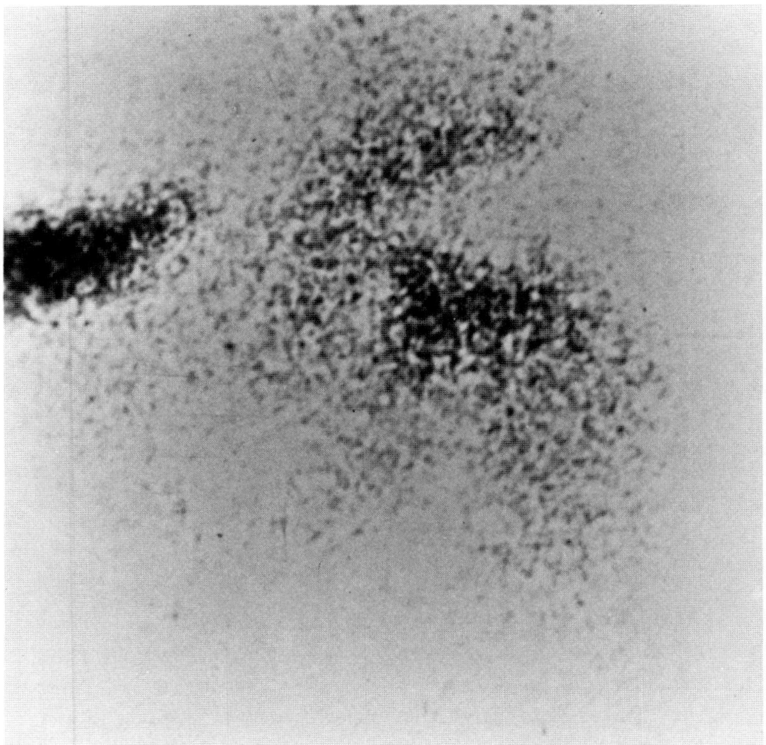

B

FIGURE 5–3
Early Legg-Calvé-Perthe's disease. Radiographs were equivocal. A. *A routine bone scan of the hips. Detail of the hips is partially obscured by the normally high uptake of radionuclide due to the growth plates of the femoral capital epiphyses.* B. *Pinhole collimator view of the left hip, showing decreased uptake of radionuclide in the left femoral head, consistent with Legg-Calvé-Perthe's disease.*

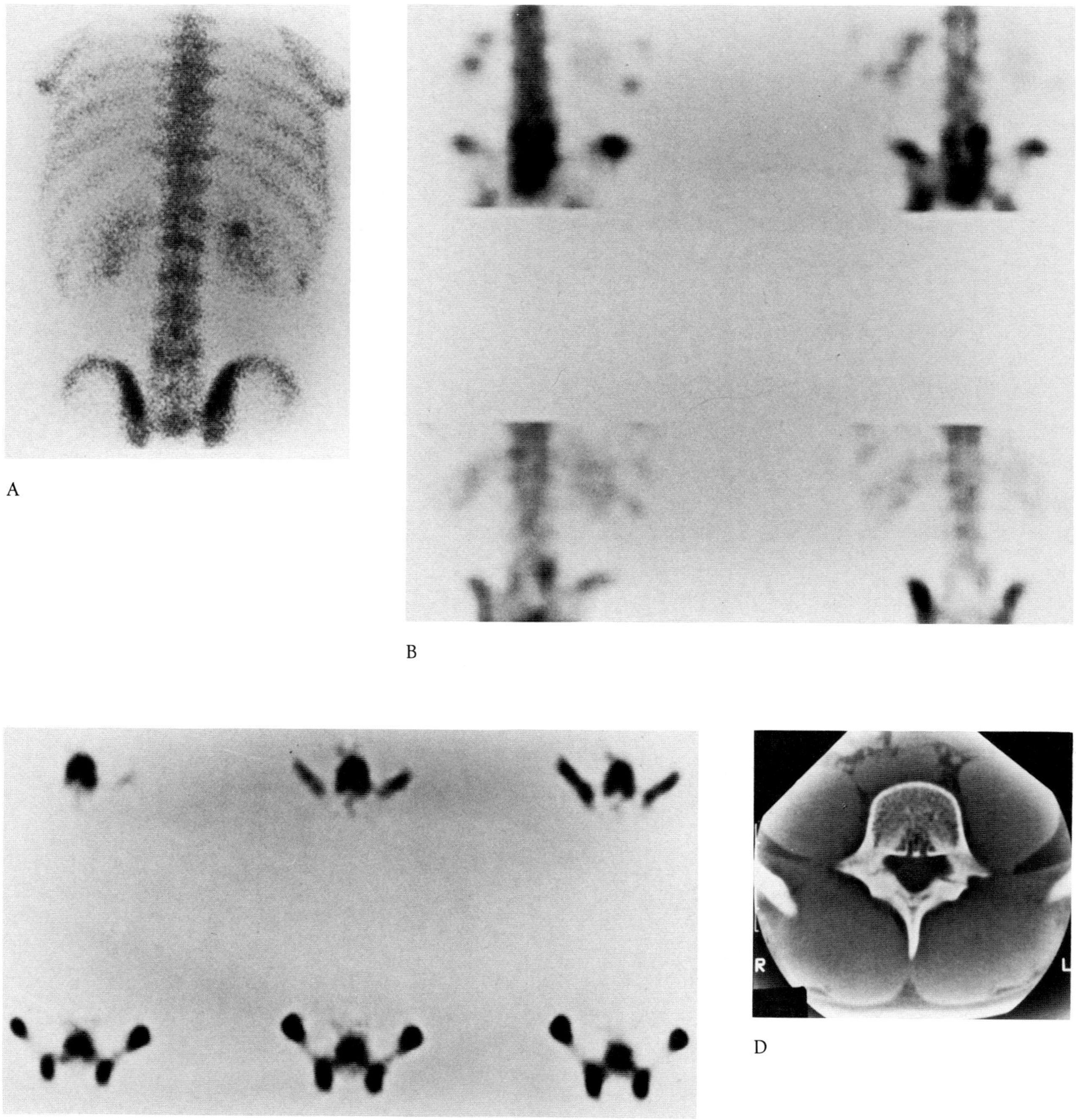

FIGURE 5–4
Stress fracture of the pars interarticularis. A 17-year-old basketball player had low back pain. Routine radiographs were normal. A. *A bone scan of the spine was normal.* B. *Coronal view of a SPECT scan of the lumbar spine shows increased uptake in the region of the pars interarticularis on the left at L-4 and less intense increased uptake on the right.* C. *An axial SPECT scan at the level of L-4 shows increased uptake in the region of the left pars.* D. *A CT scan shows a fracture through the left pars at L-4 and a less prominent fracture line through the right pars.*

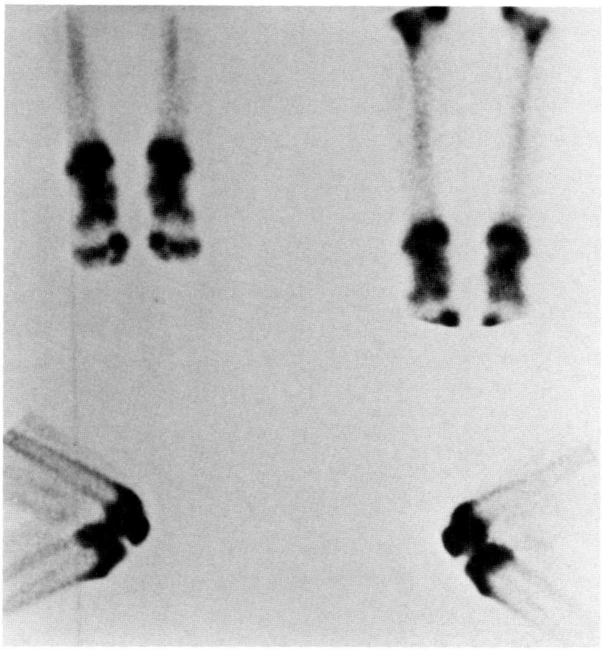

FIGURE 5–5
Renal osteodystrophy. There is increased uptake in both knees, ankles, and feet due to increased bone turnover from renal osteodystrophy. The patient did not have arthritis or joint pain.

INFLAMMATORY ARTHRITIS

When bone scanning is done in patients with rheumatoid arthritis or the other inflammatory arthritides, the characteristic patterns of joint involvement in these diseases can be seen. Rheumatoid arthritis has a tendency to affect joints bilaterally and symmetrically (Figure 5–6). Involvement of the wrist joints, metacarpophalangeal joints, proximal interphalangeal joints of the hand, and the intertarsal and metatarsophalangeal joints of the feet is common.

Severe pain in the region of a joint in a patient with arthritis may not always be due to arthritic disease, but can be due to a stress fracture (Figure 5–7) [12,13]. Infectious arthritis may also be the cause of pain in a patient with rheumatoid arthritis, but radionuclide scanning cannot differentiate infectious from noninfectious inflammatory arthritis. The seronegative spondyloarthropathies have a tendency to involve the areas of tendon insertions, such as the Achilles and plantar fascia on the calcaneus, and may show focal areas of increased uptake at these sites (Figure 5–7). The sacroiliac joints and apophyseal joints of the spine may also be involved in the seronegative spondyloarthropathies. Ankylosing spondylitis has symmetric involvement of the sacroiliac joints, whereas in psoriatic and

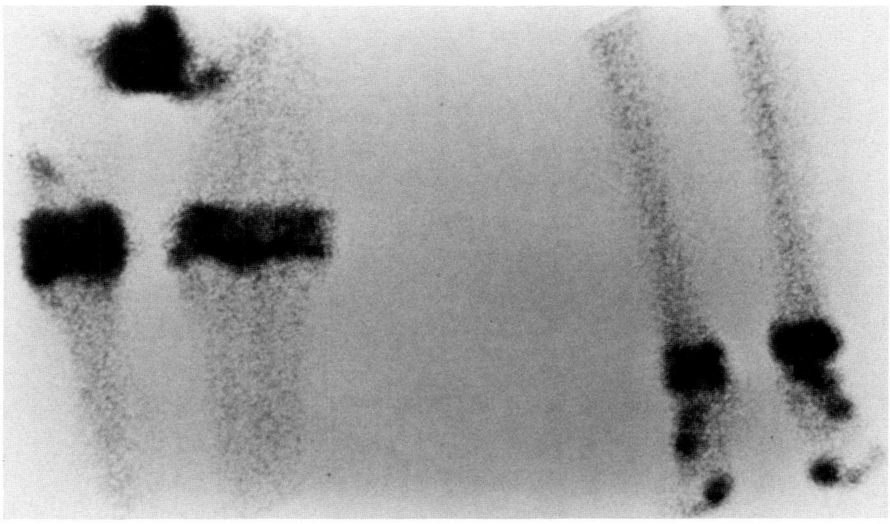

FIGURE 5–6
Rheumatoid arthritis. There is diffuse increased uptake in both knees, ankles, and feet due to rheumatoid arthritis.

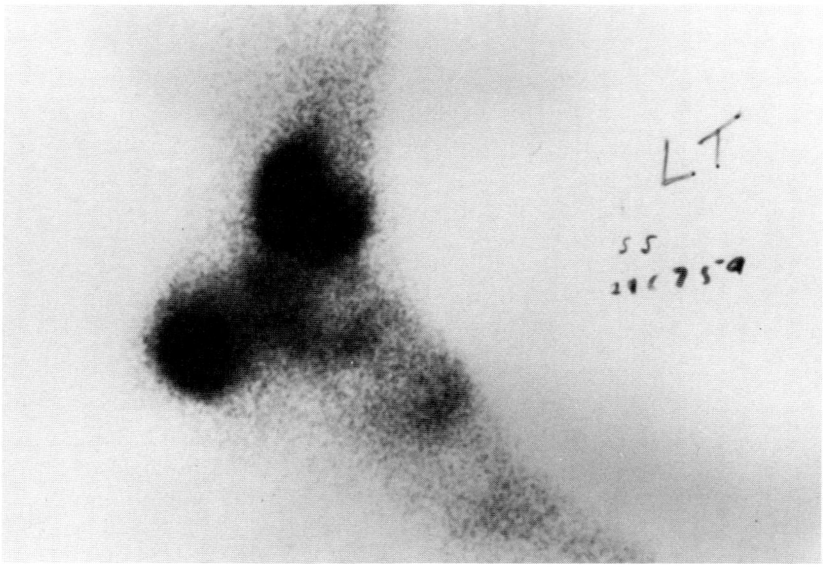

A

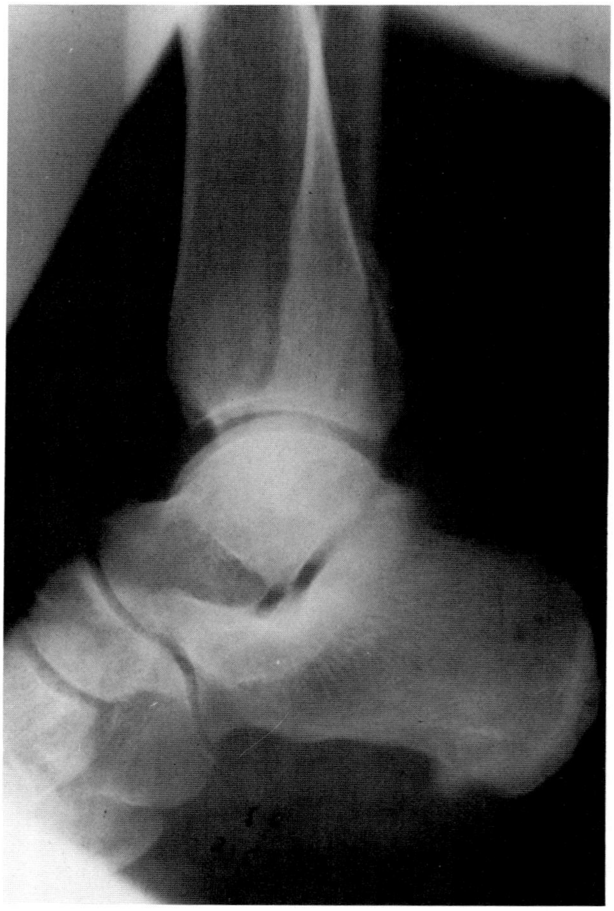

B

FIGURE 5–7
Stress fracture of the distal part of the fibula in a patient with arthritis associated with inflammatory bowel disease. A. *There is increased uptake in the plantar aspect of the calcaneus and the distal part of the fibula.* B. *A bony spur and erosion are present in the plantar aspect of the calcaneus on the radiograph. There is a stress fracture in the distal aspect of the fibula.*

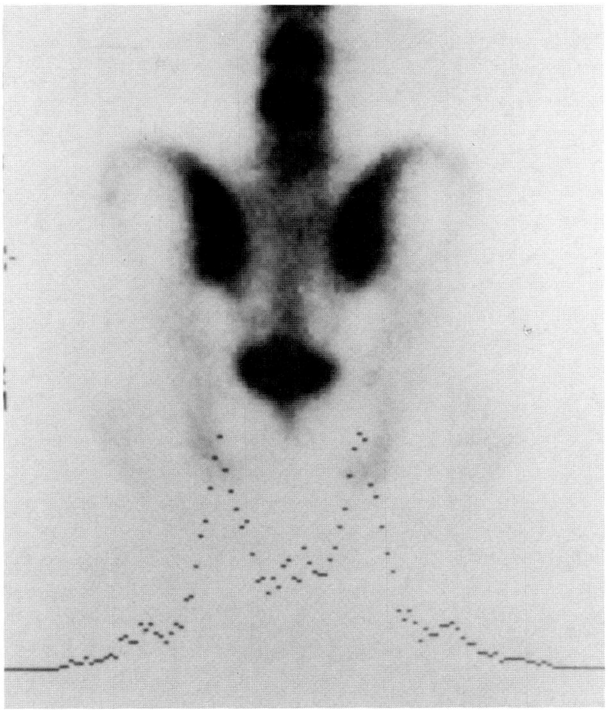

FIGURE 5–8 ▶
Quantitative bone scan of the sacroiliac joints in ankylosing spondylitis. A posterior bone scan of the pelvis was done. A region of interest across the pelvis and the region of the sacroiliac joints were assessed by computer, with a curve seen on the bottom of the scan. The counts in the region of the sacroiliac joints were more than 1.4 times the counts in the midsacrum, consistent with bilateral sacroiliitis due to ankylosing spondylitis.

Reiter's disease, there tends to be asymmetric involvement. The sacroiliac joints normally have high uptake of radionuclide, making symmetric abnormalities difficult to assess.

Quantitative bone scanning of the sacroiliac joints with computer analysis has been performed for the assessment of early ankylosing spondylitis (Figure 5–8) [14]. The method most frequently used requires a posterior scan of the pelvis, including the sacroiliac joints and the sacrum. The ratio of counts in the sacroiliac joints to the sacrum is determined by computer analysis. The average ratio has been reported to be about 1.25. A patient with sacroiliitis will tend to have higher ratios; however, an overlap between normal individuals and patients with ankylosing spondylitis occurs. This makes quantitative bone scanning for the evaluation of early ankylosing spondylitis unreliable [14].

Increased activity in the sacroiliac joints may also occur in hyperparathyroidism and degenerative arthropathy. In patients with ankylosing spondylitis, when fusion of the vertebral bodies and apophyseal joints occur, there no longer should be increased uptake of radionuclide. If a fracture or failure of fusion at one level occurs, it may allow abnormal motion to occur at a single level, causing bony erosions and vertebral destruction. This causes a focal area of high uptake of radionuclide at this level.

OSTEOARTHRITIS

Inflammatory arthritis tends to cause diffuse increased vascularity and uptake in the entire joint. Osteoarthritis may cause increased uptake in a portion of a joint. In the hips, the superior aspect of the hip joint is the most frequent site (Figure 5–9). In the knee, the tibial plateau shows in-

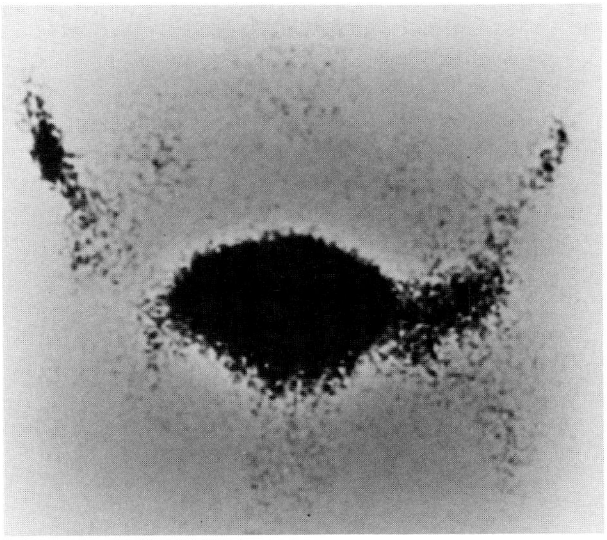

FIGURE 5–9
Osteoarthritis of the hip. There is increased uptake in the superior aspect of the left hip due to osteoarthritis.

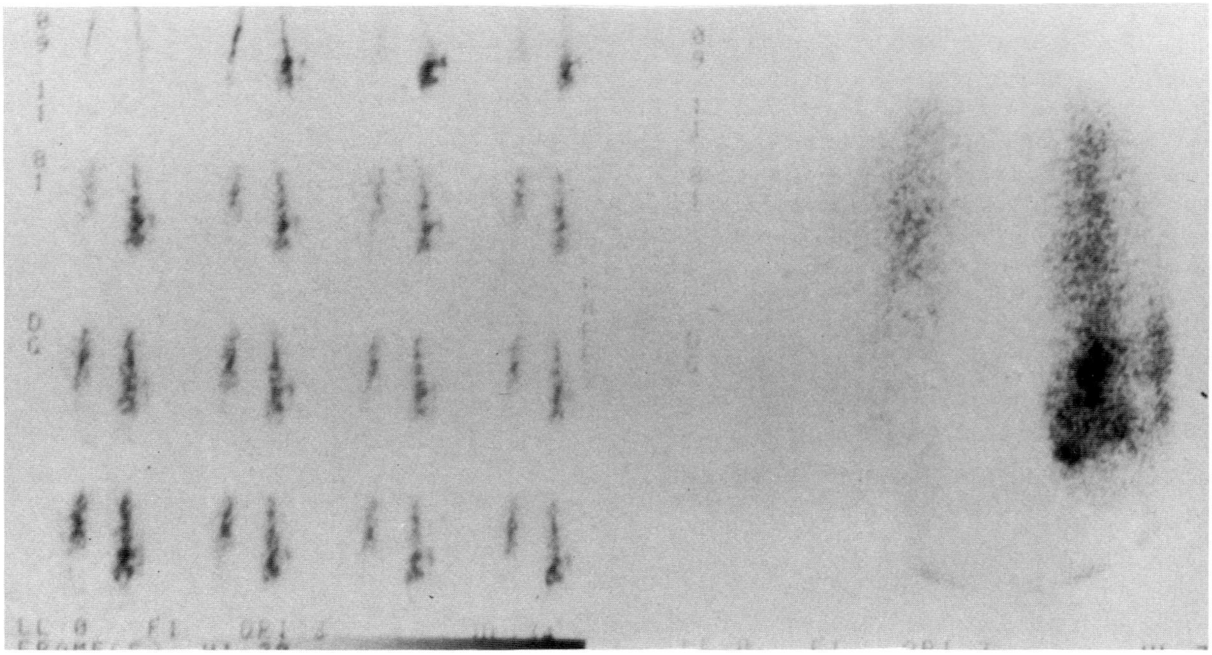

A

FIGURE 5–10
Osteoarthritis of the knees. In a patient who had radiographic changes of osteoarthritis in both knees, there was increased pain in the left knee. A. *Dynamic flow and blood-pool scans of the knees show increased vascularity in the left knee.* B. *Delayed images show diffuse increased uptake in the entire left knee, most severe in the medial compartment. There is increased uptake in the right medial tibial plateau.*

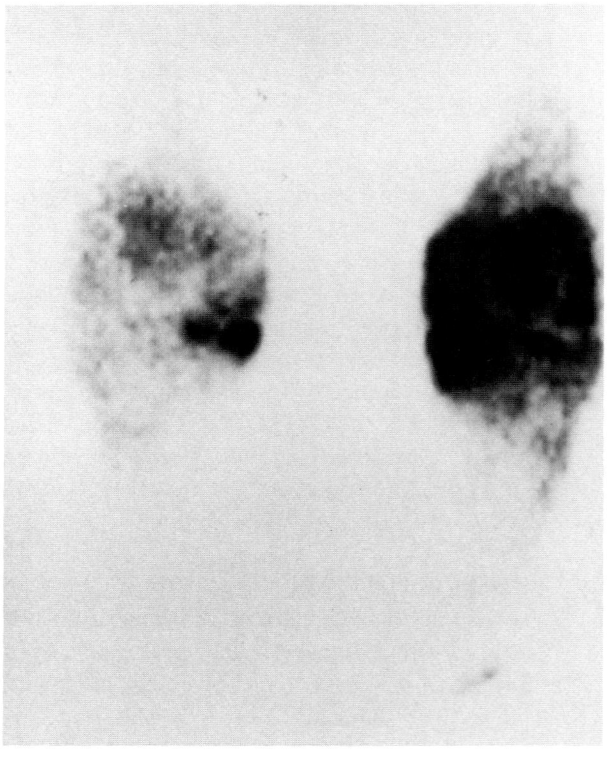

B

creased uptake earlier than does the femoral condyle (Figure 5–10). Any combination of medial, lateral, or patellofemoral compartment involvement can occur in osteoarthritis of the knee. Bone scanning can be used for determining compartmental involvement in planning surgery. In some cases of osteoarthritis, episodes of inflammation may cause increased vascularity on the early phase of the bone scan and diffuse increased uptake on the late phase (see Figure 5–10). Increased vascularity on the early phase correlates better with clinical findings of acute pain than does increased uptake on the late phase, which may be due to chronic arthritic changes of bone sclerosis or osteophyte forma-

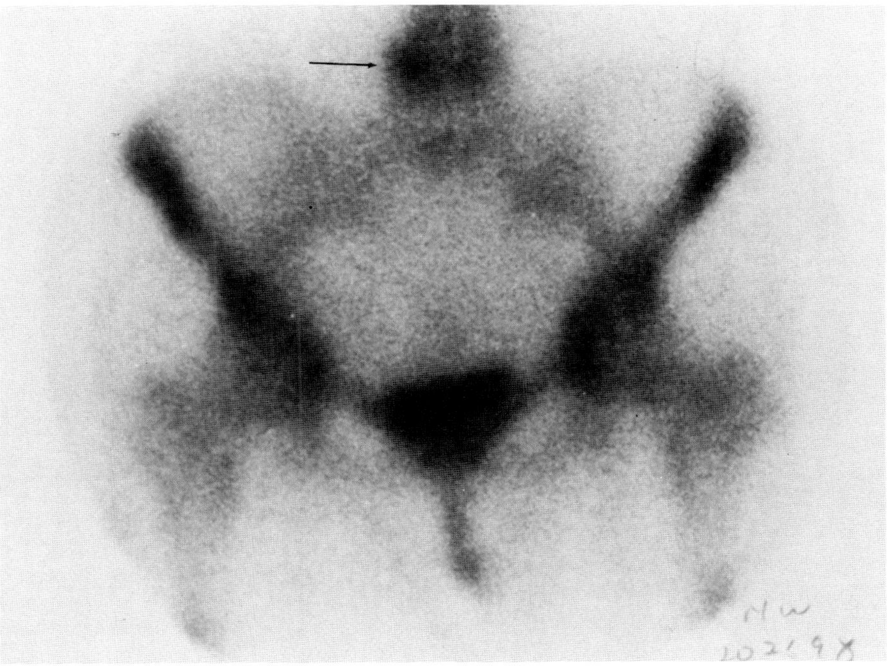

FIGURE 5–11
Osteoarthritis of the apophyseal joint at L4–5. A patient who complained of right hip pain had normal uptake of radionuclide in both hips. There is increased uptake in the right apophyseal joint at L4–5 (arrow) *due to osteoarthritis.*

tion. Patients with only mild radiographic osteoarthritis may have severe pain, whereas patients with advanced osteoarthritis may have less severe pain. Frequent sites of increased uptake due to osteoarthritis are the acromioclavicular joints, sternoclavicular joints, basal joints of the thumb, and first metatarsophalangeal joints. In the spine, increased uptake can be due to degenerative disc disease with sclerosis and osteophyte formation in the vertebral bodies. Osteoarthritis of the apophyseal joints causes focal increased uptake to the right or left of the midline (Figure 5–11). Patients with scoliosis tend to develop degenerative disease on the concavity of the curve. Multiple areas of increased uptake in the spine can be mistaken for metastatic disease. Radiographic correlation will show the degenerative changes in these cases.

NEUROPATHIC OSTEOARTHROPATHY

Neuropathic osteoarthropathy causes increased vascularity and uptake and may be difficult to differentiate from infectious or noninfectious inflammatory arthritis (Figure 5–12). Fractures, dislocations, and bony debris in the joint cause inflammatory changes and synovitis [15]. Increased uptake in the tarsometatarsal and intertarsal joints in diabetics raises the suspicion of neuropathic osteoarthropathy even when radiographic abnormalities are not apparent. Early diagnosis, leading to treatment with decreased weight bearing, may help prevent the destruction and disorganization of the joint that occur later in the course of the disease.

Stress Fractures

Stress fractures can occur when increased stress is applied to normal bone, called fatigue fractures, or when there is normal stress on osteopenic bone, called insufficiency fractures [16].

INSUFFICIENCY FRACTURES

Insufficiency fractures occur in patients with osteoporosis, osteomalacia, or hyperparathyroidism. Bone scanning is an important method of diagnosing stress fractures, as radiographs may appear normal at the onset of symptoms, and may not show abnormalities for several weeks or longer, or may never show any abnormality [17]. Increased vascularity is present on the early phase scan in recent stress fractures (Figure 5–13). The early phase scan returns to normal sooner than does the late phase.

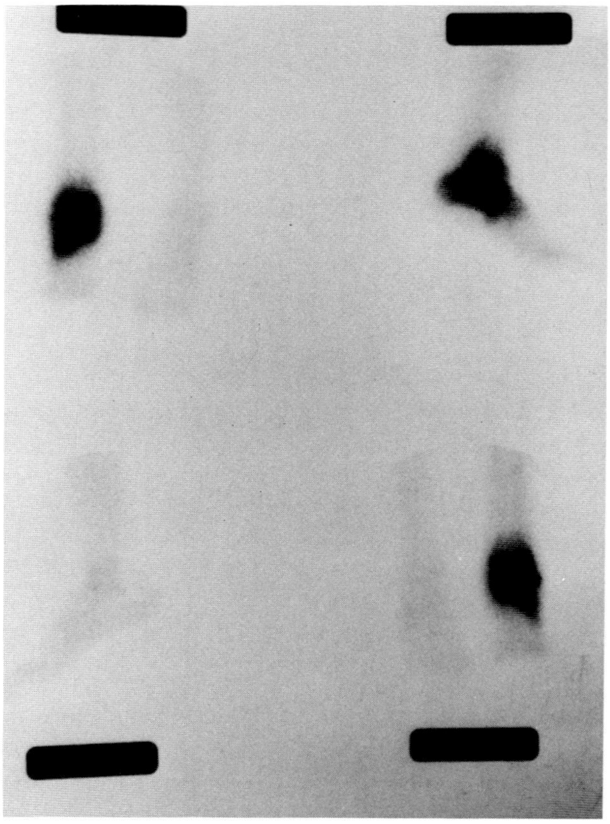

A

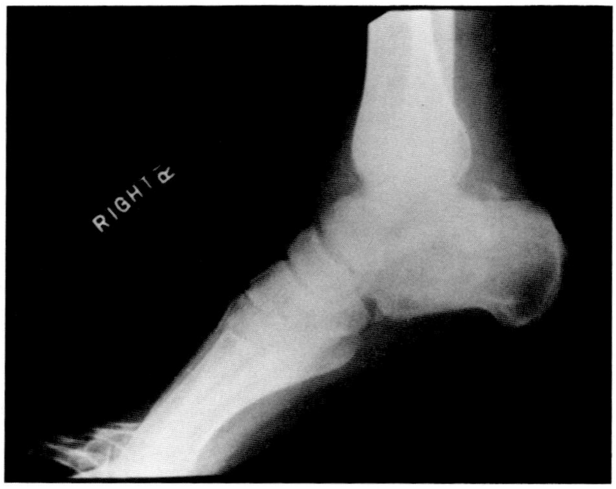

B

FIGURE 5–12
Neuropathic arthropathy. A diabetic presented with right ankle and foot swelling. A. The bone scan shows high uptake in the right ankle and hind part of the foot. B. A lateral radiograph shows soft tissue swelling and small bony fragments in the soft tissue. C. CT scan shows fractures of the talus and calcaneus, subluxation at the subtalar joint, and bone fragments within the ankle joint.

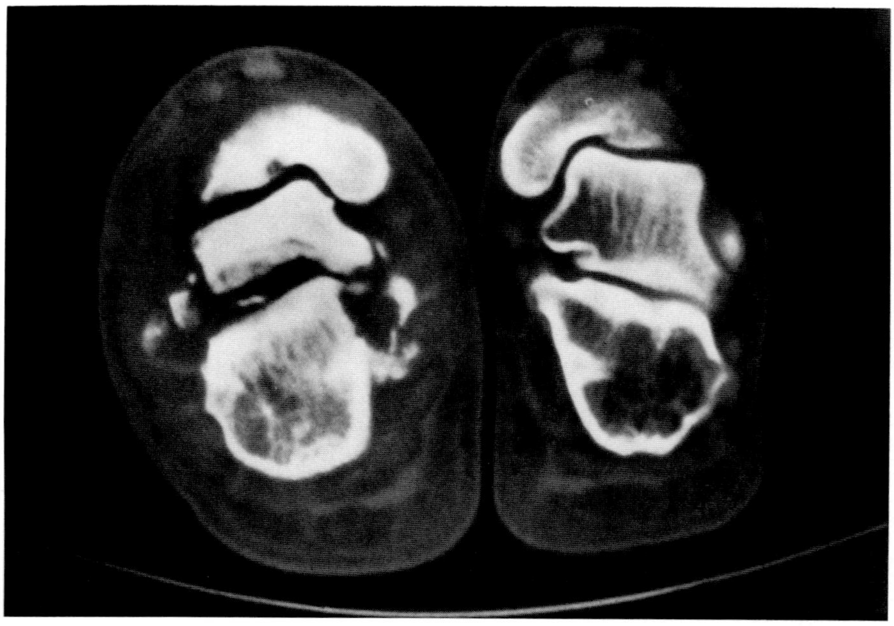

C

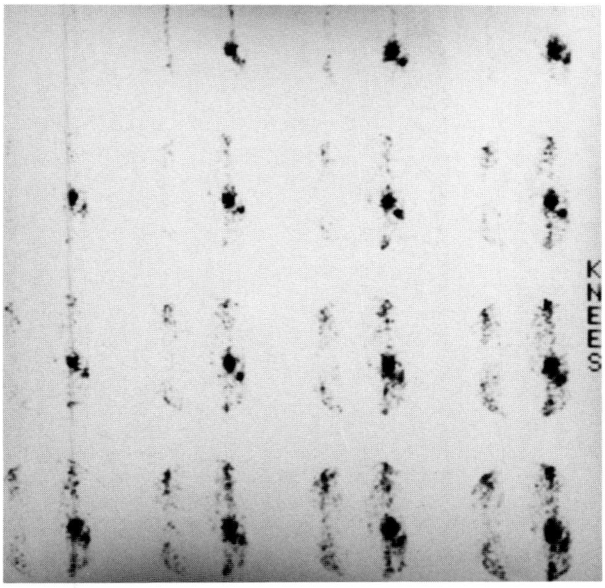

A

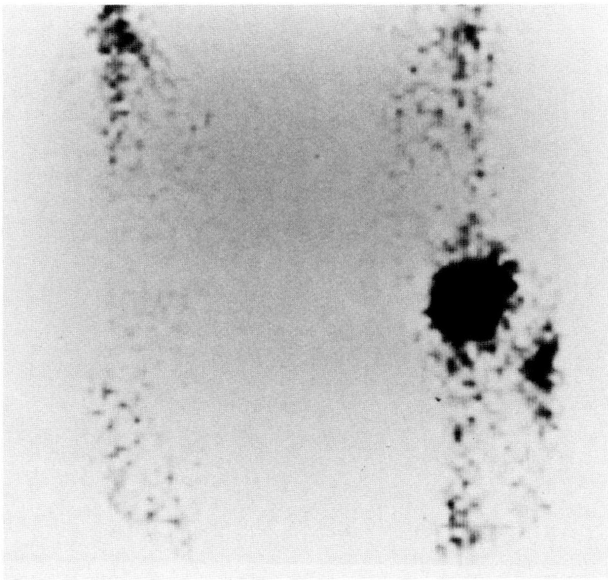

B

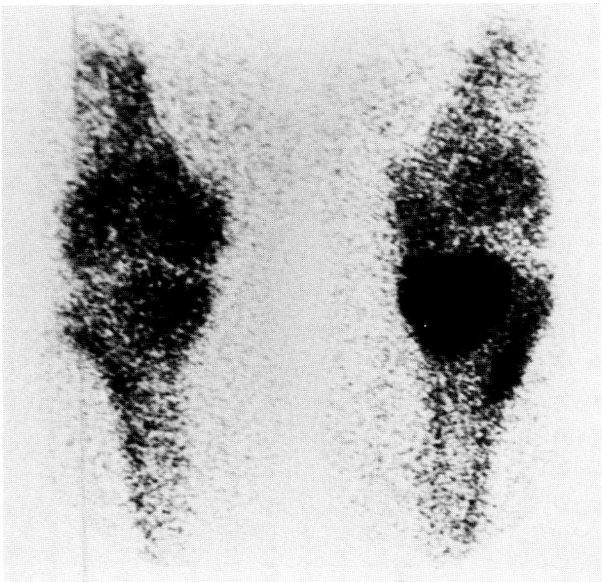

C

FIGURE 5–13
Insufficiency fracture of the tibial plateau in a woman with osteoporosis. A. Dynamic flow scan. B. Blood-pool scan. Both scans show increased vascularity in the medial compartment of the left knee. C. Delayed static image shows high uptake in the left medial tibial plateau.

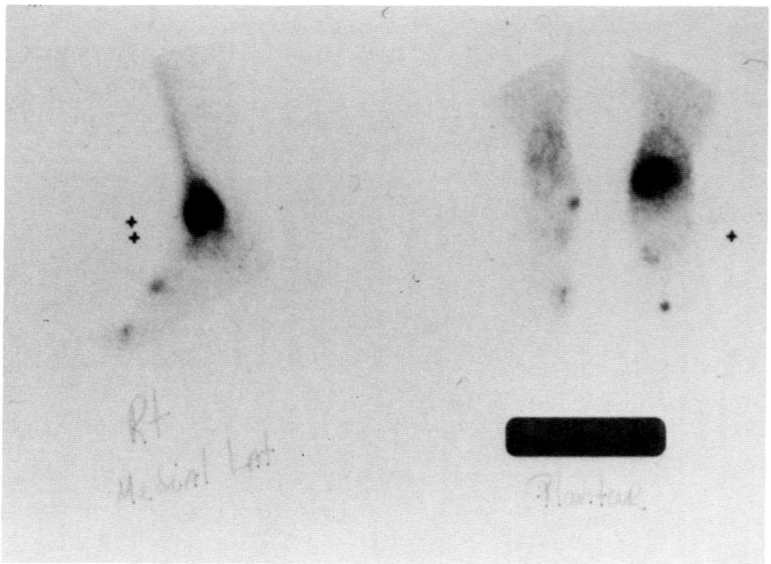

A

FIGURE 5–14
Insufficiency fracture of the distal part of the tibia. A. A bone scan shows high uptake in the distal area. B. A radiograph shows faint sclerosis (arrow) *in the distal metaphyseal region of the tibia.*

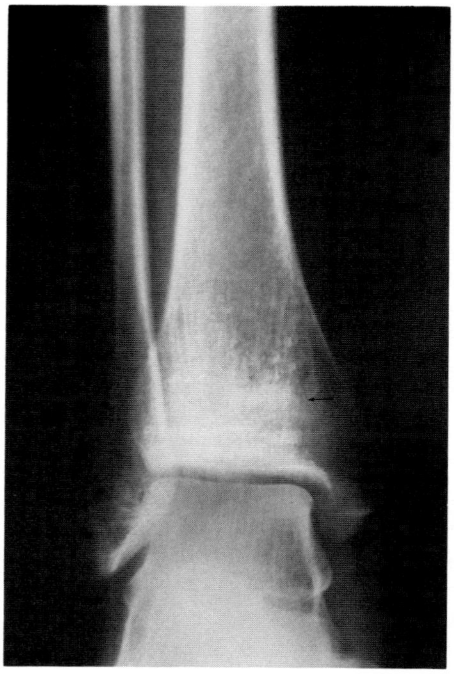

B

Insufficiency fractures often occur in the metaphyseal region of long bones, especially in the tibial plateaus and in distal aspects of the tibia, presenting with knee or ankle pain (Figures 5–13 and 5–14) [14,18]. Another frequent site of stress fracture is the calcaneus, and ankle or foot pain is the presenting symptom (Figure 5–15). Since many of the patients with insufficiency fractures do not have a history of excessive activity, this diagnosis may not be suspected. Insufficiency fractures also occur frequently in the medial aspect of the femoral neck and less frequently in the lateral aspect (Figure 5–16). Insufficiency fractures also occur in the superior and inferior pubic rami. These cause pain in the anterior part of the groin and may clinically simulate hip disease. They present as focal areas of increased uptake, but may be obscured by radionuclide in a full bladder. Radionuclide in the sides of the bladder, when the bladder is almost empty, may also give an appearance of round focal hot spots in the pubic rami and may simulate fracture. A repeat scan done with a delay of 15 minutes or half an hour will show that these hot spots are increasing in size and are due to bladder activity. If the patient cannot empty a full bladder that obscures the pubic rami, the patient may be catheterized or a scan can be done 24 hours later when the patient has voided. Insufficiency fractures simulating hip pain may also occur in the superior aspect of the acetabulum and in the sacrum (Figure 5–17) [19,20].

Superior acetabular insufficiency fractures do not show abnormalities initially on radiographs and high uptake in the acetabulum can be mistaken for neoplastic disease. Eventually, a faint area of sclerosis may develop in the superior aspect of the acetabulum. Insufficiency fractures of the sacrum can involve both sacral alae and the body of the sacrum, causing an H-shaped or a butterfly-shaped pattern of increased uptake; such a bone scan is virtually diagnostic (Figure 5–18). Insufficiency fractures of a single sacral ala present with increased uptake in the region

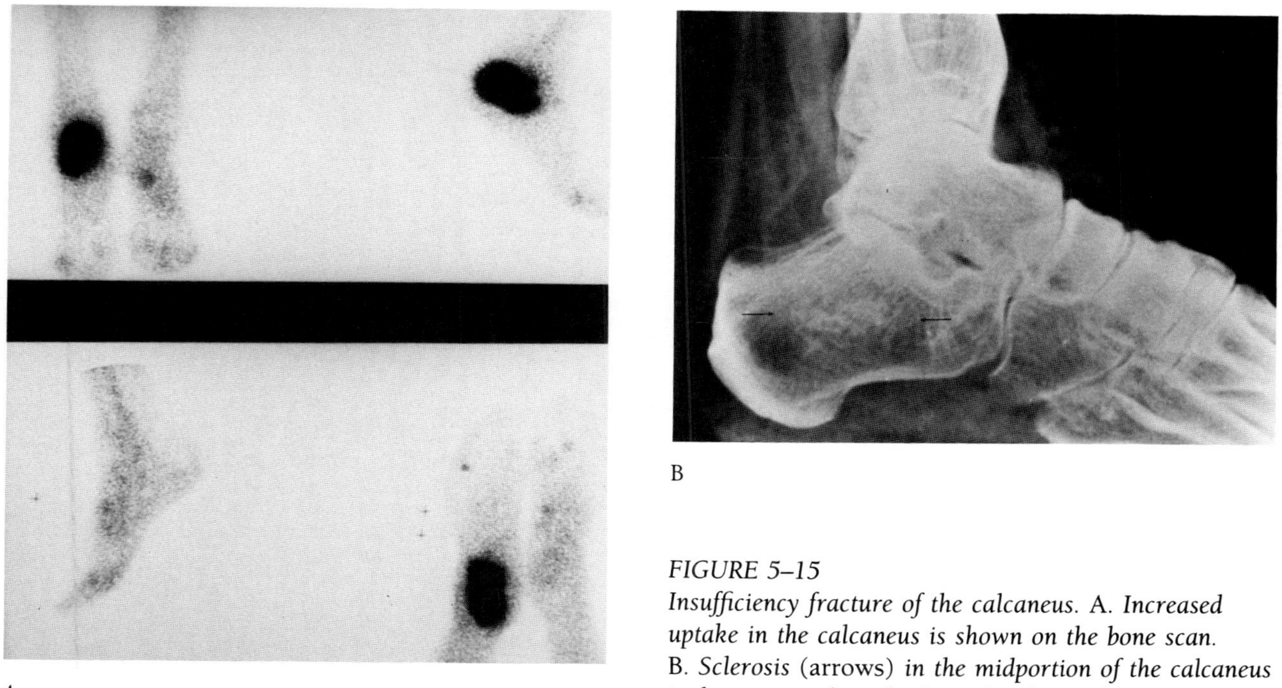

A

B

FIGURE 5–15
Insufficiency fracture of the calcaneus. A. *Increased uptake in the calcaneus is shown on the bone scan.* B. *Sclerosis* (arrows) *in the midportion of the calcaneus is demonstrated on the lateral radiograph.*

of the sacroiliac joint. They create a problem in differential diagnosis and have to be differentiated from sacroiliitis or metastatic disease (Figure 5–19). CT scans may be helpful. Initially they can be normal, but eventually, a linear area of sclerosis may develop in the sacral ala.

Knee pain may be caused by insufficiency fractures of the tibial plateau that can occur medially or laterally (see Figure 5–13). When high uptake occurs in the tibial plateau, the differential diagnosis also includes osteonecrosis and early osteoarthritis. However, osteonecrosis is relatively rare in the tibial plateaus, except in patients who have been taking corticosteroids or who have some other underlying conditions associated with osteonecrosis. Early osteoarthritis usually does not cause as marked increased vascularity and uptake in the tibial plateau, as do insufficiency fractures. Insufficiency fractures rarely involve the femoral condyles, except in patients with underlying metabolic bone disease (see Figure 5–1). Insufficiency fractures of the distal tibial metaphyses cause ankle pain. With insufficiency fractures of the metaphyseal regions of bone, a fracture line is not visible. The only radiographic findings are faint transverse areas of sclerosis that occur in the metaphysis, usually several weeks after the onset of pain (see Figure 5–14).

FATIGUE FRACTURES

Fatigue fractures due to exercise or excessive activity present less of a problem in diagnosis than do insufficiency fractures because of the suggestive history. Fatigue fractures can occur in the femoral neck or pubic rami, causing hip pain. Fatigue fractures seldom cause knee pain as they usually do not involve the femoral condyles or tibial plateaus. More frequently, fatigue fractures involve the shaft of the tibia and, less commonly, the shaft of the fibula or femur. Fatigue fractures can occur in the calcaneus, causing ankle or foot pain. One of the most common sites of fatigue fractures is the metatarsals. These are called "march fractures." High uptake is seen in the

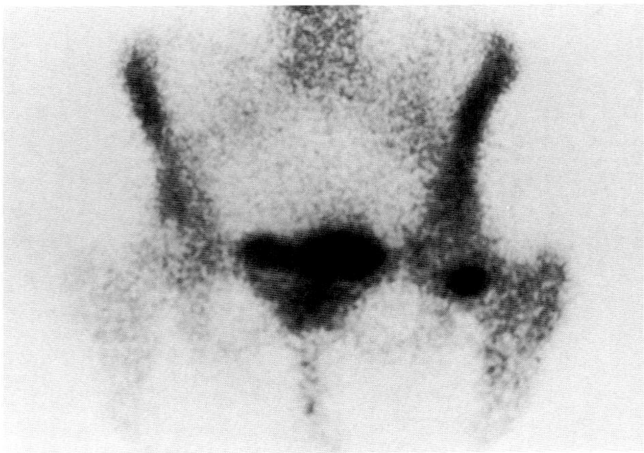

A

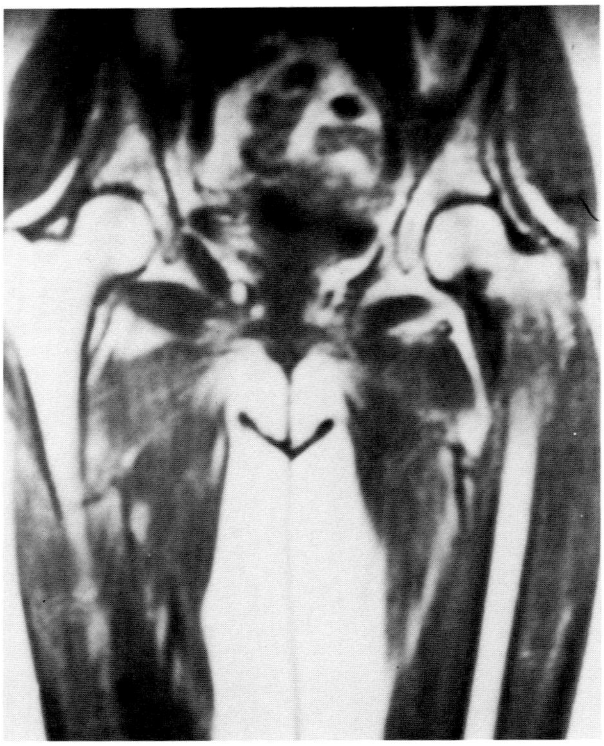

B

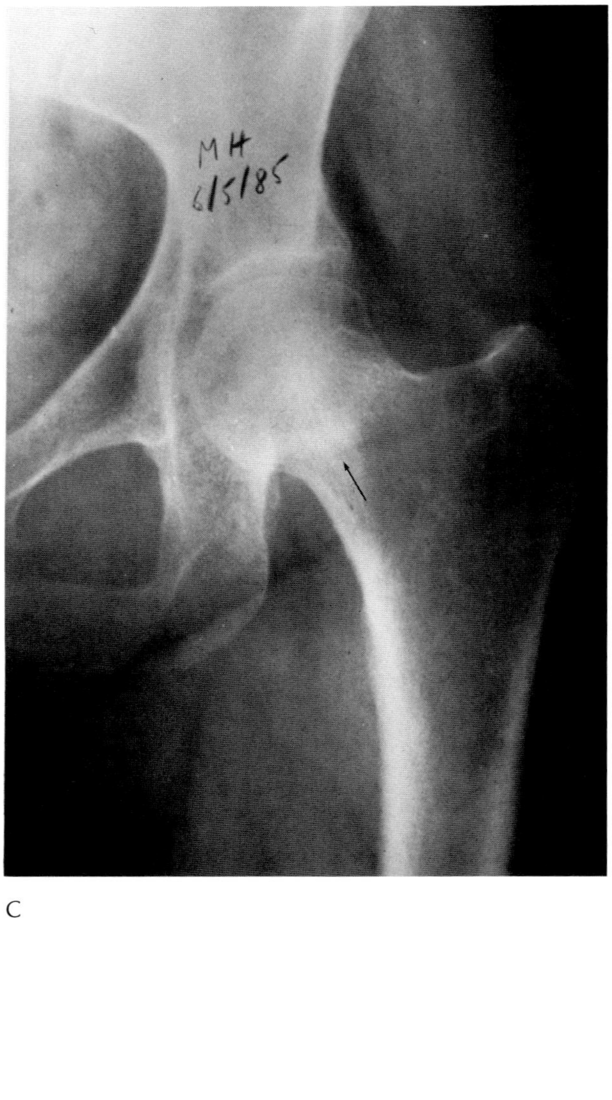

C

FIGURE 5–16
Insufficiency fracture of the femoral neck. A 64-year-old woman presented with left hip pain and normal radiographs. A. *A bone scan showed a focal area of increased uptake in the left femoral neck.* B. *A T1-weighted MRI scan showed a wedge-shaped area of decreased signal intensity in the medial aspect of the left femoral neck.* C. *A radiograph done 1 month after the onset of symptoms showed sclerosis in the medial aspect of the femoral neck* (arrow) *due to the insufficiency fracture.*

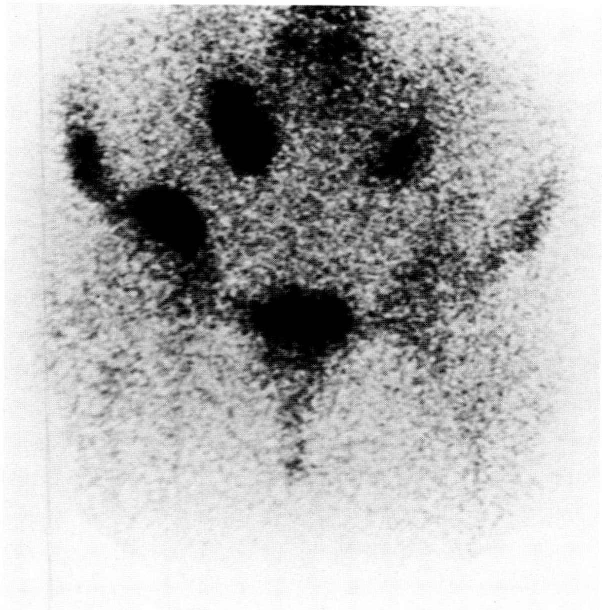

FIGURE 5–17
Insufficiency fractures of the superior aspect of the acetabulum and sacrum. There is high uptake in the superior part of the right acetabulum and the right and left sacroiliac regions.

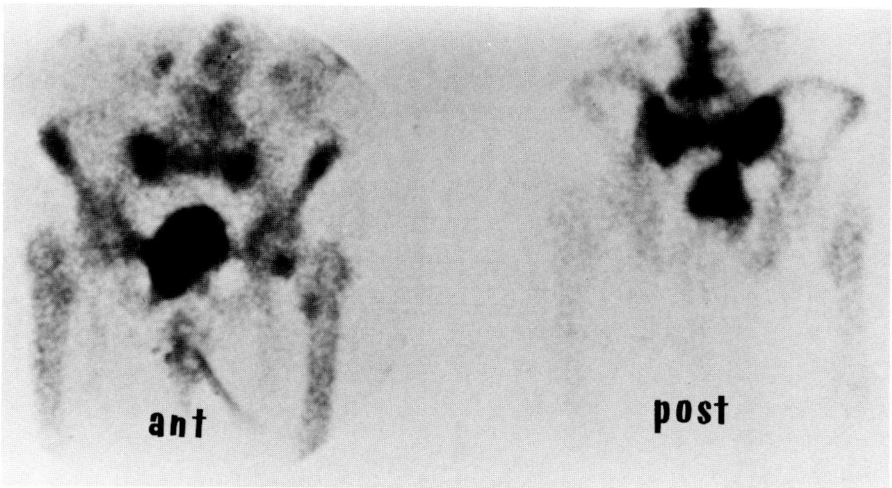

FIGURE 5–18
Insufficiency fractures of the sacrum, pubis, and femoral neck. A bone scan of the pelvis shows high uptake in both sacral alae and the body of the sacrum, the right pubis near the symphysis, and the medial aspect of the femoral neck.

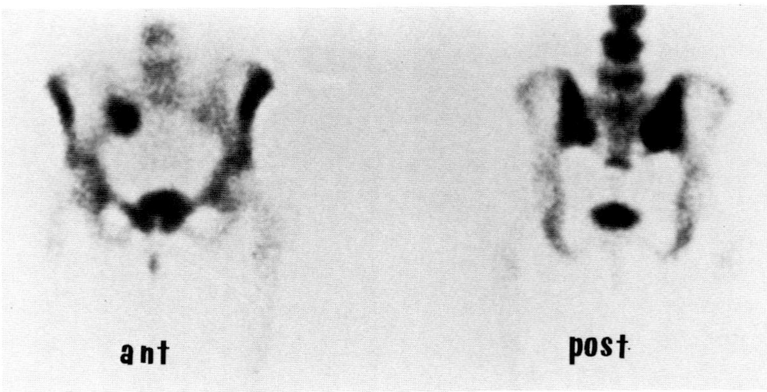

A

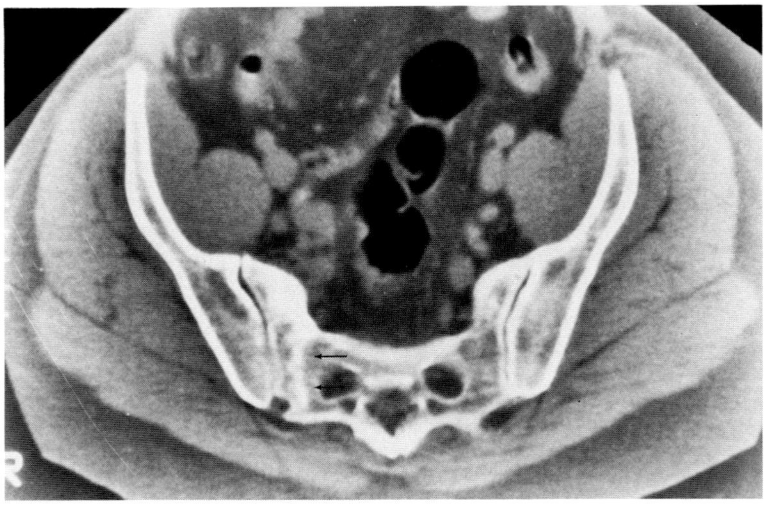

B

FIGURE 5–19
Insufficiency fracture of a sacral ala. A. Anterior and posterior scans show high uptake in the right sacroiliac region. B. A CT scan shows sclerosis in the right sacral ala (arrows) *representing a healing insufficiency fracture.*

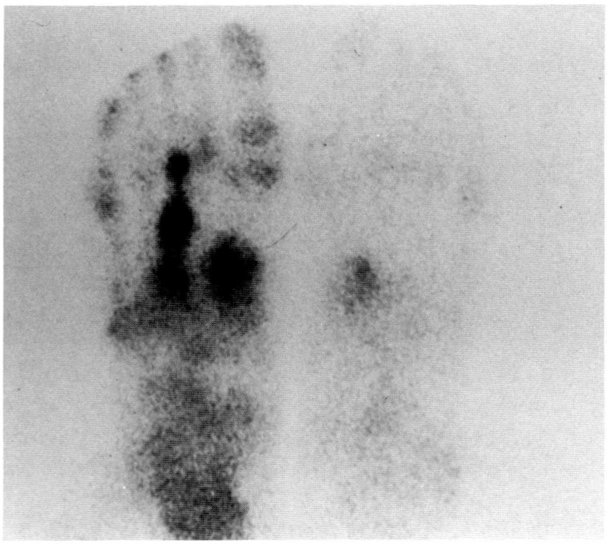

FIGURE 5–20
March fracture. A bone scan of the foot shows high uptake in the midshaft of the third metatarsal. The uptake extends along the metatarsal shaft to the metatarsal head.

shaft of the metatarsal and may extend up or down the metatarsal shaft, even though the fracture is localized (Figure 5–20). With fatigue fractures of the metatarsals, usually the fracture line or periosteal reaction is eventually seen in the shaft of the metatarsal. Tarsal navicular stress fractures may be difficult to detect on radiographs and the diagnosis can be delayed (Figure 5–21) [21]. Arthritis or osteoid osteoma may appear similar on a bone scan.

Fatigue fractures seldom occur in the upper part of the body and the upper extremities as these are not weight-bearing joints. Fatigue fractures that occur in the upper ribs or clavicle can simulate shoulder pain (Figure 5–22). Plain films or tomograms may show sclerosis in the rib or clavicle, corresponding to the hot spot on the bone scan and confirming the diagnosis. Stress fractures can occur in the elbows as a result of throwing activities. In children, stress fractures can occur through ununited apophyses. Stress fractures of the par interarticularis (traumatic spondylolysis) may be a cause of back pain (see Figure 5–4) [4,22]. Increased uptake suggests a fracture of recent origin that may benefit from immobilization [23]. Old spondylolysis that occurred in childhood does not show increased uptake.

Osteochondritis Dissecans

In osteochondritis dissecans, a fragment of bone and sometimes cartilage separates from the articular surface. It occurs in children and young adults and is associated with chronic trauma or stress. The most frequent site of osteochondritis dissecans is the medial femoral condyle. The second most common site is the ankle, affecting either the medial or the lateral articular surface of the talus. Another site is the elbow, with the capitellum involved. This occurs especially in athletes involved with throwing activities or gymnastics. On a bone scan, osteochondritis dissecans is seen as a small focal area of increased uptake at the affected site (Figure 5–23). Bone scanning is seldom needed for diagnosis, however, as radiographic abnormality is usually present when clinical symptoms occur. If increased vascularity is not demonstrated on the early phase of the scan, it suggests that the abnormality is not due to a recent osteochondral fracture. Marked increased vascularity and uptake in an area of osteochondritis dissecans suggest a recent fracture, and treatment with immobilization may cause healing with union of the fragment in some cases.

Osteonecrosis

NONTRAUMATIC OSTEONECROSIS

Nontraumatic osteonecrosis may be the cause of joint pain. It affects the hips most frequently, followed by the knees, shoulders, ankles, and elbows. In the majority of cases, some underlying cause, including corticosteroid use, alcoholism, sickle cell anemia, Gaucher's disease, and others, is associated with the condition. Bone scanning is more sensitive than radiographs for the diagnosis of osteonecrosis; however, MRI is even more sensitive (Figure 5–24) [24,25].

Nontraumatic osteonecrosis tends to be bilateral and to involve multiple joints in the body. Because of this, bone scanning may be valuable as a screening procedure that can diagnose nontraumatic osteonecrosis in asymptomatic joints. In nontraumatic osteonecrosis, the bone scan usually shows increased vascularity and uptake at the affected site. This is due to revascularization and bone repair in the region of osteonecrosis. Decreased or absent uptake may be seen better with SPECT scanning (Figure 5–25) [5]. Bone scanning is most helpful for the diagnosis in early stages of the disease: stage I, where radiographic abnormalities are absent and pain is present; or stage II, in which there are areas of sclerosis and lucency without fracture of the subchondral cortex or deformity of the articular surface. In stages I and II, core decompression biopsy has been recommended to help prevent progression of the condition. The early stages of osteonecrosis show focal high uptake usually localized to one side of the joint, helping to differentiate it from arthritis, in which there is usually uptake on both sides of the joint. With progression of osteonecrosis and collapse and deformity of bone, secondary osteoarthritis occurs and the condition may not be differentiated from primary osteoarthritis.

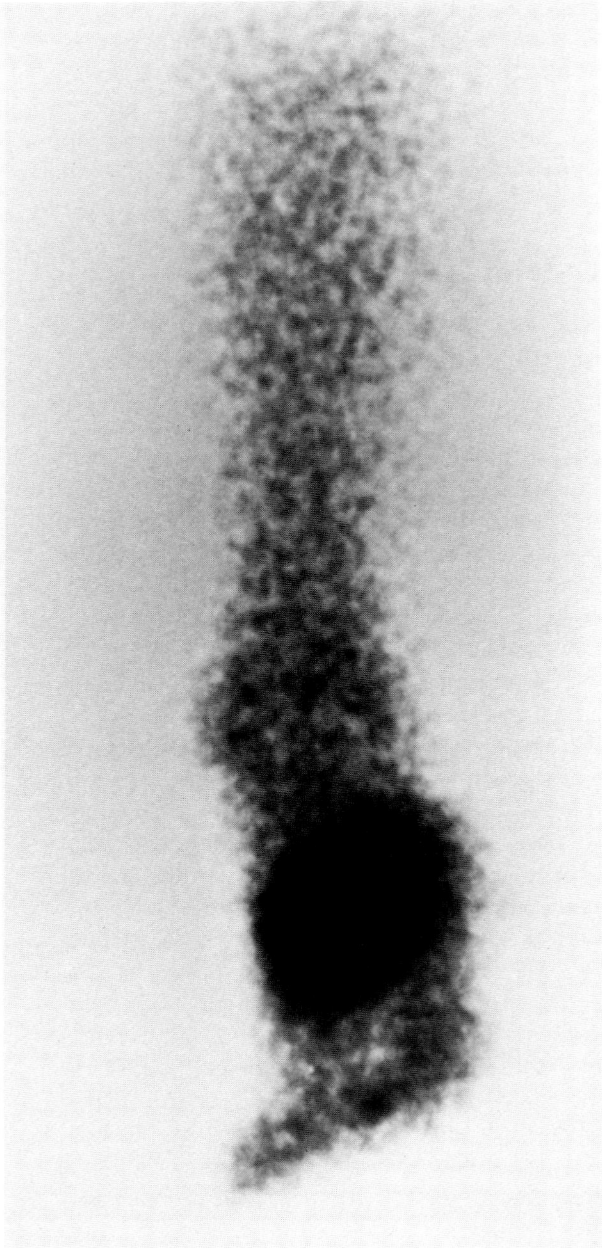

A

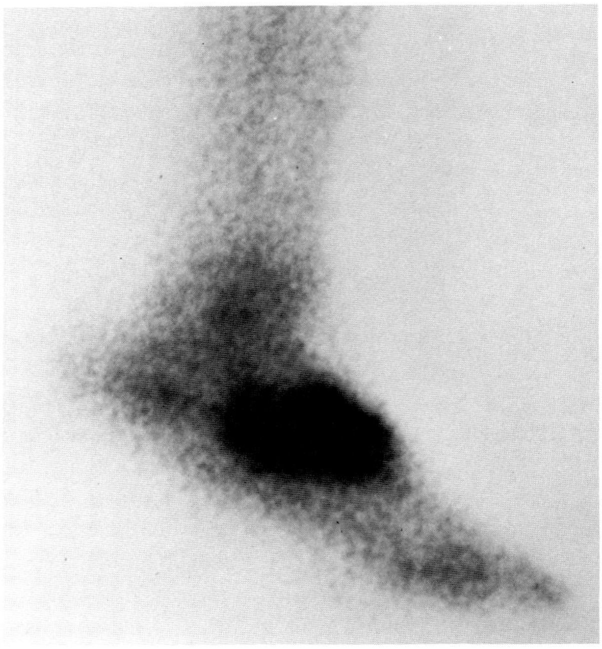

B

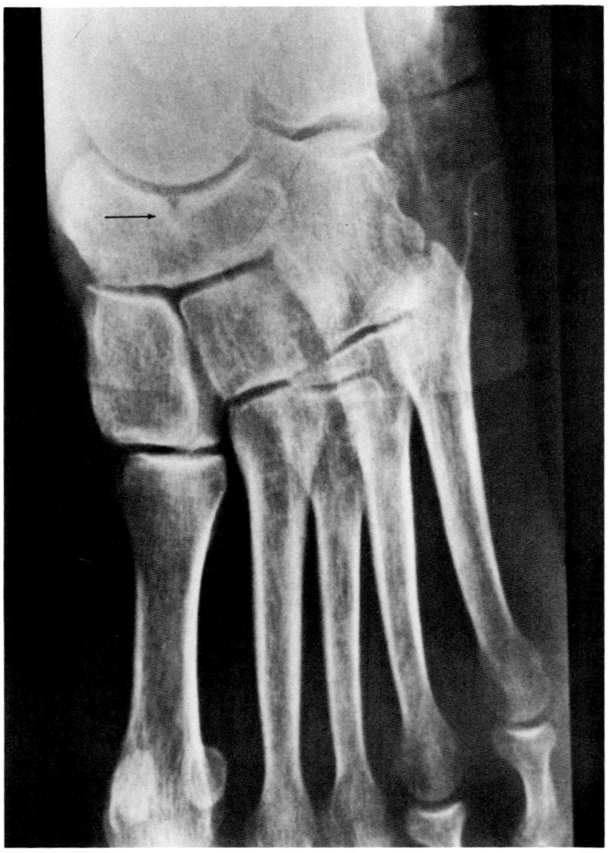

C

FIGURE 5–21
Tarsal navicular stress fracture. A, B. *Bone scan of the foot shows increased uptake in the tarsal region, most prominent around the navicular.* C. *Radiograph of the foot shows a stress fracture of the navicular* (arrow).

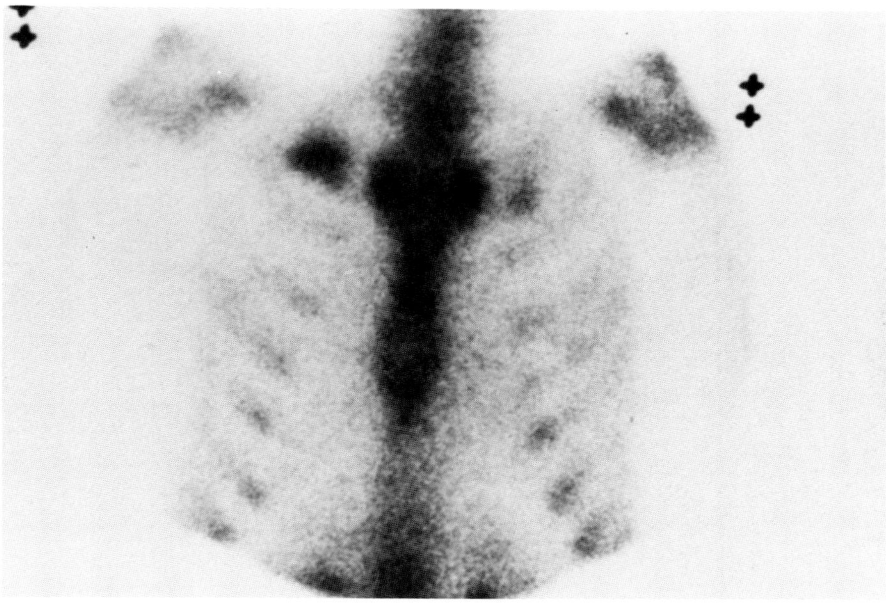

A

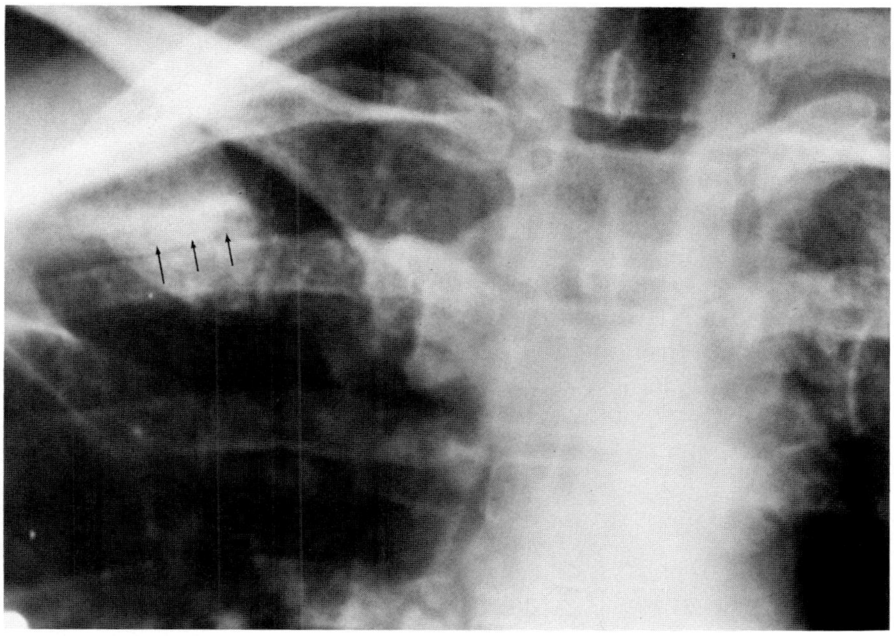

B

FIGURE 5–22

Stress fracture of the first rib. A. *A bone scan shows high uptake in the anterior aspect of the right first rib.* B. *A radiograph shows sclerosis in the anterior aspect of the rib* (arrows).

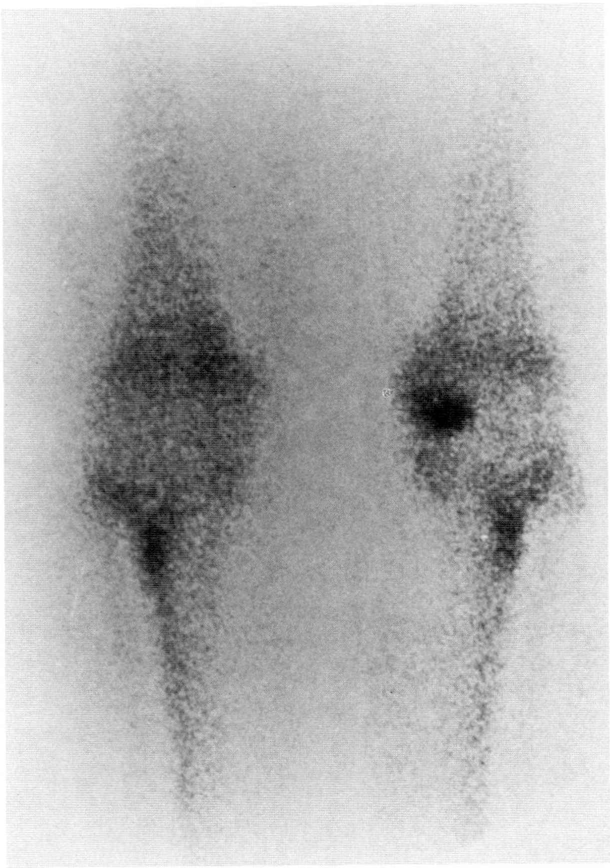

A

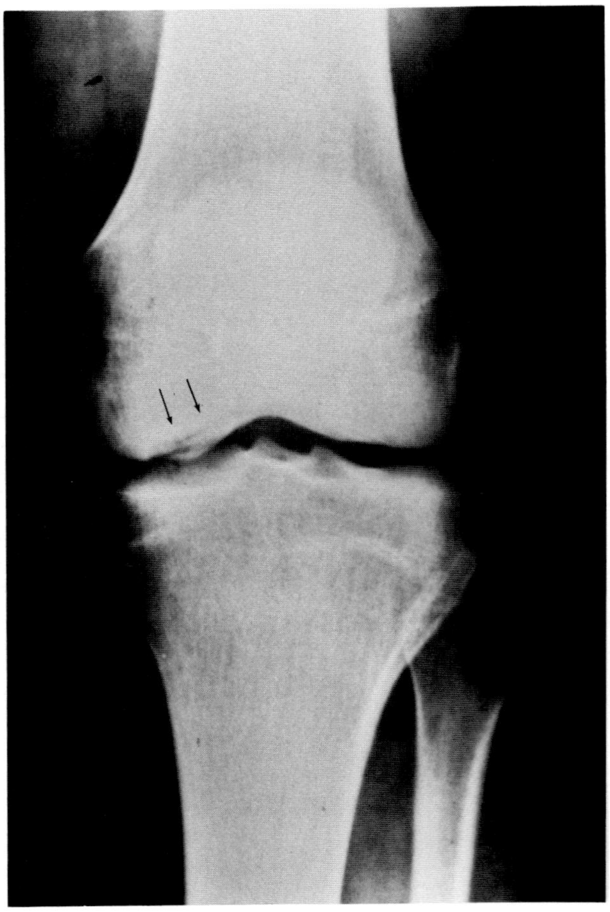

B

FIGURE 5–23
Osteochondritis dissecans of the knee. A. *The bone scan shows a small focal area of increased uptake in the left medial femoral condyle.* B. *A radiograph shows a bony fragment separated from the articular surface of the femur but not loose in the joint in the medial femoral condyle near the notch* (arrows).

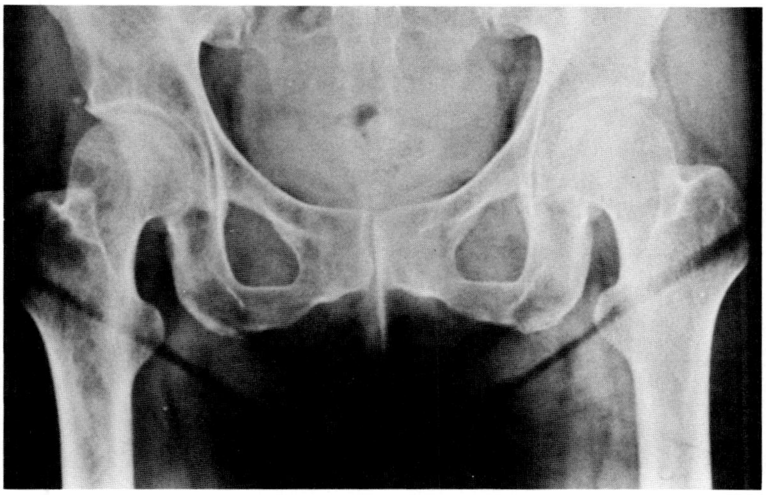

A

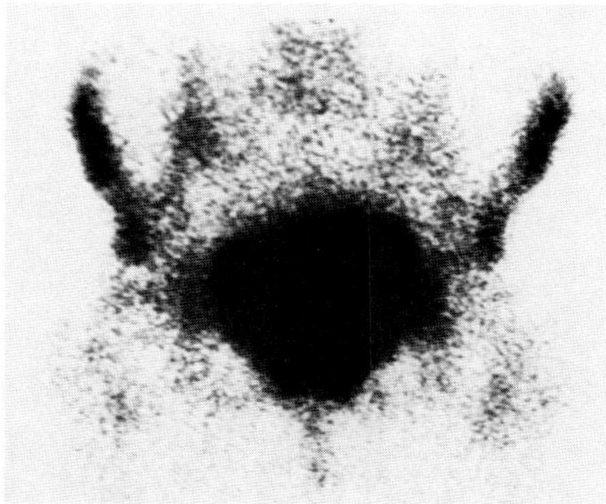

B

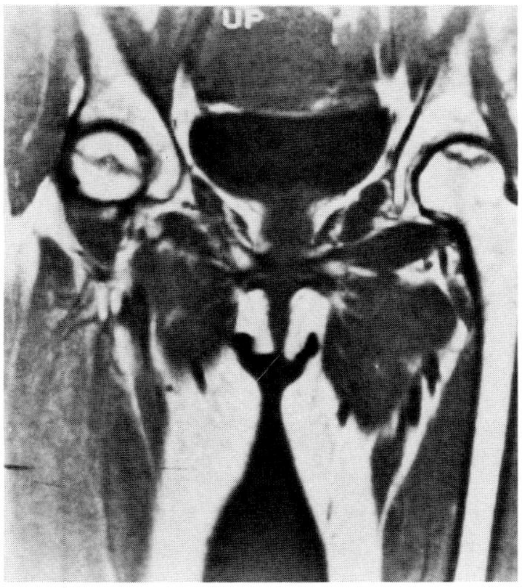

C

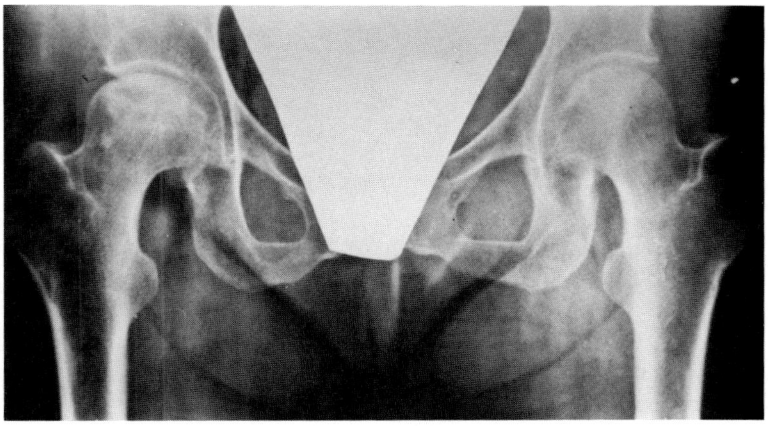

D

FIGURE 5–24
Osteonecrosis of the femoral head. A. *A radiograph of the pelvis (12/21/88) was normal.* B. *Bone scan (12/29/88) was normal.* C. *T1-weighted MRI scan (12/29/88) showed focal abnormalities in the superior aspects of the right and left femoral heads, with a rim of decreased signal intensity.* D. *Radiograph of the pelvis made 2 years later (1/29/90) showed sclerosis in the right femoral head with collapse of the articular surface, diagnostic for osteonecrosis. There is faint sclerosis and lucency in the left femoral head but no collapse of the cortex.*

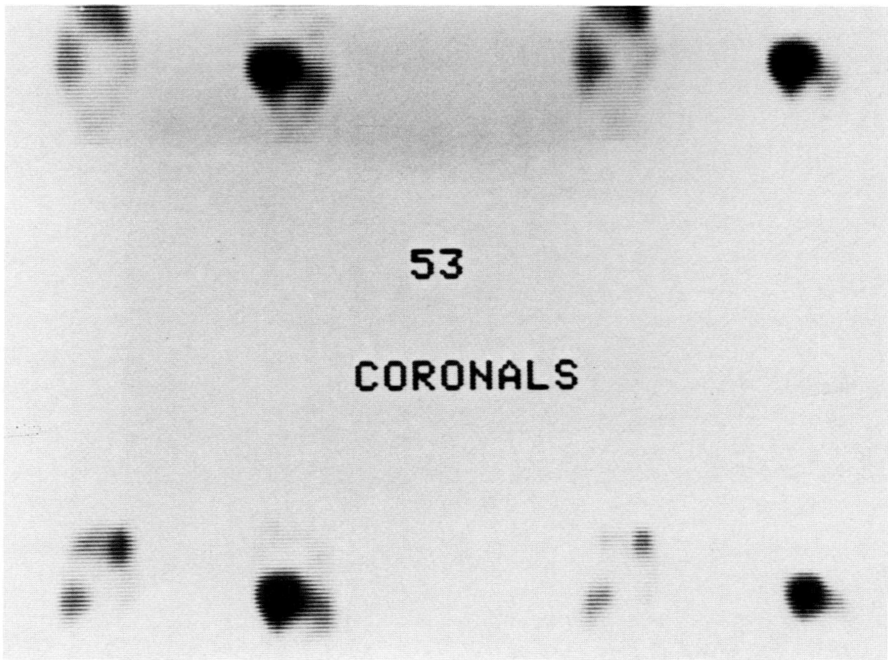

FIGURE 5–25
Osteonecrosis of the knees associated with corticosteroids. Coronal views of a SPECT scan of the knee show areas of increased and decreased uptake in the tibial plateaus and femoral condyles.

In the hip, there are several patterns of osteonecrosis (Figure 5–26). There may be a focal area of increased uptake in the superior and lateral aspect of the femoral head or in the entire femoral head, or there may be a rim of increased uptake surrounding an area of normal and perhaps decreased uptake in the femoral head. This last pattern is virtually diagnostic of osteonecrosis. The former patterns are nonspecific and may be seen in other conditions. In the knee, nontraumatic osteonecrosis is most frequent in the lateral femoral condyle, followed closely by the medial femoral condyle, and then the medial and lateral tibial plateaus. Nontraumatic osteonecrosis of the ankle and foot is most common in the proximal part of the talus and less frequent in the distal aspect and in the metatarsal heads. Osteonecrosis may affect the metacarpal heads and carpal bones (Figure 5–27).

SPONTANEOUS OSTEONECROSIS

Spontaneous osteonecrosis of the knee is a condition that differs from nontraumatic osteonecrosis in that there is no underlying cause [26]. It occurs in elderly individuals from the ages of 60 and older. It presents clinically with a sudden onset of knee pain and almost exclusively affects the weight-bearing aspect of the medial femoral condyle. Bone scans show a focal area of increased uptake in the medial femoral condyle (Figure 5–28). Due to hyperemia and synovitis, there is usually also mild diffuse increased uptake in the entire knee. Radiographs may appear normal initially; however, in most cases by 6 months, an area of sclerosis and lucency, collapse of the articular surface, or a lesion similar to that seen in osteochondritis dissecans with a small bony fragment occurs. There is often a fairly rapid progression to secondary osteoarthritis.

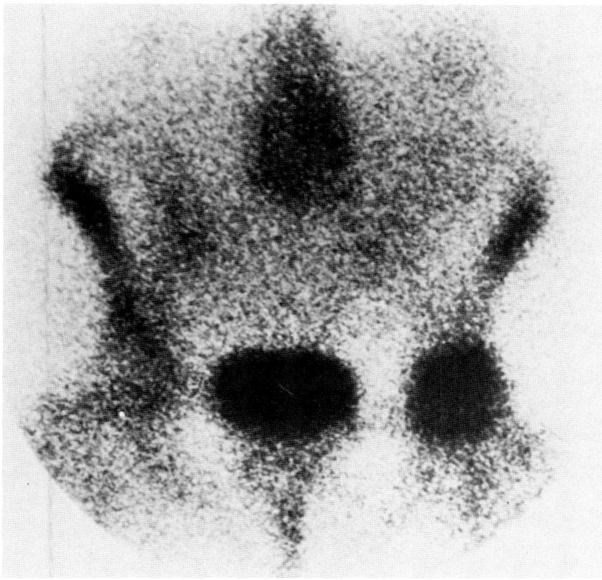

A

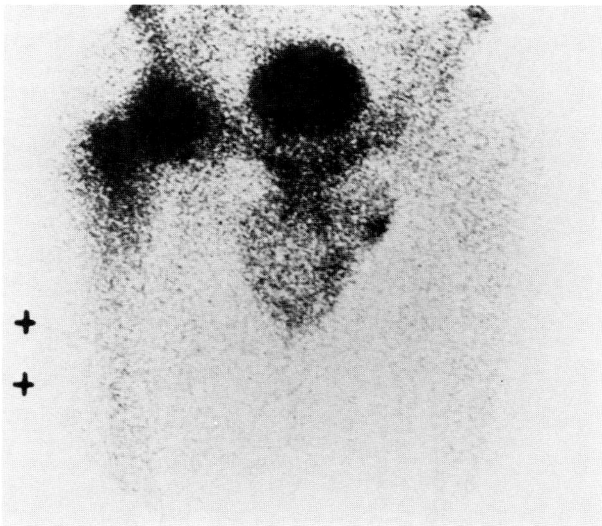

B

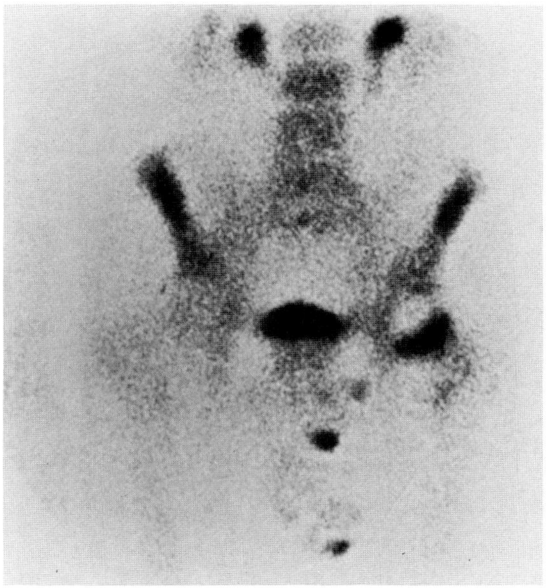

C

FIGURE 5–26
Bone scan patterns of osteonecrosis of the femoral head. A. High uptake in the superior aspect of the femoral head. B. High uptake in the entire femoral head and neck. C. Rim of increased uptake in the femoral head surrounding an area of decreased or normal uptake.

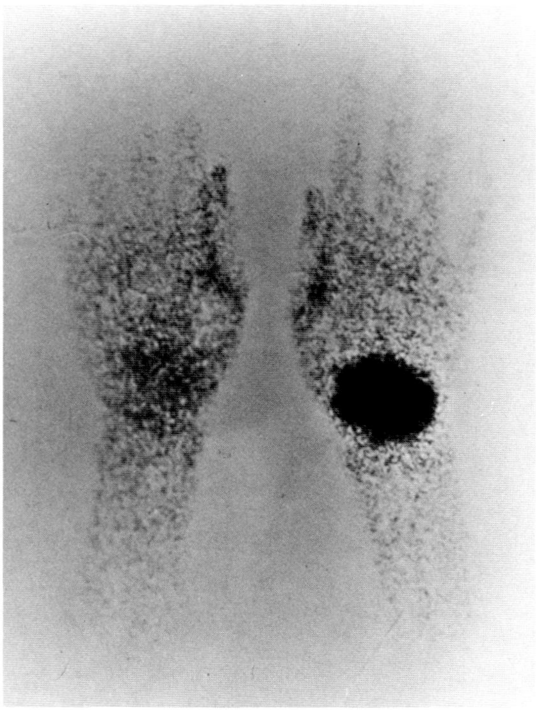

A

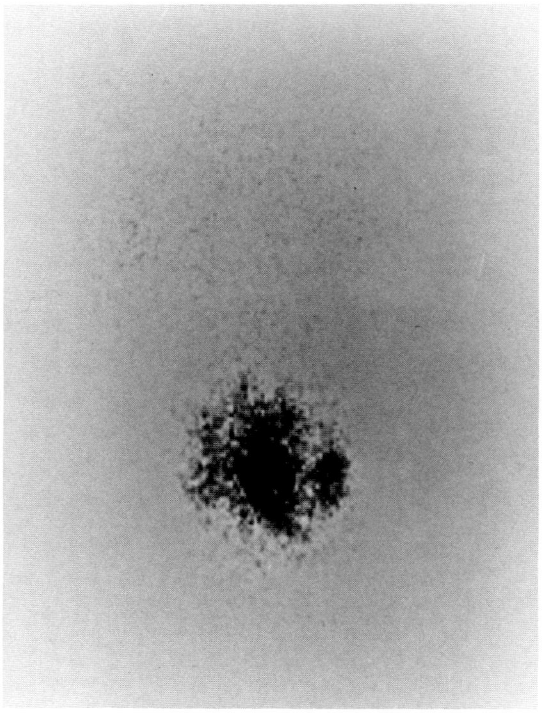

B

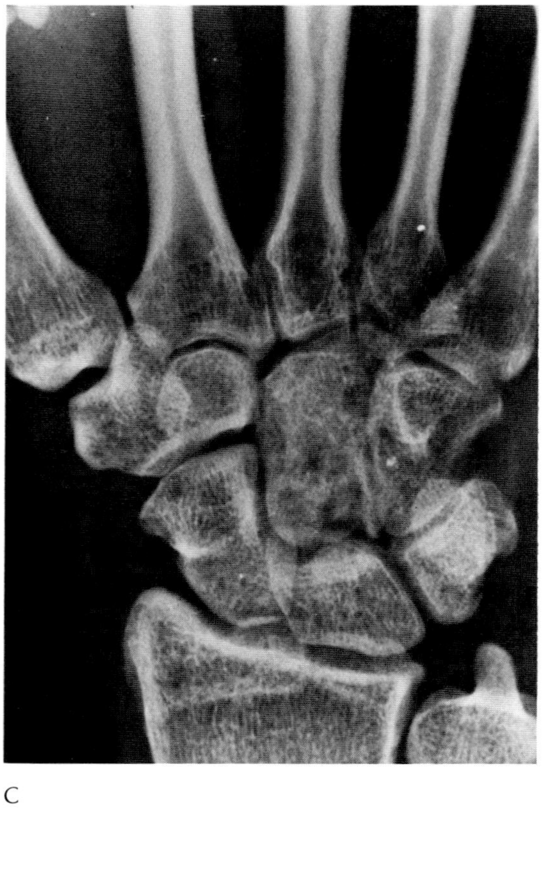

C

FIGURE 5–27
Osteonecrosis of the capitate. A. A bone scan shows high uptake in the wrist. B. A lighter view of the bone scan of the wrist shows that the area of highest uptake is in the proximal portion of the capitate. C. A radiograph shows sclerosis and deformity of the capitate.

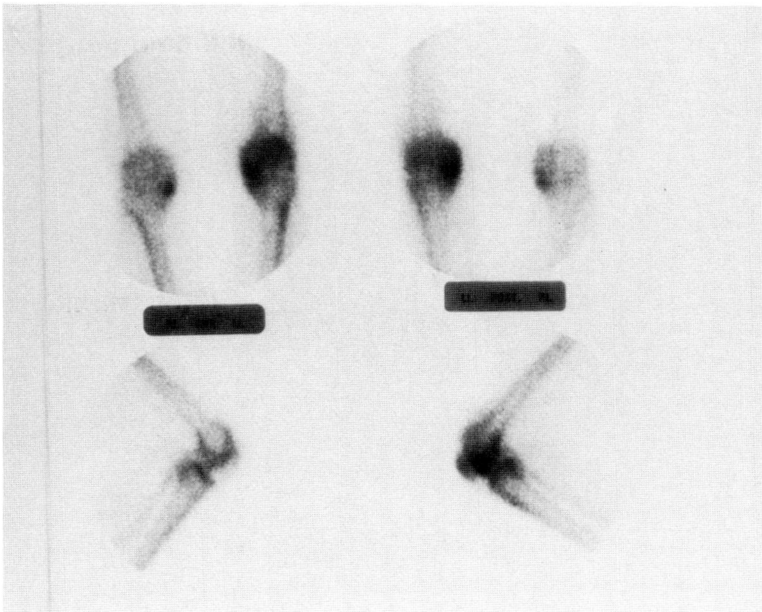

A

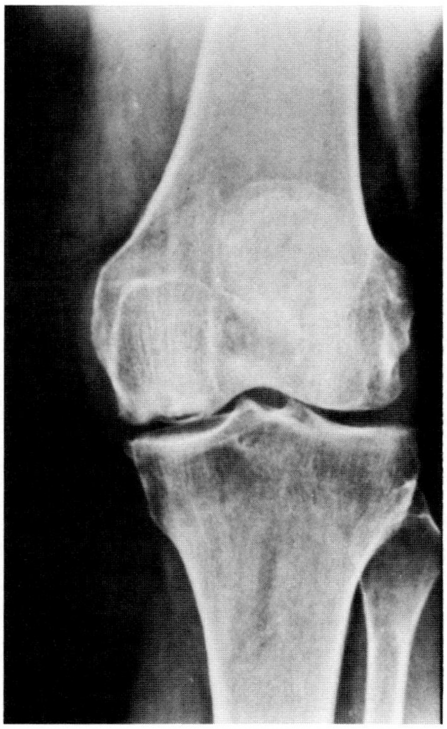

B

FIGURE 5–28
Spontaneous osteonecrosis of the left knee. A. *A bone scan of the knee shows a focal area of high uptake in the left medial femoral condyle. There is also mild diffuse increased uptake in the entire left knee, and mild increased uptake in the medial compartment of the right knee, most prominent in the right medial tibial plateau due to osteoarthritis.* B. *A radiograph of the knee shows osteonecrosis in the weight-bearing aspect of the medial femoral condyle, with flattening of the condyle and a bony fragment present.*

Spontaneous osteonecrosis can be differentiated from osteoarthritis in that the area of highest uptake of radionuclide is in the medial femoral condyle, whereas in osteoarthritis the tibial plateau usually shows higher uptake.

POSTTRAUMATIC OSTEONECROSIS

Posttraumatic osteonecrosis is due to interruption of the blood supply due to acute trauma. Acute interruption of the blood supply causing osteonecrosis can occur in dislocations of the hip and ankle or fractures of the femoral neck, talus, and scaphoid. Symptoms from posttraumatic osteonecrosis can occur months after the fracture when bone collapse occurs. The bone scan, however, can show decreased vascularity and uptake at the affected site at the time of the fracture, indicating disruption of the blood supply (Figure 5–29). Later, revascularization and bone repair occur, causing increased vascularity and uptake in the affected site.

Transient Osteoporosis

Transient osteoporosis is a condition of unknown origin and presents with sudden onset of pain in one of the joints of the lower extremity with no known underlying or predisposing abnormality. This condition occurs most frequently in middle-aged and elderly individuals, in men more frequently than in women [27]. Initial radiographs are often normal. The clinical findings are nonspecific, so the diagnosis is often first suspected only after a bone scan is done. Abnor-

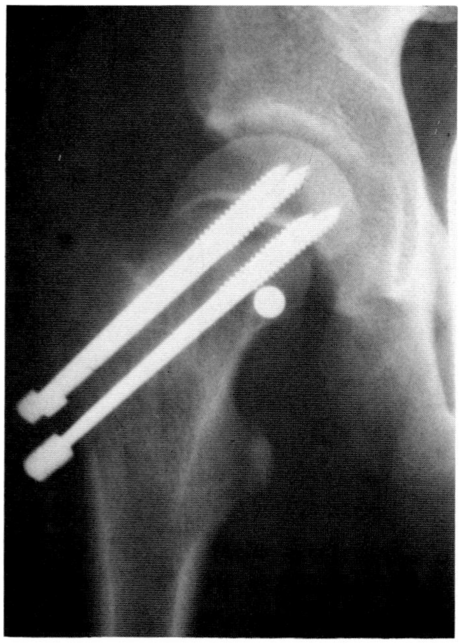

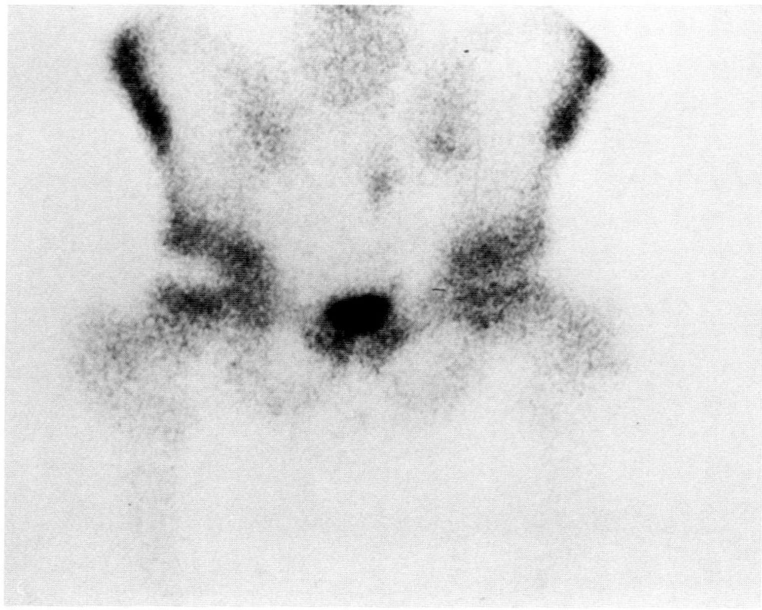

A B

FIGURE 5–29
Posttraumatic osteonecrosis of the femoral head. A. Knowles pins are present through the femoral neck in a child who had posterior dislocation of the hip and a fracture through the growth plate. B. A bone scan shows decreased uptake in the entire femoral head, indicating that it is avascular. The femoral head later went on to collapse and deformity.

mal findings are striking, showing marked increased vascularity and uptake at the affected site.

The most frequent site of involvement is the hip, with high uptake localized to the femoral head (Figure 5–30). In the knee, high uptake is localized to either the medial or the lateral femoral condyle (Figure 5–31). In the hip and knee, the bone scan findings may be identical to those of osteonecrosis. Later in the course of the disease, localized osteoporosis is present and involves the affected bone. After several months, osteoporosis becomes generalized, involving the entire joint. This helps differentiate transient osteoporosis from osteonecrosis. In MRI scans, transient osteoporosis shows decreased signal intensity on T1-weighted images, diffusely involving the femoral head and neck or femoral condyle. With progressive T2 weighting, signal intensity increases. On MRI scanning in early osteonecrosis, a segmental lesion involving a portion of the femoral head or femoral condyle occurs. Transient osteoporosis can also involve the ankle or foot. In the foot, increased uptake is usually seen in the tarsal region and in the heads of the metatarsals (Figure 5–32). A migratory pattern occurs in some patients, with symptoms developing in other joints of the lower extremity at a later date. Spontaneous remission occurs within several months to a year, with return of the bone scan to normal and reossification on radiographs.

Reflex Sympathetic Dystrophy Syndrome

Reflex sympathetic dystrophy (RSD) syndrome is a condition in which pain, swelling, autonomic dysfunction, vasomotor instability, and trophic changes in the skin, hair, and nails occur around

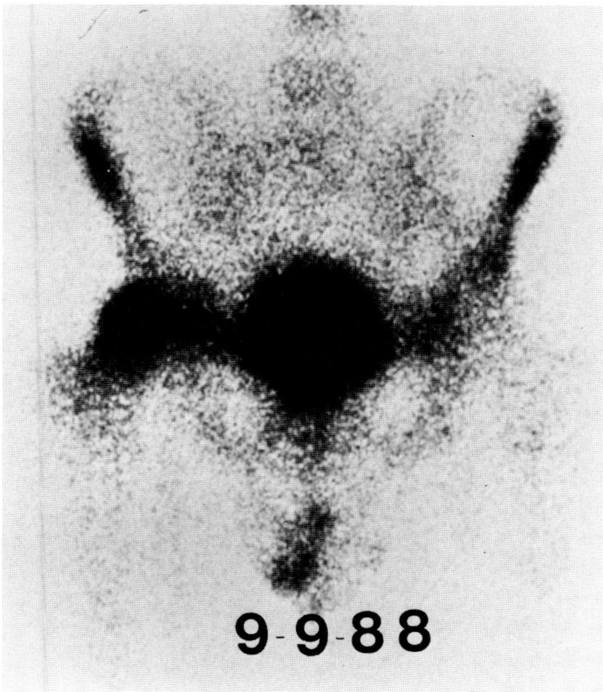

A

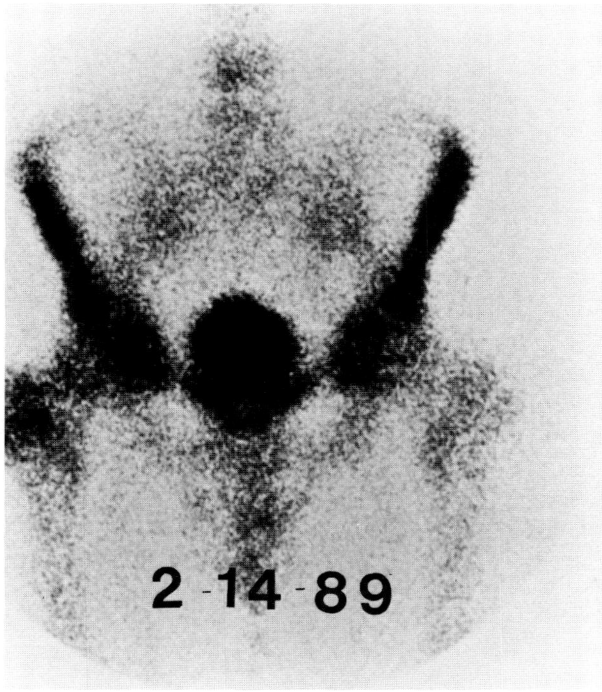

B

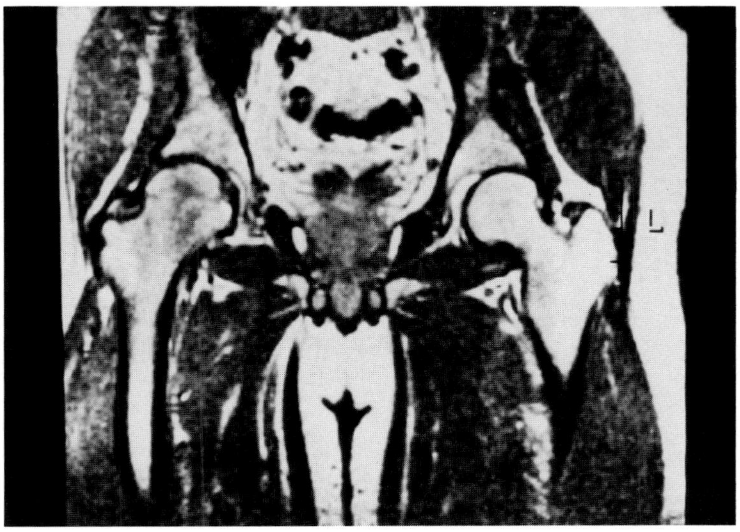

C

FIGURE 5–30

Transient osteoporosis of the hip. A. There is increased uptake in the femoral head on a bone scan (9/9/88). B. A T1-weighted MRI scan shows patchy diffuse decreased signal intensity in the right femoral head and neck. C. A bone scan (2/14/89) after the patient was asymptomatic shows normal uptake in the femoral head. There is mild increased uptake in the greater trochanter at the site of a biopsy.

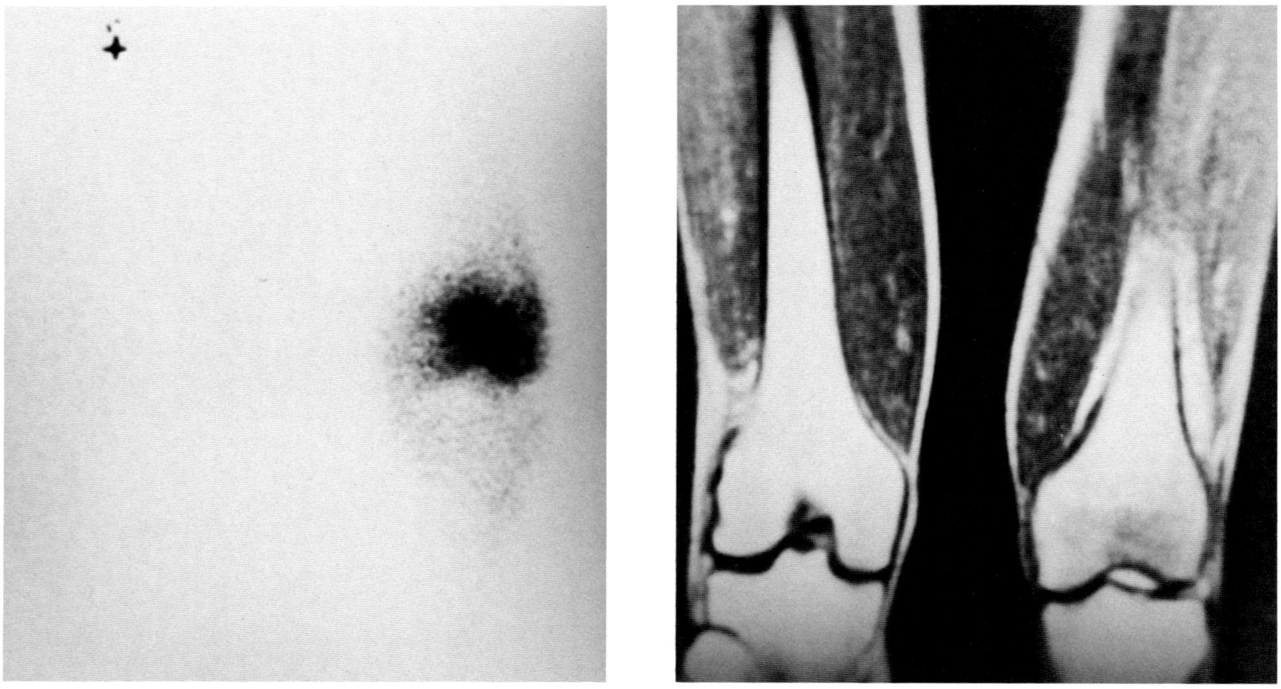

A B

FIGURE 5–31
Transient osteoporosis of the knee. A. *Bone scan shows increased uptake in the left lateral femoral condyle.* B. *T1-weighted MRI scan shows diffuse decreased signal intensity in the lateral femoral condyle.*

a joint. Osteoporosis often develops and is known as Sudeck's atrophy. It is usually associated with some underlying cause, including previous trauma, infection, myocardial infarction, peripheral neuropathy, or central nervous system disorder. The findings on bone scan in RSD syndrome vary and are nonspecific [28,29]. The findings on three-phase bone scans may vary according to the stage of the disease. In the acute phase of the disease, there is increased vascularity and uptake. In the dystrophic phase, there may be normal vascularity on the early phase scan, and increased uptake on the late phase. While in the atrophic phase, there may be decreased vascularity on the early phase scan and normal uptake on the late phase. Abnormal increased uptake affects the entire joint, diffusely (Figure 5–33). This contrasts with transient osteoporosis, which is localized, affecting only a portion of the joint. RSD syndrome also has a tendency to involve all of the joints of the affected extremity. The diagnosis of RSD is based mainly on clinical findings supported by the bone

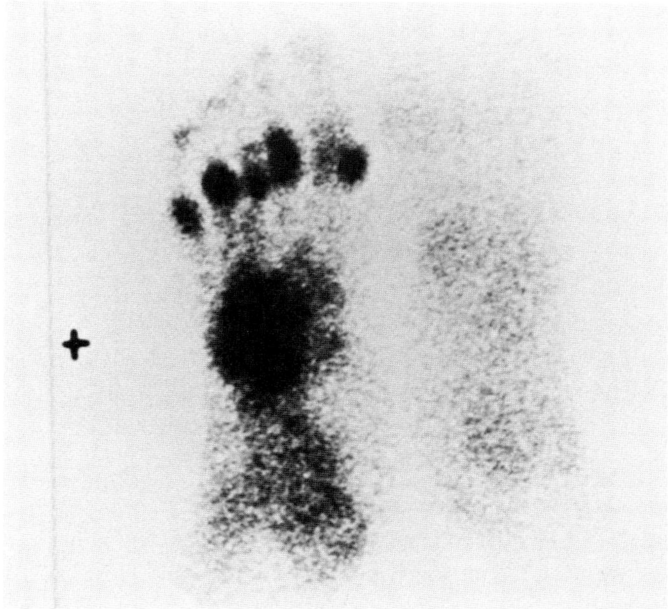

A

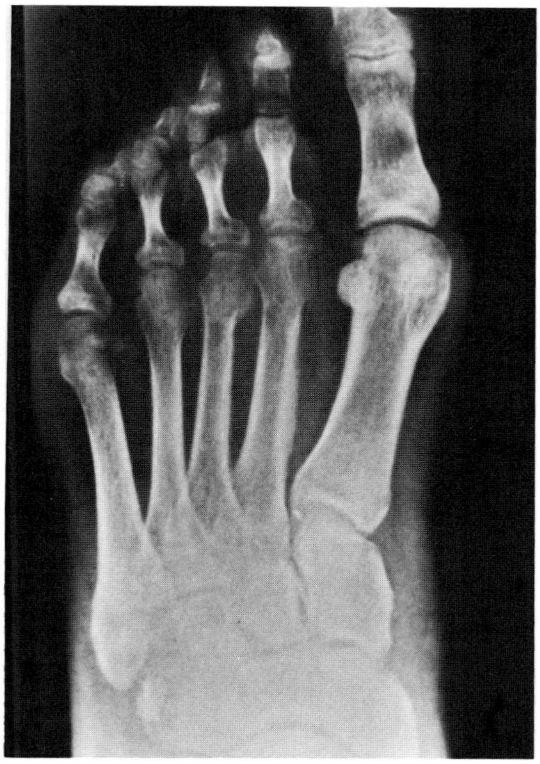

B

FIGURE 5–32
Transient osteoporosis of the foot. A. A bone scan shows increased uptake in the tarsal and metatarsal phalangeal regions of the foot. B. A radiograph shows osteoporosis of the foot.

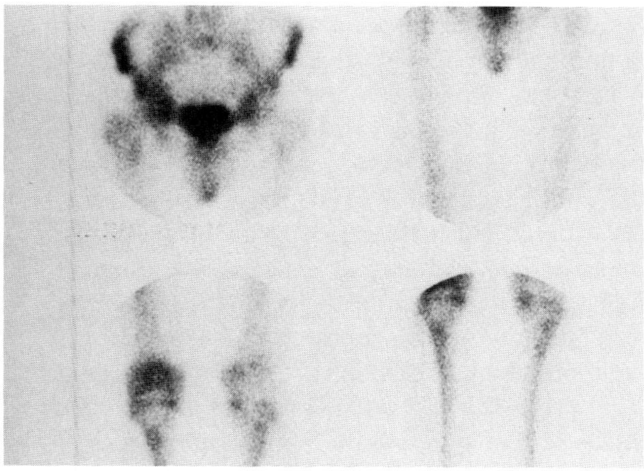

A

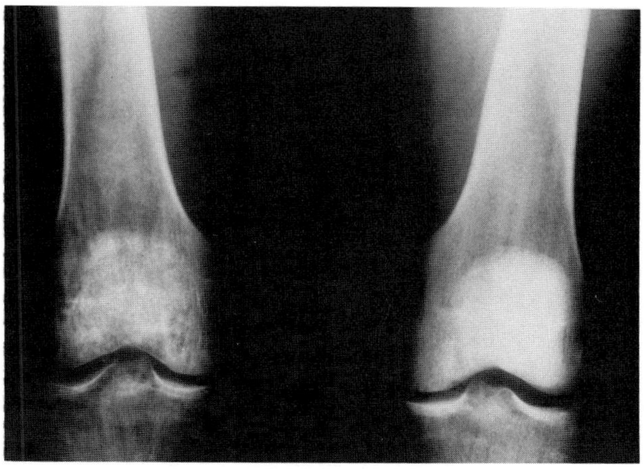

B

FIGURE 5–33
Reflex sympathetic dystrophy syndrome of the knee. A. There is increased uptake diffusely in the right knee and, to a lesser intensity, in the right hip. B. A radiograph shows osteoporosis of the right knee.

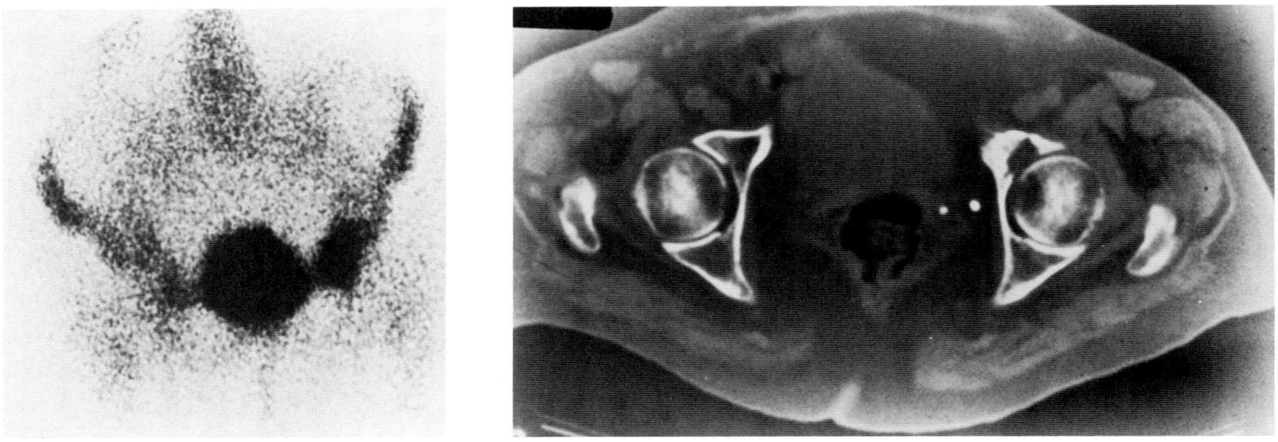

A B

FIGURE 5–34
Metastatic disease to the acetabulum. A patient with a history of breast carcinoma presented with left hip pain. A. There is high uptake in the left acetabulum on a bone scan. B. CT scan shows a mixed sclerotic and lytic lesion in the anterior aspect of the left acetabulum.

scan. The bone scan can be used to follow the course of the disease during treatment, which includes medication, exercise, and sympathetic block.

Neoplastic Disease

METASTATIC DISEASE

Metastatic disease may be the cause of joint pain and has to be ruled out in elderly patients who have severe and unremitting pain. Metastatic disease causing joint pain is most frequent in the hips and shoulders and is only rarely the cause of joint pain distally. Metastatic disease involves the axial skeleton more than the appendicular. Pain in the hips may be caused by metastatic lesions in the acetabulum, pubic, or ischial bones or proximal aspects of the femurs (Figure 5–34). Pain in the shoulders may be due to metastatic disease in the scapula, ribs, or proximal region of the humeri. Solitary metastatic lesions are not infrequent, but more frequently metastatic disease involves multiple sites in the skeleton in a random pattern [30]. Abnormal bone scan findings can identify the site for percutaneous needle or open surgical biopsy when radiographs are normal. Whole-body bone scans help in choosing the easiest site to biopsy [31].

PRIMARY TUMORS

Primary malignant bone or soft tissue tumors around the joint may simulate joint pain. Malignant bone tumors almost always cause increased vascularity and uptake. Benign tumors may have normal or increased uptake. Malignant bone tumors around the pelvic and scapular bones may be difficult to identify initially on radiographs. In elderly patients, chondrosarcomas are the most frequent primary malignant bone tumor in the pelvis. In younger patients, Ewing's sarcomas can involve the flat bones, such as the scapula and iliac bones, causing shoulder or hip pain (Figure 5–35). CT or MRI scans following the bone scan may show the bony lesion and soft tissue mass associated with these tumors. Soft tissue neoplasms that do not involve the bone can show increased vascularity on the early phase scan and only very minimal increased uptake in the soft tissues on the late phase (see Figure 5–2).

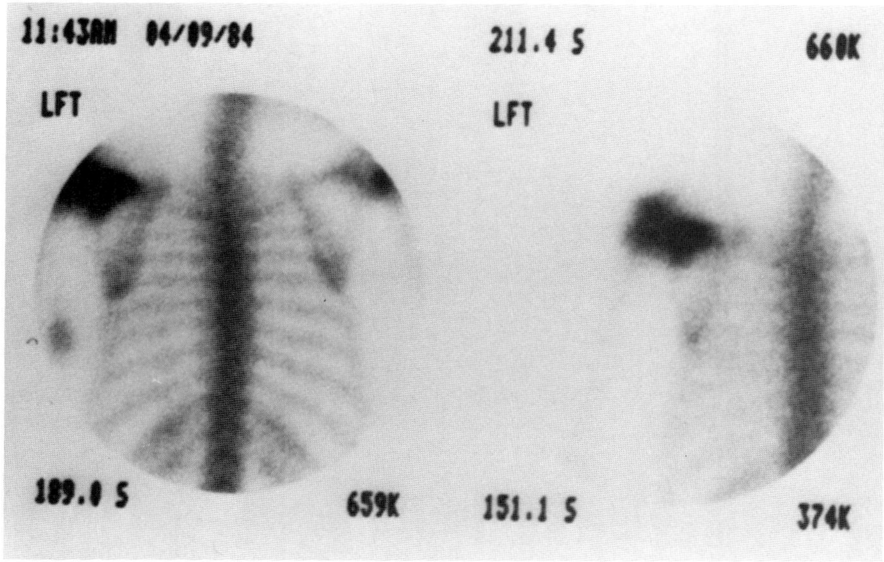

A

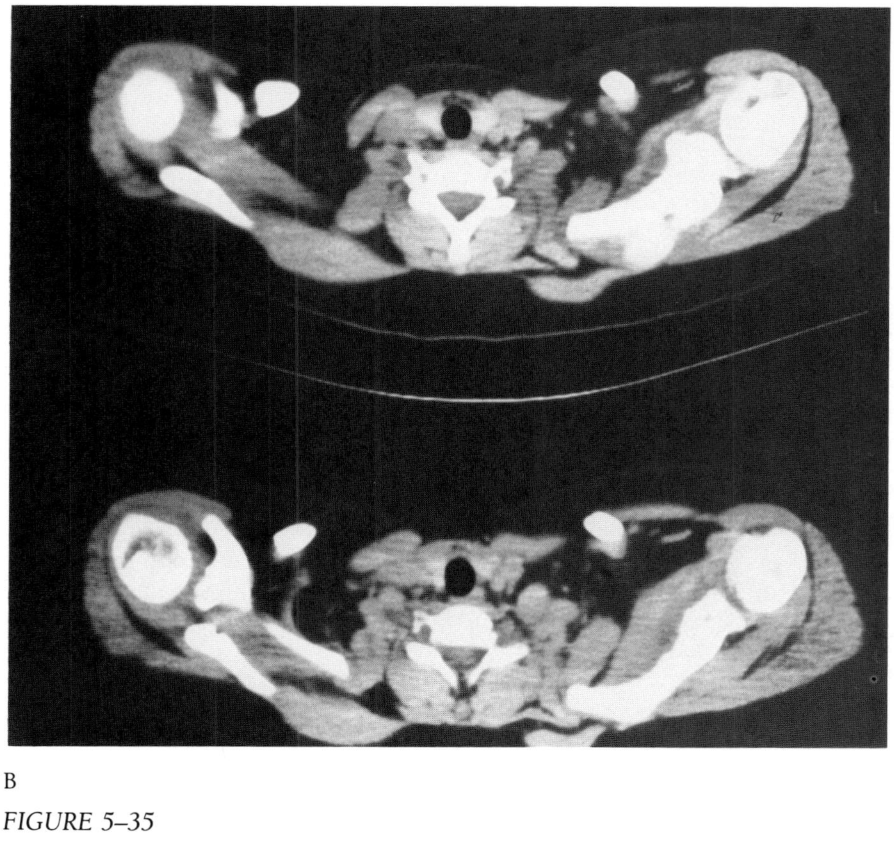

B

FIGURE 5–35

Ewing's sarcoma of the scapula. A. A bone scan shows increased uptake in the left scapula around the glenoid in a patient with normal radiographic findings. B. CT scan shows a large soft tissue mass arising from the left scapula.

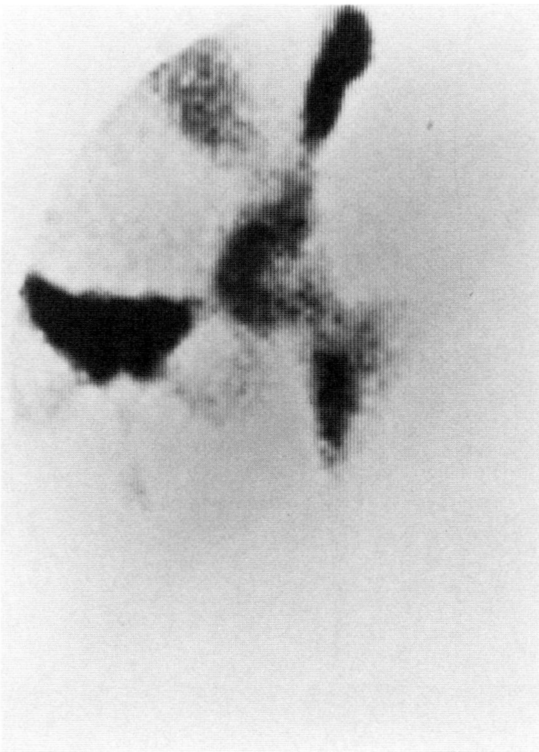

A

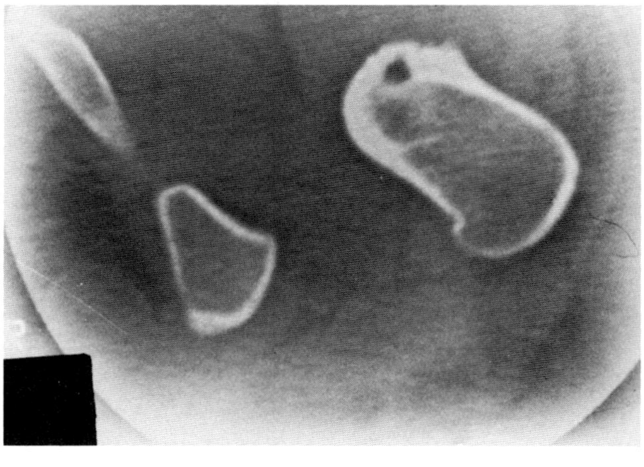

B

FIGURE 5–36
Osteoid osteoma of the femoral neck. A. *A bone scan shows increased uptake in the left femoral neck.* B. *CT scan shows small round lytic lesion in the left femoral neck, representing an osteoid osteoma.*

Benign tumors that involve the epiphyses, causing joint pain, include giant-cell tumor and chondroblastoma. Both cause increased vascularity and uptake on a bone scan but are usually easily visible on radiographs.

Intra-articular osteoid osteomas can simulate arthritis by causing synovitis, as well as bone pain. Bone scans usually show osteoid osteomas as a focal area of high uptake (Figure 5–36) [32]. The nidus of the osteoid osteoma, which is the actual tumor, shows the highest uptake, whereas the bone sclerosis that occurs due to the osteoid osteoma shows less intense increased uptake than the nidus [33]. Osteoid osteomas in the intracapsular region of the hip can cause diffuse increased uptake in the hip and proximal area of the femur, with a focal area of high uptake not readily visible. CT scanning using thin sections will help localize the exact site of an osteoid osteoma. Osteoid osteomas of the femoral neck have to be differentiated from stress fractures, which may have a similar appearance on bone scans and radiographs.

Infection

Acute hematogenous osteomyelitis is generally a disease of children and occurs in the metaphyseal region of long bones. Clinical findings of acute osteomyelitis, which include fever, high white blood cell count, toxemia, and tenderness, may or may not be present. Radiographs are usually normal at the onset of symptoms. Three-phase bone scans that show localized increased vascularity and uptake in the metaphyseal region of a long bone in a child are virtually diagnostic (Figure 5–37). Osteomyelitis should show increased uptake in the arterial as well as the capillary, venous, and blood-pool phases. Occasionally a "cold" area of decreased uptake may be seen due to ischemia (Figure 5–38) [34].

Contiguous osteomyelitis associated with soft tissue lesions is a more difficult diagnostic problem. Increased uptake may be due to periosteal reaction from irritation of bone adjacent to the soft tissue lesion, as well as to infection of the bone itself. Cellulitis without osteomyelitis

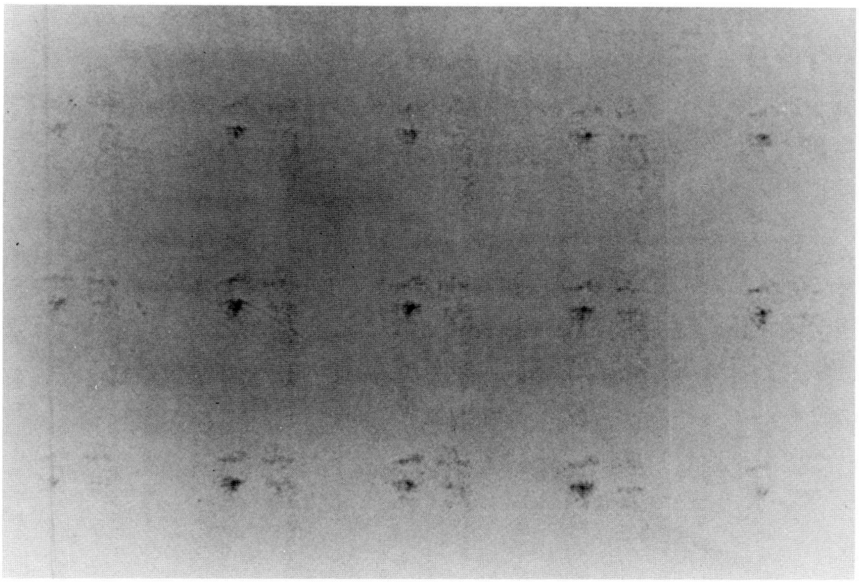

A

FIGURE 5–37
Osteomyelitis of the knee in a child. A, B, C. *Three-phase bone scan shows increased vascularity in the proximal area of the right tibia on the dynamic flow* (A) *and blood-pool* (B) *scans, and increased uptake in the metaphysis just below the growth plate* (arrow) *on the delayed image* (C). *Radiographs were normal.*

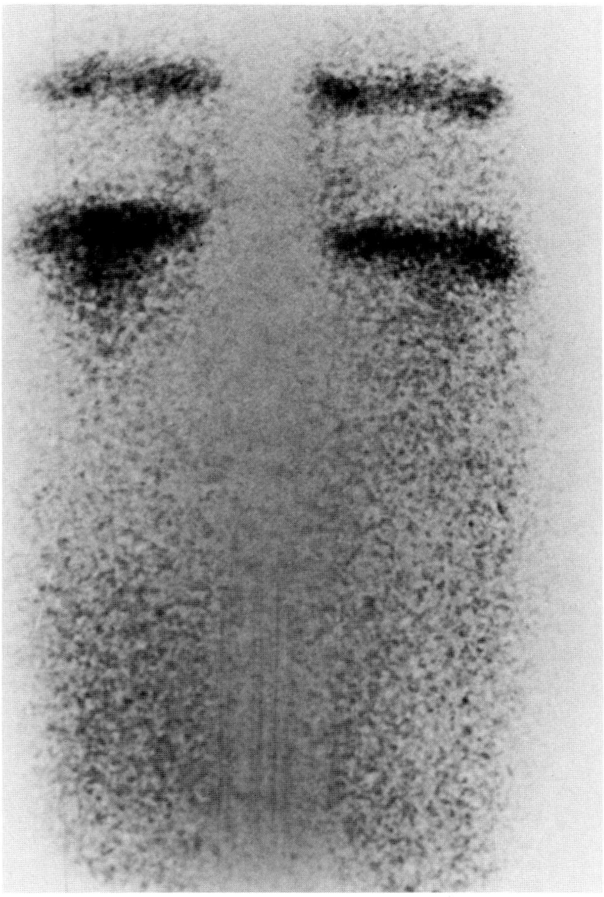

B

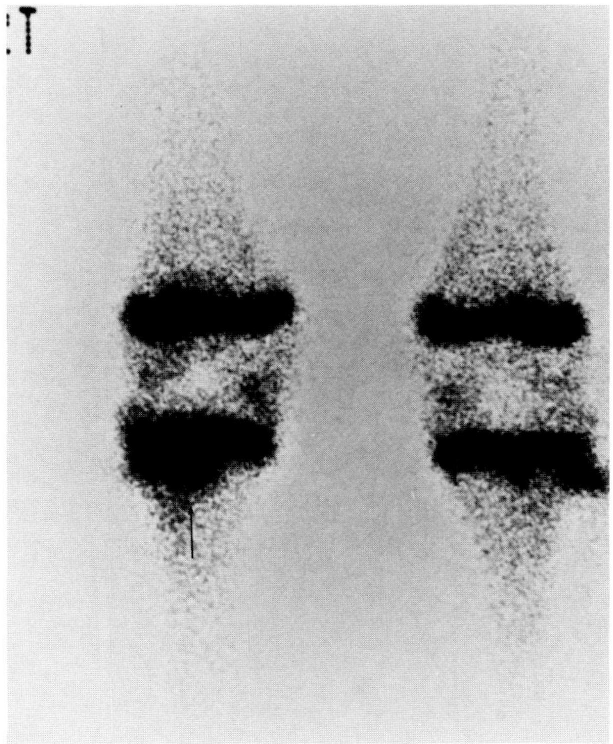

C

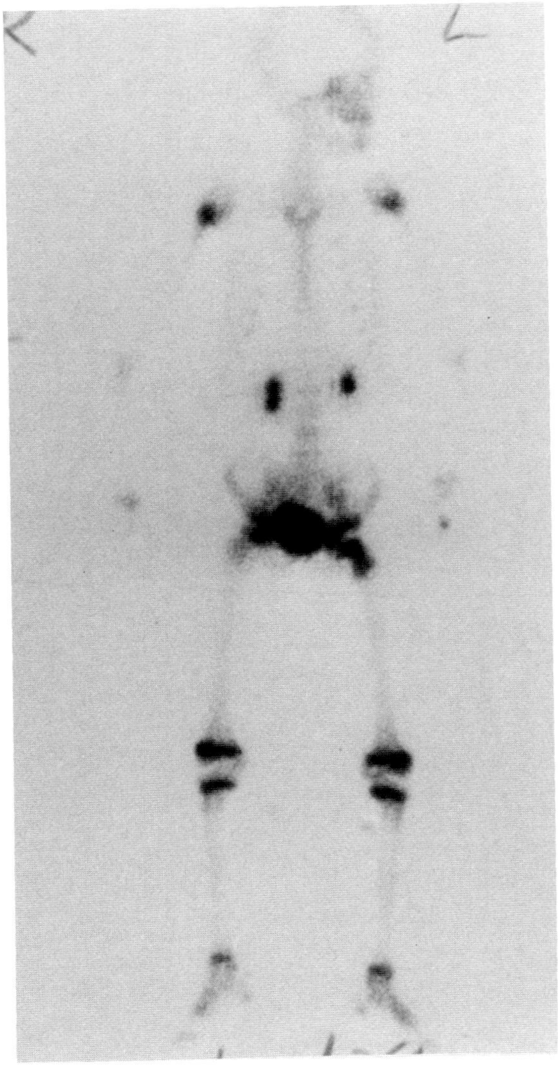

A

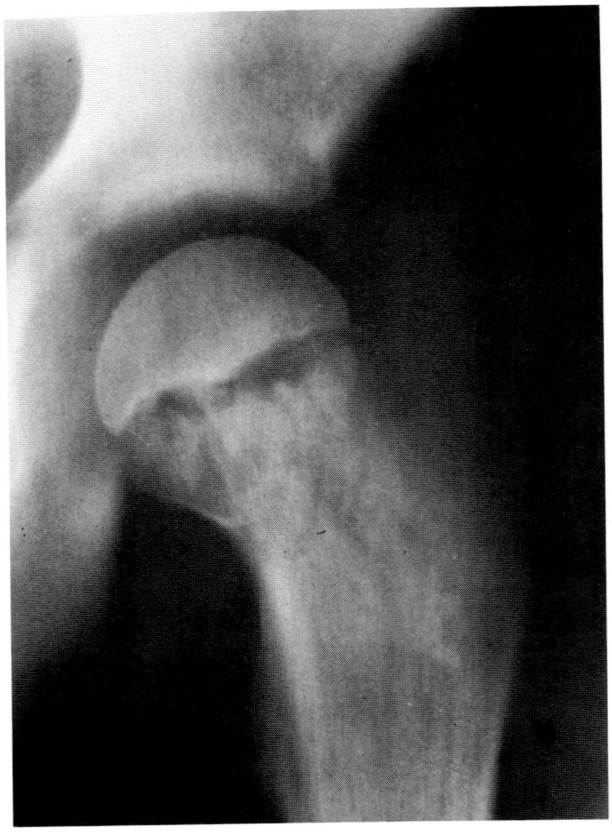

B

FIGURE 5–38
Osteomyelitis of the proximal aspect of the femur with ischemia of the femoral head. A. On a whole-body bone scan, there is increased uptake in the proximal part of the left femur. There is decreased uptake in the left femoral head due to ischemia. B. A tomogram of the left hip shows evidence of osteomyelitis in the femoral neck. There is increased density in the femoral head, indicating avascular necrosis.

causes increased vascularity in the early phase scan and either no or minimal increased uptake in the bone on the late phase scan.

Infectious arthritis causes diffuse increased vascularity and uptake involving the entire joint. This appearance is identical to that seen in noninfectious inflammatory arthritis, such as rheumatoid arthritis. Joint aspiration is needed for diagnosis. Bone scan findings in infectious sacroiliitis are abnormal even before radiographs or CT scans show erosion of the sacroiliac joint (Figure 5–39). Unilateral high uptake in the sacroiliac joint is highly suspicious for infection. An elevated erythrocyte sedimentation rate (ESR) and elevated white blood cell count also support the diagnosis of infection, but an elevated ESR can be present in noninfectious inflammatory arthritides. Noninfectious inflammatory sacroiliitis, however, rarely causes as intense increased uptake in the sacroiliac joint as seen in infection.

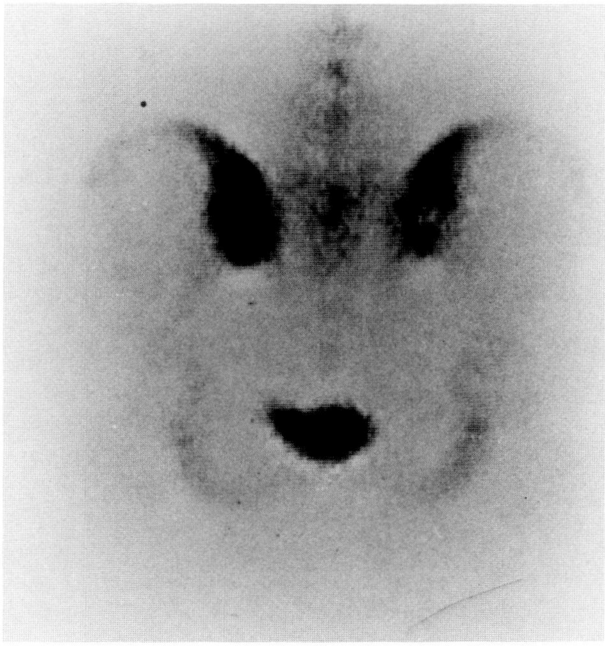

A

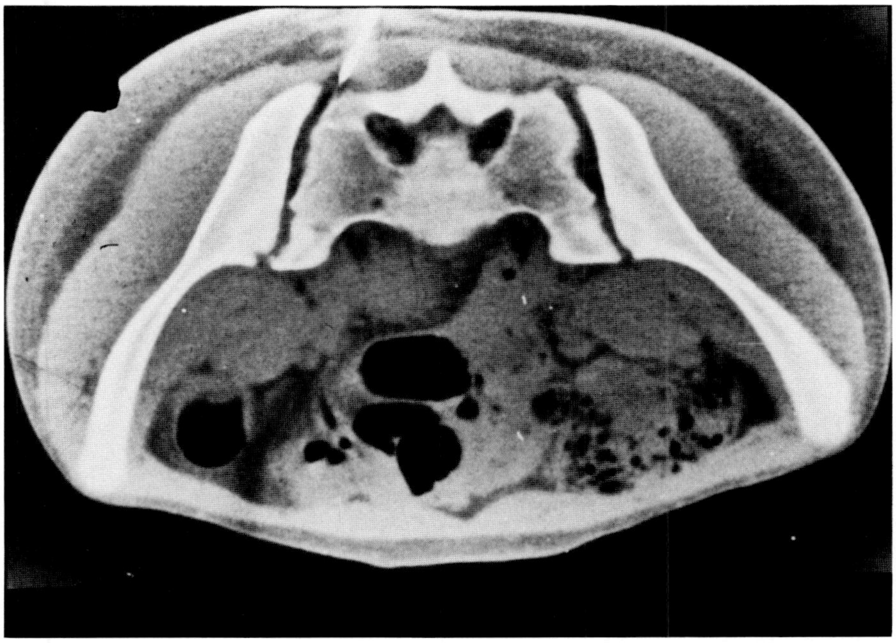

B

FIGURE 5–39
Septic sacroiliitis. A. There is increased uptake in the left sacroiliac joint on a bone scan in a patient with left hip and back pain and normal radiographs. B. CT-guided needle aspiration of the left sacroiliac joint recovered Staphylococcus aureus. *No destruction or erosion of the left sacroiliac joint is seen.*

REFERENCES

1. Maurice HD, Newman JH, Watt I. Bone scanning of the foot for unexplained pain. J Bone Joint Surg [Br] 1987; 69:448–56.
2. Seltzer SE, Weissman BN, Feinberg HJ, Markisz J. Improved diagnostic imaging in joint diseases. Semin Arthritis Rheum 1982; 11:315–30.
3. Sy WM, Bay R, Camera A. Hand images: normal and abnormal. J Nucl Med 1977; 18:419–24.
4. Collier BD, Johnson RP, Carrera GF, et al. Painful spondylolysis or spondylolisthesis studied by radiography and single-photon emission computed tomography. Radiology 1985; 154:207–11.
5. Collier BD, Carrera GF, Johnson RP, et al. Detection of femoral head avascular necrosis in adults by SPECT. J Nucl Med 1985; 26:979–87.
6. Collier BD, Johnson RP, Carrera GF, et al. Chronic pain assessed by SPECT: comparison with other modalities. Radiology 1988; 157:795–802.
7. Gillen GJ, McKillop JH, Hilditch TE, Davidson JK, Elliott AT. Digital filtering of the bladder in SPECT bone studies of the pelvis. J Nucl Med 1988; 29:1587–95.
8. Eggleston DE, Bartold K, Abghari R. Recognition of renal osteodystrophy on bone imaging. Clin Nucl Med 1985; 11:543–4.
9. Fogelman I, McKillop JH, Gray HW. The "hot patella" sign: is it of any clinical significance? J Nucl Med 1983; 24:312–5.
10. Hoffer PB, Genant HK. Radionuclide joint imaging. Semin Nucl Med 1976; 6:121–37.
11. Green FA. Editorial. Joint scintiscans: present status. J Rheumatol 1979; 6:4.
12. Schneider R, Kaye JJ. Insufficiency and stress fractures of the long bones occurring in patients with rheumatoid arthritis. Radiology 1975; 116:595–9.
13. Young A, Kinsella P, Boland P. Stress fractures of the lower limb in patients with rheumatoid arthritis. J Bone Joint Surg [Br] 1981; 63:239–43.
14. Goldberg RP, Genant HK, Shimshak R, Shames D. Applications and limitations of quantitative sacroiliac joint scintigraphy. Radiology 1978; 128:683–6.
15. Newman JH. Non-infective disease of the diabetic foot. J Bone Joint Surg [Br] 1981; 63:593–6.
16. Savoca CL. Stress fracture: a classification of the earliest radiographic signs. Radiology 1971; 100:519–24.
17. Greaney RB, Gerber FH, Laughlin RL, et al. Distribution and natural history of stress fractures in U.S. Marine recruits. Radiology 1983; 146:339–46.
18. Manco LG, Schneider R, Pavlov H. Insufficiency fractures of the tibial plateau. AJR 1983; 140:1211–5.
19. Cooper KL, Beabout JW, McLeod RA. Supraacetabular insufficiency fractures. Radiology 1985; 157:15–7.
20. Schneider R, Yacovone J, Ghelman B. Unsuspected sacral fractures: detection by radionuclide bone scanning. AJR 1985; 144:337–41.
21. Pavlov H, Torg JS, Freiberger RH. Tarsal navicular stress fractures: radiologic evaluation. Radiology 1983; 148:641–5.
22. Papanicolaou N, Wilkinson RH, Emans JB, Treves S, Micheli LJ. Bone scintigraphy and radiology in young athletes with low back pain. AJR 1985; 145:1039–43.
23. Van Den Oever M, Merrick MV, Scott JHS. Bone scintigraphy in symptomatic spondylolysis. J Bone Joint Surg [Br] 1987; 69:453–6.
24. Beltran J, Herman LJ, Burk JM, et al. Femoral head avascular necrosis: MR imaging with clinical-pathologic and radionuclide correlation. Radiology 1988; 166:215–20.
25. Gregg PJ, Walder DN. Scintigraphy versus radiography in the early diagnosis of experimental bone necrosis. J Bone Joint Surg [Br] 1980; 62: 214–20.
26. Aglietti P, Insall JN, Buzzi R, Deschamps G. Idiopathic osteonecrosis of the knee; etiology, prognosis and treatment. J Bone Joint Surg [Br] 1983; 65:588–97.
27. Lakhanpa S, Ginsburg WW, Luthra HS, Hunder GG. Transient regional osteoporosis. Ann Intern Med 1987; 106:444–50.
28. Greyson ND, Tepperman PS. Three-phase bone studies in hemiplegia with reflex sympathetic dystrophy and the effect of disuse. J Nucl Med 1984; 25:423–9.
29. Kozin F, Soin JS, Ryan LM, Carrera GF, Wortmann RL. Bone scintigraphy in the reflex sympathetic dystrophy syndrome. Radiology 1981; 138:437–43.
30. Boxer DI, Todd CEC, Coleman R, Fogelman I. Bone secondaries in breast cancer: the solitary metastasis. J Nucl Med 1989; 301:1318–20.

31. Collins JD, Bassett L, Main GD, Kagan C. Percutaneous biopsy following positive bone scans. Radiology 1979; 132:439–42.

32. Swee RG, McLeod RA, Beabout JW. Osteoid osteoma. Radiology 1979; 130:117–23.

33. Helms CA, Hattner RS, Volger JB. Osteoid osteoma: radionuclide diagnosis. Radiology 1984; 151:779–84.

34. Park HM, Rothschild PA, Kernek CB. Scintigraphic evaluation of extremity pain in children: its efficacy and pitfalls. AJR 1985; 145:1079–84.

6

Ultrasonography of the Musculoskeletal System

CAROL B. BENSON

The development of high-frequency ultrasound transducers has permitted visualization of structures just beneath the skin, with resolution of lesions as small as 1–2 mm. Subcutaneous tissues, muscles, tendons, and bursae can now be evaluated for masses, abnormal fluid collections, and areas of discontinuity. This is particularly useful in evaluating tendons and bursae around joints, but sonography is also used to localize soft tissue swelling and masses and to characterize them as cystic or solid. This real-time imaging modality provides excellent guidance for needle aspiration and biopsy.

On sonographic examination, muscles have a homogeneous echotexture with low-level echoes throughout. Fascial planes are thin, echogenic lines coursing through and around the muscles. Tendon fibers are oriented longitudinally from the body of the muscle to the tendon insertion. On transverse sonographic image, tendons appear more echogenic than muscle. On longitudinal image, when the beam is parallel with the fibers of the tendon, tendons appear slightly less echogenic, but still more echogenic than muscle [1]. Fibrous tissue and scarring within muscles are poorly defined, irregularly shaped areas that have increased echogenicity.

The sonographic beam cannot penetrate bone and is sharply reflected by it. As a result, the edge of bone appears brightly echogenic, with complete absence of echoes behind it. Sonography is occasionally useful in evaluating the surface of an osseous structure, but otherwise plays no diagnostic role in the assessment of bony structures.

Cartilage overlying articular surfaces of bone appears sonographically as a smooth, very hypoechoic band lining the bone. In most joints the articular cartilage is located in sites that are inaccessible to the ultrasound beam. The femoral condylar articular cartilage, however, is one whose anterior and inferior surface can be seen when the knee is flexed, and whose posterior surface can sometimes be seen with extension [2]. Cartilage of neonatal epiphyseal centers is also very hypoechoic and does not obscure structures deep to it. For this reason, sonography can be used to assess the location of the epiphyseal center with respect to surrounding bone and cartilaginous structures, such as the femoral head in the developing acetabulum.

Shoulder

The shoulder is the joint best suited for evaluation by ultrasound. Most of the structures involving the joint or surrounding it, such as the rotator cuff tendons, the biceps tendon, the deltoid muscle, and the subdeltoid bursa, are accessible to the sonographic beam. It is these structures that are most commonly affected by injuries and inflammatory processes. The tendons can be evaluated for tears, ruptures, or inflammation; the bursa, for inflammation.

High-frequency transducers (5–10 MHz) are applied directly to the shoulder on a layer of aqueous gel in an orientation perpendicular to the long axis of the humerus, with the beam directed toward the humeral head. Just beneath the transducer is a layer of subcutaneous tissue. Beneath that is a band of muscle, the deltoideus, of homogeneous echotexture (Figure 6–1) contained by echogenic fascial planes. Beneath the

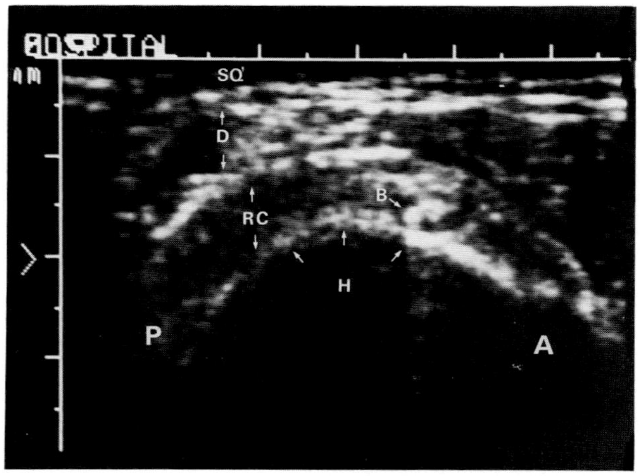

FIGURE 6–1
Transverse image of shoulder with a thin layer of subcutaneous tissue (SQ) overlying the deltoideus muscle (D), over the rotator cuff (RC), draped over the humeral head (H). The biceps tendon (B) is seen within the rotator cuff. A = anterior; P = posterior.

deltoideus, the rotator cuff is a homogeneous band of muscle and tendon draped over the humeral head (see Figure 6–1). The humerus, like bone elsewhere, has a bright edge and casts a shadow behind it.

The rotator cuff surrounds the superior-lateral surface of the humeral head. Its anterior portion is made up of the subscapularis, which inserts on the lesser tuberosity. Laterally the cuff is composed of the supraspinatus tendon and more posteriorly the infraspinatus and teres minor tendons. These three tendons insert on the greater tuberosity. On transverse image, the normal rotator cuff has homogeneous echotexture and a uniform thickness (Figure 6–2) of variable size, depending on the patient body habitus, but usually greater than 6 mm. Posteriorly, the cuff flares into the teres minor muscle (Figure 6–3).

Longitudinal evaluation of the rotator cuff involves rotating the transducer until it lies parallel with the humeral shaft. The rotator cuff has a triangular shape as it emerges from beneath the acromion to converge on its insertions on the tuberosities (Figure 6–4) [3].

The biceps tendon travels in the bicipital groove in the humerus and passes between the

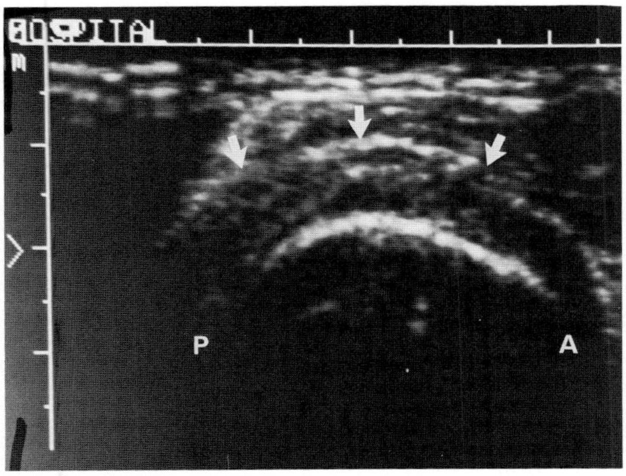

FIGURE 6–2
Rotator cuff. Transverse image showing homogeneous band (arrows) *of the cuff of uniform thickness. A = anterior; P = posterior.*

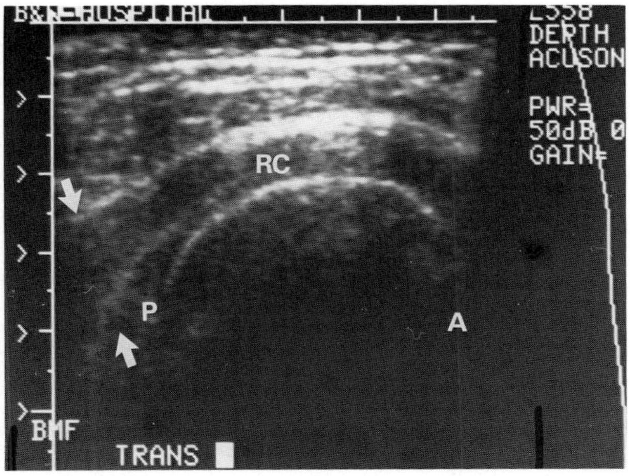

FIGURE 6–3
Posterior rotator cuff. Transverse image of rotator cuff (RC) tendon over the humeral head and flaring posteriorly into the teres minor and infraspinatus muscles (arrows) *from which it arises. A = anterior; P = posterior.*

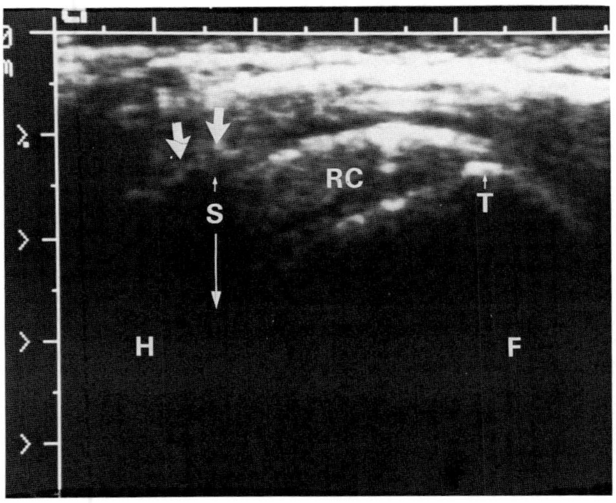

FIGURE 6–4
Rotator cuff. On longitudinal image, the rotator cuff (RC) tendon has a triangular shape converging on the tuberosity (T, arrow). *Proximally, the acromion* (arrowheads) *obscures the tendon, casting an acoustic shadow (S) leaving no echoes behind it. H = head of patient; F = foot.*

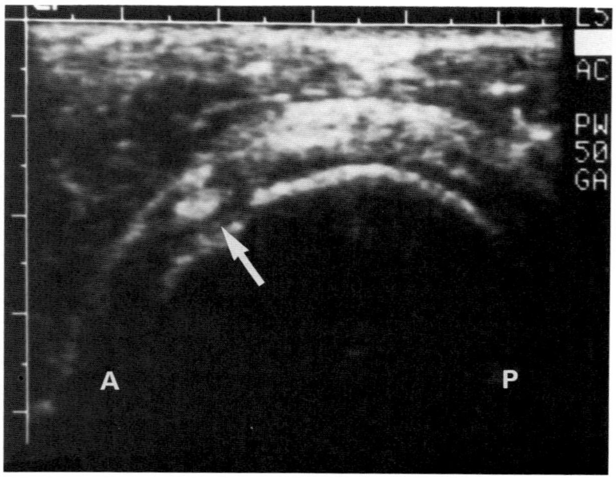

A

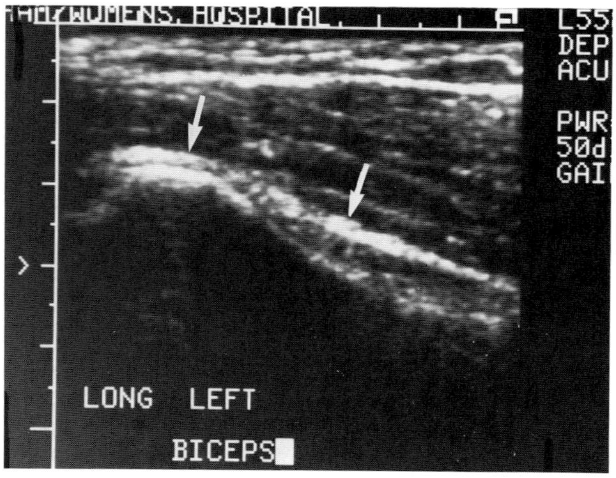

B

FIGURE 6–5
Biceps tendon. A. *Transverse image of rotator cuff with echogenic biceps tendon* (arrow) *within it, passing beneath the rotator cuff to its insertion on the glenoid labrum. A = anterior; P = posterior.* B. *Longitudinal scan of echogenic biceps tendon* (arrows) *running parallel to the humeral shaft in the bicipital groove.*

lesser and greater tuberosities before it dips below the rotator cuff to insert on the glenoid labrum. The tendon is an echogenic circle or ellipse on transverse scan and an echogenic line running parallel to the humeral shaft on longitudinal scan (Figure 6–5) [4].

Tears of the rotator cuff most commonly involve the supraspinatus tendon [5]. Partial and small full-thickness tears have similar sonographic appearances. They cause altered echogenicity in the involved portion of the cuff, usually with focal poorly defined areas of increased echogenicity (Figure 6–6) [3,6,7]. After the acute injury, cystic or hypoechoic areas due to hemorrhage may also be seen (Figure 6–7). The outer border of the cuff may be concave or flattened instead of convex [8]. Another sign of a rotator cuff tear is an area of focal thinning in the midportion of the cuff (Figure 6–8) [9]. Secondary signs of the tear may include fluid in the bicipital tendon sheath, joint effusion, and elevation of the humerus with respect to the acromion [7]. Larger tears of the rotator cuff leave visible defects or discontinuity in the cuff, often with hypoechoic areas due to hemorrhage filling the defect (Figure 6–9) [4,6]. In cases of severe or very large tears, there may be complete absence of visualization of any part of the cuff as a result of retraction of the tendon beneath the acromion [4,8,10].

In experienced hands, sonography is reported to have sensitivities for rotator cuff tears of 90–100% with specificities of 75–100% [4,6–8,10]. The technique does have limitations for diagnosing tears. Patients who have had previous shoulder surgery usually have disruption of the normal tissue planes and distortion of the muscles and tendons that cause difficulties in interpretation of the sonographic findings [11]. Tendonitis involving the rotator cuff can also mimic a rotator cuff tear. This inflammatory process, like rotator cuff tears, most commonly affects the supraspinatus tendon. The tendon may be heterogeneous in the involved area with foci of decreased echogenicity. Chronic tendonitis with calcific deposition produces an irregular echogenic focus that may not shadow, and mimics a rotator cuff tear [9]. Dense calcific deposits due to tendonitis, however, have a characteristic ap-

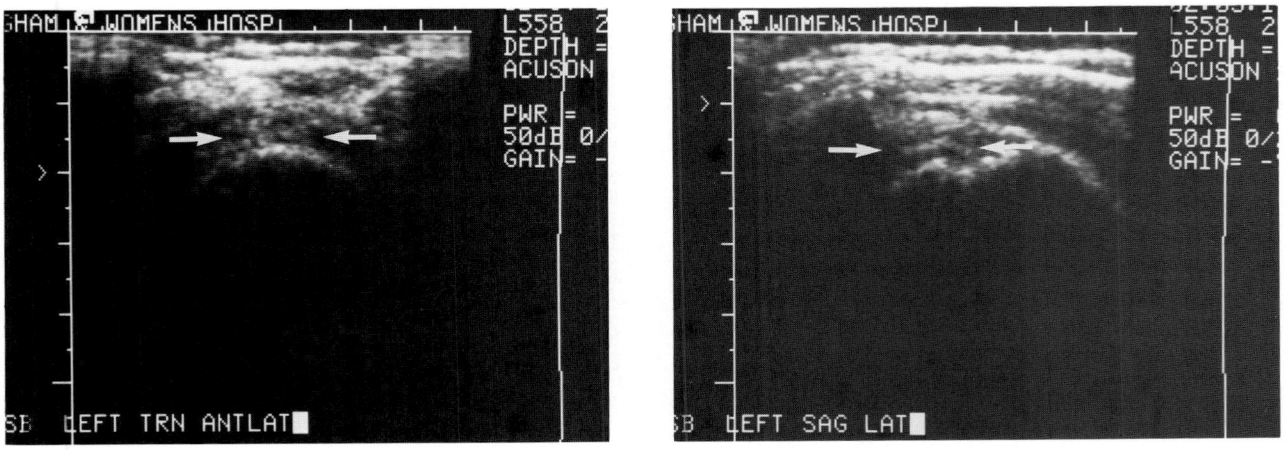

A

B

FIGURE 6–6
Partial rotator cuff tear. A. *A transverse scan shows poorly defined area of increased echogenicity* (arrows) *within rotator cuff.* B. *On longitudinal image, the area of increased echogenicity* (arrows) *is again seen.*

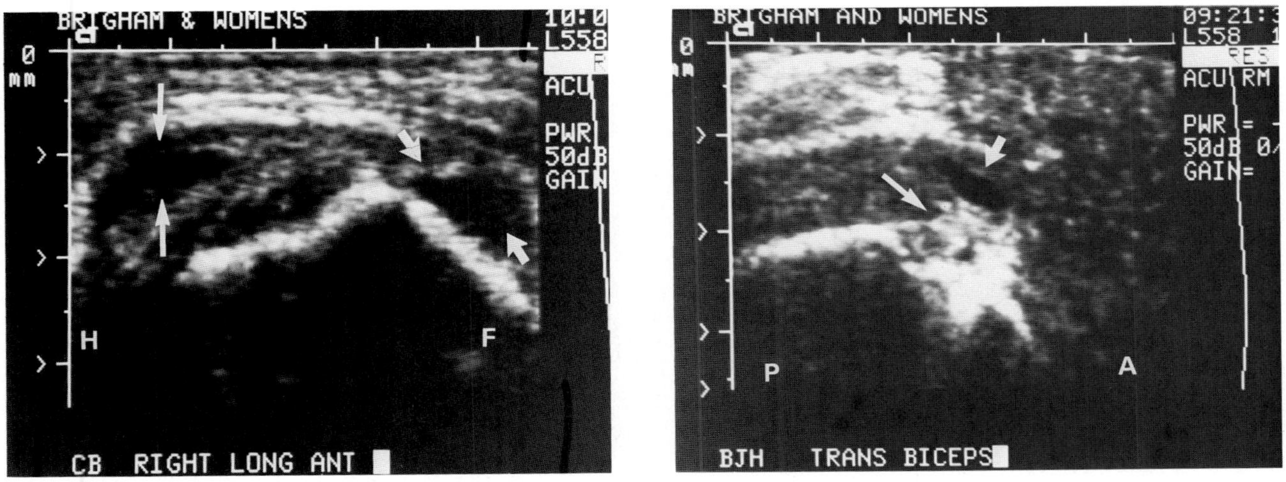

A

B

FIGURE 6–7
Acute rotator cuff tear. A. *Longitudinal scan showing an irregular fluid collection within the rotator cuff* (arrows) *as well as below the tuberosity* (arrowheads) *due to hemorrhage. The tendon of the rotator cuff is disrupted due to the tear. H = head; F = foot.* B. *Transverse image showing acute rotator cuff tear with echogenic focus within the cuff* (arrow) *as well as fluid* (arrowhead) *from hemorrhage. A = anterior; P = posterior.*

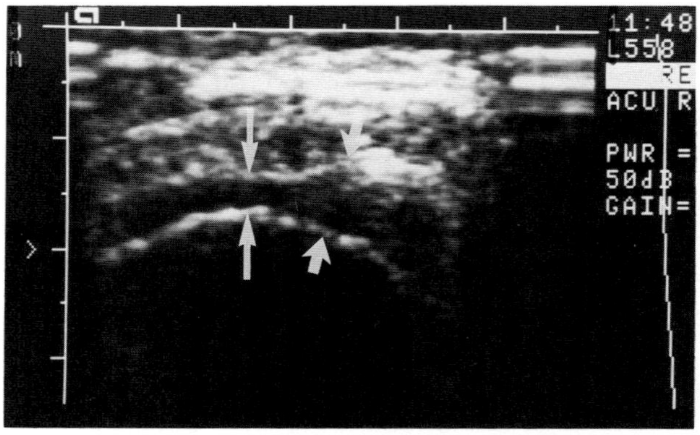

FIGURE 6–8 ▶
Rotator cuff tear. Transverse image showing abrupt thinning of the rotator cuff (arrows) *due to a tear next to a portion of the cuff of normal thickness* (arrowheads).

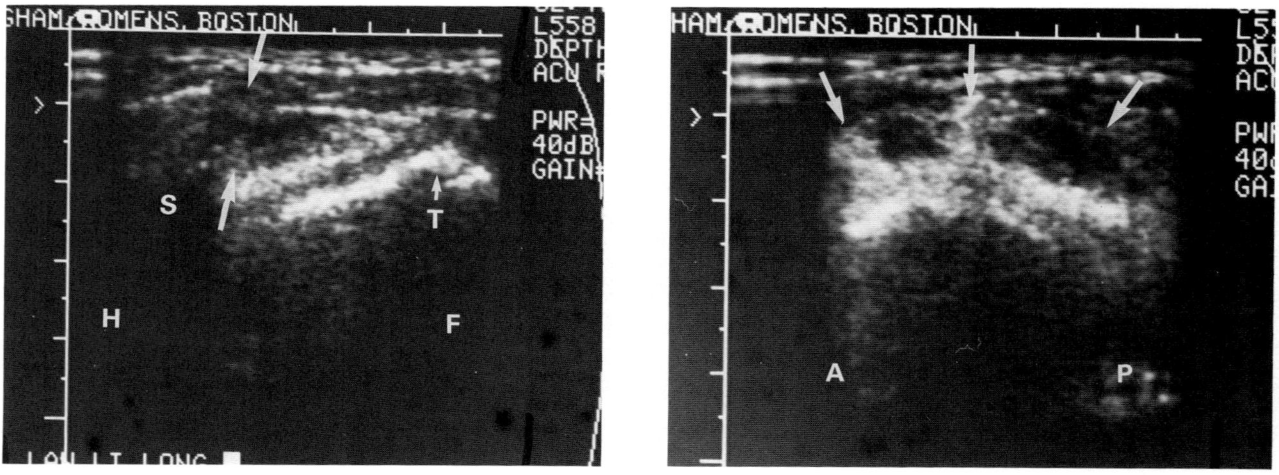

A

B

FIGURE 6–9
Large rotator cuff tear. A. *Longitudinal image showing defect* (arrows) *in rotator cuff. T = tuberosity* (arrow); *S = shadow from acromion; H = head; F = foot.* B. *Transverse image showing heterogeneous echogenicity of rotator cuff* (arrows) *with areas of increased and decreased echogenicity and poorly defined margins. A = anterior; P = posterior.*

pearance that should not be confused with a rotator cuff tear. In these cases, the calcific focus is very echogenic and casts an acoustic shadow (Figure 6–10).

Sonographic evaluation of the biceps tendon and its sheath provides information about shoulder abnormalities. Absence of the echogenic ellipse or circle of the tendon from the bicipital groove or from between the tuberosities is seen with rupture of the tendon and retraction, or severe tenosynovitis with fraying, degeneration, and atrophy of the tendon. Fluid surrounding the biceps tendon within the tendon sheath (Figure 6–11) is an indirect sign of shoulder pathology, which proves to be a rotator cuff tear in the majority of cases [3].

The subdeltoid bursa lies between the deltoideus muscle and the rotator cuff tendon. Normally the bursa is not visible to ultrasound because it is a thin echogenic band between the muscles that cannot be distinguished from the normal fascial planes. When the bursa is involved with inflammation, however, fluid collects within the bursa, which becomes visible on ultrasound as a crescentic anechoic collection bounded by echogenic lines (Figure 6–12). Chronic bursitis leads to thickening of the bursa, which on sonog-

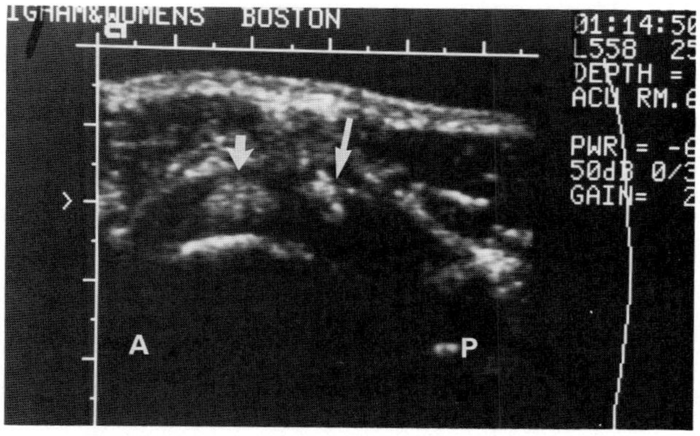

FIGURE 6–10
Calcific tendonitis. Transverse image of rotator cuff showing brightly echogenic focus due to calcification (arrow) *and second area of tendonitis with increased echogenicity* (arrowhead). *A = anterior; P = posterior.*

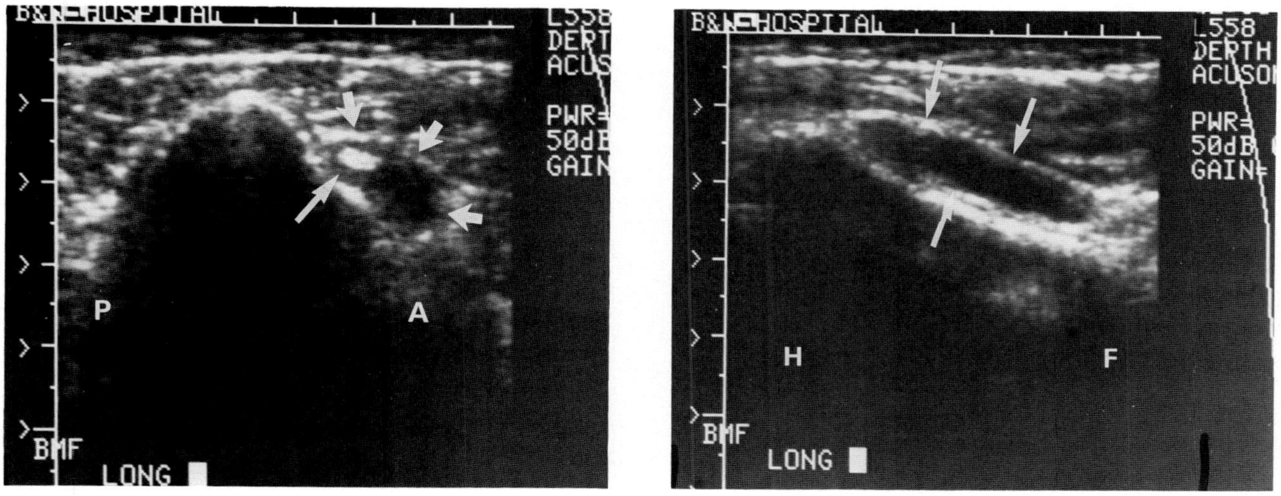

A

B

FIGURE 6–11
Fluid in bicipital tendon sheath. A. *Transverse image showing biceps tendon* (arrow) *surrounded by fluid* (arrowheads). *A = anterior; P = posterior.* B. *Longitudinal scan of tendon sheath containing fluid* (arrows). *H = head; F = foot.*

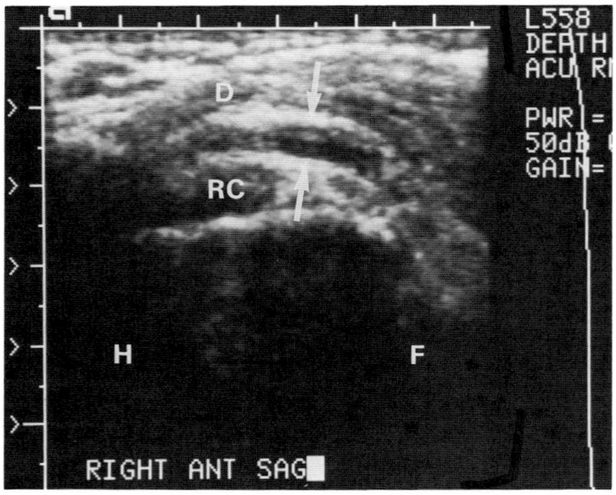

FIGURE 6–12
Bursitis. Sagittal image showing fluid in subdeltoid bursa (arrows). *D = deltoideus; RC = rotator cuff; H = head; F = foot.*

raphy is a hypoechoic area between the echogenic planes [8].

Other Tendons

Any muscle tendon can be evaluated by ultrasound provided it is located in a site accessible to the high-frequency sonographic beam. The quadriceps, patellar, and Achilles tendons are large tendons well located for evaluation for partial or complete tears or rupture. With a tear of one of these tendons, just as in the rotator cuff, focal discontinuity and altered echogenicity are seen. Complete rupture leads to discontinuity of the tendon, sometimes with abrupt irregular edges, if there is separation of the tendon fragments due to retraction. Inflammation of the tendon causes swelling and generalized decreased echogenicity in the involved area. In the hands, wrists, and feet, small tendons within tendon sheaths can be studied. Tenosynovitis causes fluid within the tendon sheath surrounding the involved echogenic tendon (Figure 6–13) [12].

Soft Tissues of the Extremities

Soft tissue masses, cysts, abscesses, hematomas, and foreign bodies can be evaluated by ultra-

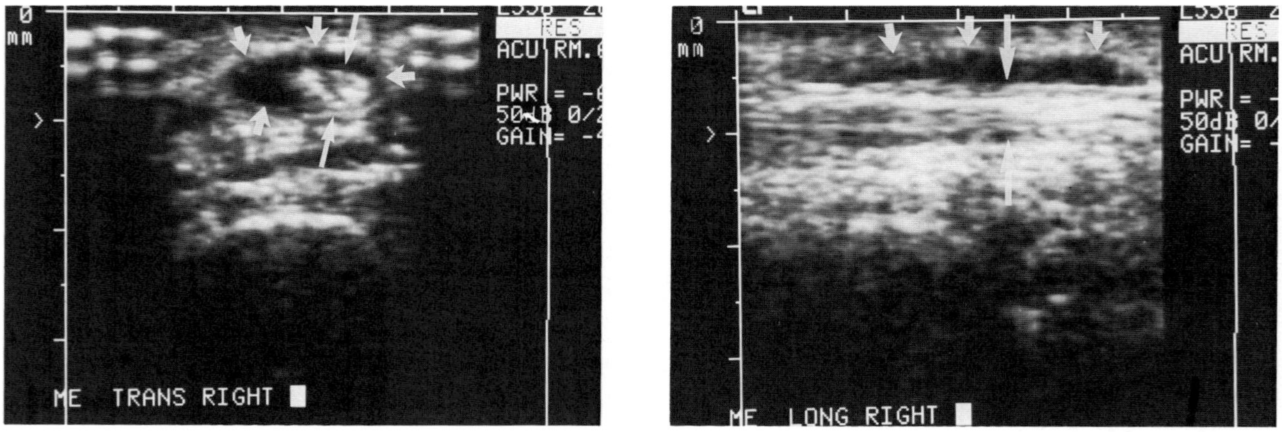

A B

FIGURE 6–13
Tenosynovitis. A. *Transverse scan of tendon* (arrows) *with fluid surrounding it* (arrowheads). B. *Longitudinal image of fluid in tendon sheath* (arrowheads) *around tendon* (arrows).

sound for size, location, and echotexture. It is particularly useful in evaluating the popliteal fossa where popliteal or Baker's cysts may be symptomatic but nonpalpable. By ultrasound, these cysts are easily located and characterized, commonly appearing as well-defined anechoic, elongated collections in the popliteal fossa, with smooth walls (Figure 6–14). Debris, septations, and internal echoes may be seen within the cysts [13]. Infections of the extremities may be limited to the subcutaneous tissue, cellulitis; involve muscle, myositis; or involve the osseous structures, osteomyelitis. Ultrasound can be used to help define the site and extent of disease. Cellulitis leads to focal thickening of the subcutaneous tissue. With abscess formation an anechoic or hypoechoic collection will be seen that has irregular borders and increased through-transmission (Figure 6–15) [14]. Myositis causes swelling and altered echogenicity of the involved muscle. The fascial planes are distorted or lost and the muscle has diffusely decreased echogenicity. As in the subcutaneous tissues, abscess formation is an irregular fluid collection with few echoes in it and acoustic enhancement [15]. In patients with acute osteomyelitis, an abnormal fluid collection may be seen adjacent to the bone, resulting from extension of the subperiosteal infection [16]. The bone itself cannot be evaluated as the sonographic beam cannot penetrate it.

Other soft tissue lesions of any size can be studied by ultrasound. Foreign bodies can be localized for surgical removal (Figure 6–16). Solid masses can be characterized and localized with respect to surrounding structures. Cystic lesions can be documented and guidance for aspiration provided.

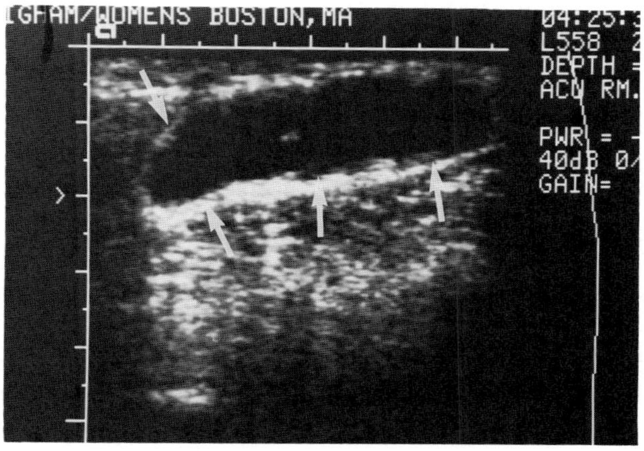

FIGURE 6–14
Popliteal cyst. Elongated hypoechoic cyst with well-defined walls in the popliteal fossa (arrows).

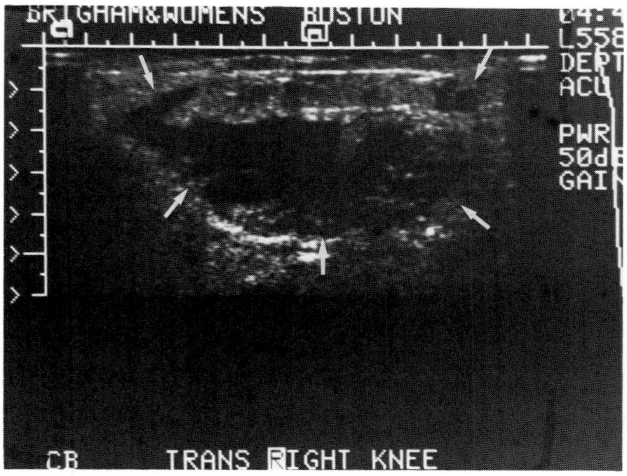

FIGURE 6–15
Soft tissue abscess. Sonogram showing large irregular mass (arrows) *containing anechoic fluid and debris localized by sonography for drainage.*

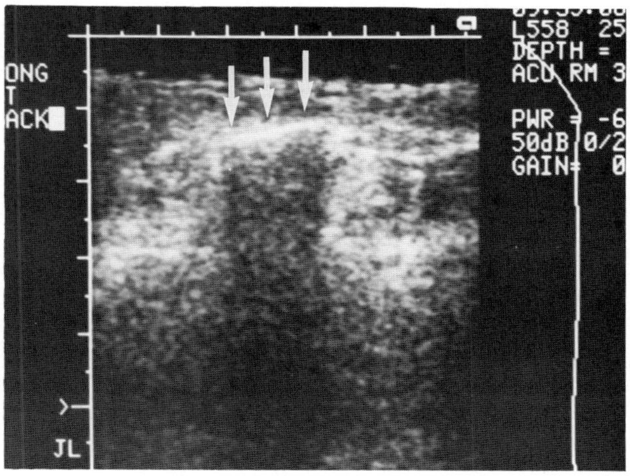

FIGURE 6–16
Bullet. Sonogram of bullet (arrows) *lodged in subcutaneous tissues of the back, localized by ultrasound for excision.*

REFERENCES

1. Fornage BD, Rifkin MD. Ultrasound examination of tendons. Radiol Clin North Am 1988; 26:87–107.
2. Richardson ML, Selby B, Montana MA, Mack LA. Ultrasonography of the knee. Radiol Clin North Am 1988; 26:63–75.
3. Middleton WD, Edelstein G, Reinus WR, Melson GL, Totty WG, Murphy WA. Sonographic detection of rotator cuff tears. AJR 1985; 144:349–53.
4. Middleton WD, Reinus WR, Tohy WG, Melson GL, Murphy WA. Ultrasound of the biceps tendon apparatus. Radiology 1985; 157:211–5.
5. Lanzer WL. Clinical aspects of shoulder injuries. Radiol Clin North Am 1988; 26:157–60.
6. Crass JR, Craig EV, Feinberg SB. Ultrasonography of rotator cuff tears: a review of 500 diagnostic studies. J Clin Ultrasound 1988; 16:313–27.
7. Mack LA, Nyberg DA, Matsen FA. Sonographic evaluation of the rotator cuff. Radiol Clin North Am 1988; 26:161–77.
8. Hodler J, Fretz CJ, Terrier F, Gerber C. Rotator cuff tears: correlation of sonographic and surgical findings. Radiology 1988; 169:791–4.
9. Middleton WD, Reinus WR, Melson GL, Totty WG, Murphy WA. Pitfalls of rotator cuff sonography. AJR 1986; 146:555–60.
10. Mack LA, Matsen FA, Kilcoyne RF, Davies PK, Sickler ME. Ultrasound evaluation of the rotator cuff. Radiology 1985; 157:205–9.
11. Crass JR, Craig EV, Feinberg SB. Sonography of the postoperative rotator cuff. AJR 1986; 146:561–4.
12. Fornage BD, Rifkin MD. Ultrasound examination of the hand and foot. Radiol Clin North Am 1988; 26:109–29.
13. Pathria MN, Zlatkin M, Sartoris DJ, Scheible W, Resnick D. Ultrasonography of the popliteal fossa and lower extremities. Radiol Clin North Am 1988; 26:77–85.
14. Vincent LM. Ultrasound of soft tissue abnormalities of the extremities. Radiol Clin North Am 1988; 26:131–44.
15. Phipps WR, Benson CB, McShane PM. Severe thigh myositis following intramuscular progesterone injections in an in vitro fertilization patient. Fertil Steril 1988; 49:536–7.
16. Abiri MM, Kirpekar M, Ablow RC. Osteomyelitis: detection with US work in progress. Radiology 1988; 169:795–7.

7

Musculoskeletal Applications of CT

RICHARD B. RAFAL

JOHN A. MARKISZ

The initial radiographic evaluation of suspected musculoskeletal abnormalities usually begins with a routine radiograph and, if necessary, continues with a radionuclide scan. Experience over 80 years has demonstrated that the conventional x-ray very often provides a diagnosis of any bone abnormality [1,2] and more recently scintigraphy has been shown to be able to survey quickly the entire body for additional bony lesions [3]. In spite of the information provided by these modalities, CT has impacted significantly on the diagnostic workup of diseases of the musculoskeletal system due to its ability to accurately image subtle abnormalities in both osseous and soft tissue structures [4,5].

High-resolution axial views, photographed using bone windows, usually supplement conventional x-ray findings for irregularities in bone (Figure 7–1). Soft tissue structures are poorly defined by standard x-ray studies and radionuclide examinations; however, it is here that CT excels. The detection and delineation of soft tissue masses and soft tissue extension of primary bone tumors are much better evaluated by CT than by plain x-rays or radionuclide studies (Figure 7–2). Pathologic processes involving joints

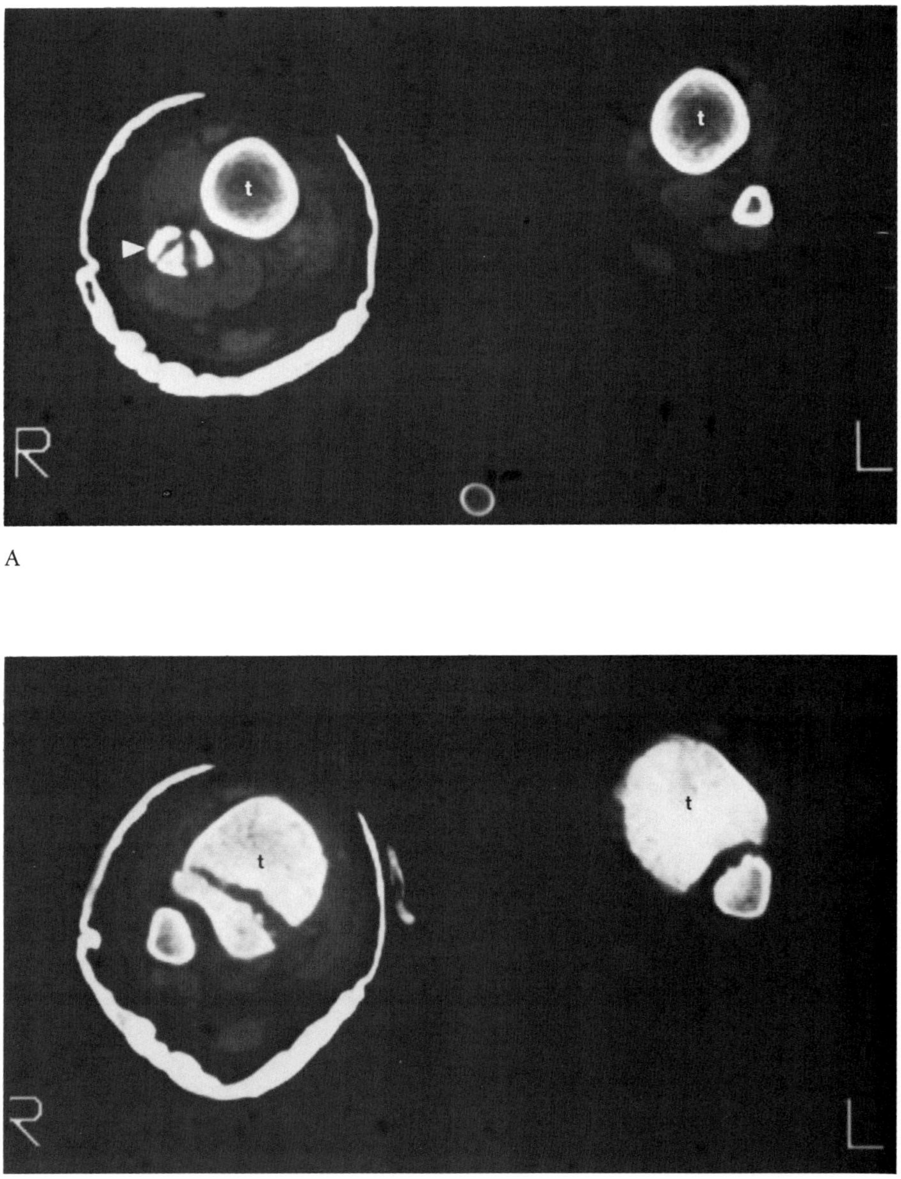

FIGURE 7–1
Axial CT scans through the lower extremities of a 15-year-old boy after a motorcycle accident. A cast surrounds an edematous right leg. A. Comminuted fracture of the fibula (arrowhead). *B. Fracture at the distal end of the tibia, with separation of fracture fragments* (arrow). *t = tibia.*

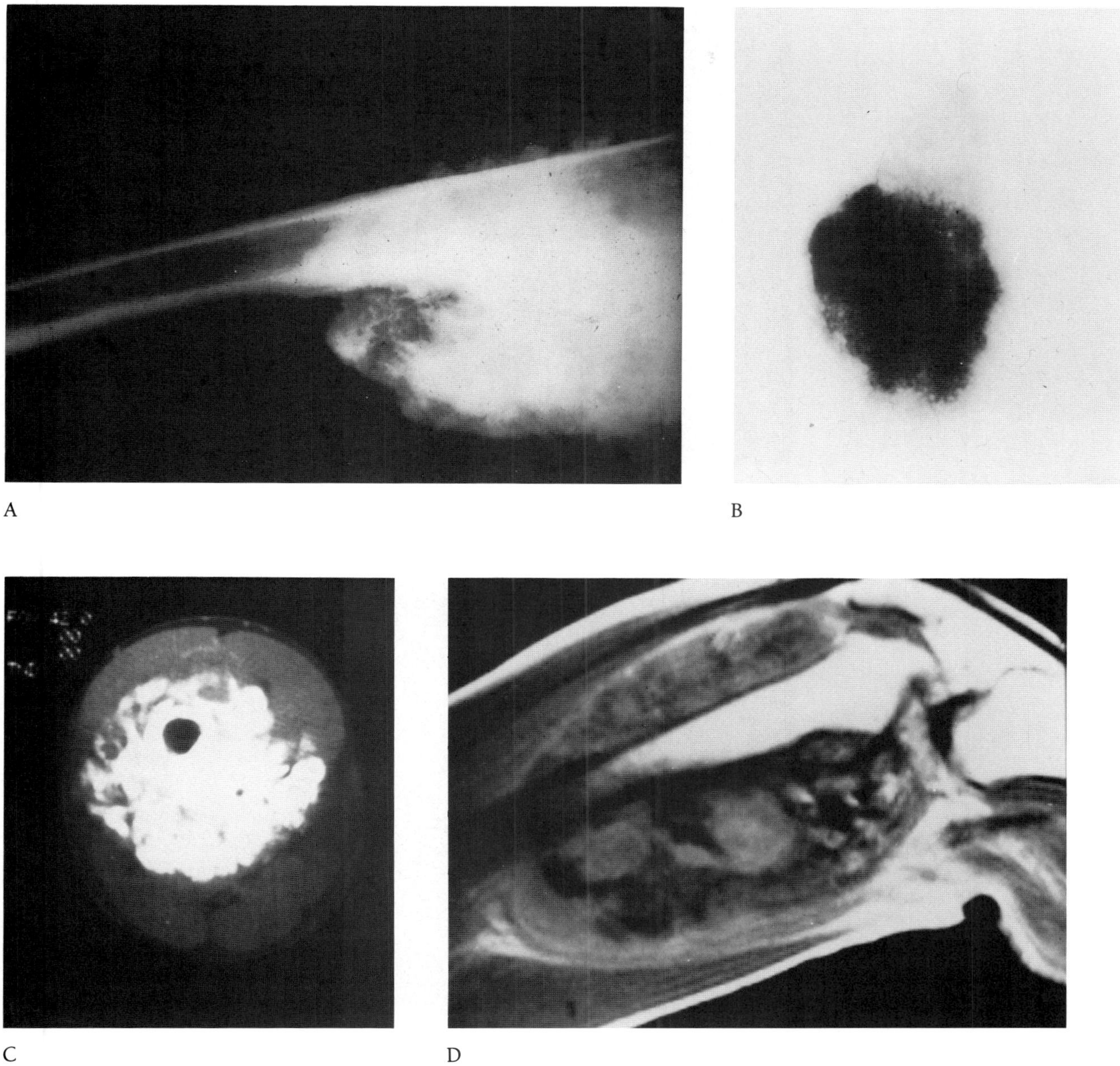

FIGURE 7–2

Osteogenic sarcoma in the distal aspect of the femur of an 18-year-old man. A. The diagnosis can be made from the plain radiograph. B. The tumor mass is obvious in the radionuclide bone scan. Even though there is inherently little detail in the scan, it demonstrates the presence of the tumor and, more importantly, the remainder of the scan was negative, eliminating the concern about skip lesions. C. The CT demonstrates the exact degree of extension into the soft tissues, with a normal-appearing medullary cavity. D. Although the tumor is well delineated as a heterogeneous mass by MRI, calcifications, which allowed a diagnosis to be made on plain film, cannot be distinguished by this method.

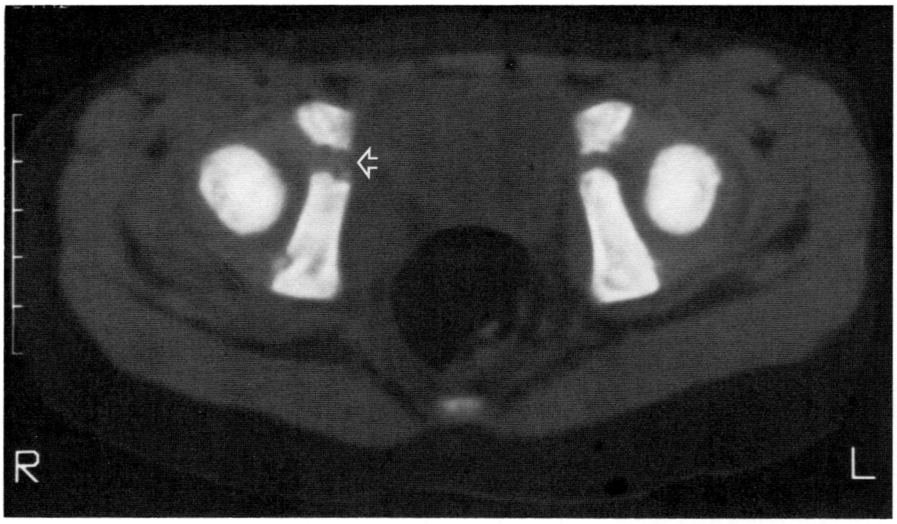

A

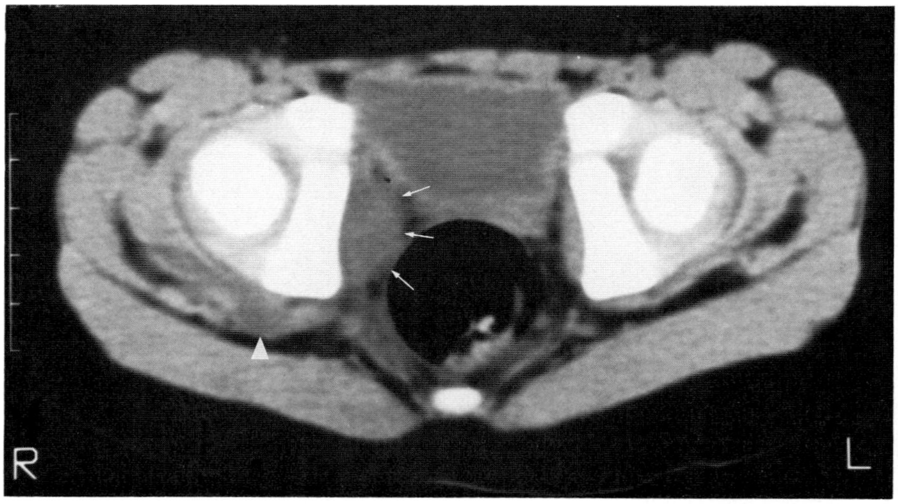

B

FIGURE 7–3

Four-year-old girl with repeated right hip dislocations. A. Bone window best depicts a widening of the growth plate between the right acetabulum and the anterior pubic ramus (open arrow), *consistent with a Salter I fracture. The irregularity of the posterior aspect of the right acetabulum is compatible with a healing fracture. This can be compared with the normal left side. B. Soft tissue technique better demonstrates the swelling of the right obturator muscle* (small arrows) *and gluteus minimus muscles* (arrowhead).

can also be better appreciated due to multiplanar image reformatting, and CT-guided biopsies ensure that the appropriate mode of entry is made and that tissue is secured from the appropriate area.

The techniques vary widely, depending on the part of the body involved, the type of abnormality suspected, and the particular CT unit employed. Contiguous or overlapping thin slices might be preferred for a complex joint if multiplanar reformatting is required, whereas slice thicknesses of 10 mm with a 15-mm increment are acceptable for the evaluation of a suspected soft tissue mass. All studies should be photographed at two different window widths and levels, one highlighting bone and the other soft tissue structures. If uncertainty persists, Hounsfield units are sometimes helpful. Both extremities can be imaged simultaneously for comparison, and to minimize the possibility of mistaking a congenital variant for an abnormality (Figure 7–3). Intravenously administered contrast material is of limited utility, being used if no mass is visualized on the noncontrast study or if precise localization of vasculature is of importance [5]. For the evaluation of joint abnormalities, injection of intra-articular contrast medium permits a CT-arthrotomographic examination [6,7].

Bone Tumors

Even though the vast majority of bone tumors are initially diagnosed by conventional x-rays, the tissue discrimination and resolution available have made CT a most useful adjunct. After the presence of a tumor has been determined radiographically, CT is extremely useful in tumor staging and in the evaluation of the precise magnitude of bone involvement and soft tissue extension. CT is very sensitive to the degree of calcification in tissue and can reveal minute changes in bone density, whereas radiographs may not demonstrate changes until there is a pronounced change in the degree of mineralization. A determination of the actual extent of disease within a bone, therefore, can be much better demonstrated by CT than with plain films alone. Questionable radiographic findings can be corroborated by CT in order to verify the presence or absence of a lesion. Radiographs are relatively insensitive to soft tissue differences; however, CT is able to precisely distinguish between fat, muscle, blood, and other fluids. This allows tissue planes to be precisely defined, and normal and abnormal structures differentiated.

Additional information about the vascularity of the tumor can be obtained by performing the CT examination with and without intravenously administered contrast medium. The increased radiographic density provided by the contrast agent generates information about the relative vascularity of the mass. The effect of administration of contrast can be variable, depending on the particular type of tumor that is involved. Although enhancement was substantiated in only 40% of soft tissue tumors, significant information can be obtained if it does occur. In this series of 50 patients, two tumors within muscle were isodense with muscle on the initial series, and could be demonstrated only after the administration of contrast material [8]. In addition, the relationship of a mass to blood vessels is more easily and more accurately illustrated on a postcontrast study. This is extremely important when major resection of the tumor is contemplated. CT is usually able to demonstrate the position of the mass relative to major blood vessels and determine whether there has been any encasement of major structures, often eliminating the necessity of angiography. Currently, with the use of CT, plain tomography and angiography are seldom indicated in preoperative staging of bone tumors [9].

ESTABLISHING A PRIMARY DIAGNOSIS

Studies have demonstrated that CT is valuable in establishing a primary diagnosis in bony lesions only if there is a soft tissue component or if the plain film is equivocal [8,10]. In over half of the cases, CT was deemed useful in assessing the actual extent of tumor involvement. Other studies have established that CT can provide unique information that will alter patient management in over 80% of the patients [11,12]. This is especially true for patients with vague radiographic indications of an abnormality. With its much higher resolution, CT can usually confirm or disprove questionable x-ray findings. When the diagnosis is in doubt and a biopsy is required, CT has been shown to benefit the vast majority of patients involved with a CT-guided biopsy [15,16]. In addition to being able to determine the best technical approach to a biopsy site, CT was found to cancel 21% of the biopsies as unnecessary because the scan established the diagnosis, and also was able to indicate when an open biopsy would be preferable.

DETERMINING EXTENT OF DISEASE

Once a diagnosis has been established, a knowledge of the exact magnitude of disease is necessary before treatment planning can begin. The range of tumor spread and the proximity to vital structures are usually the determining factors when one decides on operative or nonoperative therapy. The high sensitivity and superb resolution available with CT, as well as the ability to adjust the viewing windows to emphasize bony structures, make it especially valuable in determining the true extent of disease. Evaluation of the cortex, medullary canal, periosteal, and parosteal regions can be achieved easily and accurately. An accurate assessment of osseous detail of the matrix can be obtained in order to determine the extent of cortical destruction, involvement of the marrow space, and degree of soft tissue extension (Figure 7–4).

Osteolytic lesions demonstrate a loss of cortical and trabecular bone, and CT can determine if there has been cortical disruption (Figure 7–5). Osteoblastic lesions are viewed as areas of irregular, increased density with indistinct margins (Figure 7–6). Tumor within the medullary cavity is usually seen as an increased density, compared to the relative lucency of normal adipose tissue within the marrow space (Figure 7–7) [15]. Osteomyelitis within the marrow cavity similarly appears as a region of increased density, and must be correlated with the clinical findings (Figure 7–8) [16]. Occasionally, subperiosteal reaction can be observed, and unsuspected osteomyelitis has been diagnosed by the presence of gas within the marrow space [17].

DETECTING METASTASES

The presence of skeletal metastases from either a bone or other primary neoplasm drastically

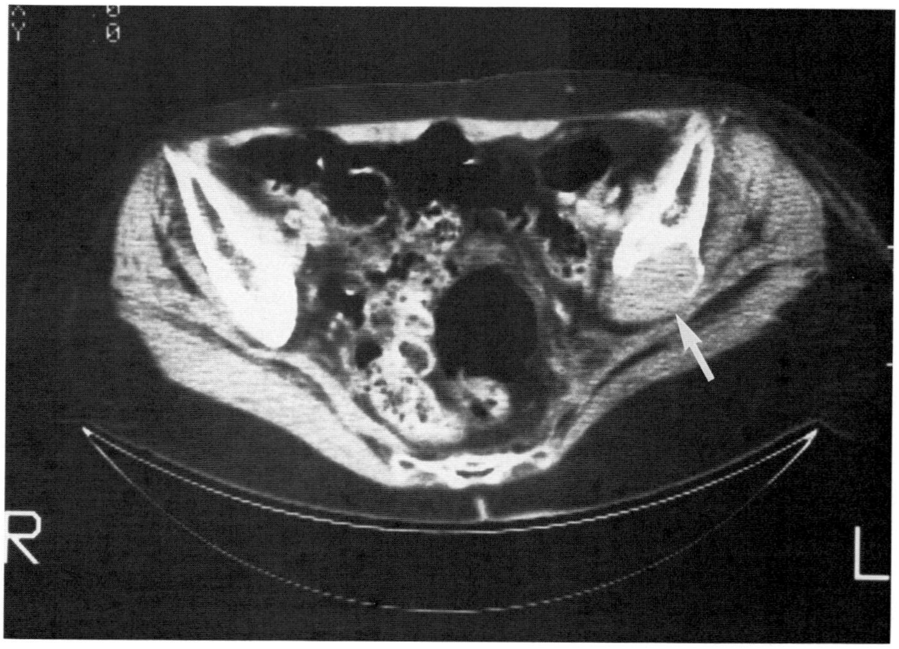

FIGURE 7–4
A plasmacytoma of the left iliac bone (arrow). *The cortical breakthrough and soft tissue extension are well demonstrated.*

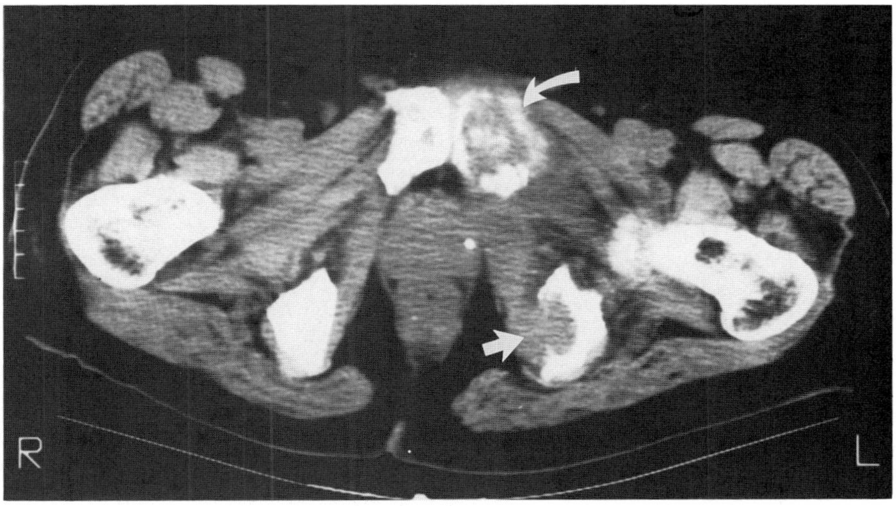

A

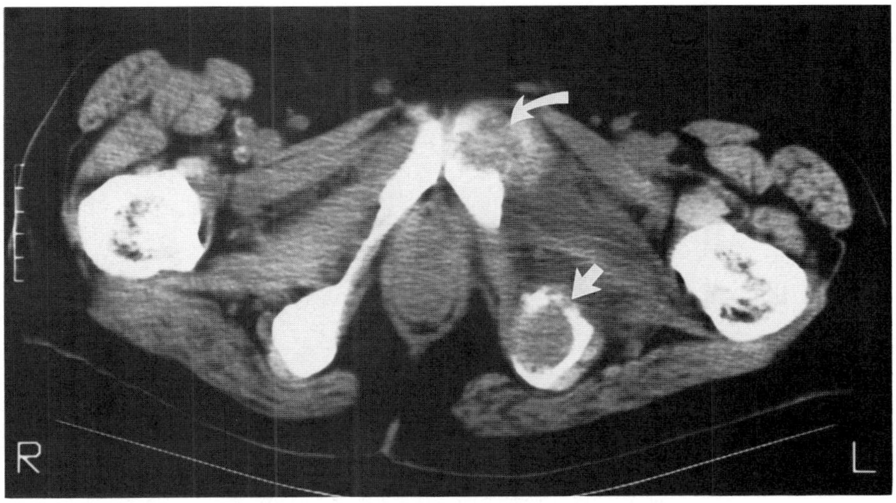

B

FIGURE 7–5
A 67-year-old man with bladder carcinoma. A, B. *CT demonstrates osteolytic bony metastases to the left superior pubic ramus and left pubic symphysis* (curved arrows), *as well as a lesion of the left ischial tuberosity* (short arrows).

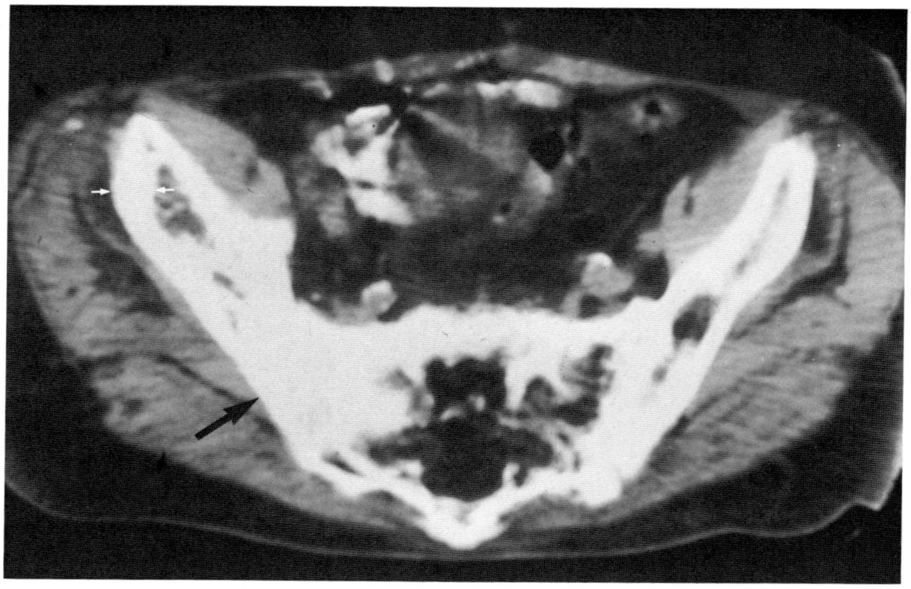

FIGURE 7–6
CT image of sarcomatous degeneration of Paget's disease (black arrow). *Note the thickened cortex of the right iliac bone* (white arrows).

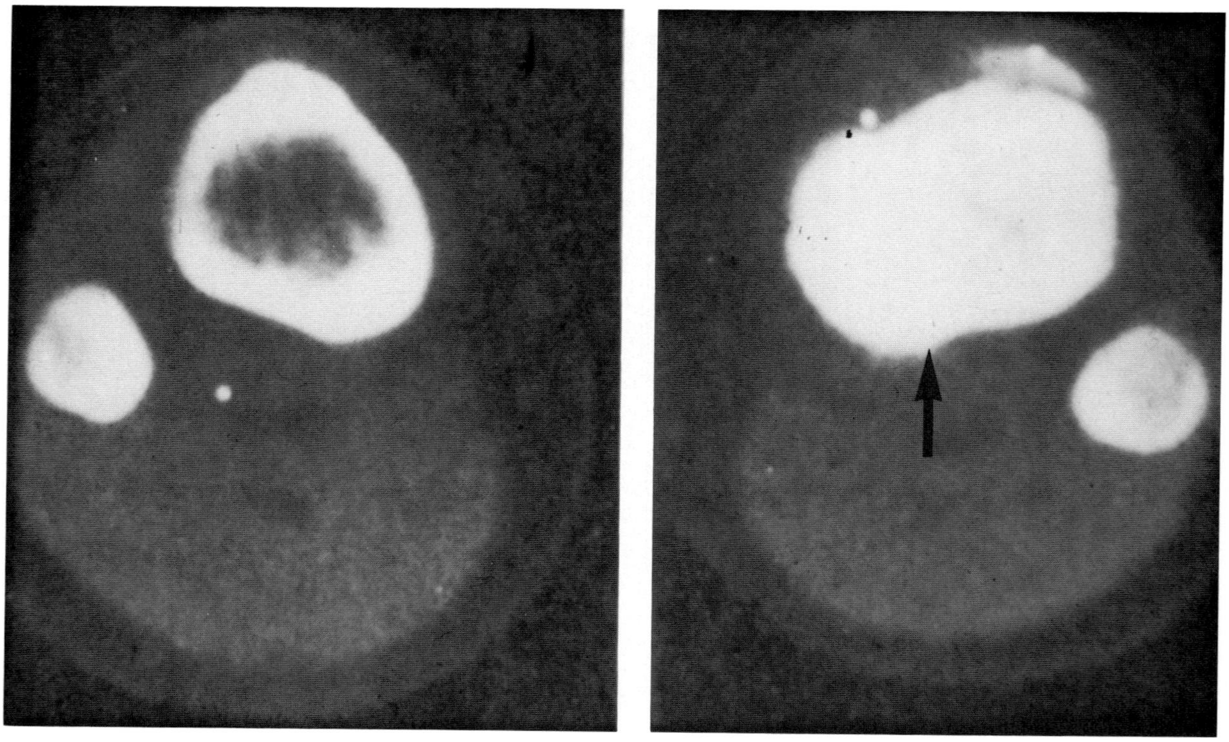

FIGURE 7–7
Transaxial CT scan revealing an osteogenic sarcoma in the proximal part of the left tibia (arrow). *The normal low-density fat-containing marrow, observed in the right tibia, has been replaced by the osteoblastic tumor.*

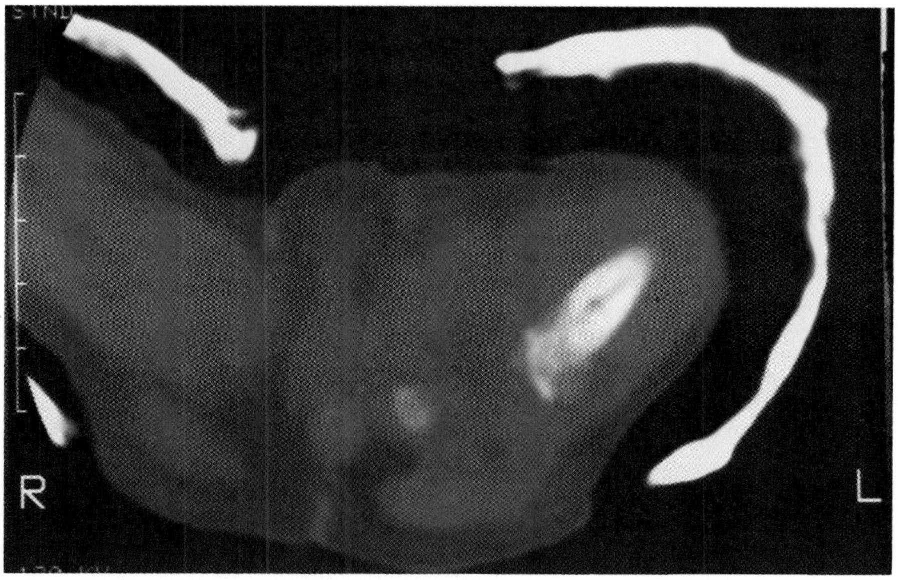

A

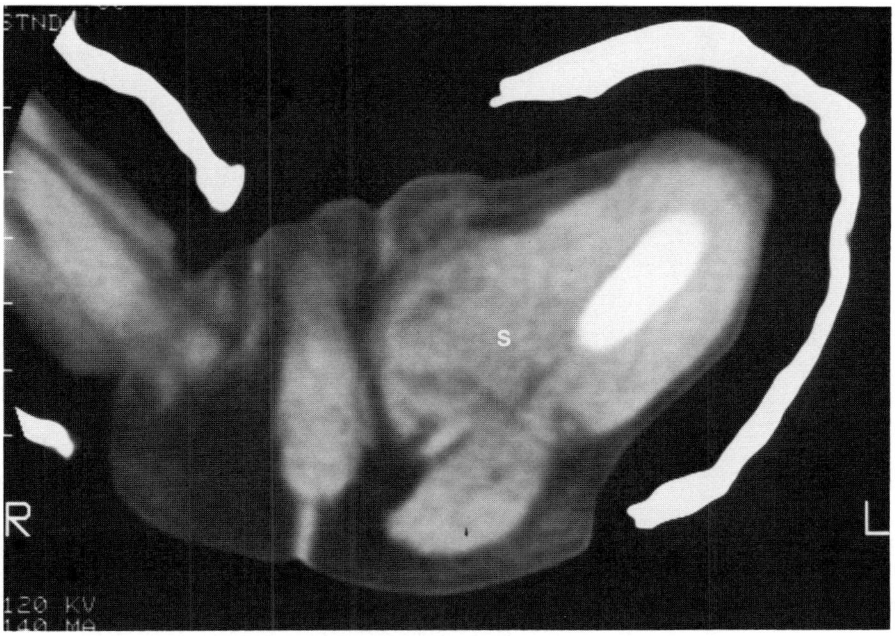

B

FIGURE 7–8
Osteomyelitis of the left femur in an infant in a cast as demonstrated by CT scan.
A. There is metaphyseal and proximal diaphyseal irregularity and expansion.
B. The same section photographed using soft tissue windows better demonstrates marked soft tissue swelling (S).

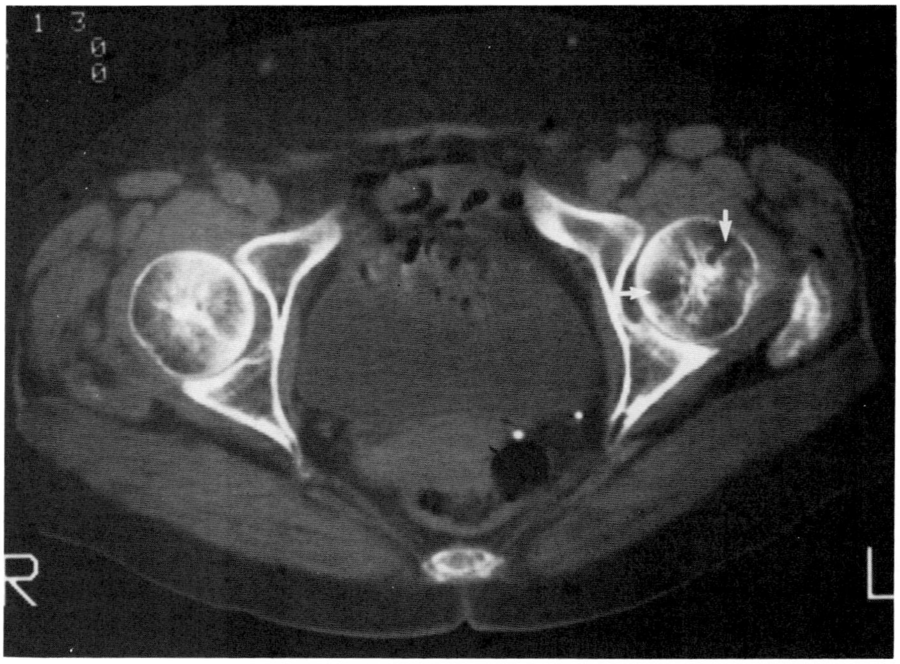

A

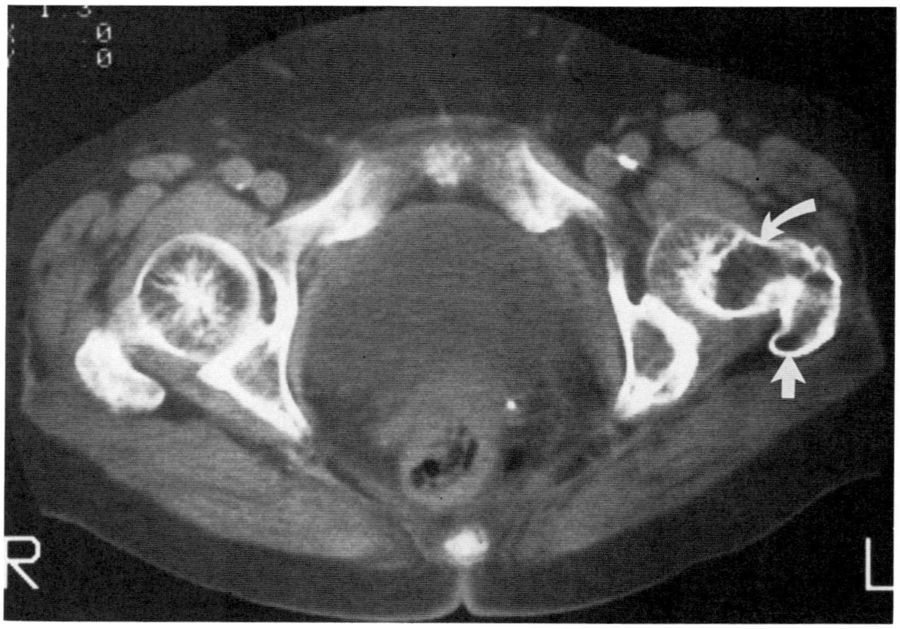

B

FIGURE 7–9

Metastatic disease from breast carcinoma in a 77-year-old woman. A. *Small lucent lesions are seen in the left femoral head* (small arrows). B. *The lytic lesion of the left femoral neck* (curved arrow) *with surrounding sclerosis, and a small lytic area in the greater trochanter* (arrow) *are also indicative of metastases.*

alters prognosis and treatment planning [10]. CT is too slow and the radiation dose would be too high to consider using it as a screening modality. This can be accomplished quickly and easily by a radionuclide bone scan (see Chapter 3). The bone scan is quite sensitive, but not specific, however, in detecting indiscriminately both metastatic disease and degenerative joint changes. It is in this setting, to evaluate suspected metastatic lesions, that CT is important (Figure 7–9). The spine and pelvis, because of the irregularly shaped and overlapping bony structures, are difficult regions in which to identify small metastatic foci by routine x-ray. Since more than half of all bony metastases are found in the spine (39%) and pelvis (12%) [3], CT becomes especially useful in differentiating between degenerative changes and tumor in these areas, when abnormalities detected by scintigraphy are evaluated.

DIFFERENTIATING LESIONS

Although obtaining a definite pathologic diagnosis by CT is very seldom possible, there are situations in which scan patterns are distinctive. CT has been found to be accurate, for example, in differentiating an osteochondroma or osteocartilaginous exostosis from a chondrosarcoma. A chondrosarcoma is an inhomogeneous, irregular mass with a calcified center, whereas an osteochondroma presents as a well-marginated, bony mass with a lucent, organized center that is contiguous with the medullary space of the adjacent bone. In the chondrosarcoma there is a thin, smooth cartilaginous cap that is thick and irregular [18]. Intraosseous lipomas are rare lesions of bone, usually being found in the calcaneus. They can be easily identified on CT scans by their homogeneous appearance consistent with fat, often with a sclerotic rim [19].

The resolution available with CT can often be helpful in patients with an osteoid osteoma, in which the nidus cannot be located by plain films or by conventional tomography. Although benign, the nidus of this painful lesion must be surgically excised. Due to the high resolution and ability to vary the bone windows, thin-section CT scans are a very sensitive method of locating the nidus, thus localizing the surgical site [16,20,21]. Most other benign lesions will be easily identified by the standard radiograph, eliminating the necessity of obtaining a CT.

Soft Tissue Lesions

Because of an impressive ability to detect and evaluate soft tissue abnormalities, especially when compared to routine plain films, CT has become the modality of choice for the evaluation of soft tissue lesions of the musculoskeletal system (Figure 7–10). Soft tissue tumors are very well visualized by CT because of density differences with normal tissue or because of distortion or destruction of normal structures. Although a specific histologic diagnosis is rarely possible, differentiating malignant from benign lesions is often feasible. Keeping in mind that there is a significant overlap in findings, benign tumors are usually homogeneous and smoothly outlined and do not invade surrounding structures or infiltrate surrounding fat planes. Malignancies are frequently heterogeneous on both nonenhanced and enhanced series, are irregular without well-defined borders, infiltrate surrounding fat, and invade surrounding structures. Weeks et al. were able to predict whether a tumor was malignant or benign, with an 88% success rate [22].

Fat-containing tumors, lipomas and liposarcomas, are the most common types of soft tissue neoplasms in the musculoskeletal system. The benign lipoma is entirely composed of fat, with smooth borders and an extremely thin capsule. It can be diagnosed quite definitively by CT as a homogeneous mass, with CT numbers consistent with fat. This is contrasted with the malignant liposarcoma, which is highly variable in fat composition, as well as in appearance, with areas of soft tissue density within them. The less fat contained within these tumors, the more aggressive they are. As with other malignancies, the appearance of liposarcomas can be widely varied [23]. They will be heterogeneous and irregular and contain indistinct margins on both pre- and postcontrast studies (Figure 7–11) [24].

COMPARTMENTS

The muscles and neurovascular bundles of the extremities are divided into compartments by ex-

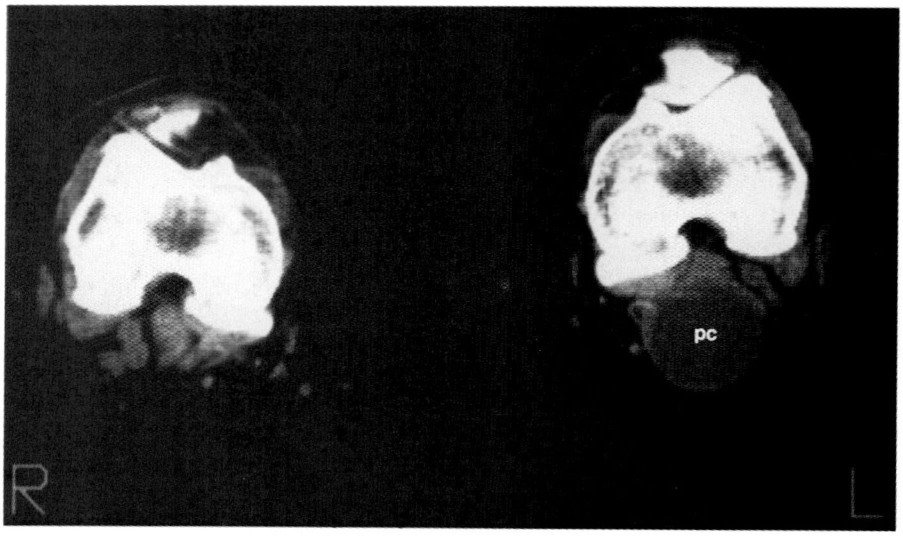

FIGURE 7–10
A lesion in the posterior aspect of the left knee is observed by CT as a homogeneous, low-density, well-circumscribed mass, consistent with a popliteal cyst (pc).

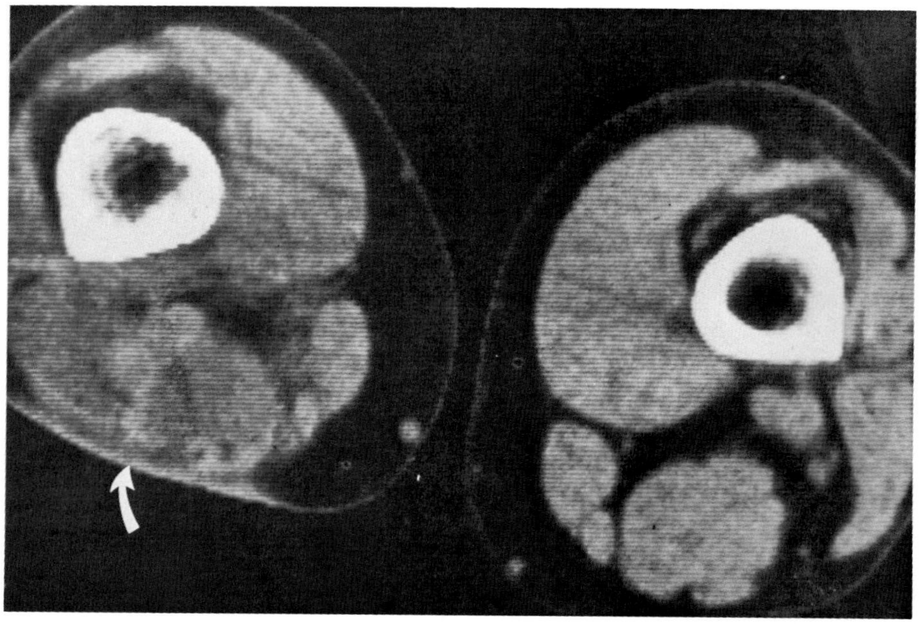

FIGURE 7–11
CT scan depicting a liposarcoma of the right thigh, confined to the posterior compartment (curved arrow). *Note the indistinct margins and the infiltration of the subcutaneous fat.*

tensions of the deep fascia and the intermuscular septa [5,25]. An examination of any soft tissue irregularity in terms of these compartments is significant since the obliteration of fascial planes and crossing of compartmental boundaries are often signs of a malignant process. In the upper arm, the brachial fascia is continuous with the fascia covering the deltoideus and pectoralis major muscles and separates the upper arm into an anterior (biceps brachii, brachialis, and coracobrachialis muscles) compartment and a posterior (triceps brachii and anconeus muscles) compartment. It continues into the forearm as the antebrachial fascia, attaching to the olecranon and posterior border of the ulna and the flexor and extensor retinacula of the wrist. Multiple intermuscular septa arise from it, both transversely and longitudinally, making a compartmental analysis of the forearm extremely involved.

In the thigh, the fascia lata extends from the sacrum and coccyx to the iliotibial band, separating the thigh into anterior (quadriceps femoris and sartorius muscles), medial (gracilis, adductors, and pectineus muscles) and posterior (semimembranosus, semitendinosus, and biceps femoris muscles) compartments. The fascia cruris is an extension of the fascia lata, and extends from the tibia, patella, and patellar tendon to the flexor and extensor retinacula of the ankle. It is continuous with the anterior and posterior crural intermuscular septa, which divide the lower leg into four compartments: anterior (tibialis anterior, extensor digitorum longus, extensor hallucis longus, and peroneus tertius muscles), lateral (peroneus longus and peroneus brevis muscles), superficial posterior (gastrocnemius, soleus, and plantaris muscles), and deep posterior (tibialis posterior, popliteus, flexor digitorum longus, and flexor hallucis longus muscles).

ABNORMALITIES

Tumor, hemorrhage, and infection usually remain confined within a particular compartment, provided trauma or surgical intervention has not disrupted the fascial planes. As a tumor increases in size and aggressiveness, however, it crosses compartmental lines, becoming extracompartmental. Hemorrhage presents in a variety of CT patterns, but typically remains within compartmental bounds, inducing edematous changes in the surrounding musculature. An acute hemorrhage appears homogeneous, with a density slightly higher than the density of muscle (Figure 7–12). With time a fluid-fluid interface is observed and eventually a pseudocapsule of low homogeneous density is also observed, decreasing in size as it becomes resorbed. Initially an intramuscular bleed may be difficult to distinguish from a neoplasm. An abscess usually initially respects compartmental boundaries. Unless gas is present within the abscess, this too can be confused with neoplasm. A CT-guided aspiration allows proper therapy to be initiated.

Disorders of the Joints

The continued evolution of CT toward higher resolution, decreased radiation exposure, and multiplanar reformatting techniques has been responsible for the increased utilization of CT for the evaluation of disorders of the joints (Figure 7–13). Although many new techniques are continually being tested, CT has established itself as a major orthopedic modality and offers a proven method of obtaining high-resolution, multiplanar views of the bony and soft tissue structures of the joints.

The traumatic hip presents a particular problem to a radiologist, because the pain that the patient experiences often limits the examination to a single view, or suboptimal study at best. Subtle acetabular fractures are often not visualized, with one series reporting 29% being missed by routine examinations (Figure 7–14) [31]. CT has been used very successfully to evaluate the hip subsequent to trauma to the acetabulum, as well as to evaluate fractures and the presence of fracture fragments in the shoulder, foot, and ankle (Figure 7–15). CT offers a relatively easy and accurate method to demonstrate fracture lines and identify fracture fragments within the joint space [32]. It has been reported to have changed patient management in up to one-third of suspected pelvic fractures (Figure 7–16) [33].

Even though fractures of the bones of the shoulder are readily demonstrated on routine x-rays, subtle dislocations, involvement of the labrum, and the size and position of fracture

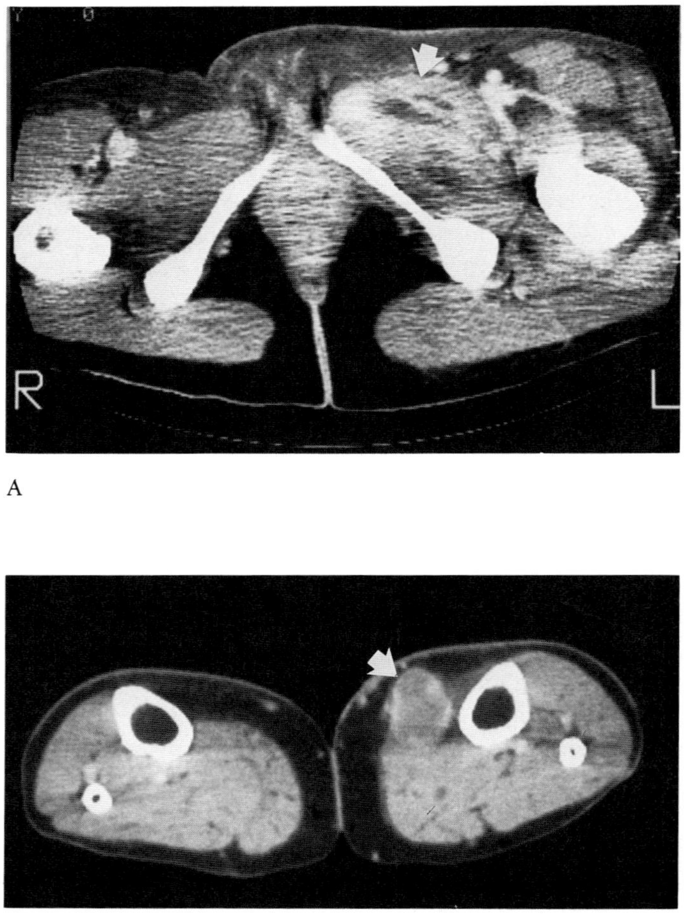

A

B

FIGURE 7–12
CT evaluation of two different patients with hematomas. A. Hemorrhage anterior to the left ischium (arrow) *displays a higher density than surrounding muscle. B. Localized hematoma is present in the thigh. Note the scattered hyperdense areas in the periphery of the lesion* (arrow).

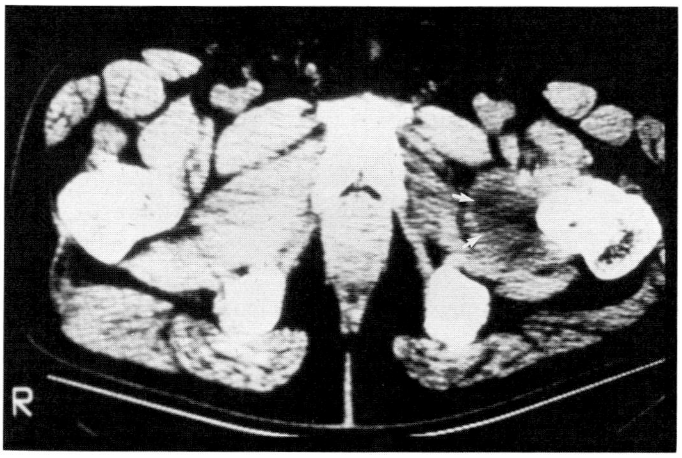

FIGURE 7–13
CT section at the level of the femoral necks in a patient with a septic left hip. A low-attenuation fluid is distending the left hip joint (arrows).

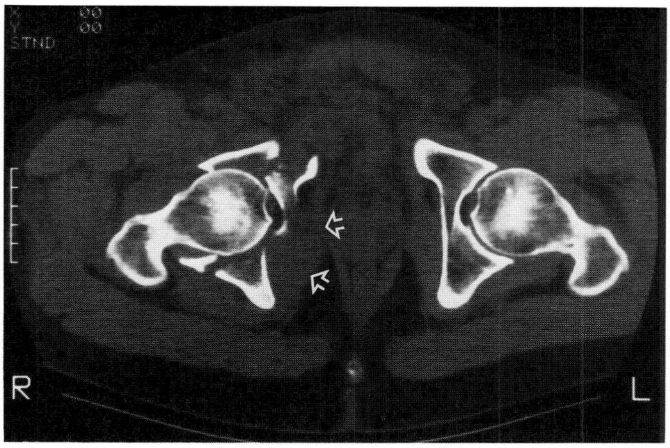

A

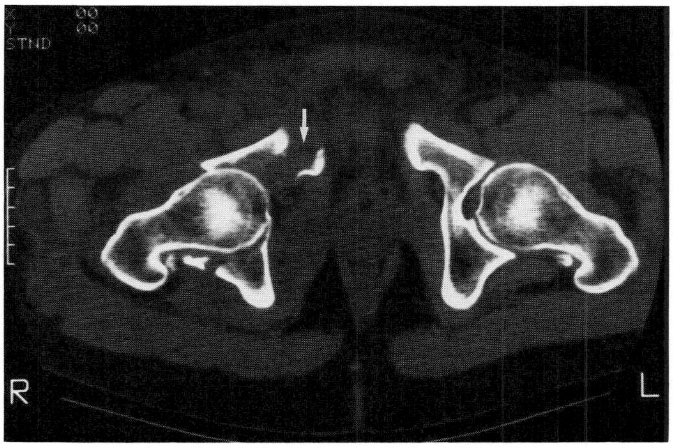

B

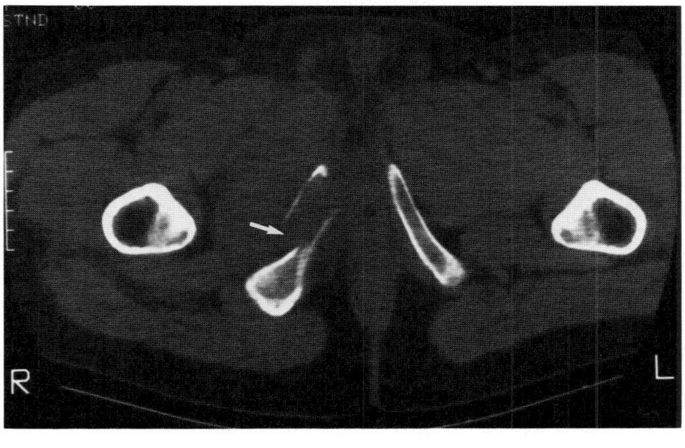

C

FIGURE 7–14

A. *There is excellent demarcation of a right acetabular fracture involving the medial aspect and the anterior and posterior columns. Note the adjacent hematoma* (open arrows). B, C. *There are associated fractures of the right superior and right inferior pubic rami, respectively.*

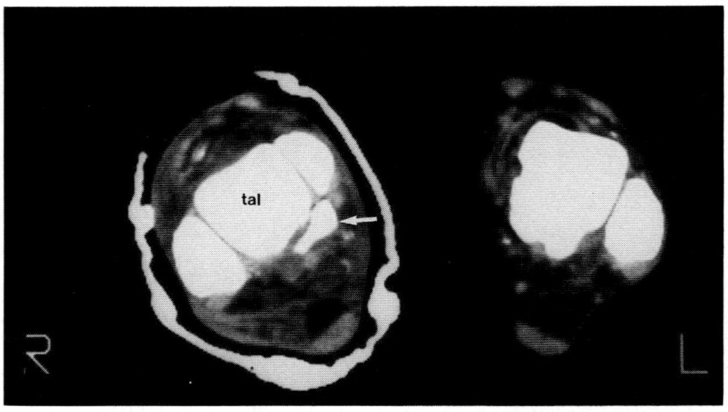

A

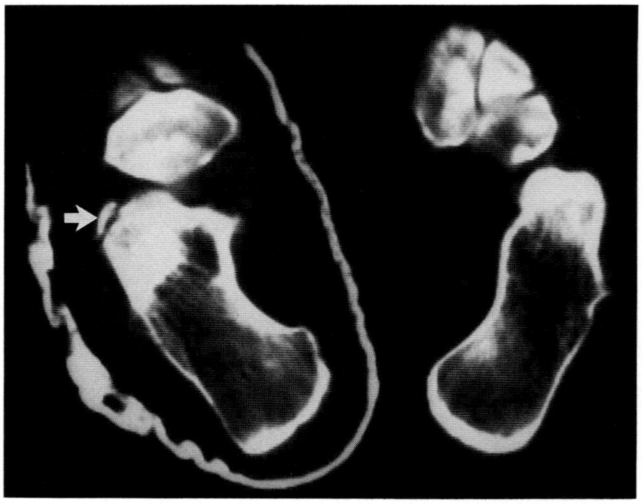

B

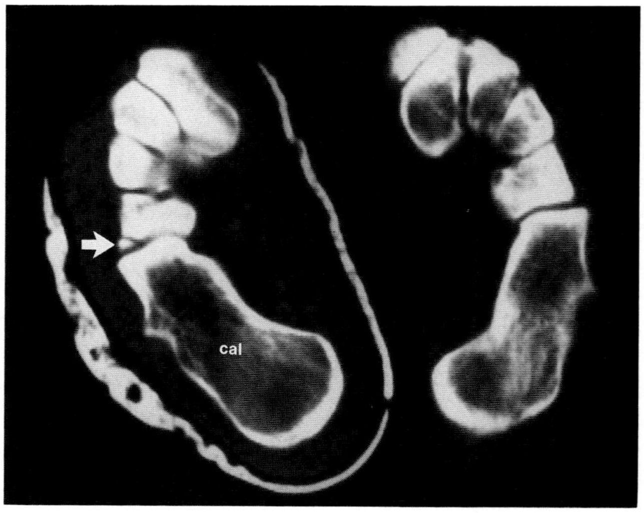

C

FIGURE 7–15
Right ankle trauma. A. *A displaced, comminuted fracture* (arrow) *of the posteromedial aspect of the right talus (tal) is seen, with extensive soft tissue swelling and a surrounding cast.* B. *A small chip fracture can be observed along the anterolateral aspect of the right calcaneus.* C. *A slightly displaced fracture* (arrow) *of the posterolateral surface of the right cuboid is seen to be extending into the joint space between the calcaneus (cal) and the cuboid.*

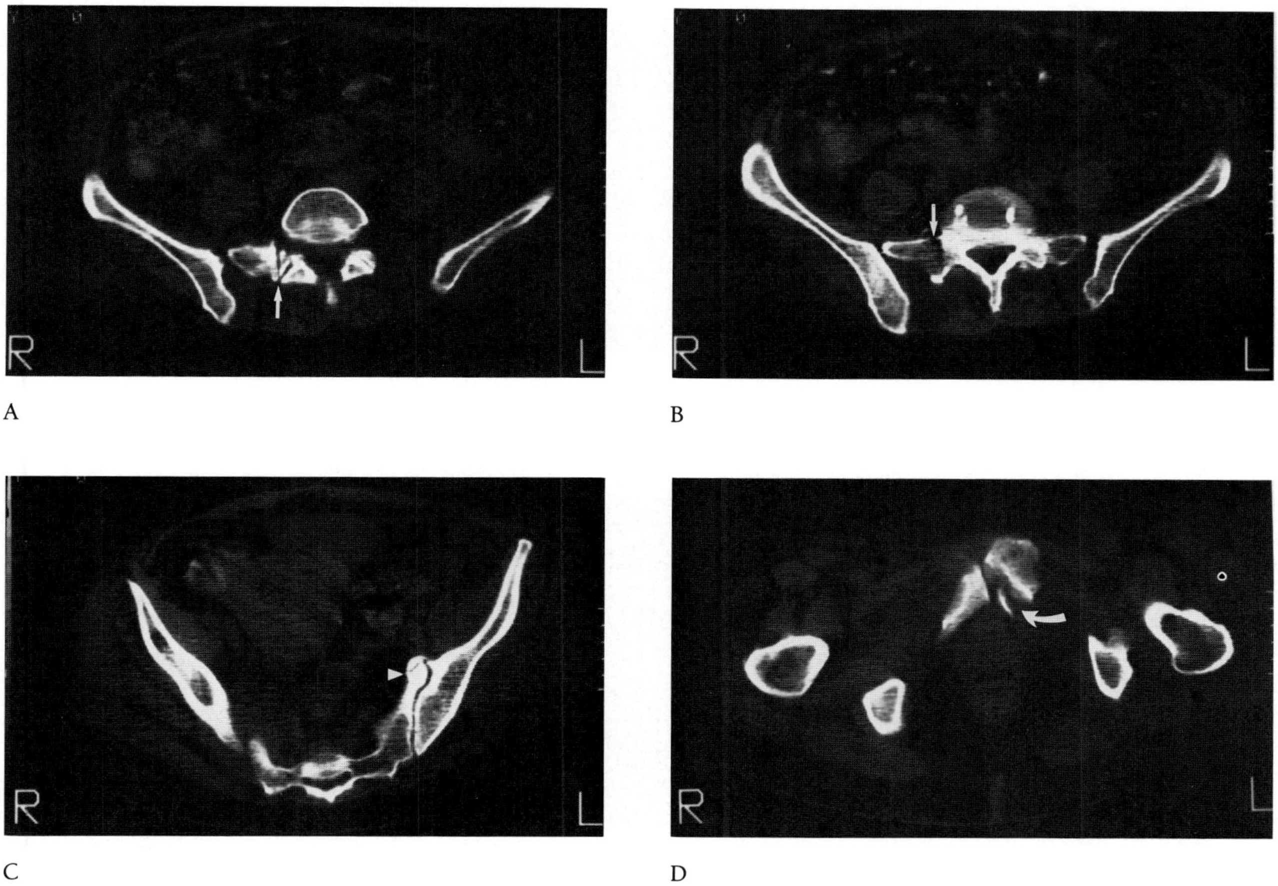

FIGURE 7–16

Trauma patient with multiple fractures. A, B. *A comminuted fracture is seen on the right side of the sacrum* (arrows). C. *Small avulsion off of the anterior sacrum, near the sacroiliac joint, is well visualized* (arrowhead). D. *A displaced fracture of the left pubic bone* (curved arrows) *is shown.*

fragments are often missed, but can be well demonstrated by CT arthrotomography. After the injection of air and contrast material into the joint space, thin sections can easily demonstrate pathologic changes [34]. Multiple traumatic abnormalities of the wrist and carpal tunnel [35,36], complex ankle fractures [37], and even sternoclavicular joint trauma [38] are best evaluated by CT. Calcaneal fractures are extremely difficult to document with routine radiographs, and account for almost 60% of all tarsal fractures. CT has been able to consistently demonstrate fractures and fracture fragments, as well as the posttrauma trapping and impingement of tendons [39–41].

CT Compared with Other Modalities

Many multimodality evaluations have been performed in order to determine the precise role for CT in the evaluation of abnormalities of the musculoskeletal system. One comparison of CT, conventional radiography and tomography, and scintigraphy was undertaken to evaluate the utility of each technique in the treatment planning of

128 patients with confirmed primary bone malignancies [10]. Assessed was the ability to accurately define the extent of disease, soft tissue extension, and spatial relationships to major vessels. CT was found to provide the most useful and complete information, although as a primary diagnostic tool it was necessary in only 4% of the cases. One major weakness is the difficulty in distinguishing hyperemia and edema in reactive soft tissues from actual tumor. CT also accurately evaluates the extent of bony involvement, which was often overestimated by scintigraphy [26]. Many similar studies have echoed comparable results: Although CT is not usually necessary to establish a diagnosis, it does demonstrate a marked effect on the determination of the extent of disease and treatment planning [27,28].

Postoperative and posttherapy evaluations of patients with malignancies rely heavily on serial CT studies. Sequential scans are able to detect the recurrence of tumor postoperatively much earlier than plain films or conventional tomography do. Bone scans are quite reliable for detecting distant metastases; however, these may remain positive in the operative region for a considerable length of time, making evaluation of an early recurrence difficult. The evaluation of tumor response to therapy has been quite effective when monitored by CT. Changes in the size and consistency (e.g., necrosis) of tumor masses can easily and accurately be observed by CT, in order to monitor response to chemotherapy and radiation therapy [9,29,30].

Conclusions

CT has become an effective and essential modality in the evaluation of diseases of the musculoskeletal system. Even though it infrequently aids in the diagnosis of bone neoplasms, it has become an integral part of determinating the exact magnitude of disease within bone, as well as evaluating extension into the soft tissues. It is significantly better than radiographs or scintigraphy for the evaluation of soft tissue tumors, and can accurately monitor tumor sites for recurrence, or for therapeutic response. CT has also become an indispensable tool for evaluating joints and possible occult fractures within the pelvis.

REFERENCES

1. Moser RP Jr, Madewell JE. An approach to primary bone tumors. Radiol Clin North Am 1987; 25:1049–93.
2. Levine E. Computed tomography of musculoskeletal tumors. CRC Crit Rev Diagn Imaging 1981; 19:279–309.
3. McNeil BJ. Value of bone scanning in neoplastic disease. Semin Nucl Med 1984; 14:277–286.
4. Arger PH, Coleman B, Dalinka MK. Computed tomography in orthopedics. Orthop Clin North Am 1983; 14:217–32.
5. Petasnick JP. CT of the extremities. Semin Ultrasound CT MR 1986; 7:308–19.
6. Passariello R, Trecco F, DePaulis F, DeAmicis R, Bonanni G, Masciocchi C. Computed tomography of the knee joint: clinical results. J Comput Assist Tomogr 1983; 7:1043–9.
7. Deutsch AL, Resnick D, Mink JH. Computed and conventional arthrotomography of the glenohumeral joint: normal anatomy and clinical experience. Radiology 1984; 153:603–9.
8. Levine E, Lee KR, Neff JR, Maklad NF, Robinson RG, Preston DF. Comparison of computed tomography and other imaging modalities in the evaluation of musculoskeletal tumors. Radiology 1979; 131:431–7.
9. Heare TC, Enneking WF, Heare MM. Staging techniques and biopsy of bone tumors. Orthop Clin North Am 1989; 20:273–85.
10. Lukens JA, McLeod RA, Sim FH. Computed tomographic evaluation of primary osseous malignant neoplasms. AJR 1982; 139:45–8.
11. Marcove RC, Mike V, Hajek JV, Levin AG, Hutter RV. Osteogenic sarcoma under the age of twenty-one. A review of 145 operative cases. J Bone Joint Surg [Am] 1970; 52:411–23.
12. Levine E, Neff JR. Dynamic computed tomography scanning of benign bone lesions: preliminary results. Skeletal Radiol 1983; 9:238–45.
13. Haaga JR, Alfidi RJ. Precise biopsy localization by computed tomography. Radiology 1976; 118:603–7.
14. Hardy DC, Murphy WA, Gilula LA. Computed tomography in planning percutaneous bone biopsy. Radiology 1980; 134:447–50.

15. Helms CA, Cann CE, Brunelle FO, Gilula LA, Chafetz N, Genant HK. Detection of bone-marrow metastases using quantitative computed tomography. Radiology 1981; 140:745–50.
16. Hermann G, Rose JS. Computed tomography in bone and soft tissue pathology of the extremities. J Comput Assist Tomogr 1979; 3:58.
17. Ram PC, Martinez S, Korobkin M, Breiman RS, Gallis HR, Harrelson JM. CT detection of intraosseous gas: a new sign of osteomyelitis. AJR 1981; 137:721–3.
18. Kenney PJ, Gilula LA, Murphy WA. Computed tomography to distinguish osteochondroma and chondrosarcoma. Radiology 1981; 139:129–39.
19. Reig-Boix V, Guinot-Tormo J, Risent-Martinez F, Aparisi-Rodriguez F, Ferrer-Jimenez R. Computed tomography of intraosseous lipoma of os calcis. Clin Orthop 1987; 221:286–91.
20. Gamba JL, Martinez S, Apple J, Harrelson JM, Nunley JA. Computed tomography of axial skeletal osteoid osteomas. AJR 1984; 142:769–72.
21. McConnell B, Dell PC. Localization of an osteoid osteoma nidus in a finger by use of computed tomography: a case report. J Hand Surg 1984; 9A:139–41.
22. Weeks RG, McLeod RA, Reiman HM, Pritchard DJ. CT of soft tissue neoplasms. AJR 1985; 144:355–60.
23. Weinberger G, Levinsohn EM. Computed tomography in the evaluation of sarcomatous tumors of the thigh. AJR 1978; 130:115–8.
24. Hunter JC, Johnston WH, Genant HK. Computed tomography evaluation of fatty tumors of the somatic soft tissues: clinical utility and radiologic-pathologic correlation. Skeletal Radiol 1979; 4:79–91.
25. Warwick R, Williams PL. Gray's anatomy. 35th British ed. Philadelphia: WB Saunders, 1973:535–77.
26. Hudson TM, Scheibler M, Springfield DS, Enneking WF, Hawkins IF Jr, Spanier SS. Radiology of giant cell tumors of bone: computed tomography, arthro-tomography, and scintigraphy. Skeletal Radiol 1984; 11:85–95.
27. Destouet JM, Gilula LA, Murphy WA. Computed tomography of long-bone osteosarcoma. Radiology 1979; 131:439–45.
28. deSantos LA, Bernadino ME, Murray WA. Computed tomography in the evaluation of osteosarcoma: experience with 25 cases. AJR 1979; 132:535–40.
29. Shrikhoda A, Jaffe N, Wallace S, Ayala A, Lindell MM, Zornoza J. Computed tomography of osteosarcoma after intraarterial chemotherapy. AJR 1985; 144:95–9.
30. Mail JT, Cohen MD, Mirkin LD, Provisor AJ. Response of osteosarcoma to preoperative intravenous high-dose methotrexate chemotherapy: CT evaluation. AJR 1985; 144:89–93.
31. Pearson JB, Hargadon EJ. Fractures of the pelvis involving the floor of the acetabulum. J Bone Joint Surg [Am] 1962; 44:550.
32. Gilula LA, Murphy WA, Tailor CC, Patel RB. Computed tomography of the osseous pelvis. Radiology 1979; 132:107–14.
33. Griffiths HJ, Standertskjöld-Nordenstam CG, Burke J, Lamont B, Kimmel J. Computed tomography in the management of acetabular fractures. Skeletal Radiol 1984; 11:22–31.
34. Deutsch AL, Resnick D, Mink JH, et al. Computed and conventional arthrotomography of the glenohumeral joint: normal anatomy and clinical experience. Radiology 1984; 153:603–9.
35. Wechsler RJ, Webbe MA, Rifkin MD, Edeiken J, Branch HM. Computed tomography diagnosis of distal radioulnar subluxation. Skeletal Radiol 1987; 16:1–5.
36. Patel RB. Evaluation of complex carpal trauma: thin-section direct longitudinal computed tomography scanning through a plaster cast. Comput Tomogr 1985; 9:107–9.
37. Cooperman DR, Spiegel PG, Laros GS. Tibial fractures involving the ankle in children. The so-called triplane epiphyseal fracture. J Bone Joint Surg [Am] 1978; 60:1040–6.
38. Stark P. Computed tomography of the sternum. CRC Crit Rev Radiol 1987; 27:321–49.
39. Rosenberg ZS, Feldman F, Singson RD, Price GJ. Peroneal tendon injury associated with calcaneal fractures: CT findings. AJR 1987; 149:125–9.
40. Rosenberg ZS, Feldman F, Singson RD. Intra-articular calcaneal fractures: computed tomographic analysis. Skeletal Radiol 1987; 16:105–13.
41. Guyer BH, Levinsohn EM, Fredrickson BE, Bailey GL, Formikell M. Computed tomography of calcaneal fractures: anatomy, pathology, dosimetry, and clinical relevance. AJR 1985; 145:911–9.

8

Principles of MRI

R. JAMES R. KNOWLES

MRI is generally considered much more esoteric than CT or radionuclide studies. CT uses x-ray attenuation measurements to determine high-resolution cross-sectional anatomy (along with some ancillary physiologic information inferred from the behavior of injected contrast materials). Radiotracer studies, on the other hand, tend to mirror selected physiologic processes rather closely but are somewhat lacking in anatomic detail. MRI provides both anatomic detail and a broad physiologic sensitivity to subtle soft tissue and functional differences. Furthermore, MRI uses only magnetic fields and radiowaves: There is no ionizing radiation as is required by both CT and nuclear medicine. The theory underlying MRI is still relatively unfamiliar to those in medicine but is not inherently harder than many other theories already mastered. Here we only consider the highlights of MRI theory. Fortunately, there are now many excellent books [1–8], as well as a recent review article [9], for those who wish to pursue these matters further.

Contributions to the MR Image

Many factors in both the patient and the instrumentation make important contributions to the MR image. It is easier to understand imaging sequences if one has first analyzed the basic building blocks of the imaging process.

MAGNETIC NUCLEI

Atomic nuclei are composed of protons and neutrons, both of which possess intrinsic spins and magnetic moments. Protons and neutrons also have a strong tendency to pair; that is, two protons or two neutrons will line up opposite to each other so that their spins and magnetic moments cancel out. Unpaired protons or neutrons produce net nuclear spins and net nuclear moments; that is, they produce magnetic nuclei.

Magnetic nuclei of biologic interest include hydrogen (^{1}H), carbon (^{13}C), fluorine (^{19}F), sodium (^{23}Na), and phosphorus (^{31}P). Note that all of these nuclei have odd mass numbers, indicating an unpaired nucleon. For example, ^{13}C is magnetic, but neither ^{12}C (the more common isotope) nor ^{14}C (a radioactive isotope) is magnetic. Conventional MRI images hydrogen (^{1}H) and is often called proton imaging. Because of its presence in water and lipids, hydrogen has the highest biologic abundance of any magnetic isotope. Moreover, hydrogen also has the highest intrinsic sensitivity to NMR techniques.

On the other hand, one should be aware that while the hydrogen in most water and lipids makes large contributions to the MR image, the hydrogen in proteins makes essentially no contribution. This difference in behavior is a result of differences in bonding. More restrictive bonding reduces contributions to the image, essentially by shortening the relaxation time T2 (relaxation times are discussed later). Such considerations are often expressed by saying that only "mobile" hydrogen is imaged.

STATIC MAGNETIC FIELD

Magnetic field strength can be expressed in either Gauss or Teslas (1 T = 10,000 G). The earth's magnetic field averages about 0.5 G. Most MRI units produce magnetic fields between 0.15 and 1.5 T (1,500–15,000 G).

When placed in an external magnetic field, magnetic nuclei align along the direction of the magnetic field (by convention, the static magnetic field is taken as the z-direction). This alignment process is complicated by the intrinsic nuclear spin. Because of their spins, magnetic nuclei actually precess about the direction of the external magnetic field, like spinning tops precess about the direction of the gravitational field. The frequency of this precession is given by Larmor's equation:

$$\omega = \gamma B$$

where ω is the angular frequency, γ is a constant (the gyromagnetic ratio), and B is the magnetic field strength. Note that γ is different for different magnetic nuclei (e.g., 42.58 MHz/T for ^{1}H but 17.25 MHz/T for ^{31}P). For proton imaging between 0.15 and 1.5 T, the Larmor frequency range is 6.4–64 MHz (1 MHz = 1,000,000 cycles/sec), that is, from below the CB radio bands through the lower VHF TV channels.

Increasing the static magnetic field strength not only increases the Larmor frequency, but also increases the available signal because the higher fields align more nuclei. In addition, higher magnetic fields increase T1 but leave T2 essentially unchanged [10–12].

RADIOFREQUENCY PULSES AND RELAXATION TIMES

If radiofrequency (RF) energy at the Larmor frequency is provided to magnetic nuclei precessing at the Larmor frequency, resonance occurs. Resonance facilitates the addition of energy to the precessing nuclei. This additional energy causes the angle of precession to increase so that the nuclei tip further and further away from the z-axis (Figure 8–1). If the RF energy is sufficient to cause the precessing nuclei to tip 90 degrees away from the z-axis and into the x-y or transverse plane, the associated RF pulse is called a 90-degree pulse. Similarly, if the RF energy is sufficient to tip the precessing nuclei 180 degrees so that they rotate in the opposite direction about the z-axis, the associated RF pulse is called a 180-degree pulse.

Magnetic nuclei precessing so as to produce transverse magnetic moments are especially important because it is this rotating transverse mag-

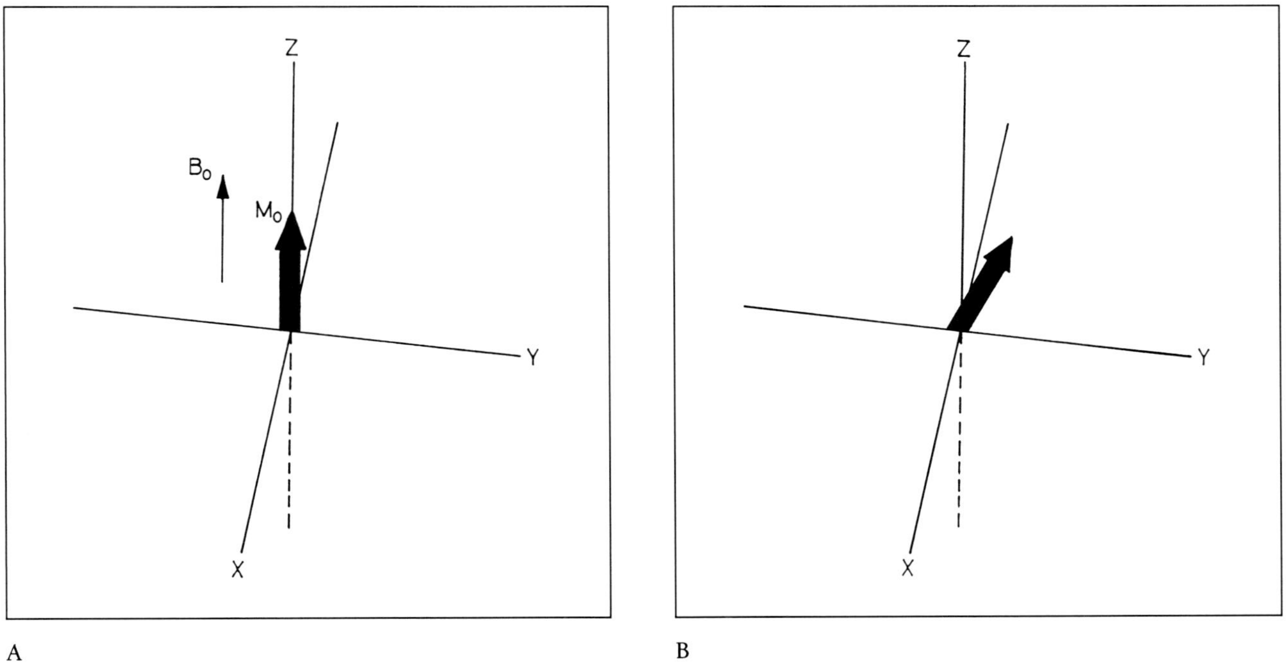

FIGURE 8–1
By convention, the direction of the static magnetic field is taken as the z-direction. A. *Magnetic nuclei tend to align with the z-axis and precess about it.* B. *Adding RF energy to the precessing nuclei causes the axis of precession to rotate away from the z-axis. The angle between the axis of precession and the z-axis is the tip (flip) angle.*

netization that induces signal from which the MR image is constructed. Since this signal is generated by precession at the Larmor frequency, it is also a Larmor frequency radiowave.

Once the RF pulse is over, various relaxation mechanisms begin to alter the precession of the tipped nuclei. Initially, all of the tipped nuclei precess together in phase (Figure 8–2A). This changes rapidly because the nuclei experience different microscopic and macroscopic magnetic fields. Those in higher fields precess faster; those in lower fields precess slower; phase coherence is lost. Transverse magnetization disappears as the randomized moments cancel each other out (Figure 8–2B). The time constant of this dephasing process is T2. In the time T2, 63% of the initial phase coherence is lost. T2 is also known as the transverse relaxation time.

Another set of relaxation mechanisms cause the tipped nuclei to realign with the static magnetic field (Figures 8–2C and D). The time constant of this realigning process (as well as of the initial aligning process) is T1. In the time T1, 63% of the nuclei align or realign with the external magnetic field. T1 is also known as the longitudinal relaxation time.

In simple systems such as some crystal lattices (where T1 is known as the spin-lattice relaxation time and T2 is known as the spin-spin relaxation time), relaxation mechanisms are well understood. In biologic systems, relaxation mechanisms are much more of a mystery. T1 and T2 tell us something about the physical and chemical environment of the magnetic nuclei. In biologic systems, proton relaxation times are probably largely determined by interactions between water and various macromolecules [12]. Theoretically, T1 must always be greater than or equal to T2. In biologic systems, T1 (typically 200–800 milliseconds for soft tissues, depending on field strength) is substantially greater than T2 (typically 50–150 milliseconds) [11,12].

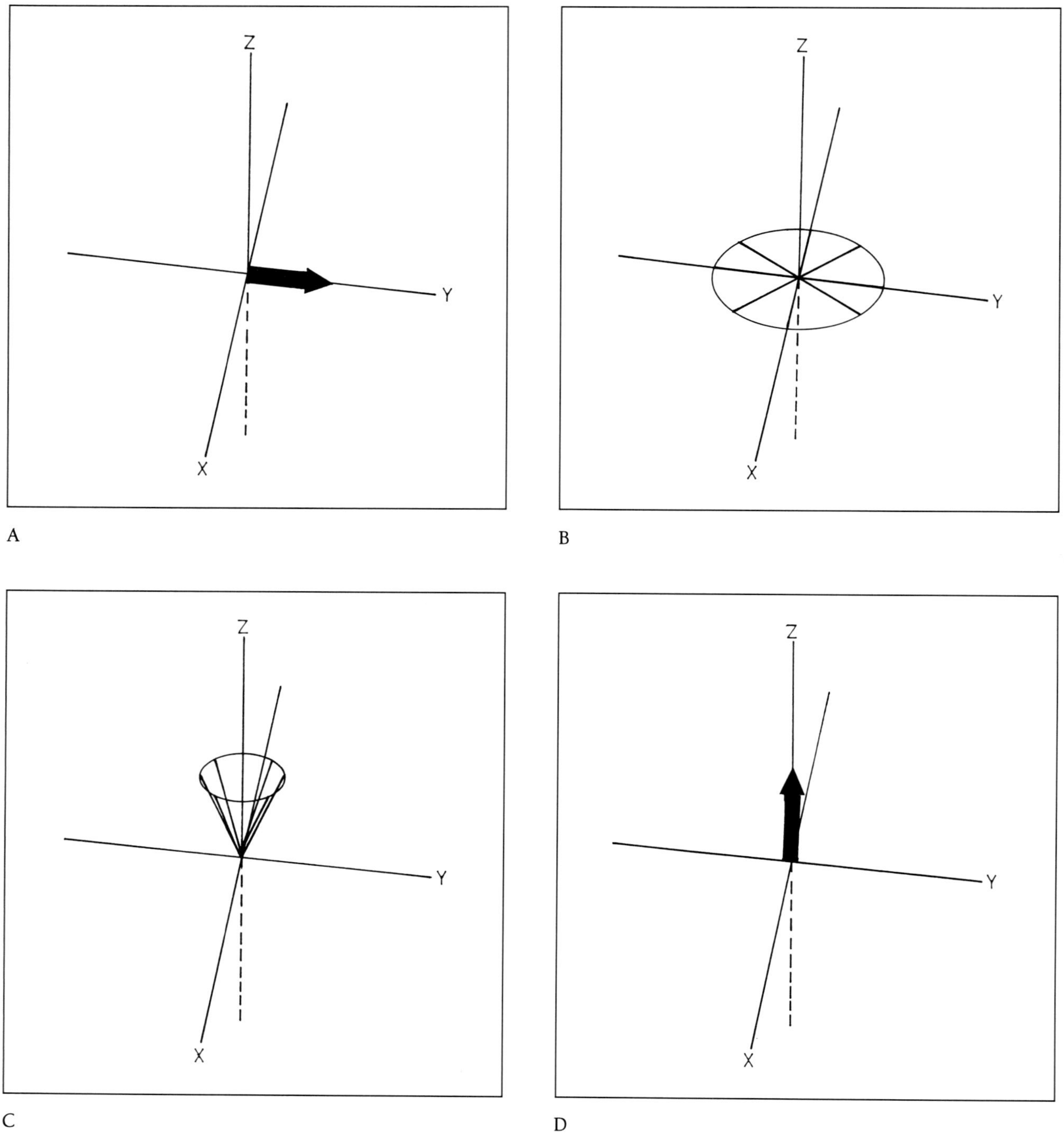

FIGURE 8–2
A 90-degree pulse is followed by T2 and T1 relaxation. A. *A 90-degree pulse tips the axis of precession into the transverse plane where the nuclei initially precess together in phase.* B. *T2 relaxation mechanisms rapidly dephase the spins.* C. *After T2 relaxation is well under way (because T2 is much shorter than T1 in biologic materials), T1 relaxation mechanisms cause substantial realignment of the nuclei with the static magnetic field in the z-direction.* D. *T1 relaxation is complete and the z-magnetization is restored.*

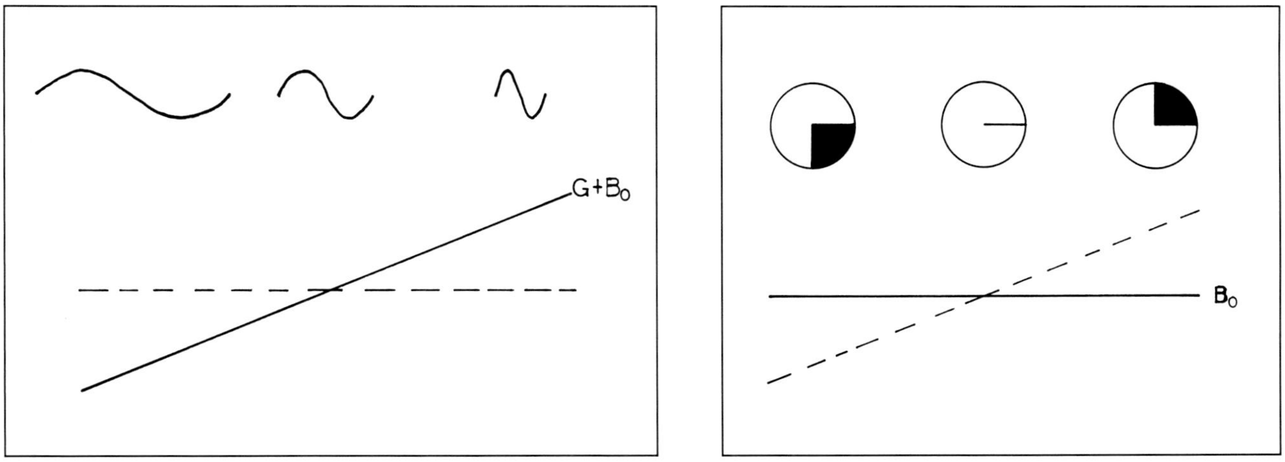

A B

FIGURE 8–3
Magnetic gradients produce linear changes in the z-magnetic field as a function of position. A. When the gradient is on, nuclei in higher magnetic fields have higher Larmor frequencies than nuclei in lower magnetic fields. B. When the gradient has been turned off, frequency differences disappear but phase differences remain because the nuclei that experienced higher magnetic fields precessed faster than the nuclei that experienced lower magnetic fields when the gradient was on.

MAGNETIC GRADIENTS AND SIGNAL LOCALIZATION

Magnetic gradients are small systematic changes in the magnetic field strength as a function of position. For example, a z-gradient implies a systematic increase or decrease in the z-magnetic field as one moves in the z-direction, and a y-gradient implies a systematic increase or decrease in the z-magnetic field as one moves in the y-direction. Note that in MRI an x-gradient does not imply an x-magnetic field, and a y-gradient does not imply a y-magnetic field, but both imply changes in the z-magnetic field as one moves in a particular direction.

When a gradient is turned on, some nuclei experience a higher magnetic field and some experience a lower magnetic field as a result of their positions (Figure 8–3). By Larmor's equation, the nuclei in the higher field have a higher precessional frequency whereas those in the lower field have a lower precessional frequency. Thus there is a one-to-one correspondence between signal frequency and position along the gradient direction (see Figure 8–3A). This association of frequency and location is termed frequency encoding. Frequency encoding can be used during the reading of a signal to determine its location relative to a read gradient. Similarly, frequency encoding can be used during RF excitation to restrict the excitation to nuclei in a certain limited region along the slice selective gradient.

After a gradient has been turned off, all the nuclei return to the same precessional frequency, but an effect of the gradient lingers on. When the gradient was on, some nuclei precessed faster and ran ahead while others precessed slower and lagged behind. These differences in phase (phase = frequency × time) remain after the gradient has been turned off (see Figure 8–3B). Such an association of phase and position is termed phase encoding. Phase differences can be detected by a phase-sensitive or quadrature detector. Since all gradients, including frequency encoding gradients, produce phase encoding, phase encoding is sometimes an undesirable side effect. In such cases, phase encoding can be removed with an equal but opposite gradient called a compensatory or rephasing gradient.

COMPUTER CONTROL AND IMAGE RECONSTRUCTION

RF pulses must be precisely tailored and delivered at precisely the right time. Similar constraints exist for gradient pulses and for the reading of signals. Such processes are computer-controlled. Furthermore, signals are measured as functions of frequency and phase. A computer is required to convert this frequency and phase information into an image that shows signal intensity as a function of position. This reconstruction is most commonly accomplished by using Fourier transforms, which mathematically link the spatial domain to the frequency domain (and vice versa). The resulting image matrix is typically 256 × 256, but lower (and occasionally higher) resolutions as well as unequal vertical and horizontal resolutions (e.g., 128 × 256) are possible. Each element of the image matrix is a picture element (pixel) that represents a volume element (voxel) in the object (patient or phantom).

Because of the almost universal use of quadrature detection in MRI, there are actually two raw images reconstructed. These are separated by 90 degrees of phase and are usually termed the real (cosine) and imaginary (sine) images. The real and imaginary images are sensitive to phase differences and can be used to calculate other phase-sensitive images. Most commonly, however, they are used to calculate magnitude images. A magnitude image is calculated on a computer by taking (pixel by pixel) the square root of the sum of the real squared value and the imaginary squared value. Note that magnitude images can have only positive valued pixels while phase-sensitive images can have positive or negative pixels (even though most displays fail to distinguish negative from positive values). Computers are also necessary for further image processing, the generation of video displays, and both short-term and long-term archiving of images.

Imaging Sequences

RF pulses of various kinds can be combined in many different ways. Similarly, magnetic gradients can be used in many different ways to locate signals in three dimensions. There are even different ways to form the signals from which MR images are reconstructed. Most practical imaging systems do not read the free induction decay (FID) that forms immediately after transverse magnetization is created; instead they create some form of echo that is read at a later time.

EXAMPLES OF PULSE SEQUENCES

Perhaps the simplest example of an RF sequence is a train of equally spaced 90-degree pulses (Figure 8–4). After each pulse, a certain amount of T1 recovery occurs before the next pulse. If this repetition time (TR) is smaller than or equal to T1, the sequence is termed a partial saturation sequence [13]. If TR is much larger than T1 (or if an additional dephasing pulse [13] is inserted), the sequence is termed a saturation recovery sequence.

If a 180-degree pulse is placed before the 90-

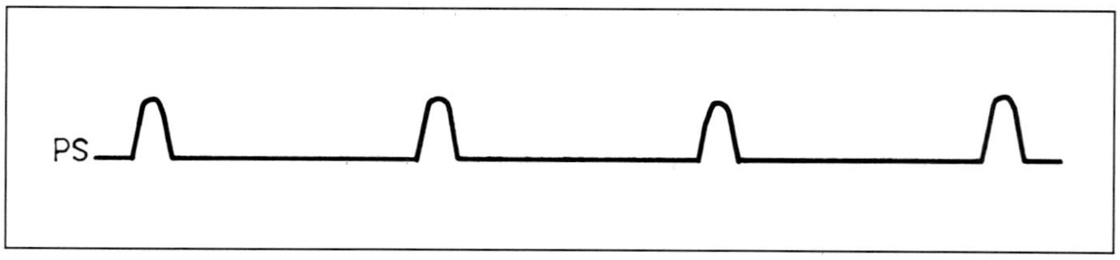

FIGURE 8–4
A train of 90-degree pulses spaced a time TR apart is termed a partial saturation sequence if TR is less than or equal to T1 and termed a saturation recovery sequence if TR is much greater than T1.

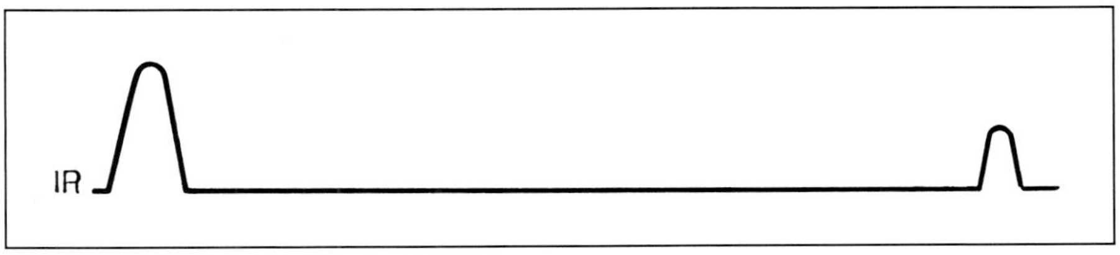

FIGURE 8–5
Placing a 180-degree pulse a time TI before a 90-degree pulse creates an inversion recovery sequence. TR is now the time between successive 180-degree pulses or the time between successive 90-degree pulses.

degree pulse, the sequence is termed an inversion recovery (IR) sequence (Figure 8–5). By initially inverting the nuclei and allowing recovery for an inversion time (TI) before applying the 90-degree pulse, signals with given T1 values can be either accented or suppressed. For example, if TI is chosen so that nuclei with a certain T1 have just recovered to the transverse plane when the 90-degree pulse is given, these nuclei will be inverted so that they have no transverse component and thus make no contribution to the image. Thus, the presence of the initial 180-degree pulse increases the T1 weighting of the resulting images.

If the 180-degree pulse is placed after the 90-degree pulse, a spin-echo sequence is created (Figure 8–6). Here the 180-degree inversion occurs after the nuclei are already precessing in the transverse plane. Hence the 180-degree pulse just flips the nuclei from one part of the transverse plane to another part, but it flips them so that the direction of rotation is reversed. That is equivalent to flipping the more rapidly precessing nuclei behind the more slowly precessing ones. As the more rapidly precessing nuclei catch up with the more slowly precessing ones, the nonrandom part of the dephasing is eliminated and the rephasing spins produce a spin echo which is like two FIDs placed back to back. This spin echo occurs at a time (TE) after the 90-degree pulse and at a time TE/2 after the 180-degree pulse. Moreover, the 180-degree pulse can be repeated at times (3/2)TE, (5/2)TE, and so on, to produce additional echoes at 2TE, 3TE, etc. Each succeeding echo in such a multiecho sequence will be smaller than its predecessors because of T2 decay (the random component of dephasing that is not affected by 180-degree pulses), and each successive echo is said to be more highly T2-weighted.

BASIC SPIN-ECHO SPIN-WARP SEQUENCE

Most MRI systems employ spin-echo RF sequences with a mixture of frequency and phase

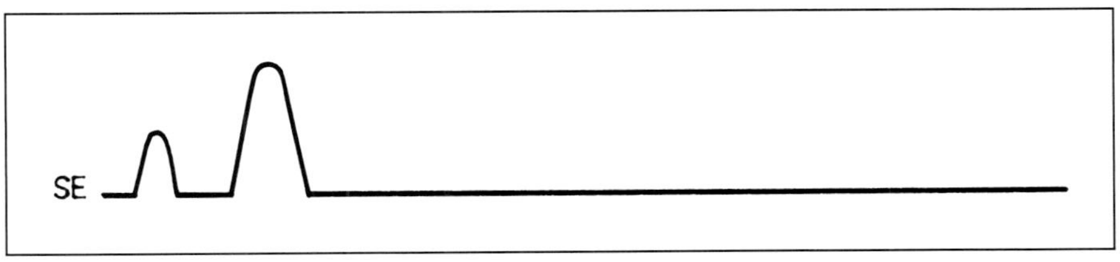

FIGURE 8–6
Placing a 180-degree pulse a time TE/2 after a 90-degree pulse creates a spin-echo sequence. The echo occurs at a time TE after the 90-degree pulse, and the entire sequence is repeated after a time TR.

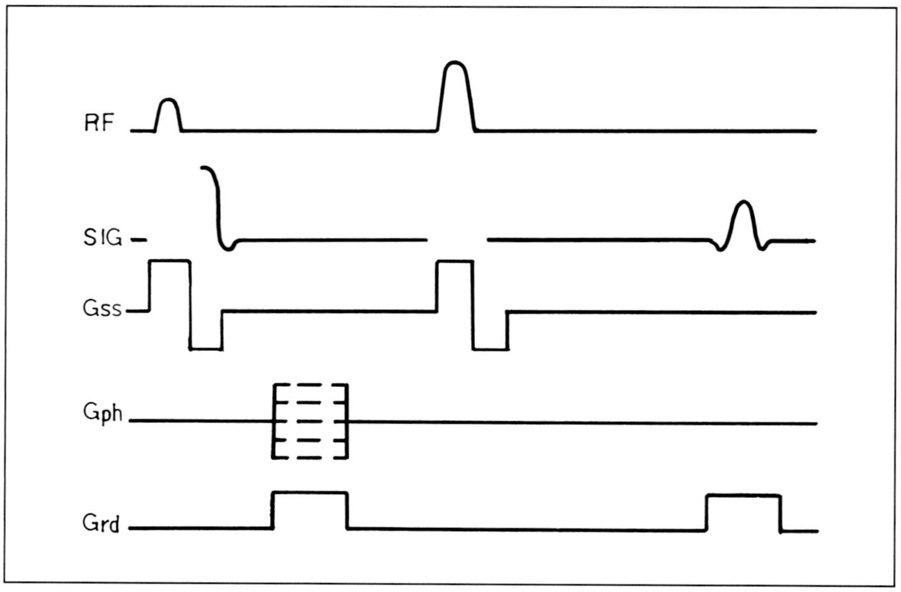

FIGURE 8–7
This timing diagram of the basic spin-echo spin-warp sequence graphs amplitude vertically and time horizontally. There are separate lines for RF transmission, signal (SIG), read gradient (Grd), phase encoding gradient (Gph), and slice selection gradient (Gss). Note that the signal line is gated off during the RF transmissions and that both RF pulses are slice selective.

encoding gradients termed spin-warp. Figure 8–7 presents this combination of RF and gradient pulses in the form of a timing diagram. In this diagram, time runs horizontally and amplitude is presented vertically. Separate lines are provided for the RF, signals, and three gradients.

In Figure 8–7, Grd is a read gradient, a frequency encoding gradient that is turned on during the reading of a spin echo to provide localization in one direction. Spatial resolution in this direction simply depends on the accuracy with which signal intensity is measured as a function of frequency. Gph is a phase encoding gradient used to provide localization in a direction orthogonal to the Grd. Because conventional detectors can only measure the sum of the phases produced by a gradient, Gph must be repeated at a sufficient number of different positive and negative values to provide the required spatial resolution in the direction of the gradient (e.g., Gph must be turned on and off at 128 different values to produce a matrix resolution of 128 in the phase encoding direction). Gss is a slice selective gradient in the third orthogonal direction. Gss is turned on during both RF pulses. Because the RF pulses are selective (i.e., they contain only a narrow band of frequencies), the frequency encoding of Gss restricts excitation to a narrow band in the direction of the gradient and thus provides localization in the third direction.

Three other features of Figure 8–7 are worth noting. First, the signal line is gated off during RF pulses since kilowatts of RF power would overload circuits designed to detect microwatts of signal. Secondly, there is an FID immediately after the 90-degree pulse, but no attempt is made to read it. Thirdly, both Gss and Grd have compensatory (rephasing) gradients to remove undesirable phase encoding. The compensatory lobe of Gss is opposite in sign to the main lobe and equal in area to the part of the main gradient after the effective RF pulse (about half of the main gradient area). The compensatory gradient for Grd is equal to the area of the main gradient

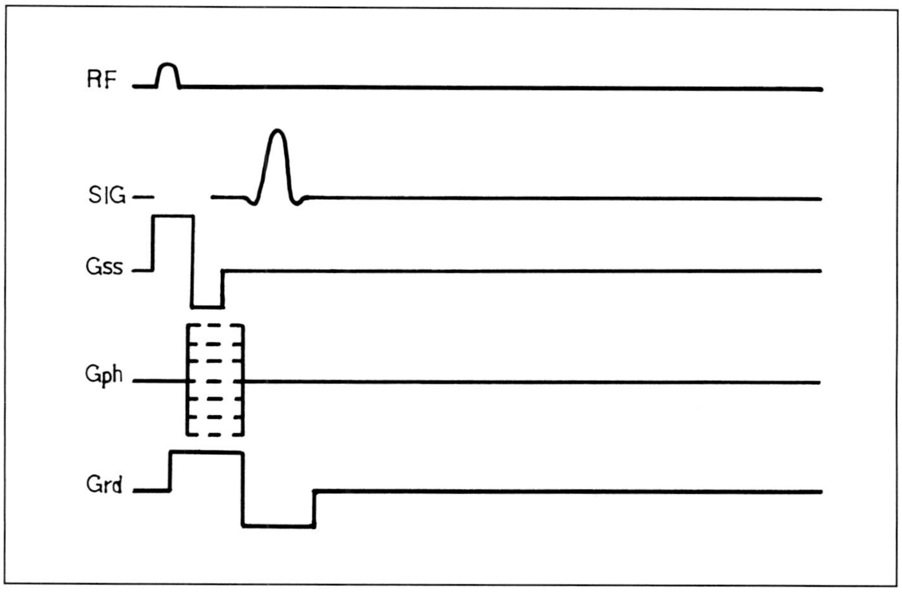

FIGURE 8–8
This timing diagram of the basic gradient-echo spin-warp sequence graphs amplitude vertically and time horizontally. There are separate lines for RF transmission, signal (SIG), read gradient (Grd), phase encoding gradient (Gph), and slice selection gradient (Gss). Note that the signal line is gated off during the RF transmissions and that the slice selective pulse is a small-angle pulse. TE is essentially the time between the center of the RF pulse and the center of the echo.

and of the same sign since the 180-degree pulse between the two gradients acts as an operational minus sign by reversing the precessional direction.

BASIC GRADIENT-ECHO SPIN-WARP SEQUENCE

Sequences employing gradient echoes are becoming more common because of their use in rapid scanning sequences [7–9,14]. Figure 8–8 is a timing diagram for a gradient-echo spin-warp sequence. Comparing Figures 8–7 and 8–8 reveals two principal differences between spin-echo and gradient-echo sequences. First, the gradient-echo sequence has no 180-degree pulse (and no associated slice selective gradient). Secondly, Grd has been moved up near the RF pulse and made bipolar, like Gss. The first lobe of the Grd dephases the FID; the second lobe rephases the dephased nuclei to produce the gradient echo.

Gradient echoes are similar to spin echoes, but they are not identical. There are dephasing effects that are reversed by 180-degree pulses that are not eliminated by gradient reversals. This can be illustrated by examining the equation for T2* (pronounced as "T2-star"):

$$\frac{1}{T2^*} = \frac{1}{T2} + \gamma Gd + \gamma M \frac{(X2 - X1)}{X1X2} + \cdots$$

where T2 is the relaxation time discussed above, γ is the gyromagnetic ratio found in the Larmor equation, G is the gradient strength, d is the distance along G, M is the magnetization, and X1 and X2 are two different magnetic susceptibilities. In the present context, the main point of this equation is that there are many other microscopic and macroscopic sources of dephasing that act like T2 mechanisms by changing the local magnetic field, and these other sources of dephasing add inversely to T2 (the random component of dephasing that cannot be eliminated without violating the Second Law of Thermodynamics) to create T2* (which is always shorter

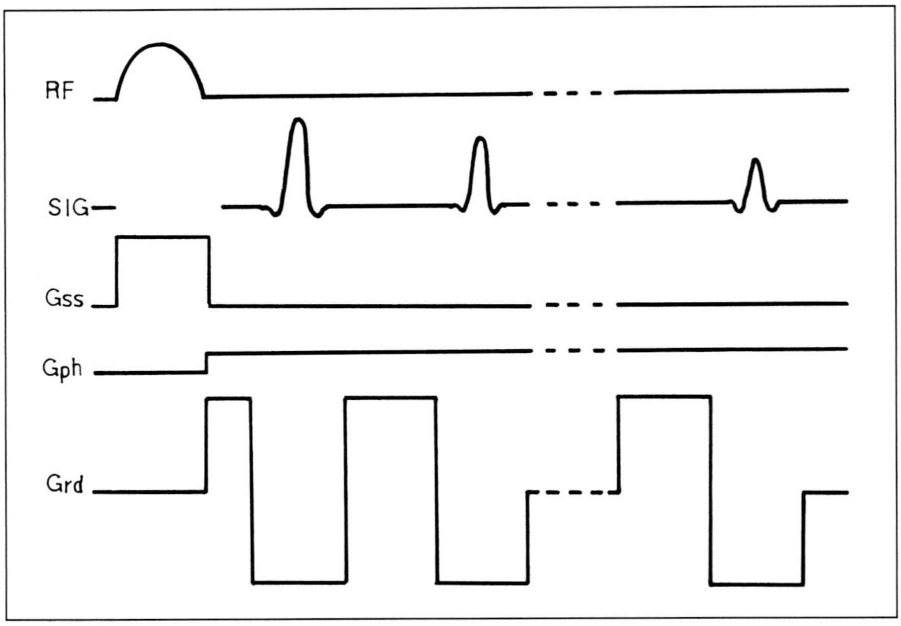

FIGURE 8–9
This timing diagram of the basic echo-planar sequence graphs amplitude vertically and time horizontally. There are separate lines for RF transmission, signal (SIG), read gradient (Grd), phase encoding gradient (Gph), and slice selection gradient (Gss). Note that the signal line is gated off during the RF transmissions and that the single RF pulse is a 90-degree pulse. The number of echoes determines the spatial resolution.

than T2). A 180-degree pulse reverses all the magnetic terms to the right of T2 (by reversing the direction of precession) so that they can be cancelled out; a gradient reversal only allows the term with the gradient in it to be cancelled out. Thus gradient echoes decay according to T2*, not just T2, which makes them more sensitive to effects such as microscopic changes in susceptibility from hemorrhage than spin echoes are [15].

One other aspect of fast gradient-echo sequences that is not immediately evident from Figure 8–8 is that the RF pulse used is generally much less than 90 degrees. Such small-angle pulses allow for faster T1 recovery (and hence more signal) when TR is made very short in order to reduce scan time.

BASIC ECHO-PLANAR SEQUENCE

In addition to these two basic imaging sequences, one other type of sequence is briefly examined here. Echo-planar techniques can be used to decrease scan time, and one variant is the basis for the first commercial scanner to image in the millisecond range [16,17]. Figure 8–9 presents the basic echo-planar sequence timing diagram. While multiple echoes are being generated by rapid reversals of Grd, Gph is continuously advancing the phase. (Note that it is also possible to advance the phase discontinuously [18].) Thus the phase encoding is different for each echo, and spatial resolution is determined by the number of echoes. The number of echoes is essentially limited by T2 or T2* (though more advanced variants can improve on this limitation by various rephasing techniques). For these reasons, signal strength can severely limit spatial resolution (the sequence in Figure 8–9 requires 64 echoes for a resolution of 64 × 64), and high magnetic fields may be helpful. Unlike spin-echo or gradient-echo sequences that repeat RF pulses with different phase encoding gradients and thus allow varying amounts of T1 recovery, there is only one 90-degree pulse in the typical echo-planar sequence. This implies that there is no inherent T1 dependence in echo-planar images, but a T1 dependence can be introduced through additional RF pulses (e.g., an initial 180-degree pulse can be used to produce an IR variant).

Many other special purpose sequences are in use, and more are proposed daily. Variants of the sequences detailed here, however, account for much of current imaging.

Controlling Image Contrast

Most imaging sequences allow the operator to vary TR. Both spin-echo and gradient-echo sequences also allow the timing of the echo (TE) to be varied, and small-angle scanning makes the tip (flip) angle an adjustable parameter.

REPETITION TIME

Adjusting TR varies the amount of T1 recovery allowed between excitations (Figure 8–10). Increasing TR increases the total amount of signal but reduces T1 discrimination. Decreasing TR decreases the total signal but increases T1 discrimination.

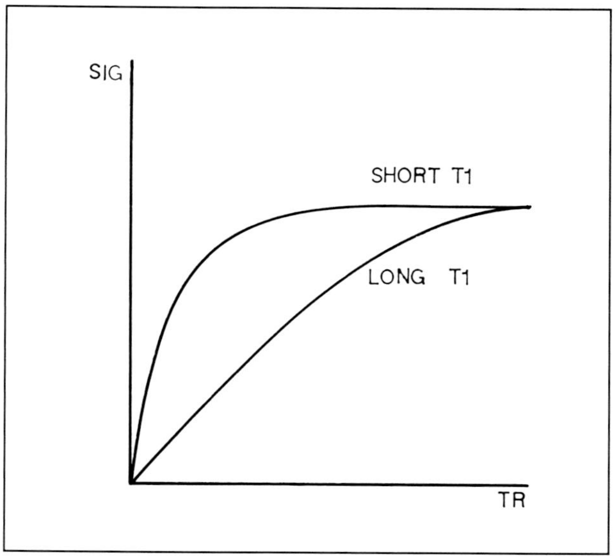

FIGURE 8–10
Signal strength is graphed vertically, and TR is graphed horizontally. Short T1 substances recover more rapidly than long T1 substances. Here both substances are assumed to have the same spin densities.

ECHO TIME

Adjusting TE varies the amount of T2 decay that occurs before the echo is read (Figure 8–11). Increasing TE decreases the total amount of signal but increases T2 discrimination. Decreasing TE increases the total signal but also decreases T2 discrimination.

In the limit of very long TRs and very short TEs, the MR image reduces to a spin density image. In the limit of very long TRs and very long TEs, maximum T2 weighting is obtained. In the limit of very short TRs and TEs, a small-angle gradient-echo sequence is generally used.

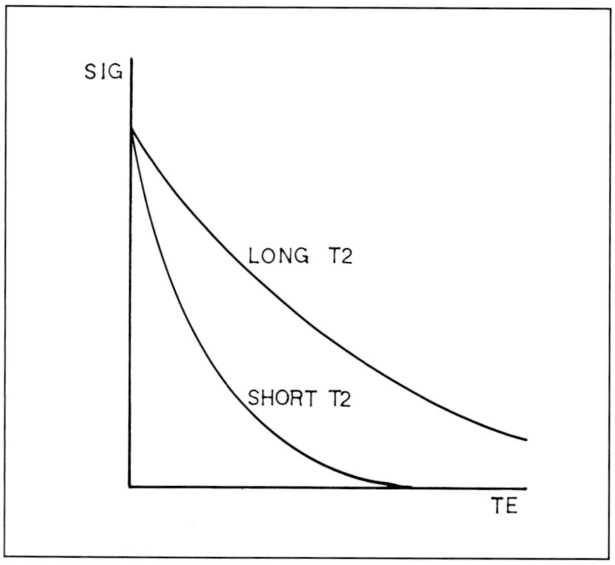

FIGURE 8–11
Signal strength is graphed vertically, and TE is graphed horizontally. Short T2 substances decay more rapidly than long T2 substances. Here both substances are assumed to have the same spin densities.

TIP (FLIP) ANGLES

In the limit of very small tip angles, the MR signal strength becomes dependent only on the density of the magnetic nuclei (Figure 8–12). For somewhat larger tip angles, the signal strength depends on the magnitude of TR relative to the relaxation times of the object imaged. For TRs near T1, short T1 components provide maximum signals at larger tip angles than do longer T1 components (see Figure 8–12A). For TRs near

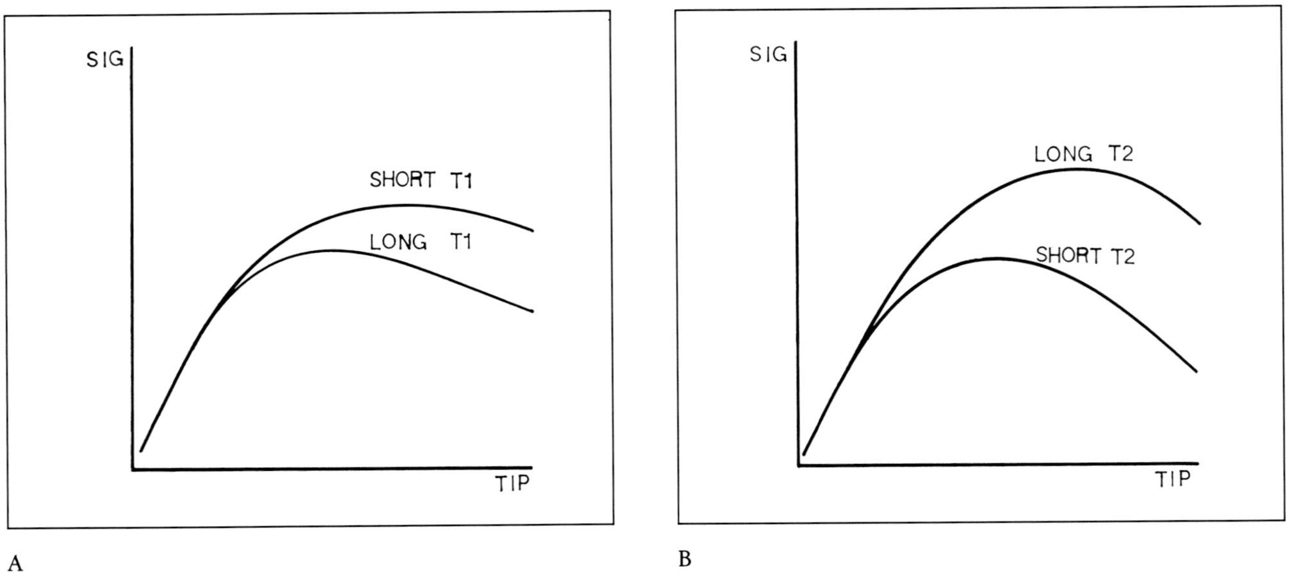

FIGURE 8–12
Signal strength is graphed vertically, and tip angles from 0 to 90 degrees are graphed horizontally. For very small tip angles, the signal intensity is determined by spin densities (assumed equal in these graphs). A. *As TR approaches T1, the signal strength at larger tip angles is determined by the relative size of T1.* B. *As TR approaches T2*, the signal strength at larger tip angles is determined by the relative size of T2*. A short T2* has a similar effect to a long T1 (and vice versa) because T2* represents a decay process while T1 represents a recovery process. In all cases, the tip angles that produce maximum signals are known as Ernst angles.*

T2 (or T2*), short T2 components provide maximum signals at smaller tip angles than do longer T2 components (see Figure 8–12B). This reciprocity between T1 and T2 results from the way T1 enters as a recovery process while T2 enters as a decay process. Moreover, because small-angle techniques are generally used with gradient echoes, T2 dependence is often replaced by T2* dependence. In all cases, the tip angles at which maximum signal strength is obtained are known as Ernst angles [19].

Many useful effects can be achieved by adjusting tip angles. For example, to make a myelographic image of the spine (i.e., to make the CSF signal greater than the spinal cord signal), a traditional spin-echo sequence with a 90-degree pulse requires a long TE to decay away the cord signal; in turn, this requires a very long TR to produce adequate signal in the CSF with its long T1 and T2. An even better myelographic effect can be achieved with a short TR gradient echo by adjusting the tip angle to the Ernst angle for CSF. This maximization of the CSF signal is aided by the slightly higher proton density of CSF.

Scan Time

During the last few years, much effort has been devoted to reducing scan time. These reductions allow sicker patients (e.g., trauma patients) to be scanned, eliminate the effects of some physiologic and voluntary motions, and may improve the economics of MRI by increasing patient throughput (even though many other factors beside machine time enter into throughput considerations).

THREE BASIC COMPONENTS

Total scan time has three basic components:

$$T = (n)(N)(TR)$$

where n is the number of times the signal is read, N is the number of times the sequence is repeated for phase encoding, and TR is the repetition time. If the signal-to-noise ratio (S/N, or SNR) is low, each signal may have to be read many times (e.g., n = 4 or 8) so that the results can be added together to achieve an adequate image. Both higher magnetic fields and better designed RF coils can improve the S/N and thus reduce n. Such improvements were among the first made, and currently most high-field scans are done with n = 1. Reductions in N and TR are discussed in a moment. First, however, it is helpful to examine these parameters in a little more detail.

For a traditional two-dimensional spin-warp technique with

$$Gx = Grd$$
$$Gy = Gph$$
$$Gz = Gss$$

N = Ny is simply the number of times the sequence is repeated with different y-gradients to improve the spatial resolution in the y-direction. For example, N = 128 is required for a matrix resolution of 128 in the y-direction, and N = 256 is required for a resolution of 256. One phase encoding gradient and one frequency encoding gradient allow signal to be obtained from an entire projected column of the object and thus produce a better S/N than two orthogonal frequency encoding gradients that could only obtain signal from a small volume element (voxel). (The noise is mostly electronic and therefore essentially the same for both techniques.) A closely related S/N advantage is also evident in three-dimensional volume scans where a second phase encoding gradient (Gz = Gph in our example) is applied in the slice selective direction. Then N = NyNz and the S/N of the volume image is improved by the square root of Nz over the corresponding two-dimensional image. Volume techniques often use this gain in S/N to achieve very thin slices (as well as to eliminate the gaps between slices). Thus, for spin-warp techniques, reducing N implies reducing spatial resolution or S/N, or both, so that a different localization scheme (e.g., echo planar) is necessary to fully exploit decreases in N.

TIME USAGE, MULTI-ECHO, AND MULTI-SLICE TECHNIQUES

In a typical spin-echo spin-warp sequence, most of the scan time is spent waiting for T1 recovery (Figure 8–13). These relatively long TRs ensure adequate total signal and are critical if T1 is long or if long TEs are utilized. Furthermore, reducing TR alters image contrast.

Long recovery periods, however, can be put to other uses. During this time, multiple echoes and data from other slices can be collected. Multiecho techniques were considered earlier. Multislice techniques are most commonly achieved by varying the frequency of the selective RF pulses. This allows (theoretically) the excitation and recovery of each slice to proceed independently. A multislice sequence can contain as many slices as can have their excitations and data collections fitted into the time period TR [20]. Thus long recovery times are not necessarily a complete waste. Furthermore, fast and ultrafast scan sequences generally do not allow such multislice/multiecho techniques.

DECREASING SCAN TIME

To speed up scanning by reducing TR, gradient-echo sequences compress both the excitation and the read aspects of the sequence relative to the traditional spin-echo sequence (see Figures 8–7 and 8–8). In addition, small-angle techniques are utilized to increase effective recovery as compensation for shorter TRs, and TEs are usually made much shorter to allow less T2 decay so that total signal is high despite short TRs. These changes are sufficient to maintain adequate signal while reducing scan times from the minutes characteristic of traditional spin-echo sequences to the seconds characteristic of gradient-echo fast scans (see Figure 8–13).

To speed up scanning by reducing N, echo-planar sequences generate multiple echoes, each with a different phase encoding, from a single RF pulse (or single group of RF pulses). In the basic ultrafast sequence (see Figure 8–9), this allows N to be reduced to one (see Figure 8–13), and the scan time to be reduced to the millisecond range.

As we have already seen, such fast scan and ultrafast scan techniques can produce very dif-

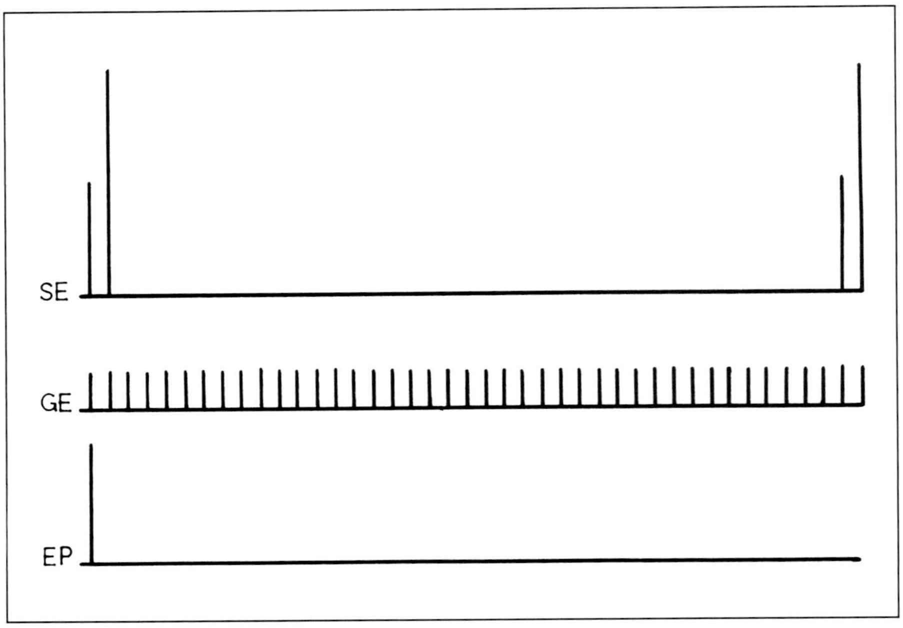

FIGURE 8–13
This reduced timing diagram compares a spin-echo (SE) sequence, a small-angle gradient-echo (GE) sequence, and an echo-planar (EP) sequence by presenting typical RF pulse sequences for each on the same time scale. The relative size of the RF pulses is also typical.

ferent image contrasts than those found in traditional spin-echo images. As described in the next section, such techniques can also substantially alter image artifacts such as those resulting from flow and chemical shift.

Common Image Artifacts

In addition to being exquisitely sensitive to patient physiology and anatomy, MRI is highly sensitive to factors in both objects (patients or phantoms) and instrumentation that produce artifacts [21]. These artifacts can either mask or mimic pathologic features, and some artifacts are even useful as diagnostic signs [21]. All artifacts must be recognized as such, and undesirable artifacts should be eliminated or reduced by a conscientiously applied quality assurance program [21,22].

MOTION ARTIFACTS

Perhaps the most commonly encountered MRI artifacts result from patient motion (Figure 8–14). Basically, these artifacts consist of displaced signal. In addition to whatever displacements occur in the direction of motion, displaced signal tends to propagate in the direction of phase encoding gradients as so-called ghosting even if the motion is not in the phase encoding direction. This ghosting is more pronounced at higher magnetic fields and is similar to that produced by various instrumental instabilities [23].

Voluntary motions can usually be eliminated by patient education and proper attention to patient comfort (although in some cases sedation or even anesthesia [24] may be necessary). Physiologic motions, on the other hand, often require gating, restricting, or restructuring of data acquisition [25–29]. Fast scan techniques can reduce or eliminate bowel and respiratory motions directly while ultrafast scan techniques can even eliminate cardiac motions.

By far the most complicated motion-related phenomena are associated with flow [4,8]. Spin-echo multislice/multiecho sequences have two or

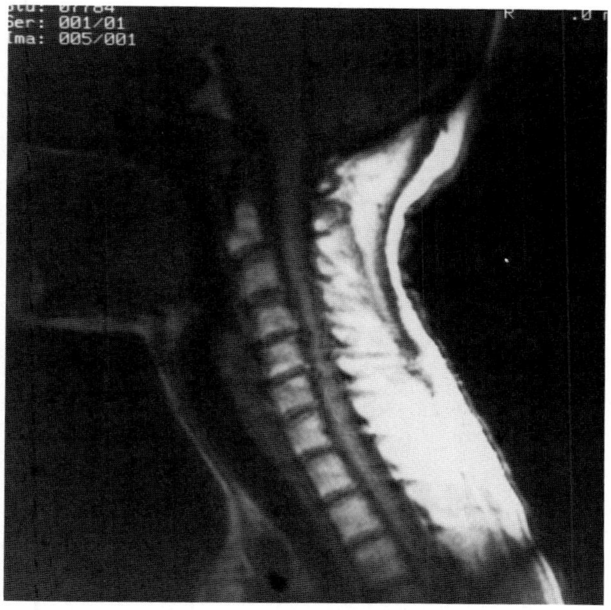

A

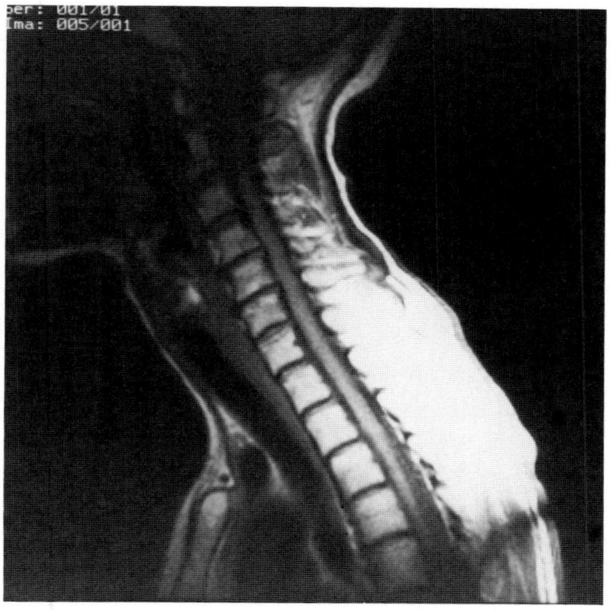

B

FIGURE 8–14
Sagittal sections through the cervical section of the spine were obtained with a posterior surface coil and a posterior-anterior phase encoding gradient. A. Hiccups produced a rather subtle series of motion artifacts that might (in part) be misinterpreted as irregularities and marked thinning of the spinal cord. B. The same slice was also imaged without the motion artifacts.

more slice-selective RF pulses. Slow flow entering the slice brings in nuclei that have not experienced previous RF pulses; these nuclei produce enhanced signals because they are not constrained by the same T1 recovery as stationary nuclei that have experienced previous RF pulses. As flow velocity increases, more of the nuclei leave the slice before the next selective pulse can create an echo; this washout reduces signal intensity. Thus, depending on flow velocity perpendicular to the slice, signal intensity can be increased or decreased by flow. This is particularly important because of the turbulence often found in actual blood flow [30]. These effects are also more pronounced where blood flows into a multislice series than where it flows out. Furthermore, there is an enhancement of even-numbered echoes because of a diffusion term that reduces the intensity of odd-numbered echoes but cancels out of even-numbered echoes [31]. Lastly, flow along gradients introduces phase modulations that are dramatic in phase-sensitive images and that may or may not be evident in conventional magnitude images.

In gradient-echo fast scan sequences, multislice techniques are seldom used, and data are usually acquired one slice at a time (even though volume techniques are available). Thus, perpendicular flow into the slice from either side brings in nuclei that have not experienced previous RF pulses and whose signal is not limited by T1 recovery. Moreover, there is only one selective RF pulse per sequence so that there is only flow enhancement and no washout effect. Thus, all flow produces very intense signals in fast scan sequences, and signal displaced along phase encoding gradients can obscure other image details. For this reason, fast scan techniques often incorporate flow compensatory gradients to reduce flow artifacts [8,32].

Ultrafast scan techniques can avoid most common flow artifacts. This can be accomplished best by gating the sequence so that it occurs during cardiac diastole.

Flow artifacts in MRI are often used to determine the patency of blood vessels. It should be clear from the preceding discussion that a diagnosis of patency is more reliable than a diagnosis of occlusion. In fact, the topic of stationary blood is rather complex [33]. Similarly, quantitative estimates of flow are complicated and require special sequences [34–37].

MAGNETIC, CONDUCTIVITY, AND SUSCEPTIBILITY ARTIFACTS

Nearly as common as motion artifacts are the artifacts induced by foreign bodies found on the outside and the inside of patients. Both metallic and nonmetallic objects can produce susceptibility artifacts [38–40]. In fact, even plastic phantoms and air spaces in the body can generate such artifacts [38]. If the object is conductive (e.g., an ionic solution in a phantom or a bit of metal), there can be induced currents and interactions with the RF pulses that produce far less subtle artifacts. The most spectacular artifacts result from magnetic materials that severely distort the local magnetic field (Figure 8–15).

Since even a hairpin or a bit of mascara containing iron oxides can ruin a scan, it is important to properly educate and prepare patients for scanning. Gowning can help eliminate snaps, clips, zippers, buckles, and objects carried in pockets, but additional care must be taken to remove jewelry since wearing it often becomes almost unconscious. Furthermore, magnetic objects can become projectiles if brought into most magnetic fields.

Internal foreign bodies, implants, and surgical appliances present more difficult choices. Magnetic foreign bodies in critical locations (e.g., in the eye) or near critical structures (e.g., blood vessels or nerves) should not even be allowed in the scan room since the magnetic field may cause such bodies to move. Similarly, prudence probably dictates that patients with magnetic aneurysm clips should not be scanned. (Note many

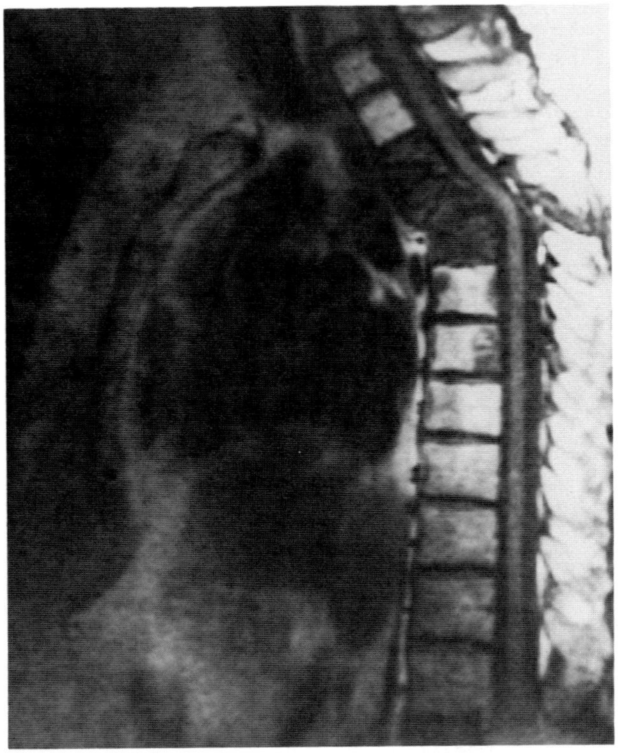

A

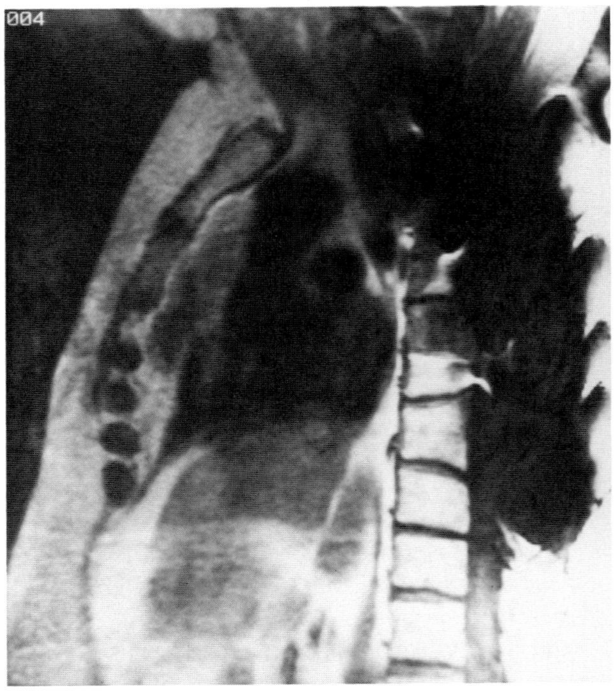

B

FIGURE 8–15 ▶
Sagittal sections through the thoracic region of the spine were obtained from a patient with widespread metastases to the vertebral bodies. A. *Surface coil images demonstrate a Gibbus deformity with T-3 collapse.* B. *Volume coil images taken after the installation of Harrington rods show metallic artifacts obscuring the upper thoracic and lower cervical regions.*

surgical clips and staples are nonmagnetic, and the number is growing [41].) In addition, cardiac pacemakers and neurostimulatory devices can be severely affected by the changing gradient and RF fields [42]. Most surgical implants and appliances, however, can be scanned, and valuable diagnostic results can be obtained despite some artifacts [21,43].

CHEMICAL SHIFT ARTIFACTS

MRI utilizes gradient-induced frequency shifts to localize signals. Similar frequency shifts can occur through microscopic magnetic fields generated by the electrons in chemical bonds surrounding magnetic nuclei. Most of the proton frequency shifts are too small to be resolved by the typical MRI unit, but the shift of 3.7 parts per million (ppm) between lipids and water is so large that it regularly causes artifacts (especially at the higher field strengths). The most common form of chemical shift artifact occurs at boundaries between soft tissues that are mostly water and tissues with a high fat content [44,45]. Here bands of increased and decreased signal follow the tissue boundaries perpendicular to the frequency encoding gradient direction (Figure 8–16). In all cases, these bands result from the aliphatic signal appearing 3.7 ppm downfield from the tissue with which it is physically associated, but the location of bands of altered signal depends on whether there is a fatty inclusion in soft tissue or whether there is a soft tissue inclusion in fat.

When fat is surrounded by soft tissue, a dark band of reduced signal appears at the upfield boundary, and a bright band of increased signal appears at the downfield boundary (see Figure 8–16A). The dark band results from the fat signal being shifted downfield and leaving behind a signal void. The bright band results from the fat signal being shifted downfield and re-enforcing the soft tissue signal already present.

When soft tissues are surrounded by fat, a bright band of increased signal appears at the upfield boundary and a dark band of reduced signal appears at the downfield boundary of the soft tissue region (see Figure 8–16B). The bright band upfield results from the fat signal being shifted downfield to add to the soft tissue signal already present. The dark band downfield results from the fat signal being shifted downfield and leaving behind a signal void. This is demonstrated in the abdomen at the fat–soft tissue interfaces of the liver and kidneys in Figure 8–16C.

Another type of artifact can result from chemical shifts. This second type of artifact is a variation in signal intensity found when TE is changed (Figure 8–17). When TE is increased, a monotonic decrease in signal is expected from T2 decay (see Figure 8–17A). The chemical shift causes the fat and water signals to rephase at different times and interact with each other so that the signal intensity oscillates as TE is increased (see Figure 8–17B). This second type of chemical shift is most commonly found in gradient echoes [46] but can be found in spin echoes [47].

At times (and especially at high magnetic fields), chemical shifts have been seen as a nuisance and special pulse sequences have been developed to suppress the artifact. At other times, chemical shift has been seen as a diagnostic tool, and special pulse sequences have been developed to produce separate water and fat images [48]. Chemical shift imaging and its role in diagnosis continue to evolve.

TRUNCATION, ALIASING, AND OTHER RECONSTRUCTION ARTIFACTS

Several related artifacts result from the way MRI uses a finite number of measurements (particularly in phase encoding measurements) to represent a continuous object. A finite number of measurements in the spatial domain corresponds to truncating a Fourier series so that no terms above a certain frequency appear. Frequencies in the object above this limit can modulate frequencies below the limit to produce aliasing artifacts (i.e., changes in signal intensity that do not correlate with changes in the object). One example of such aliasing is the ringing (termed Gibbs phenomena) that occurs at sudden transitions in the object (e.g., at the interfaces between soft tissues and CSF in the brain). Gibbs phenomena manifest themselves in the image as a repetition (in the phase encoding direction) of the edge of such an abrupt transition. This type of ghosting is often masked in noisy images and becomes

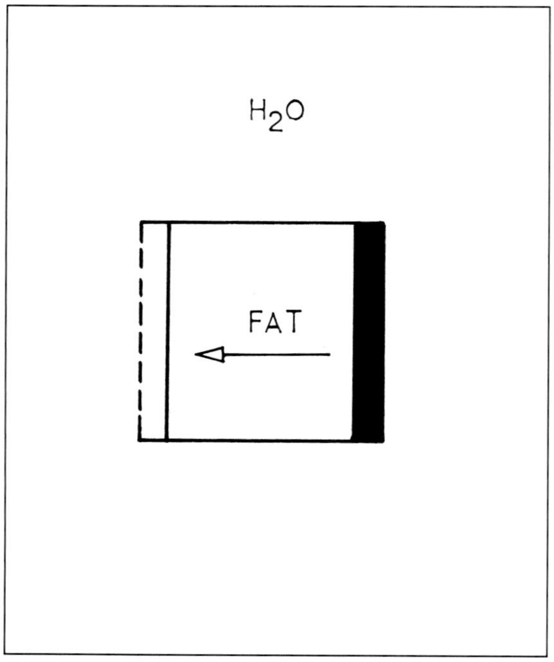

A

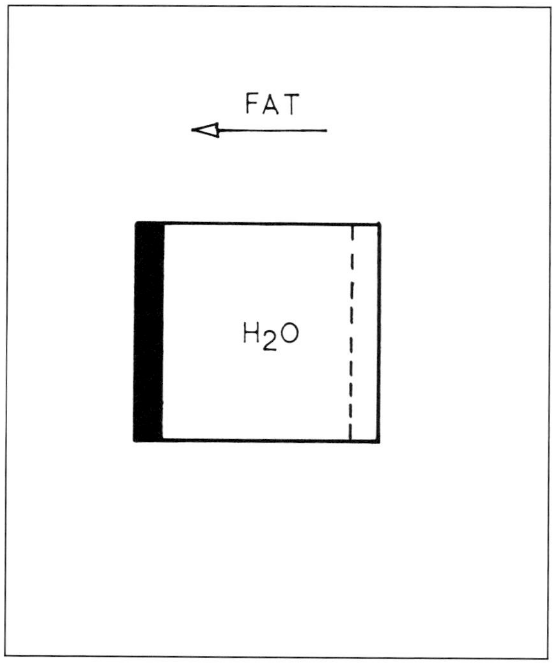

B

FIGURE 8–16
One type of chemical shift artifact occurs at boundaries between soft tissue and fat. This type of artifact is seen as a bright or dark band parallel to the tissue boundary and perpendicular to the frequency encoding gradient. A. When fat is surrounded by soft tissue, a dark band appears at the upfield boundary (image right) *and a bright band appears at the downfield boundary* (image left). *B. When soft tissue is surrounded by fat, a bright band appears at the upfield boundary* (image right) *and a dark band appears at the downfield boundary* (image left). *All of these effects result from the fat signal being shifted downfield 3.7 ppm relative to the water signal. C. The bright bands* (arrows) *at the upfield boundary and the dark bands* (arrowheads) *at the downfield boundary can often be demonstrated at the fat-liver and fat-kidney boundaries.*

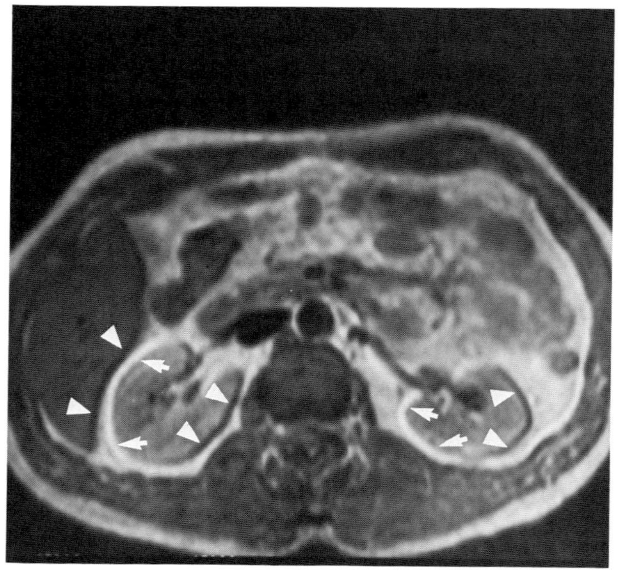

C

A

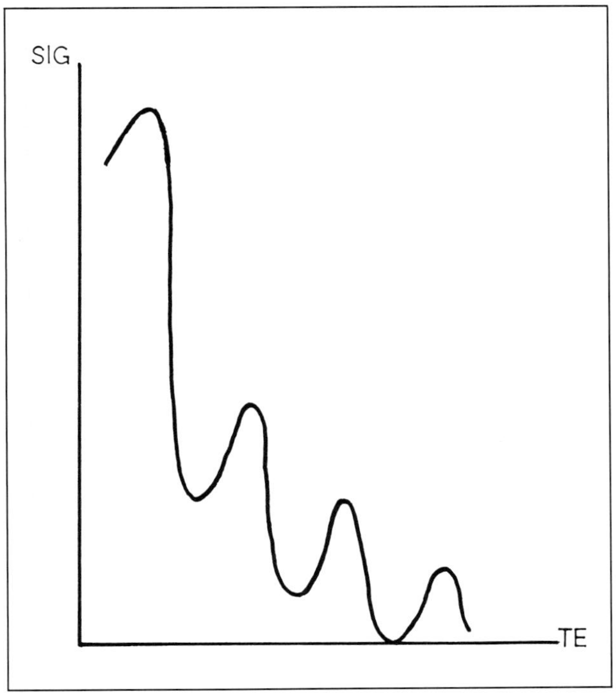

B

FIGURE 8–17
Another type of chemical shift artifact is evident in graphs of signal strength versus TE. A. *T2 decay produces a monotonic decrease in signal as TE is increased even if there are two or more substances with different T2s.* B. *Two substances with different chemical shifts can rephase at different times and interact with each other to produce oscillations in the signal strength as TE is varied.*

more apparent in higher signal images. Such artifacts can be reduced either by taking more measurements in the phase encoding direction or by processing the signal before reconstruction [49–51].

Wraparound artifacts are closely related to aliasing in that they result from a limited number of phase measurements (Figure 8–18). This can allow signals from outside the field of view to have the same phase as signals inside the field of view. Such outside signals then become superimposed on the field of view. To eliminate wraparound artifacts, the outside signals must be eliminated mathematically or physically [52].

Processing before, during, and after reconstruction can reduce artifacts and change image quality; however, the reconstruction process itself also profoundly affects image quality. For example, the Fourier reconstruction techniques now in almost universal use are more sensitive to motion but less sensitive to magnetic field inhomogeneities than are the older back-projection techniques. These Fourier techniques also take the Poisson electronic noise arriving from the detectors and redistribute it throughout the image so that the image noise itself is non-Poisson. This implies that the S/N lost in data collection through slice thickness and pixelization cannot be simply recovered by postprocessing filters [53].

INSTRUMENTAL ARTIFACTS

Many different classes of artifacts can be traced to instrumental limitations and failures [21]. Geometric distortions can clearly result from magnetic inhomogeneities. Such inhomogenei-

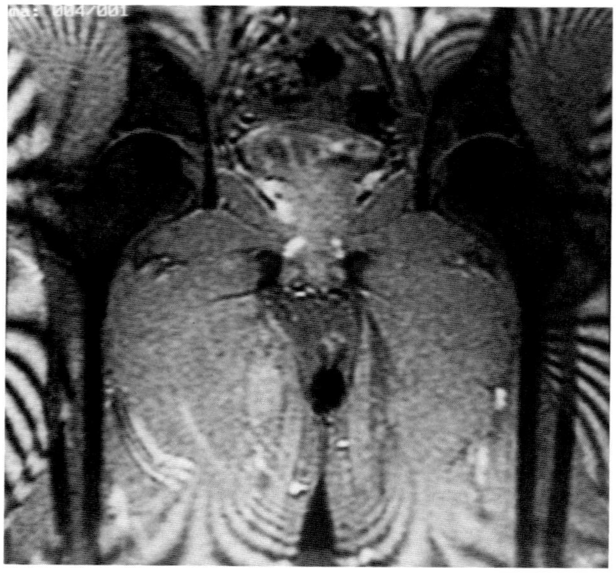

A

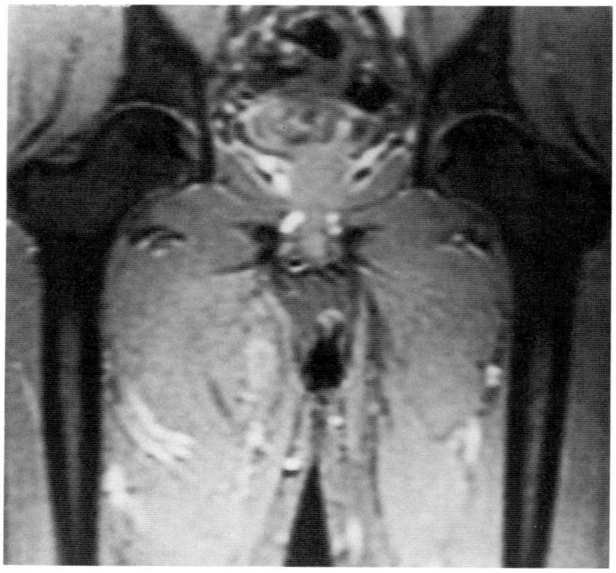

B

FIGURE 8–18
Coronal sections were obtained of both femoral heads. A. A 32-cm field of view produced substantial wraparound and zebra-stripe phase artifacts. B. Imaging the same region with a 48-cm field of view removed both artifacts.

ties can be reduced by improved shimming and linearization of gradients, but static inhomogeneities can also be removed mathematically [54,55]. It is also possible to create gradient sequences to reduce inhomogeneities due to chemical shift and susceptibility changes as well as those due to static inhomogeneities [56]. Slice geometry in MRI, however, is much more complicated than in other medical imaging modalities because it depends on object relaxation times and TR as well as on instrumental factors such as magnetic homogeneity, gradient pulse shape, and RF pulse shape [57–59]. This greatly complicates partial volume effects and implies that image geometry can change with sequence parameters such as echo number and TR [21].

Image nonuniformities can result from RF inhomogeneities as well as from geometric factors. The most common cause of RF inhomogeneities is coil design. Surface coils in particular can produce very nonuniform images because of the strong signal originating in subcutaneous fat. This effect can be reduced by coil positioning and by image processing [60]. Other common RF problems include poor RF pulse shape and inappropriate tip angles. On the receiver side, proper gain settings are important, as is the reduction or exclusion of noise sources.

To control instrumental artifacts and make rational use of service, it is essential to have a quality assurance program [21,22]. Routine quality assurance can be accomplished with relatively simple phantoms and requires no more than half an hour per day. This small investment can return substantial dividends in terms of improved image quality and cost-effective use of service.

REFERENCES

1. Mansfield P, Morris PG. NMR imaging in biomedicine. New York: Academic Press, 1982.
2. Partain CL, James AE, Rollo FD, Price RR, eds. Nuclear magnetic resonance (NMR) imaging. Philadelphia: WB Saunders, 1983.
3. Wolf GL, Popp C. NMR: a primer for medical imaging. Thorofare, NJ: Slack, 1984.
4. Esser PD, Johnston RE, eds. Technology of nuclear magnetic resonance. New York: Society of Nuclear Medicine, 1984.
5. Thomas SR, Dixon RL, eds. NMR in medicine: the instrumentation and clinical applications. New York: American Institute of Physics, 1986.
6. Morris PG. Nuclear magnetic resonance imaging in medicine and biology. Oxford: Clarendon Press, 1986.
7. Bushong SC. Magnetic resonance imaging: physical and biological principles. St. Louis: CV Mosby, 1988.
8. Wehrli FW, Shaw D, Kneeland JB, eds. Biomedical magnetic resonance imaging: principles, methodology, and applications. New York: VCH Publishers, 1988.
9. Taylor DG, Inamdar R, Bushell M-C. NMR imaging in theory and practice. Phys Med Biol 1988; 33:635–70.
10. Fullerton GD, Cameron IL, Ord VA. Frequency dependence of magnetic resonance spin-lattice relaxation of protons in biological materials. Radiology 1984; 151:135–8.
11. Bottomley PA, Foster TH, Argersinger RE, Pfeifer LM. A review of normal tissue hydrogen NMR relaxation times and relaxation mechanisms from 1-100 MHz: dependence on tissue type, NMR frequency, temperature, species, excision, and age. Med Phys 1984; 11:425–48.
12. Bottomley PA, Hardy CJ, Argersinger RE, Allen-Moore G. A review of H-1 nuclear magnetic resonance relaxation in pathology: are T1 and T2 diagnostic? Med Phys 1987; 14:1–37.
13. Bydder GM, Young IR. Clinical use of the partial saturation and saturation recovery sequences in MR imaging. J Comput Assist Tomogr 1985; 9:1020–32.
14. Frahm J, Haase A, Matthaei D. Rapid NMR imaging of dynamic processes using the FLASH technique. Magn Reson Med 1986; 3:321–7.
15. Edelman RR, Johnson K, Buxton R, et al. MR of hemorrhage: a new approach. Am J Neuroradiol 1986; 7:751–6.
16. Pykett IL, Rzedzian RR. Instant images of the body by magnetic resonance. Magn Reson Med 1987; 5:563–71.
17. Rzedzian RR, Pykett IL. Instant images of the human heart using a new, whole-body MR imaging system. AJR 1987; 149:245–50.
18. Ordidge RJ, Coxon R, Howseman A, et al. Snapshot head imaging at 0.5T using the echo planar technique. Magn Reson Med 1988; 8:110–5.
19. Ernst RR, Anderson WA. Application of Fourier transform spectroscopy to magnetic resonance. Rev Sci Instruments 1966; 37:93–102.
20. Kneeland JB, Knowles RJR, Cahill PT. Multi-section multi-echo pulse magnetic resonance techniques: optimization in a clinical setting. Radiology 1985; 155:159–62.
21. Knowles RJR, Markisz JA. Quality assurance and image artifacts in magnetic resonance imaging. Boston: Little, Brown, 1988.
22. Dixon RL, ed. MRI acceptance testing and quality control: the role of the clinical medical physicist. Madison, WI: Medical Physics Publishing, 1988.
23. Lerski RA, Straughan K, Williams JL. Practical aspects of ghosting in resistive nuclear magnetic resonance imaging systems. Phys Med Biol 1986; 31:721–35.
24. Karlik SJ, Heatherley T, Pavan F, et al. Patient anesthesia and monitoring at 1.5-T MRI installation. Magn Reson Med 1988; 7:210–21.
25. Haacke EM, Patrick JL. Reducing motion artifacts in two-dimensional Fourier transform imaging. Magn Reson Imaging 1986; 4:359–76.
26. Wood ML, Henkelman RM. Suppression of respiratory motion artifacts in magnetic resonance imaging. Med Phys 1986; 13:794–805.
27. Haacke EM, Lenz GW, Nelson AD. Pseudo-gating: elimination of periodic motion artifacts in magnetic resonance imaging without gating. Magn Reson Med 1987; 4:162–74.
28. Wood ML, Runge VM, Henkelman RM. Overcoming motion in abdominal MR imaging. AJR 1988; 150:513–22.
29. Dixon WT, Brummer ME, Malko JA. Acquisition order and motional artifacts in spin warp images. Magn Reson Med 1988; 6:74–83.

30. Axel L. Review: blood flow effects in magnetic resonance imaging. AJR 1984; 143:1157–66.
31. Carr HY, Purcell EM. Effects of diffusion on free precession in nuclear magnetic resonance experiments. Phys Rev 1954; 94:630–8.
32. Haacke EM, Lenz GW. Improving MR image quality in the presence of motion by using rephasing gradients. AJR 1987; 148:1251–8.
33. Brooks RA, Di Chiro G. Magnetic resonance imaging of stationary blood: a review. Med Phys 1987; 14:903–13.
34. Moran PR. A flow velocity zeugmatographic interlace for NMR imaging in humans. Mag Reson Imaging 1982; 1:197–203.
35. Moran PR, Moran RA, Karstaedt N. Verification and evaluation of internal flow and motion: true magnetic resonance imaging by the phase gradient modulation method. Radiology 1985; 154:433–41.
36. George CR, Jacobs G, MacIntyre WJ, et al. Magnetic resonance signal intensity patterns obtained from continuous and pulsatile flow models. Radiology 1984; 151:421–8.
37. Meier D, Maier S, Bosiger P. Quantitative flow measurements on phantoms and on blood vessels with MR. Magn Reson Med 1988; 8:25–34.
38. Ludeke KM, Roschmann P, Tischler R. Susceptibility artefacts in NMR imaging. Magn Reson Imaging 1985; 3:329–43.
39. Wendt RE, Wilcott MR, Nitz W, Murphy PH, Bryan RN. MR imaging of susceptibility-induced magnetic field inhomogeneities. Radiology 1988; 168:837–41.
40. Ericsson A, Hemmingsson A, Jung B, Sperber GO. Calculation of MRI artifacts caused by static field disturbances. Phys Med Biol 1988; 33:1103–12.
41. Shellock FG. MR imaging of metallic implants and materials: a compilation of the literature. AJR 1988; 151:811–4.
42. Erlebacher JA, Cahill PT, Pannizzo F, Knowles RJR. Effect of magnetic resonance imaging on DDD pacemakers. Am J Cardiol 1986; 57:437–40.
43. Gangarosa RE, Minnis JE, Nobbe J, Praschan D, Genberg RW. Operational safety issues in MRI. Magn Reson Imaging 1987; 5:287–92.
44. Dwyer AJ, Knop RH, Hoult DI. Frequency shift artifacts in MR imaging. J Comput Assist Tomogr 1985; 9:16–8.
45. Babcock EE, Brateman L, Weinreb JC, Horner SD, Nunnally RL. Edge artifacts in MR images: chemical shift effect. J Comput Assist Tomogr 1985; 9:252–7.
46. Wehrli FW, Perkins TG, Shimakawa A, Roberts F. Chemical shift-induced amplitude modulations in images obtained with gradient refocusing. Magn Reson Imaging 1987; 5:157–8.
47. Hinks RS, Henkelman RM. Problems with organic materials for magnetic resonance imaging phantoms. Med Phys 1988; 15:61–3.
48. Brateman L. Chemical shift imaging: a review. AJR 1986; 146:971–80.
49. Wood ML, Henkelman RM. Truncation artifacts in magnetic resonance imaging. Magn Reson Med 1985; 2:517–26.
50. Smith MR, Nichols ST, Henkelman RM, Wood ML. Application of autoregressive modelling in magnetic resonance imaging to remove noise and truncation artifacts. Magn Reson Imaging 1986; 4:257–61.
51. Parker DL, Gullberg GT, Frederick PR. Gibbs artifact removal in magnetic resonance imaging. Med Phys 1987; 14:640–5.
52. Edelman RR, Atkinson DJ, Silver MS, Loaiza FL, Warren WS. FRODO pulse sequences: a new means of eliminating motion, flow, and wraparound artifacts. Radiology 1988; 166:231–6.
53. Edelstein WA, Glover GH, Hardy CJ, Redington RW. The intrinsic signal-to-noise ratio in NMR imaging. Magn Reson Med 1986; 3:604–18.
54. Sekihara K, Kuroda M, Kohno H. Image restoration from non-uniform magnetic field influence for direct Fourier NMR imaging. Phys Med Biol 1984; 29:15–24.
55. Sekihara K, Kohno H. Image restoration from nonuniform static magnetic field influence in modified echo-planar imaging. Med Phys 1987; 14:1087–9.
56. Cho ZH, Kim DJ, Kim YK. Total inhomogeneity correction including chemical shifts and susceptibility by view angle tilting. Med Phys 1988; 15:7–11.
57. Young IR, Bryant DJ, Payne JA. Variations in slice shape and absorption as artifacts in the determination of tissue parameters in NMR imaging. Magn Reson Med 1985; 2:355–89.

58. McRobbie DW, Lerski RA, Straughan K, Quilter P, Orr JS. Investigation of slice characteristics in nuclear magnetic resonance imaging. Phys Med Biol 1986; 31:613–26.

59. Robinson EM, Hickey DS, Aspden RM, Machin M. Computer simulation of the slice profile in magnetic-resonance imaging. Phys Med Biol 1987; 32:1531–44.

60. Brey WW, Narayana PA. Correction for intensity falloff in surface coil magnetic resonance imaging. Med Phys 1988; 15:241–5.

9

MRI of Bone Disease

JOHN A. MARKISZ

The evaluation of suspected bone pathology should always begin with a standard radiograph. Osseous lesions can often be so well categorized by routine plane films that no further workup is needed [1]. The value of MRI lies in its ability to detect infiltrative abnormalities at a much earlier stage of disease, as well as to determine associated problems, such as extension into the soft tissues and neurovascular involvement, using direct, high-resolution imaging in multiple planes [2,3]. The natural contrast inherent in MRI eliminates the need for intravenous injections of contrast material. Although recent studies have demonstrated the potential usefulness of magnetic vascular contrast agents for evaluating specific musculoskeletal problems, the clinical utility of these methods has yet to be established [4–8].

Imaging of bone ranks second only to neurologic studies in the total number of MR procedures performed. The absence of respiratory, peristaltic, and pulsation artifacts in much of the skeletal system makes MRI perfectly suited for the evaluation of bone, joints, and the surrounding soft tissues. The lack of beam-hardening ar-

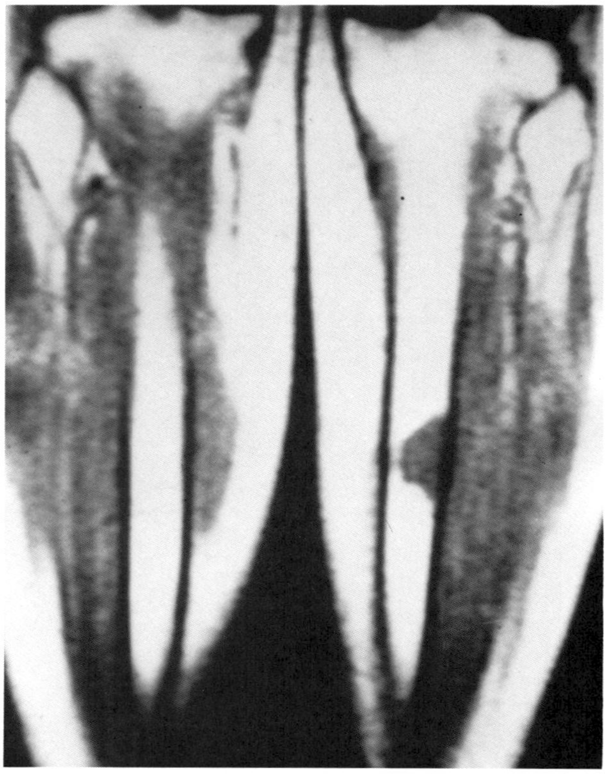

A

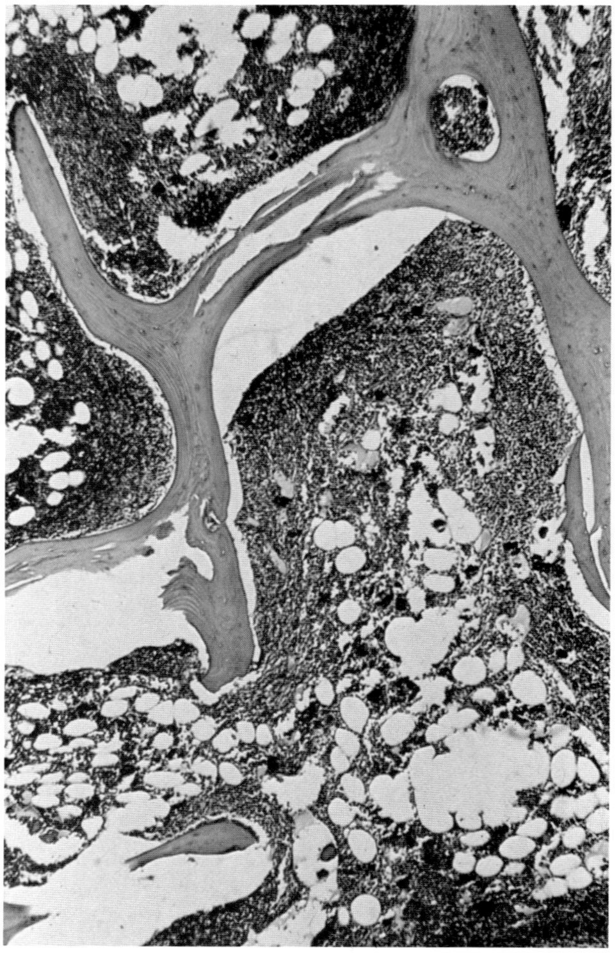

B

Figure 9–1
A. MRI demonstrates leukemic infiltrate within the left tibial shaft. Findings on x-rays and radionuclide bone scan were negative. B. Biopsy specimen demonstrates leukemic infiltrate within the marrow cavity. Trabecular bone is not yet affected.

tifacts associated with CT, continued advances in imaging software and techniques, and the significantly increased resolution now available with the development of surface coils are technical elements that have led to the very rapid rise in the number of MR bone studies [2,3].

MRI can be useful in many different clinical situations and can reveal bony abnormalities earlier than other modalities do. A significant loss of calcium is necessary before changes appear on routine radiographs, and increased osteoblastic activity is needed for radionuclide bone scintigraphy to show positive findings. MRI, however, can detect infiltrative processes even before trabecular bone is involved (Figure 9–1) [9–12]. The natural contrast of this technique delineates structures very well, so that the extent of soft tissue involvement can be readily demonstrated. Structural alterations and occult fractures, not detectable on routine plain films, can also be documented by MRI (Figures 9–2 and 9–3) [13–15].

Signal Characteristics of Bone Tissues

Cortical bone, without sufficient mobile hydrogen atoms, does not generate an MR signal and appears as a dark region on all pulse sequences. A signal can, however, be obtained from the con-

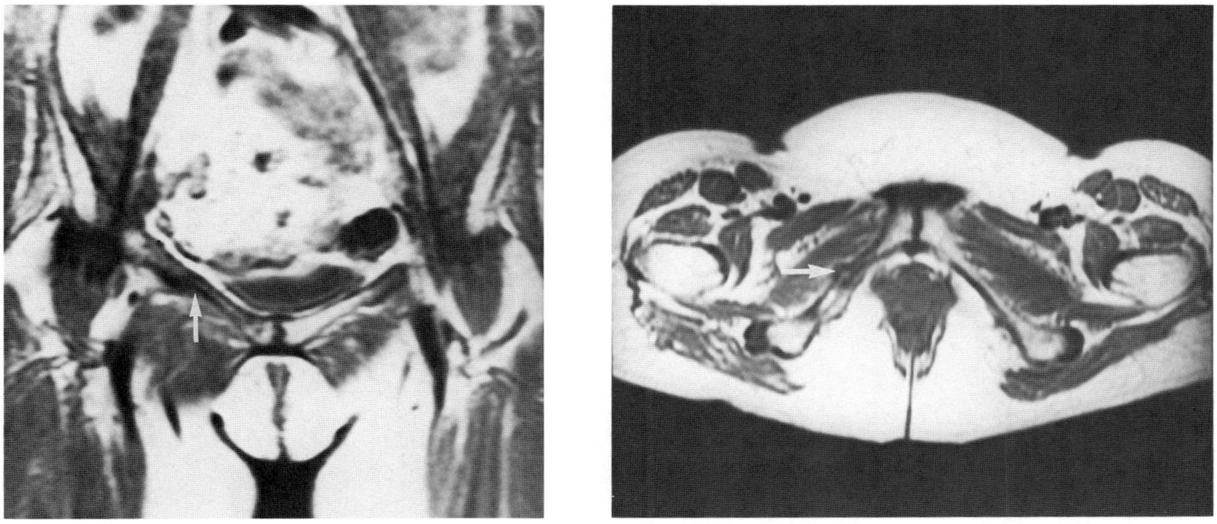

A B

FIGURE 9–2
Fracture of the pubic ramus (arrow) *was not detected on plain films, but are seen on MR images.* A. *T1-weighted coronal view.* B. *T1-weighted axial view.*

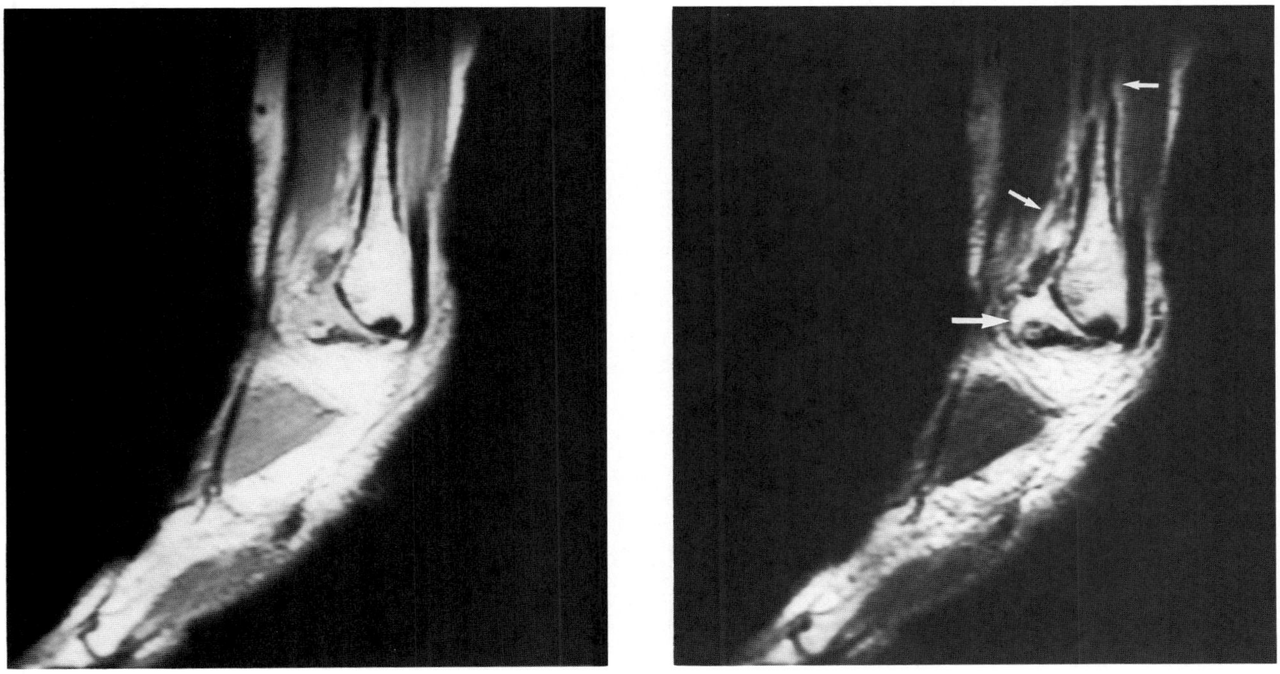

A B

FIGURE 9–3
A. *T1-weighted image. Displaced fibular fracture, diagnosed by MRI obtained for an ankle "sprain," demonstrates the value of first obtaining plain films.* B. *T2-weighted image. Surrounding area of edema within muscles* (small arrows) *is seen as increased signal as is a region of subcutaneous edema* (large arrow).

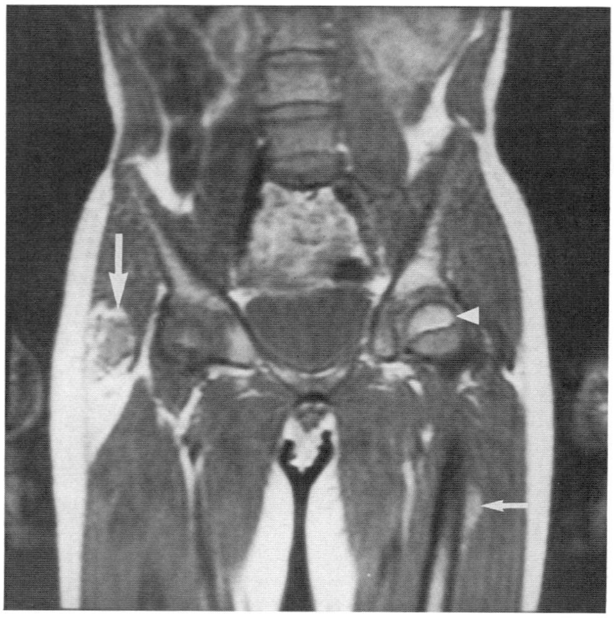

A

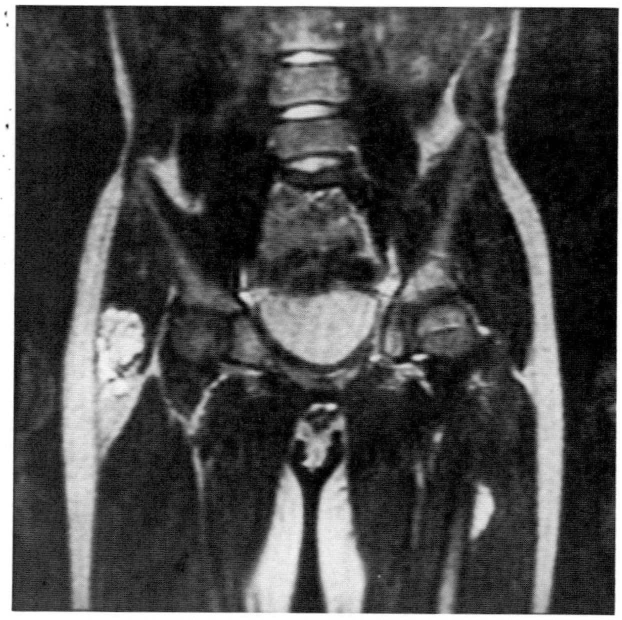

B

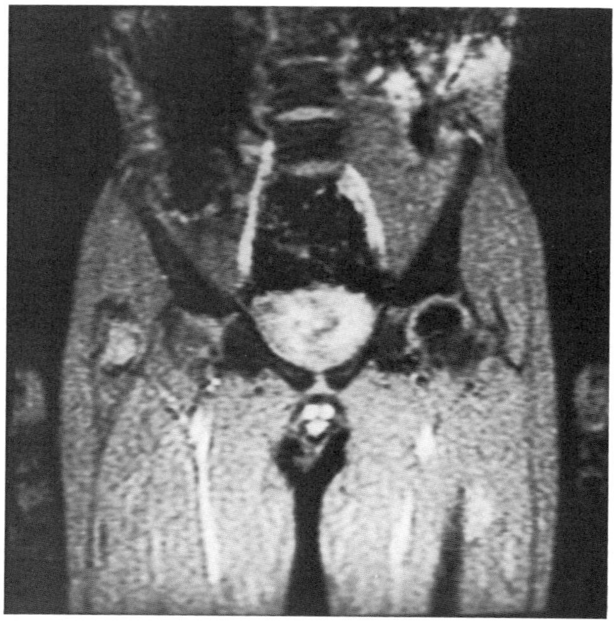

C

FIGURE 9–4

A. T1-weighted coronal images through the pelvis of a 6-year-old with a palpable mass (large arrow) *in the right hip. B, C. The mass is demonstrated to be a hemangioma on the T2-weighted images and gradient-echo images, respectively. An incidental finding was a second hemangioma in the left vastus lateralis muscle* (small arrow). *Note the decreased signal within the marrow cavity of all of the bony structures, relative to the brighter epiphysis* (arrowhead) *on the T1-weighted image (A). Gradient-echo images (C) exhibit decreased signal from fat and increased signal from vessels, fluid collections, and intervertebral discs.*

tents of the medullary cavity, which typically is filled with marrow. Marrow can normally exist in two basic forms, red marrow and yellow marrow. Red marrow consists of a network of reticular connective tissue, hematopoietic tissue, and some fat cells. The concentration of fat cells in red marrow varies from 25–50%, depending on age and sex [16]. Yellow marrow is primarily composed of fat cells (80%) with some hematopoietic tissue in a connective tissue framework. The medullary cavity with yellow marrow appears bright due to the intense signal of fat. The decreased concentration of fat in red marrow leads to a less intense image.

In infants and children, red marrow is found throughout the skeleton, with the exception of the epiphyses, which normally have a high fat content (Figure 9–4). This sometimes continues into adulthood, and it is not unusual to see epiphyses in an adult that are brighter than the rest of the bone (Figure 9–5). At about the age of 5, yellow marrow begins to supplant the red marrow. The replacement begins in the distal skeleton, proceeding centrally to the more proximal bones. Some investigators claim that replacement occurs distal to proximal within a given bone as well, while others maintain that it begins centrally, but proceeds faster distally. This latter interpretation is more compatible with the MR findings of patchy areas suggestive of red marrow in the distal aspects of the femurs of otherwise healthy young adults, especially young women. By early adulthood, red marrow is usually no longer found in metaphyses of long bones, settling in the vertebrae, proximal aspects of the femurs and humeri, pelvic bones, sternum, ribs, clavicles, scapulae, and skull. As the need for hematopoiesis decreases, more slowly in women than in men, yellow marrow continues to replace red marrow. Even at age 70, however, 50% of the marrow space in the sternum and vertebrae continues to be filled with red marrow. When reviewing a study, physicians should always remember that there *are* normal areas of red marrow in the bones of adults, so that this is not mistaken for pathologic conditions. It is not unusual to observe red marrow in the proximal part of femoral shafts and necks in women well into their 50s. Marrow within the skull can undergo

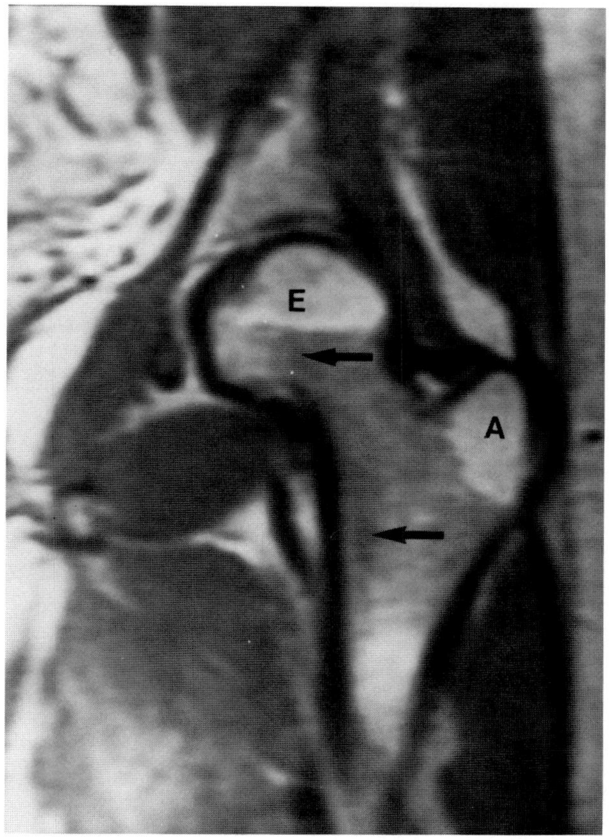

FIGURE 9–5
Femur of a 27-year-old woman with increased fat in the epiphysis (E) and apophysis (A). The arrow indicates vertical linear regions of decreased signal, attributed to increased trabecular formation along stress lines.

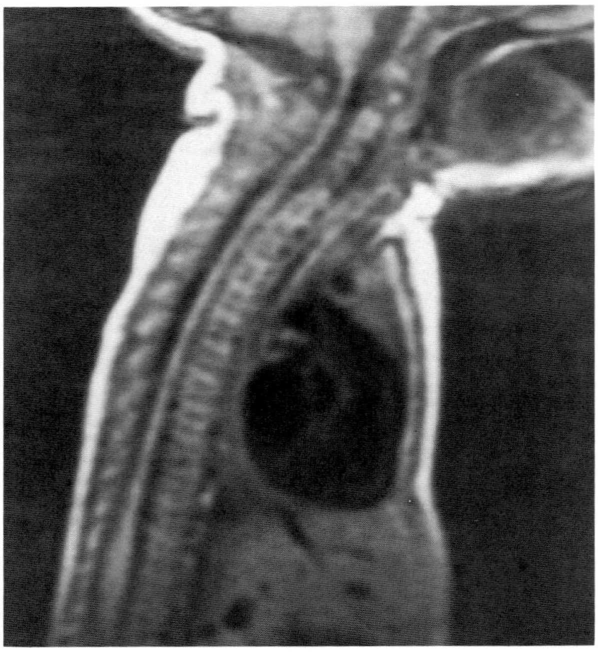

FIGURE 9–6
Vertebral column in a 1-month-old infant, with alternating discs and vertebrae of approximately equal size.

further degeneration into gelatinous marrow in extreme old age, a change not yet described by MRI. Due to the differences in fat content, then, the MR appearance of the bones in an adult is brighter than that in a child [17]. Because of the decreased signal in the marrow space of children, infiltrative diseases are more difficult to detect by MRI than they would be in adults [18].

Ongoing developmental changes in vertebral body formation result in significant differences in the MR appearance of vertebral bodies of adults and children, especially in those less than 1 year old. As a consequence of the embryonic evolution of the vertebral body from the fusion of two mesenchymal sclerotomes, vertebral bodies are observed as biconcave structures with the fusion plane of the sclerotomes, a cartilaginous plate, visualized as a dark central line [19]. Cells from the cephalic part of the sclerotome fill in the space between the vertebral bodies, forming the intervertebral discs. These discs are relatively larger and more rounded than those of adults, and of low to intermediate signal intensity (Figure 9–6). As the child grows older, the discs gradually flatten, the cartilaginous plate disappears, and the contours of the vertebral bodies approach those of the adult. The spine is noticeably straighter than it is after the child matures, with a significantly decreased thoracic kyphosis and lumbar lordosis. This allows coronal scans to include much more of the spinal column on a single view, making this imaging plane much more useful in examining small children than it is in adults [20].

The image of the medullary space is very dependent on the pulse sequence selected (see Figure 9–4) [21]. T1-weighted and proton density images yield the best signal, and excellent anatomic resolution, while T2-weighted images provide information relative to pathology. The fat in the marrow cavity does not produce a signal on gradient-echo images. For these reasons, spin-echo T1 and T2 imagings are almost exclusively used for the evaluation of bone disease. Gradient-echo images are useful in joint evaluation, but do not provide much information for purely bone problems. Inversion recovery (IR) sequences provide heavily T1-weighted images, but are time-consuming and yield little in the way of pathologic information to identify active disease processes. IR techniques that utilize short inversion time sequences in which the signal from fat is suppressed, however, permit evaluation of marrow abnormalities comparable to spin-echo T2-weighted sequences (Figure 9–7).

Fat, due to its short T1 and long T2, exhibits a bright signal on all spin-echo sequences—bright on T1-weighted images, fading slightly, but still appearing bright on relatively T2-weighted views. The medullary cavity in adults usually contains a high concentration of fat, so that the contents are visualized extremely well as bright structures. Since the signal obtained is actually an average of a large amount of the fat within the marrow space and a much smaller amount of the normally dark, trabecular bone, the signal observed is slightly less intense than that of subcutaneous fat. It is usually homogeneous; however, there is a considerable variation found in the normal population. Areas of increased red marrow, especially in women and athletes, cause patchy decreased marrow signal, as do areas of increased trabeculae formation, such as along regions of increased stress or fracture healing. These regions are usually distinctive in both position and appearance. Increased red marrow usually appears in the pelvic bones, vertebrae, and proximal aspects of the humeri and femurs. The appearance of the marrow is slightly decreased on the T1-weighted images, light to medium gray, and is not brighter on the T2-weighted views, but rather becomes isointense with the rest of the marrow space. Increased bone formation is usually darker and more linear, corresponding to stress lines and healed fracture sites [22].

Signal Characteristics of Bone Disease

MRI can detect an infiltrative process in bone earlier than any other modality, since it can observe a disease process that has invaded the medullary space, even before bone itself is affected (see Figure 9–1). Several studies [9–12] have demonstrated the earlier detection of infiltrative diseases of bone by MRI than by bone scintigraphy, CT, or plain radiography. Since neither

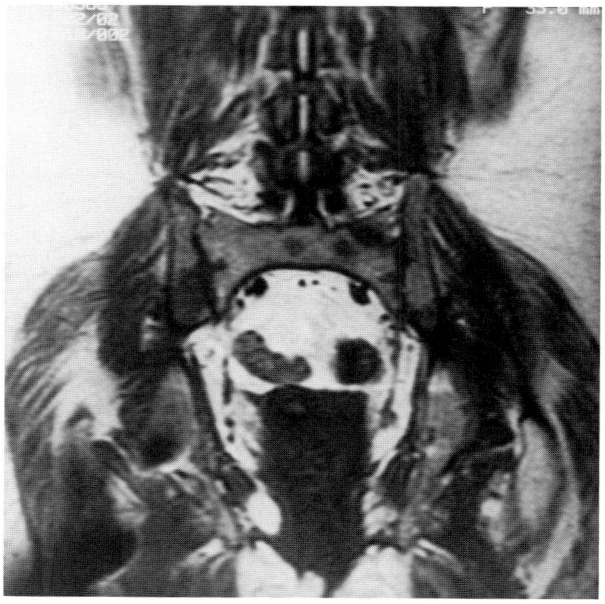

A

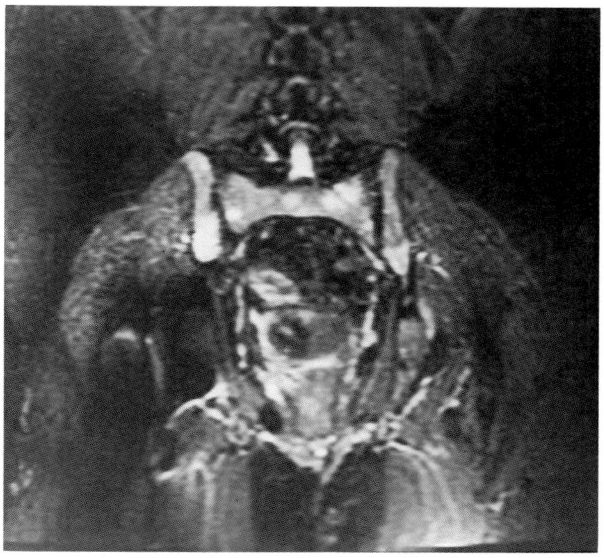

B

FIGURE 9–7
Inversion recovery (IR) sequences utilizing fat suppression techniques demonstrate areas of active disease. A. *In this patient with active sarcoidosis, a T1-weighted coronal image demonstrates nonspecific finding of multiple areas of decreased signal, compatible with either active disease or healed, fibrosed regions.* B. *IR sequence indicates areas of active disease as bright areas in darker normal marrow, from which fat signal is suppressed.*

increased osteoblastic activity (necessary for a positive finding on bone scintigram) nor significant changes in calcification (necessary for routine x-ray observation) are necessary for MRI detection of changes, infiltrative disease in bone produces detectable scan alterations significantly earlier than can be obtained by radiographic or scintigraphic methods. While CT can detect changes in fat density within the medullary space, the sensitivity is not as high as with MRI. As a routine screening procedure for metastatic disease, however, MRI is not practical. Bone scintigraphy is so well suited for screening the entire skeletal system within a relatively short period of time that MRI should not even be considered for this purpose. Scanning the entire body properly by MRI is a very time-consuming operation, and the ribs and skull, relatively common sites of disease, routinely are not well visualized. For these reasons, MRI is primarily utilized when a bony abnormality of a specific area of the skeletal system is suspected.

The majority of infiltrative disorders present as focal or diffuse regions of decreased signal within the marrow space. These are visible on MRI as dark areas when the normally bright adipose tissue within the marrow cavity has been replaced, or even partially replaced, by the disease process. Unfortunately, although MRI is highly sensitive at detecting abnormalities of bone, it is not at all specific. Malignant, benign, infectious, and traumatic conditions can present as low-signal T1 areas within the normally bright marrow space. Areas of normal red marrow, fibrosis at sites of old disease, and increased calcification due to degenerative changes can also be observed as regions of focally decreased signal, relative to the bright background of fat. During the very early days of MRI, the promise of highly specific diagnoses was made, by using calculated T1 and T2 relaxation times. Although an occasional series has found these values to be useful, experience has demonstrated that there is too large an overlap in these quantities to maintain any degree of accuracy [23]. The heterogeneity in calcified and/or necrotic neoplasms further limits the utility of this approach.

While MRI is not at all specific for the evaluation of disease within the marrow cavity, the patterns exhibited by different pathologic entities often yield clues that narrow the differential diagnosis. The signal patterns are a composite of three basic findings: whether the lesions are diffuse or focal; whether the signal is very dark, intermediate, or bright in appearance; and whether the signal on the T2-weighted images increases or decreases in intensity. This information, along with a clinical history, usually permits an accurate diagnosis to be made.

On occasion, focal areas that are brighter than normal marrow fat can be seen within bone. Areas of increased signal within bone on T1-weighted images can be caused by a relatively limited number of entities: focal fatty deposits within bone, intraosseous lipomas or hemangiomas, hemorrhage, areas of necrosis within tumors, and rarely, a telangiectatic hemangiosarcoma. Usually, except for the last, very rare neoplasm, a signal more intense than the rest of the marrow space on a T1-weighted image is very suggestive of a benign lesion. Focal fatty deposits appear to be a relatively common occurrence, due to a natural replacement of red marrow with yellow marrow [24]. These focal regions are age-related, increasing with age, and are found in over half of patients between 31 and 40 years old and in 93% of patients between 51 and 60 years old (Figure 9–8).

Hemangiomas within bone, especially within vertebrae, can often be confused with focal fatty deposits on the T1-weighted images [25]. The capillaries in this heavily vascularized lesion are covered with fat, accounting for the high T1 signal intensity. In T2-weighted images, the appearance becomes even brighter due to the long T1 of the pooled blood within the lesion (Figure 9–9). Intraosseous lipomas are an uncommon bone lesion, most often observed in the calcaneus, and can usually be distinguished by a dark sclerotic rim around them. These also behave as does fat, decreasing slightly in intensity on the more T2-weighted images. Hemorrhagic aneurysmal bone cysts, containing blood within them, also appear relatively bright on the T1-weighted images, increasing in intensity on the T2-weighted images (Figure 9–10) [26,27]. Necrotic or hemorrhagic tumors rarely, if ever, are homogeneous, which helps to distinguish them

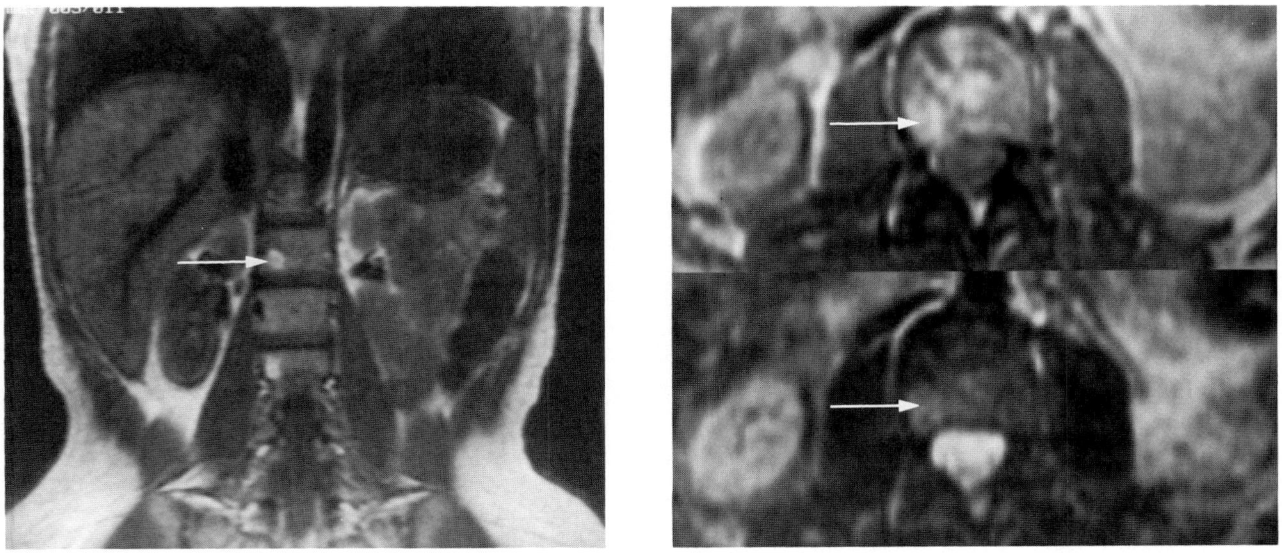

A B

FIGURE 9–8
A. *T1-weighted coronal image.* B. Top, *T1-weighted axial image.* Bottom, *T2-weighted axial image. Focal fatty deposits* (arrow) *can appear as bright regions within bone on T1-weighted images and fade, typical of fat, on T2-weighted images.*

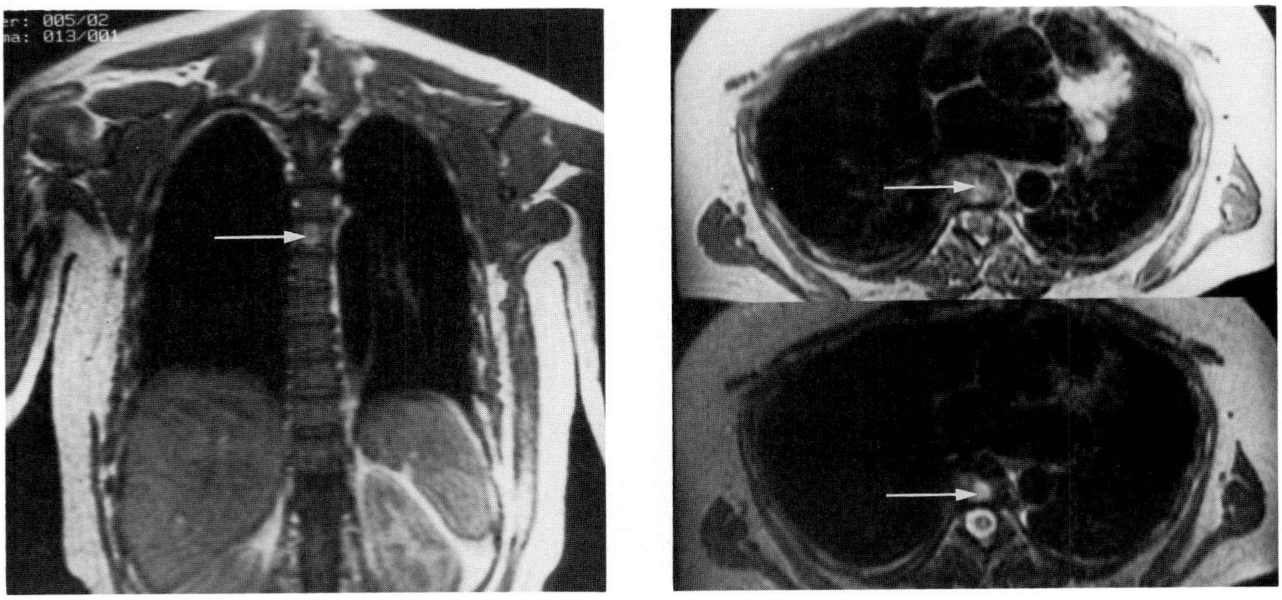

A B

FIGURE 9–9
A. *T1-weighted coronal image.* B. Top, *T1-weighted axial image.* Bottom, *T2-weighted axial image. Vertebral hemangiomas* (arrow) *are similar to focal fatty deposits on T1-weighted images, but increase in signal on the T2-weighted images.*

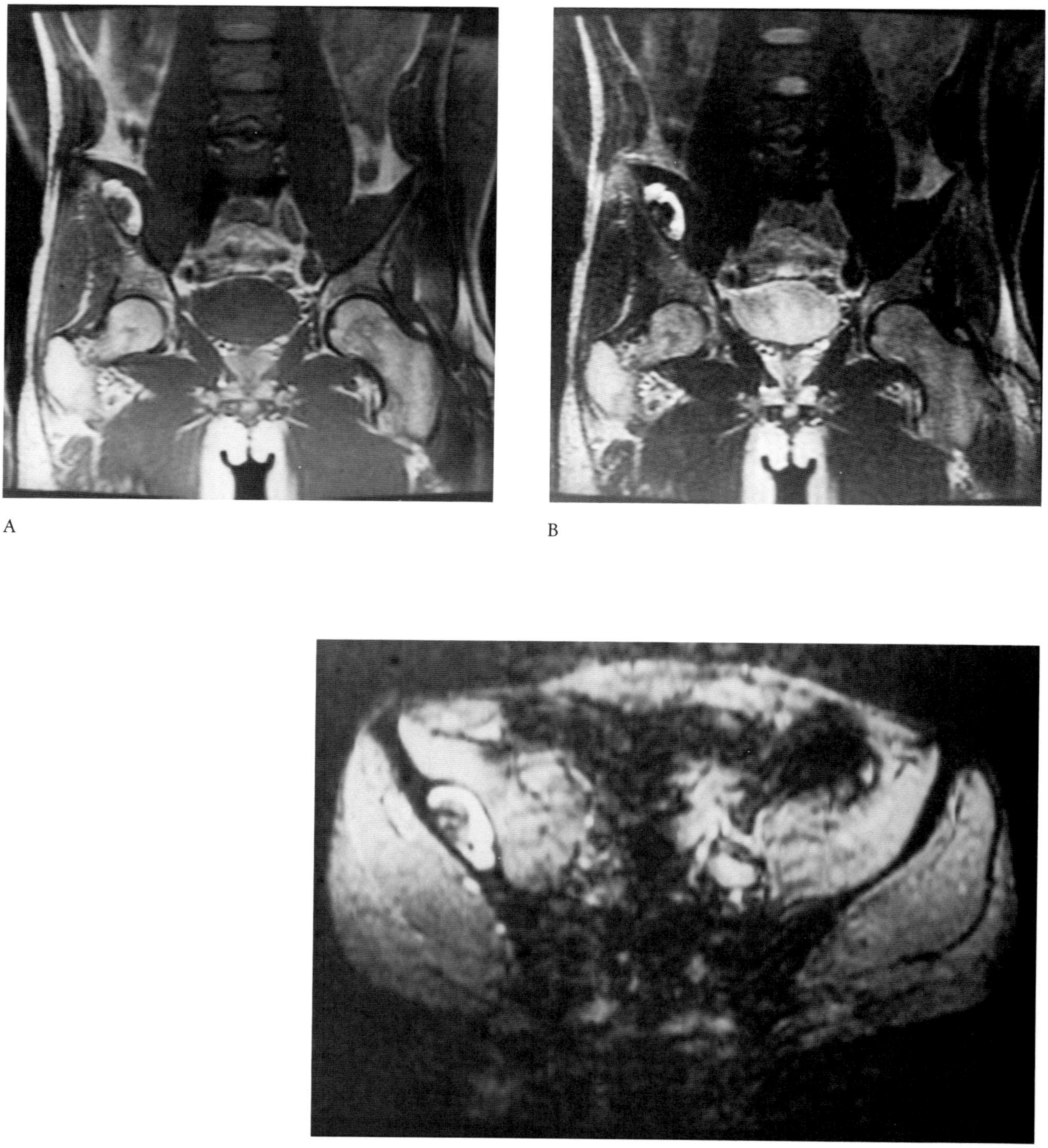

A B C

FIGURE 9–10

A, B. *T1-weighted and T2-weighted axial images, respectively, of an aneurysmal bone cyst, bright in* A *and brighter in* B. C. *Tumor is also well demonstrated on a gradient-echo image. The central dark area represents hemosiderin deposition.*

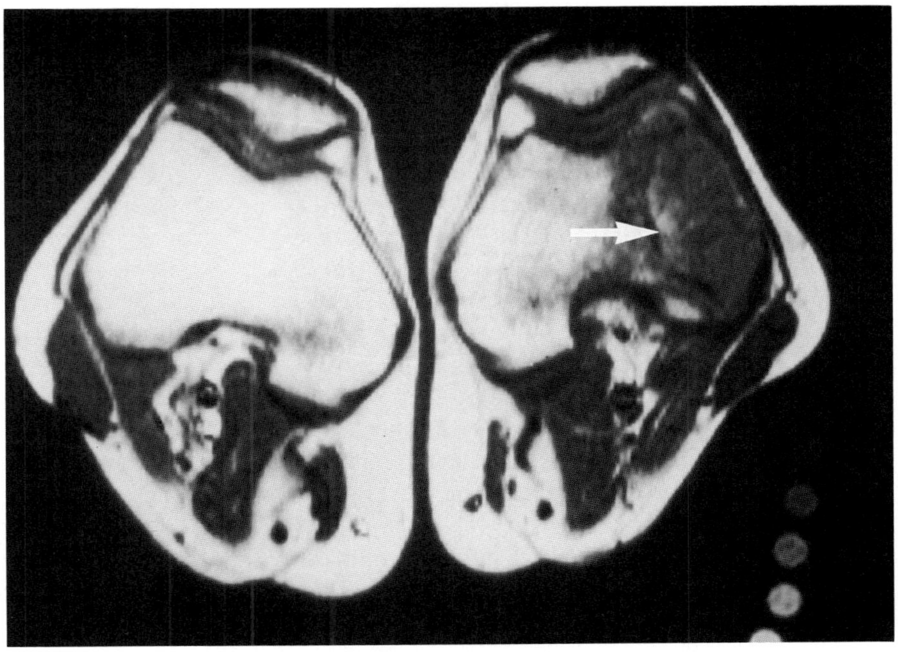

A

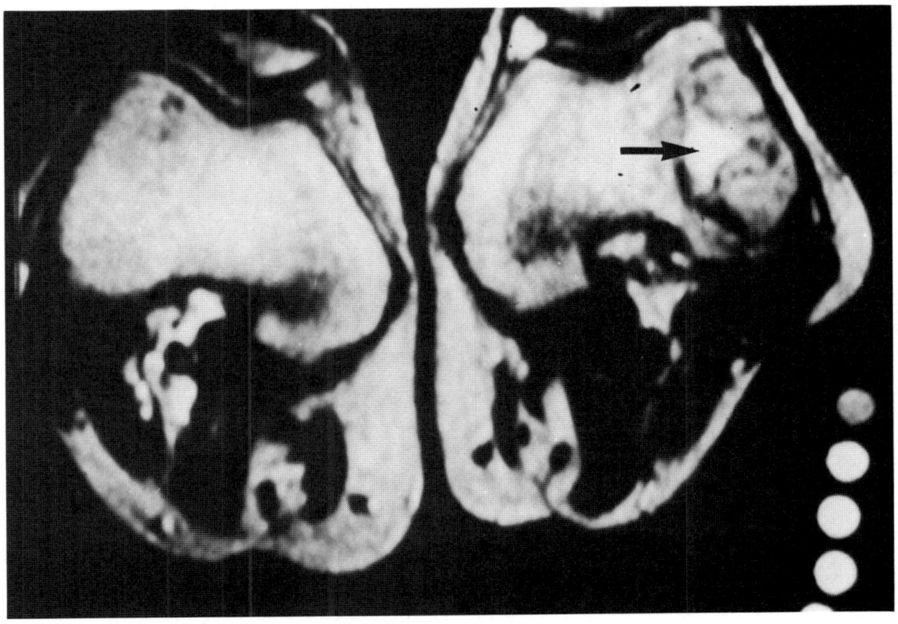

B

FIGURE 9–11

A, B. *T1-weighted and T2-weighted axial images, respectively, of a giant-cell tumor of the left lateral femoral condyle. On the T1-weighted image, the tumor is primarily dark with bright areas representing necrosis* (arrow). *On the T2-weighted image, the entire tumor brightens; however, the necrotic area becomes particularly bright.*

from a benign lesion (Figure 9–11). Degenerative changes in the spine can cause increased signal in the vertebral end-plates, as discussed in Chapter 11.

Abnormally bright marrow throughout the bones can also be observed in regions included in radiation therapy portals (Figure 9–12) [28,29]. Increased fat deposition is presumably secondary to replacement of cellular elements destroyed by radiation [30]. Effects have been noted as early as 2 weeks after treatment, and we have observed these same effects in patients 15 years after therapy has been completed. A similar pattern of diffusely increased marrow signal can also be found in patients with aplastic anemia, also secondary to the replacement of active hematopoietic cells by fat cells [31].

Diffusely decreased signal from within the medullary space on the T1-weighted sequences can be due to both benign and malignant processes. A diffuse or patchy signal that is moderately decreased can be obtained from a diffuse infiltrative process, hypercellular states, or increased trabecular formation. An active disease that has infiltrated the marrow cavity causes increased signal intensity on the T2-weighted images. Focal areas of decreased signal can also represent benign or malignant processes (Figures 9–13 through 9–18). Except for sclerotic metastases, malignancies are bright on the T2-weighted images. Areas of ischemic necrosis, bone islands, healed fracture sites, and heavily calcified regions or areas of fibrosis following therapy or trauma usually remain signal-poor on these sequences. Bone cysts usually are identified as smoothly contoured, homogeneous regions that increase uniformly in signal intensity on the T2-weighted images (see Figure 9–13).

Malignancies, either primary or metastatic, usually appear as low to intermediate signal intensity on the T1-weighted images, and can be homogeneous or heterogeneous (Figures 9–19 and 9–20). Areas of calcification appear dark, while hemorrhagic or necrotic areas can be bright. On the T2-weighted images, there is usually an increase in signal for active tumor. The very high sensitivity of MRI makes it well suited as a means of following suspected disease or response of disease to therapy (Figure 9–21). Treated malignancies may undergo fibrosis, and the marrow cavity will remain of lower signal than the remainder of the unaffected bone. In these cases, determination of regression of disease is decided by the lack of an increased signal on the T2-weighted image, or by the lack of progression of disease over time, since replacement by adipose tissue may not occur. The major exception to the interpretation of these signal intensity changes is sclerotic metastases. These heavily calcified lesions, mimicking a benign process, usually give little, if any, signal on any echo sequence (Figure 9–22). The presence of edema in adjacent musculature may be helpful, but an analysis of standard x-rays or CT studies is more reliable in this case.

Areas of infarction have a characteristic appearance on MR images, as a dark enclosed area, surrounding a bright region of fat ("ring" sign) (Figure 9–23), unchanged on T2-weighted views. An acute event may also demonstrate an area of surrounding edema within the marrow cavity, which becomes brighter on the T2 sequences. Osteonecrosis may demonstrate the ring sign, or may appear as a very dark area on T1-weighted images. Parts of this region of decreased signal may show increased signal on T2-weighted images, especially if the event is acute, with nonacute changes usually remaining low in signal (Figure 9–24; see also Figures 12–14, 12–15, and 12–31). Transient osteoporosis can exhibit a similar clinical presentation to osteonecrosis. The MR appearance of transient osteoporosis, however, is quite different [32,33]. The decrease in signal from within the marrow cavity is diffuse and intermediate in intensity, rather than dark; the signal becomes isointense with or brighter than normal marrow signal on the T2-weighted images.

Major trauma to the bony structures can almost certainly be documented by plain radiographs (see Figure 9–3). When plain films show negative findings, but there is a high clinical suspicion of injury, the exceptional sensitivity of MRI can lead to a diagnosis (see Figure 9–2). Occult fractures can be observed on high-resolution scans, surrounded by an area of decreased marrow signal, caused by surrounding edema (see Figure 12–2). Bone contusions, with

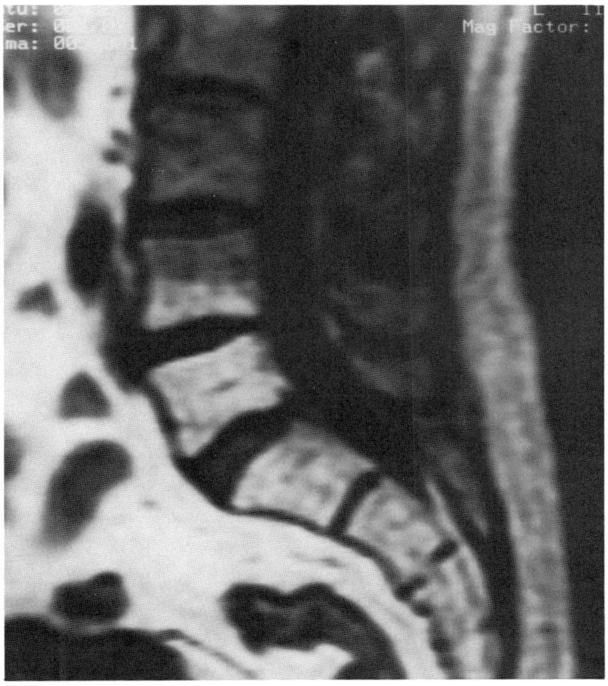

FIGURE 9–12 ▶
Radiation therapy induces increased fat deposition within bone, secondary to destroying cellular elements, causing an increased signal intensity from within the marrow space. The radiation portal for this patient with prostatic carcinoma extended midway through the L-4 vertebral body.

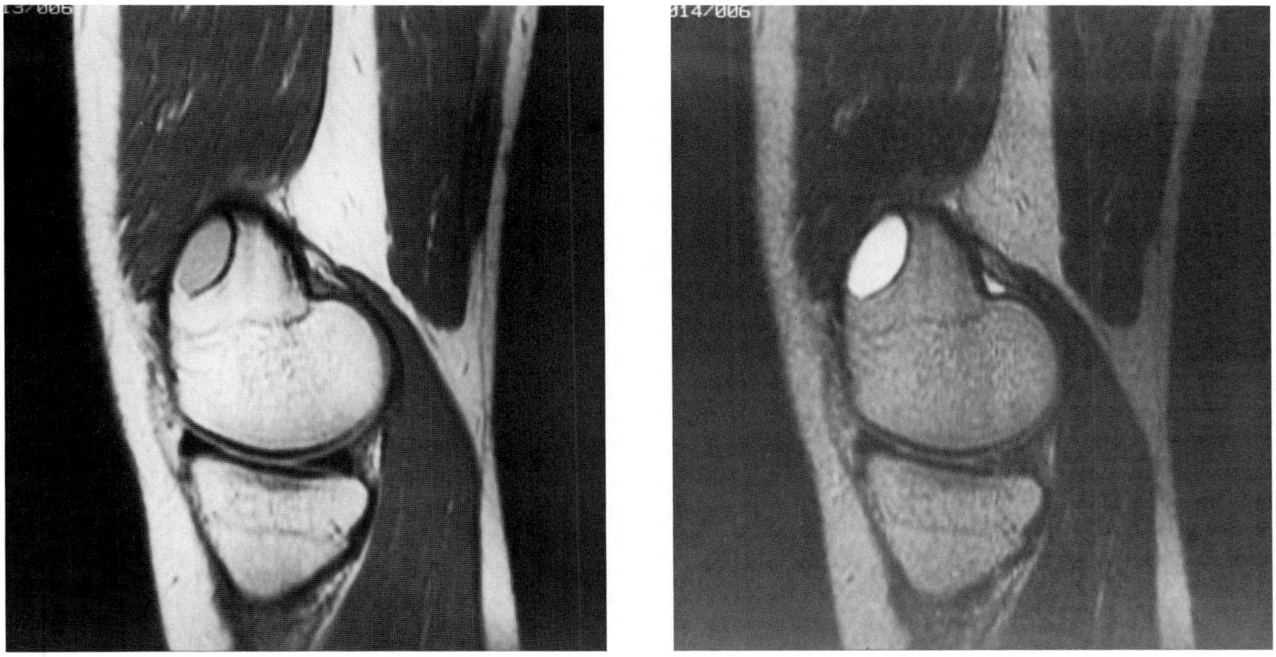

A B

FIGURE 9–13
A traumatic bone cyst is seen as a well-circumscribed homogeneous region on the T1-weighted images, increasing in brightness on the T2-weighted image. A. T1-weighted images. B. T2-weighted image.

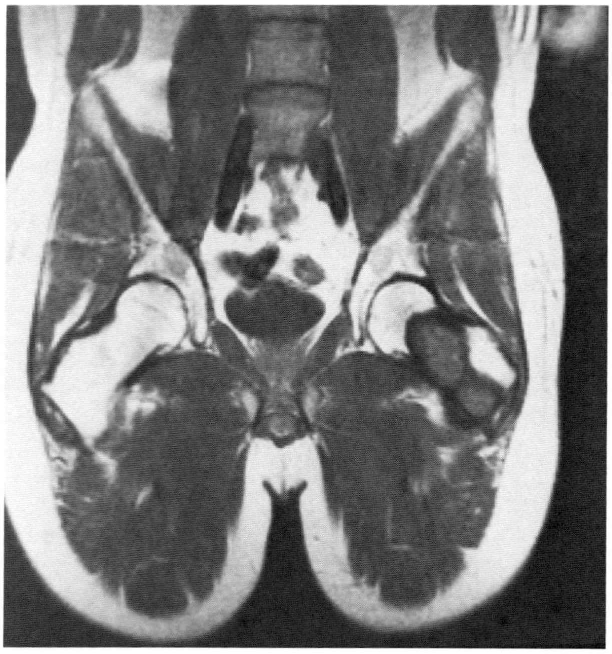

A

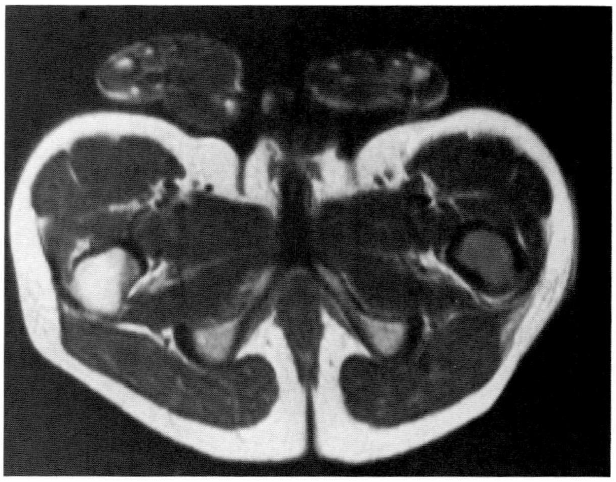

B

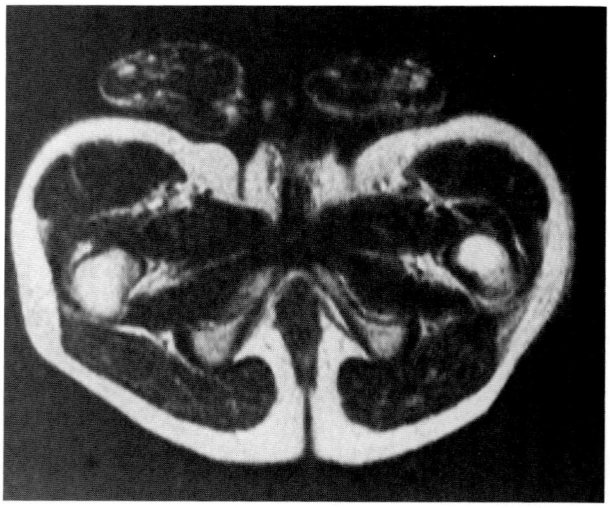

C

FIGURE 9–14
A, B. *Coronal and axial T1-weighted images, respectively.* C. *Axial T2-weighted images of fibrous dysplasia, showing increased intensity.*

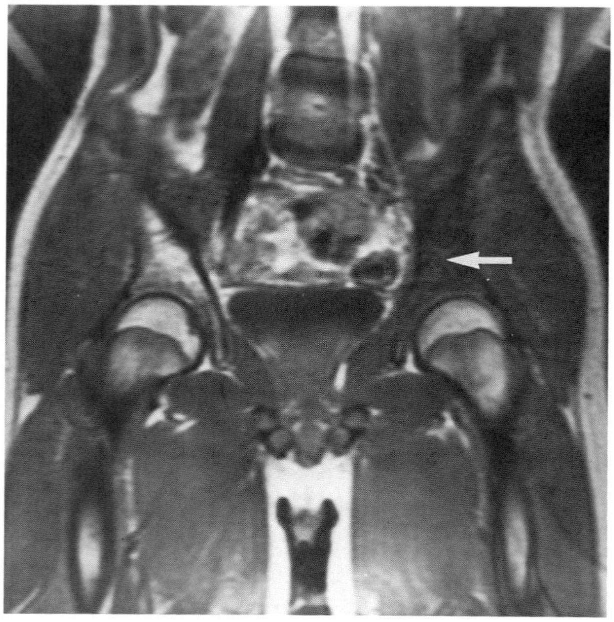

◀ *FIGURE 9–15*
T1-weighted coronal image of Ewing's sarcoma of the left ilium (arrow) *seen as an area of decreased signal.*

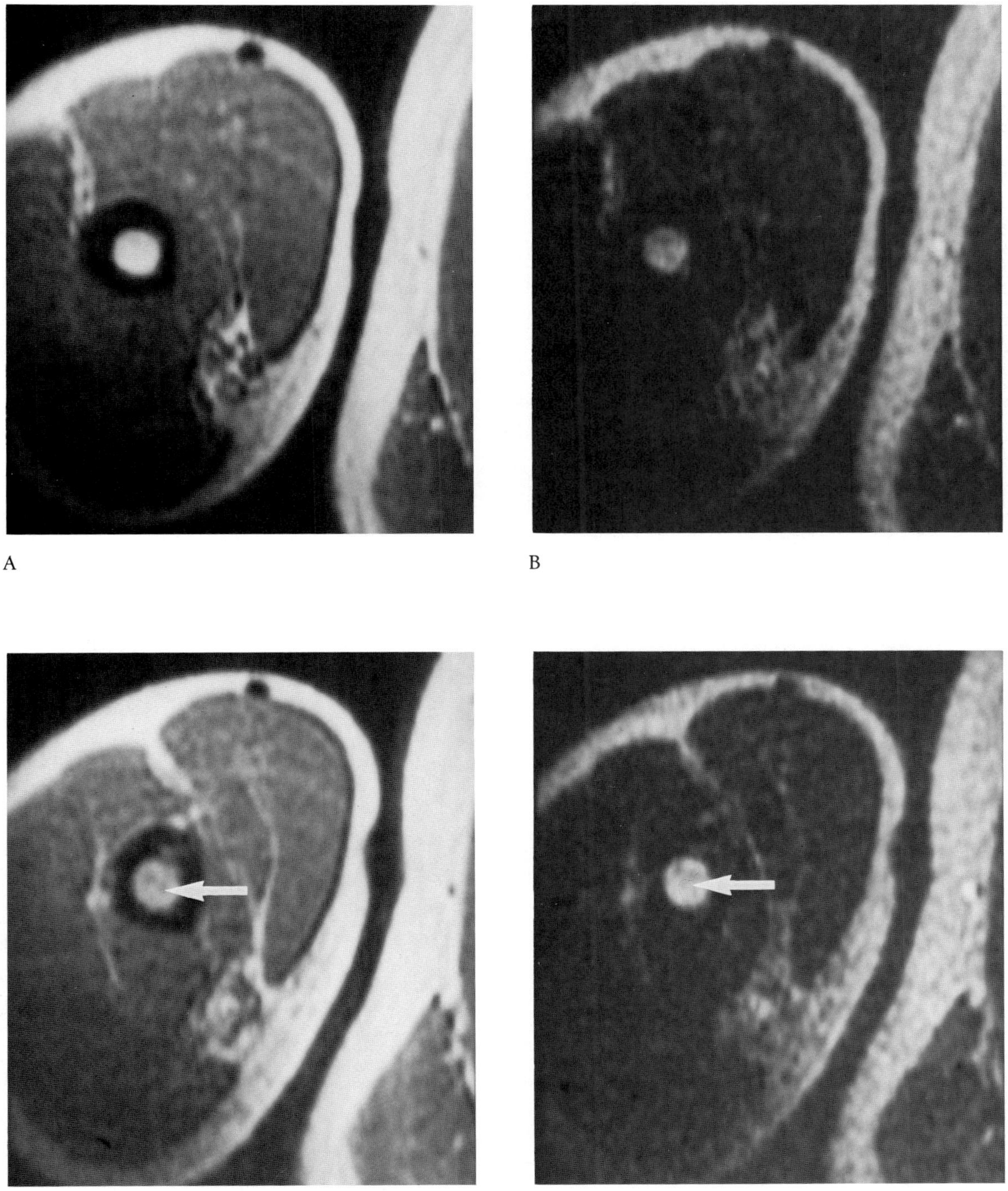

A B C D

FIGURE 9–16

A, B. *Normal marrow in patient with lymphoma in T1-weighted and T2-weighted images, respectively. Note the normal decrease in signal from fat within marrow cavity.* C, D. *The area of lymphoma infiltrating marrow cavity* (arrows) *is depicted in T1-weighted and T2-weighted images, respectively. Notice the increase in signal in the T2-weighted image due to tumor involvement.*

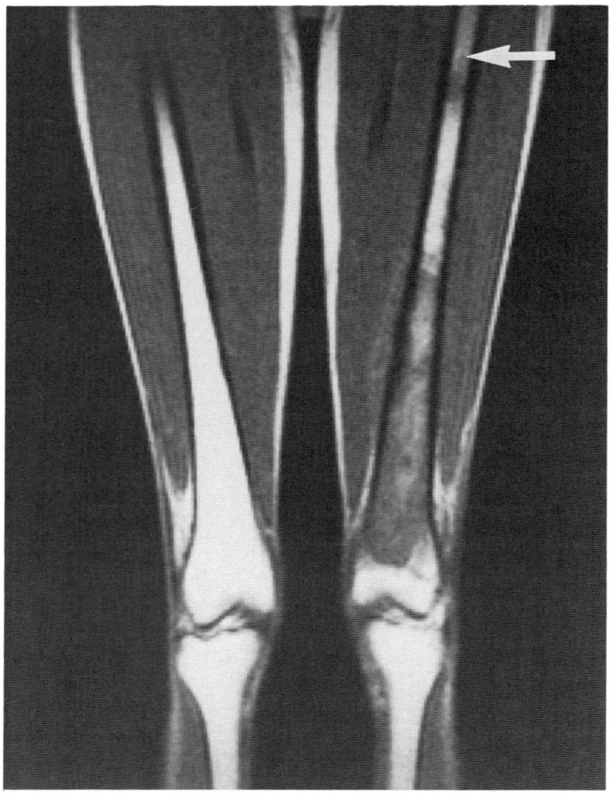

FIGURE 9–17
Osteosarcoma with evidence of skip metastases (arrow) *on a T1-weighted coronal view.*

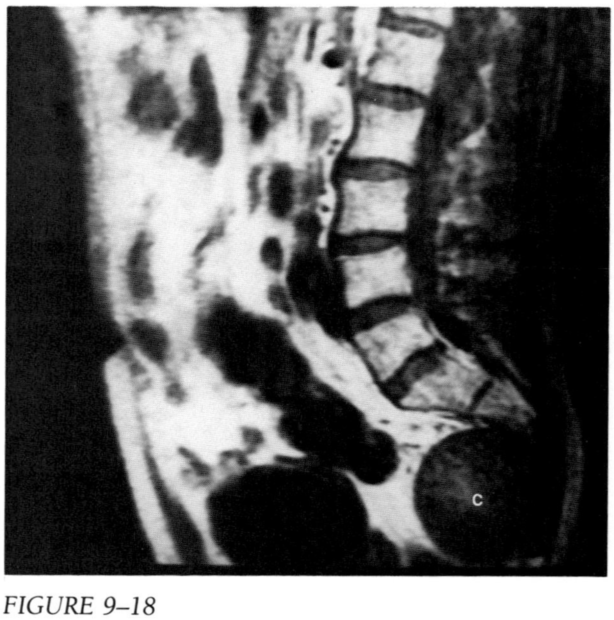

FIGURE 9–18
A large recurrent sacral chondrosarcoma (C) presents as a smooth spherical mass extending anteriorly from the sacrum.

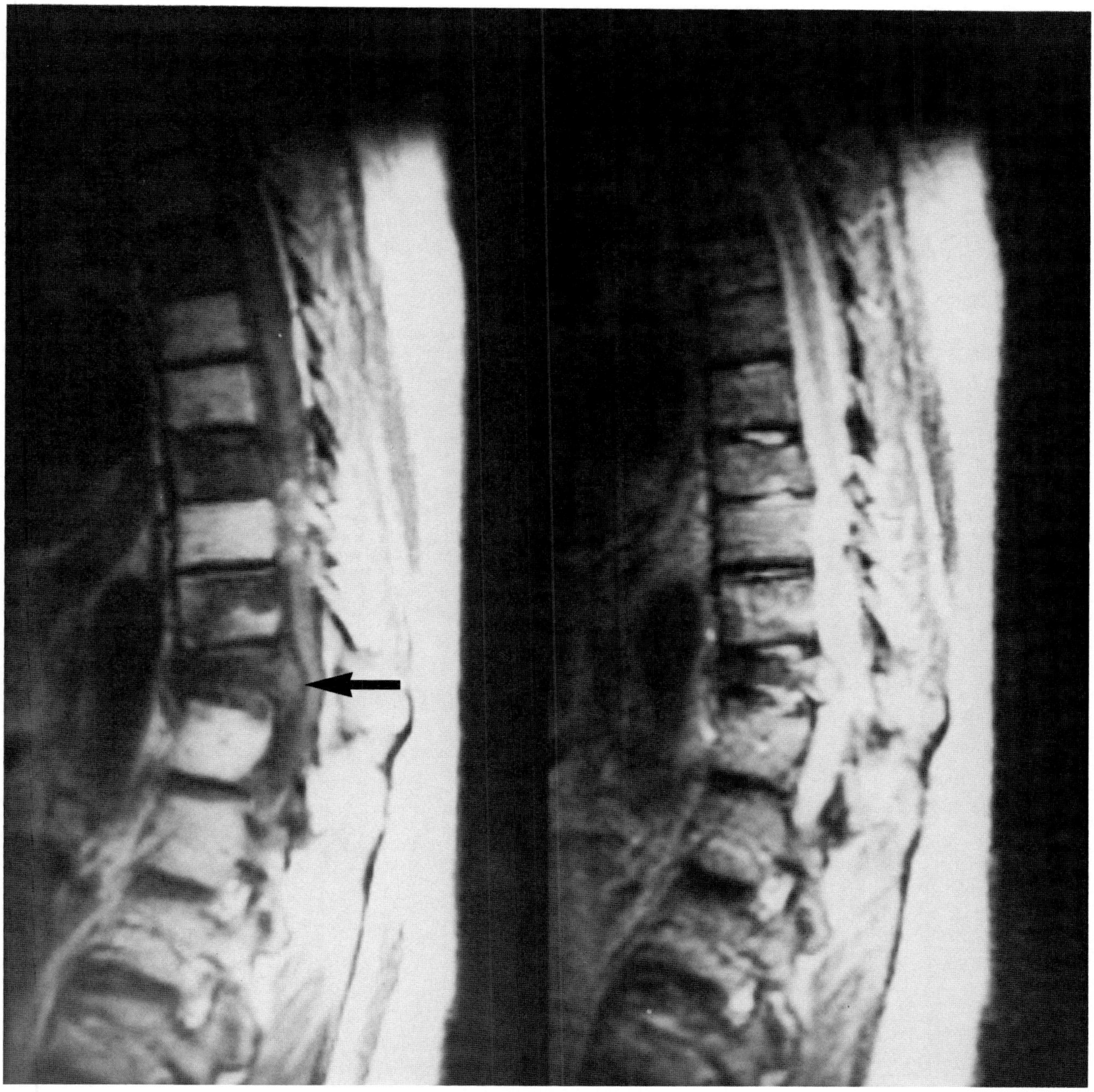

FIGURE 9–19
T1-weighted image (left) *demonstrates multiple areas of metastases from colon cancer to spine. There is an epidural mass extending from compressed vertebral body* (arrow).

FIGURE 9–20
Plasmacytoma of left ilium presents as expansile lesion in the T1-weighted image, increasing uniformly in signal intensity on the T2-weighted image. A. *T1-weighted image.* B. *T2-weighted image.*

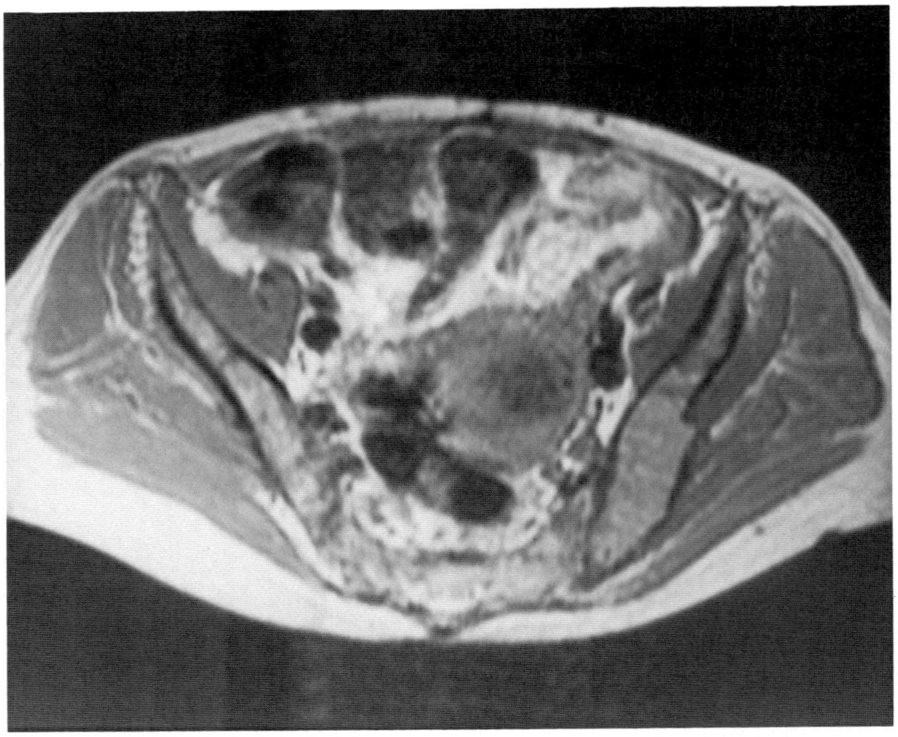

A

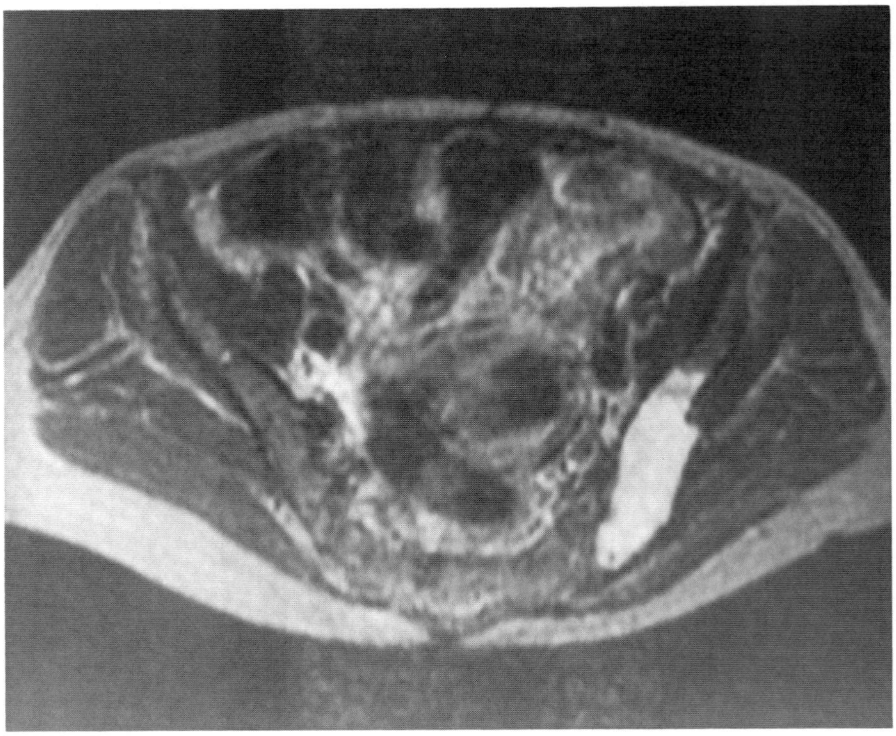

B

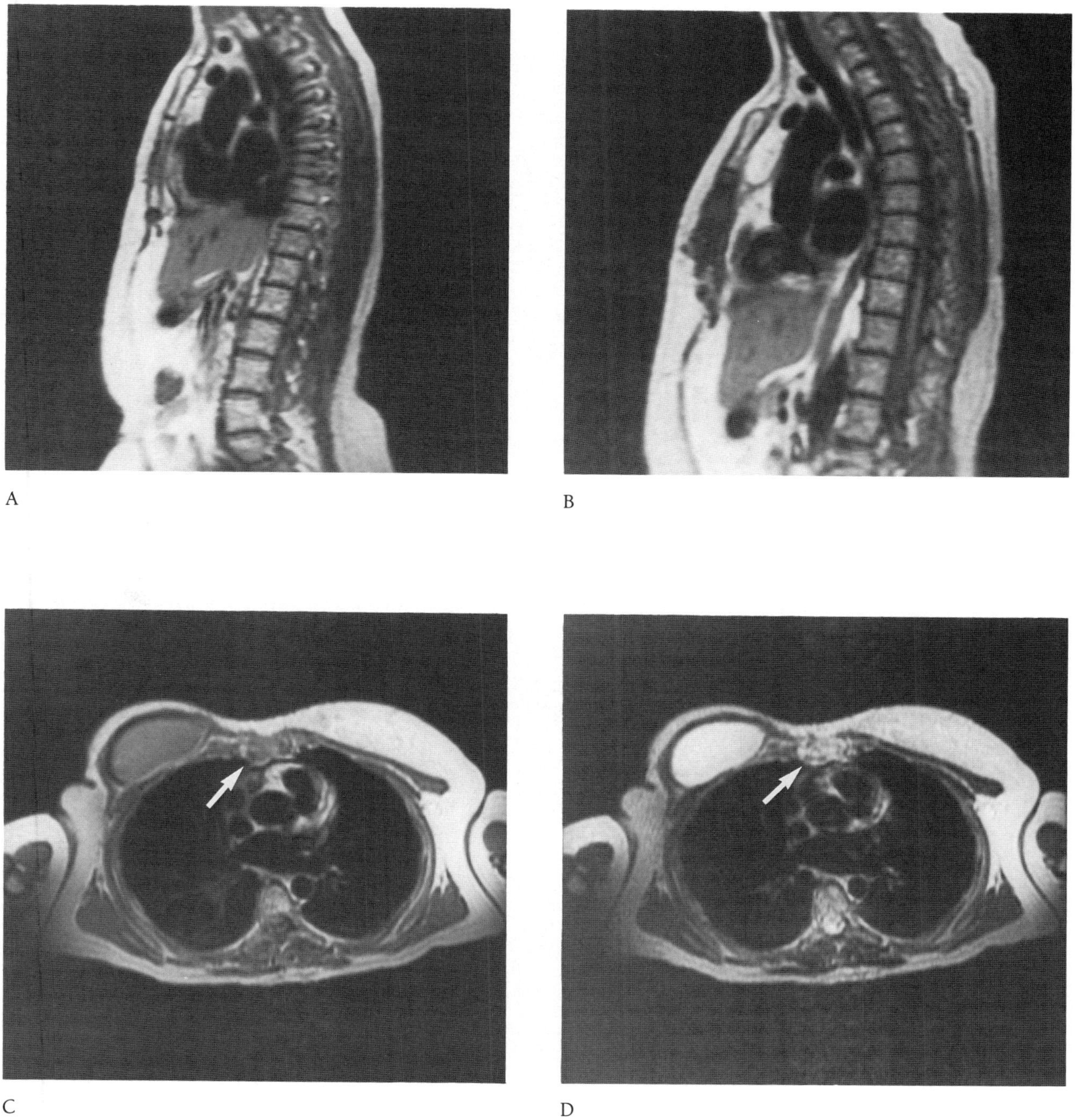

A B C D

FIGURE 9–21

A. Sagittal T1-weighted image of a patient with breast carcinoma demonstrates abnormally low signal in the sternum. B. *Two months later, a sagittal image demonstrates significant spread of the tumor through the cortex of the sternum and into soft tissues.* C, D. *Proton density and T2-weighted axial images, respectively, demonstrate increasing signal intensity from active tumor* (arrows). *Note increased signal from breast prosthesis.*

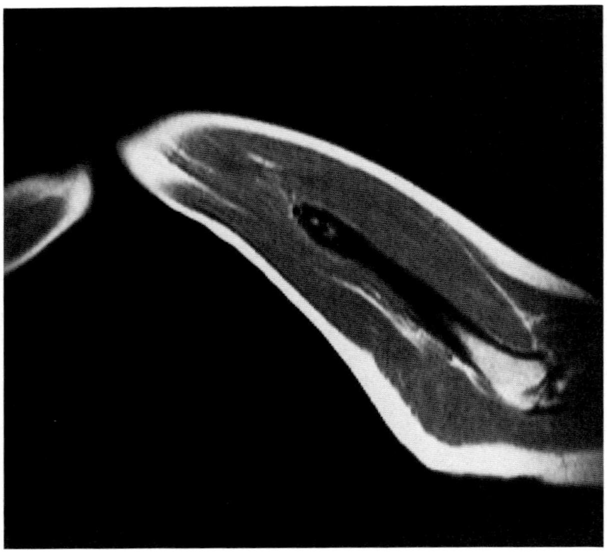

A

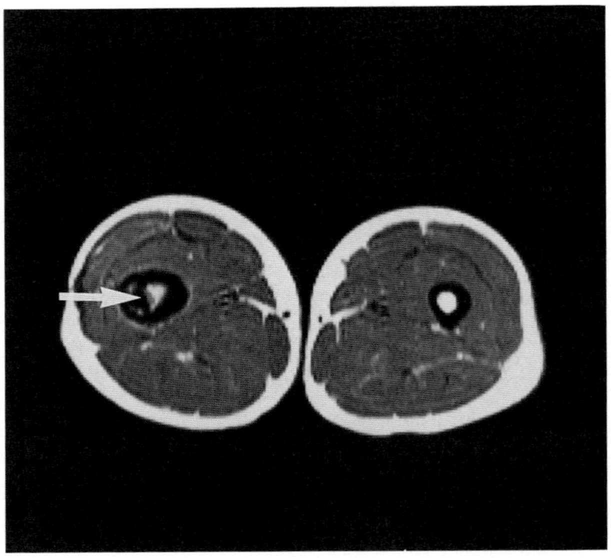

B

FIGURE 9–22
A, B, C. *T1-weighted coronal, T1-weighted axial, and T2-weighted axial images, respectively, of the mid femur with an area of consistently decreased signal due to sclerotic metastasis from breast carcinoma* (large arrow). *Note the rim of inflammation around the lesion in* B (small arrows).

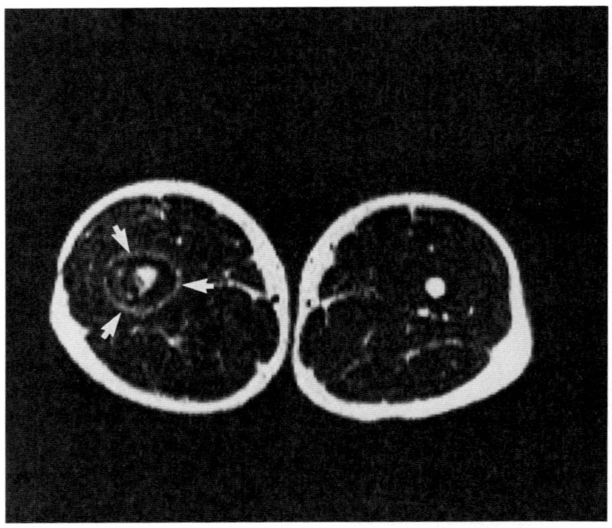

C

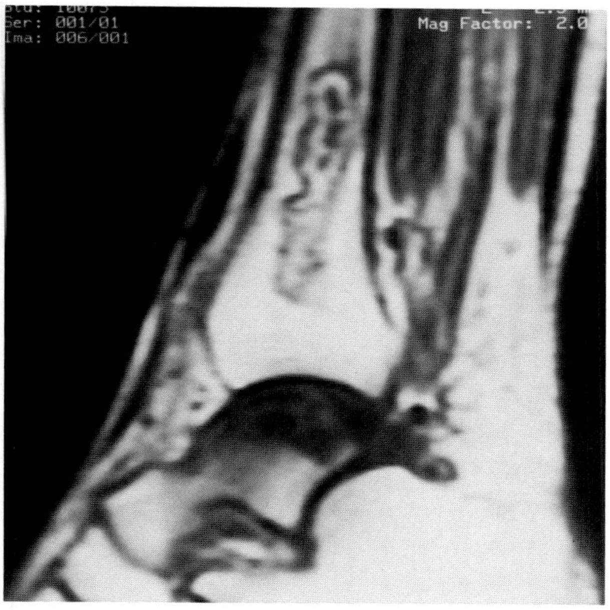

FIGURE 9–23
The ring sign of a bone infarction in the distal tibia.

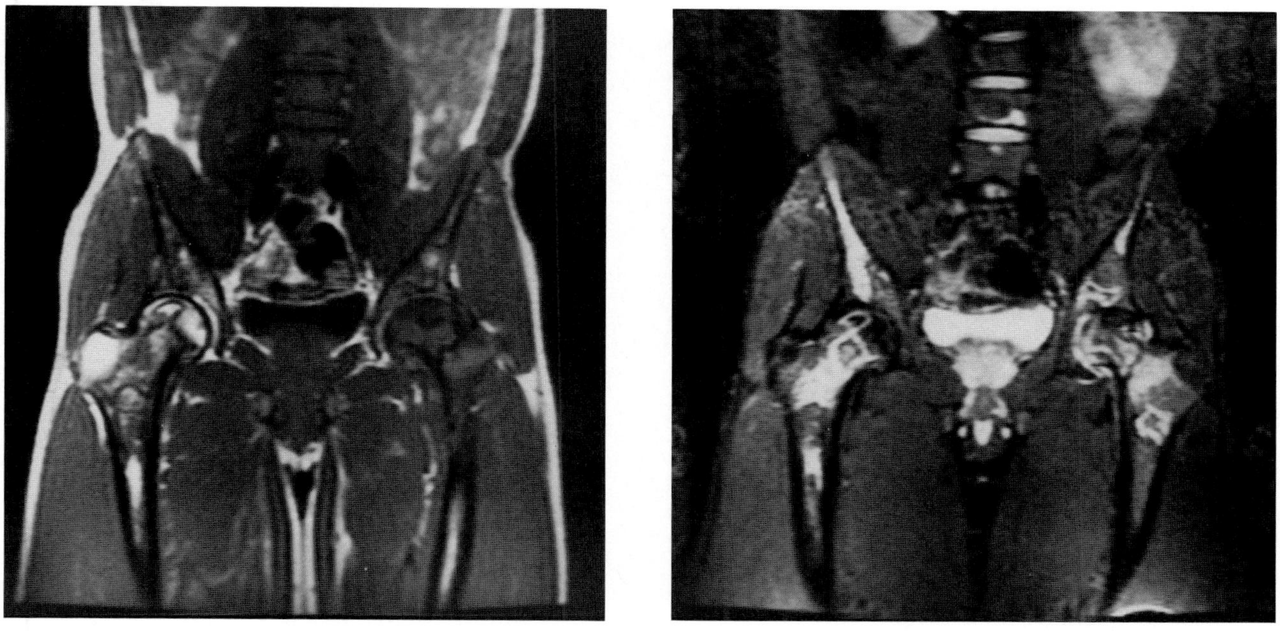

A B

FIGURE 9–24
Multiple areas of infarction with bilateral avascular necrosis of the femoral heads, in a patient with sickle cell disease. A. T1-weighted image exhibiting a ring pattern in the right femoral head and a region of diffusely decreased signal in the left, along with decreased signal throughout the bony structures due to areas of infarction. B. Fat-suppression IR image demonstrates areas of active pathology.

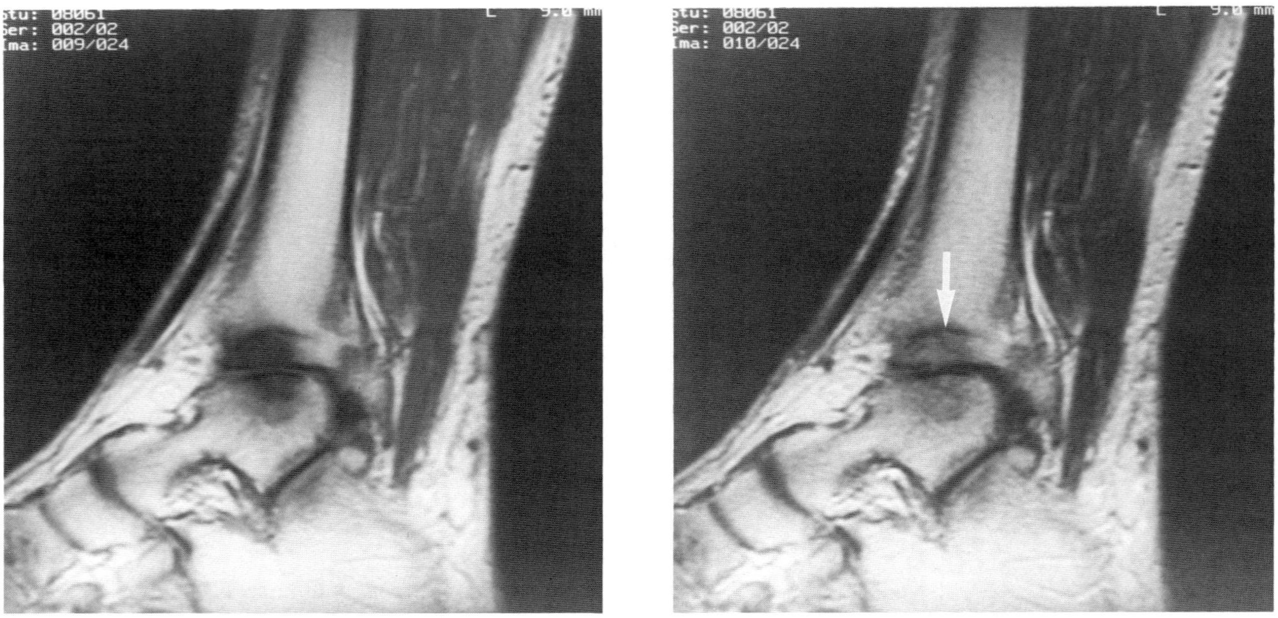

A B

FIGURE 9–25
A. *There is a large area of decreased signal in the distal aspect of the tibia and the talus on the T1-weighted image.* B. *The T2-weighted image demonstrates these areas to be almost isointense with the rest of the marrow cavity, and reveals an occult fracture at the distal part of the tibia* (arrow).

trauma-induced edema within the marrow cavity, present as areas of decreased signal on the T1-weighted scans, becoming isointense to slightly brighter than normal marrow on the T2-weighted sequences (Figure 9–25). Edematous changes within musculature, secondary to trauma, can also be readily identified [34].

Infection is an area where MRI is of great clinical assistance, permitting an early and accurate diagnosis of osteomyelitis [35–38]. The marrow signal is decreased in these cases, but is not totally absent (being dark gray, but not black). It becomes very bright on the T2-weighted image, secondary to edema within the marrow space (Figure 9–26). The MR signal of the surrounding musculature may increase slightly because of edematous changes, and the presence of an abscess suspected if the soft tissue increase becomes very bright (Figure 9–27). The changes due to osteomyelitis, again, are not specific; however, the MR findings together with the clinical picture enable a diagnosis to be made.

Conclusions

MRI is an extremely sensitive method of determining abnormalities within the marrow space. As has been shown, it is not nearly as specific as

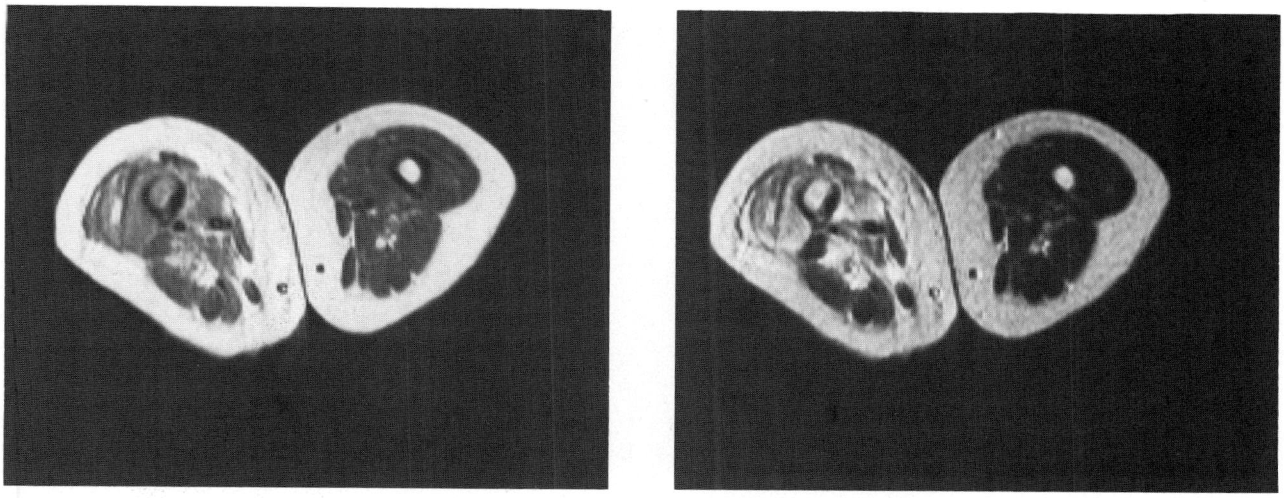

A B

FIGURE 9–26
A, B. T1-weighted and T2-weighted axial images, respectively, through the midpart of the thighs. A region of decreased signal in the marrow space of the right femur is seen in A, and increases in signal intensity in B due to osteomyelitis. The surrounding musculature is only slightly increased in intensity over the normal left side, consistent with edema.

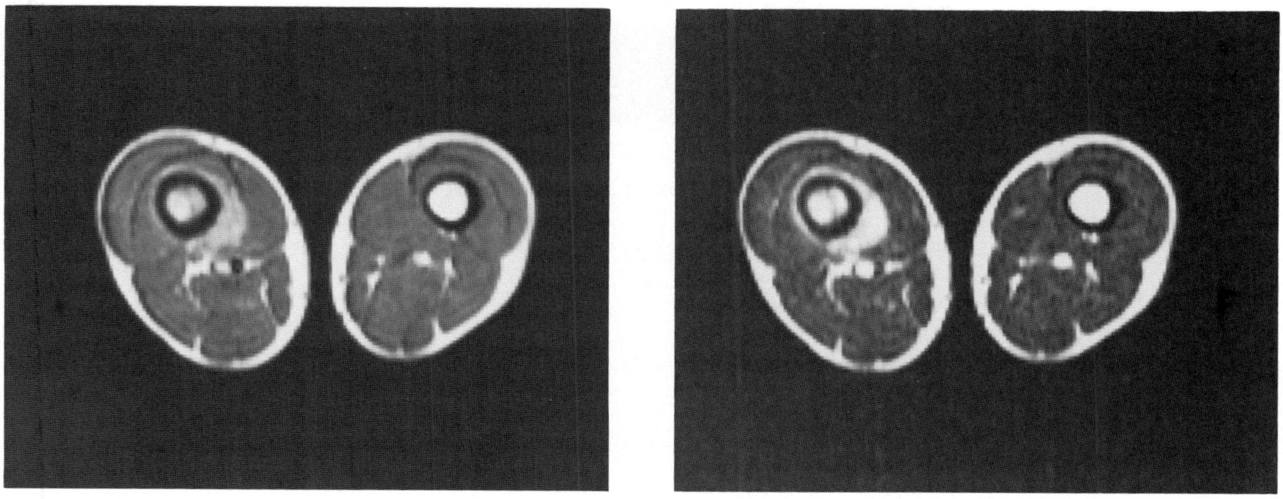

A B

FIGURE 9–27
A, B. T1-weighted and T2-weighted axial images, respectively, through the midpart of the thighs. A region of decreased signal in the marrow space of the right femur is seen in A, and increases in signal intensity in B due to osteomyelitis. The surrounding abnormal signal becomes extremely bright in B, consistent with an abscess.

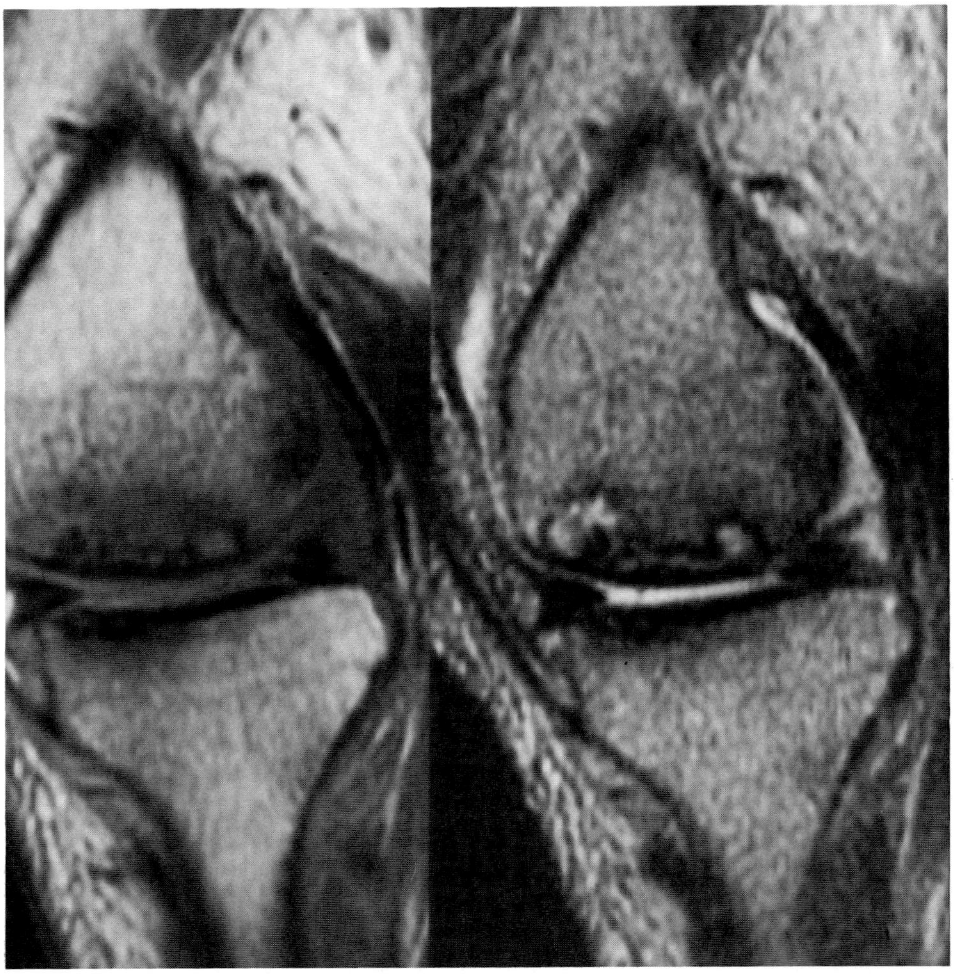

FIGURE 9–28
Subchondral cyst formation in degenerative arthritis is identified as irregular regions below the cortical surface on the T1-weighted images (left)*, which increase in signal on the T2-weighted image* (right).

plain x-rays; however, continued experience and new scanning techniques may increase specificity in the future. In certain clinical settings, however, diagnoses can be made at the present time. From a knowledge of anatomy and the signal characteristics of different tissues, many different types of particular bone abnormalities can be identified on MRI. Certain types of bone disease are much better evaluated on plain roentgenograms than on MRI, although these diseases do have a distinct MR appearance (Figures 9–28 and 9–29). Areas of calcification, arthritic and erosive changes, and anatomic anomalies should always first be documented by x-ray, with MRI studies obtained if additional information is desired [39,40].

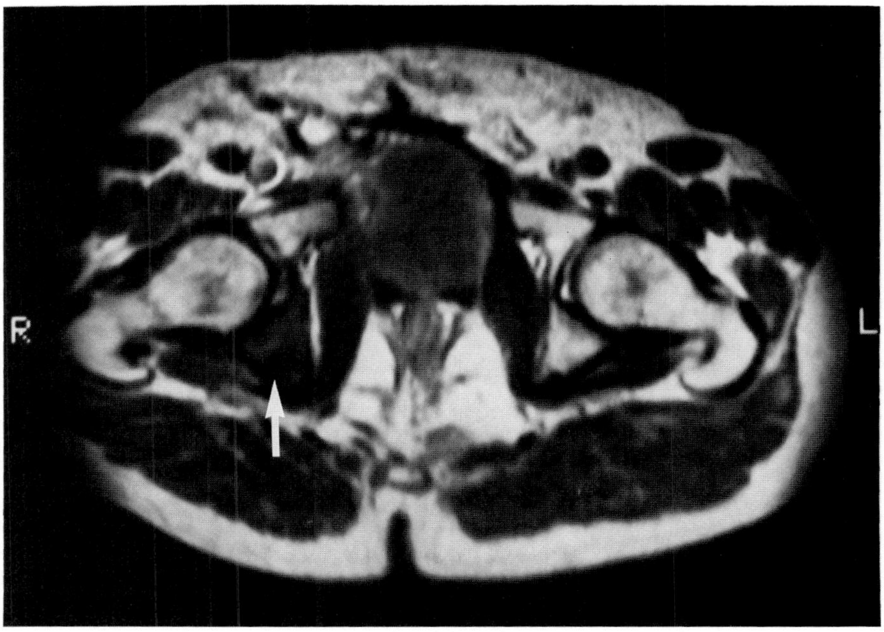

A

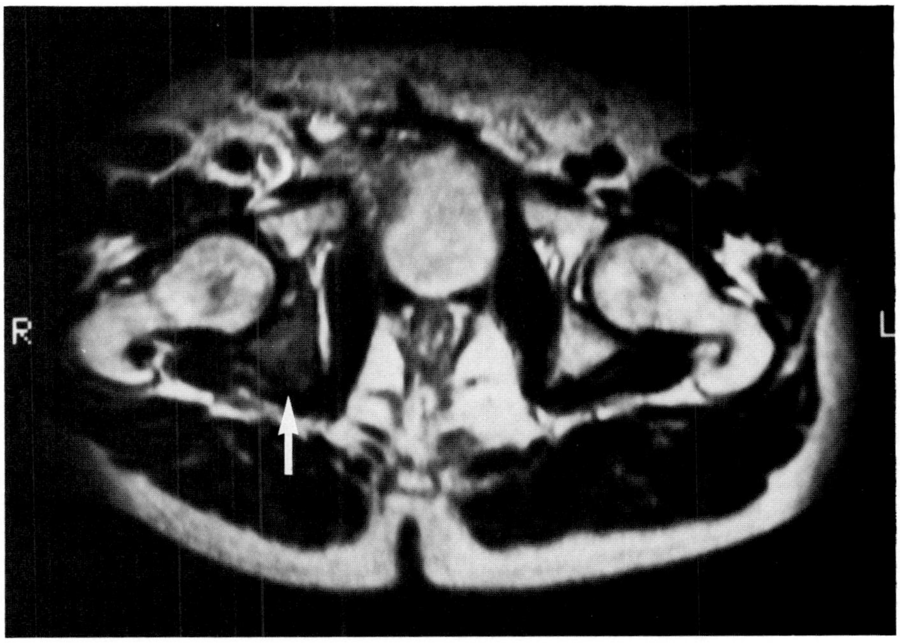

B

FIGURE 9–29

A, B. *Paget's disease of the right ischium is signal-poor on both the T1-weighted image and the T2-weighted image, respectively, due to the increased calcification. (Courtesy of Middletown Radiology Group, Middletown, NY)*

REFERENCES

1. Edeiken J. Roentgen diagnosis of diseases of bone. Baltimore: Williams & Wilkins, 1981.
2. Ehmann RL, Berquist TH, McLeod RA. MR imaging of the musculoskeletal system: a 5-year appraisal. Radiology 1988; 166:313–20.
3. Sartoris DJ, Resnick D. MR imaging of the musculoskeletal system: current and future status. AJR 1987; 149:457–67.
4. Kursunoglu-Brahme S, Riccio T, Weisman MH, et al. Rheumatoid knee: role of gadopentetate-enhanced MR imaging. Radiology 1990; 176:831–5.
5. Erlemann R, Reiser MF, Peters PE, et al. Musculoskeletal neoplasms: static and dynamic Gd-DTPA-enhanced MR imaging. Radiology 1989; 171:767–73.
6. Millen SJ, Daniels DL, Meyer GA. Gadolinium-enhanced magnetic resonance imaging in temporal bone lesions. Laryngoscope 1989; 99:257–60.
7. Hajek PC, Sartoris DJ, Neumann CH, Resnick D. Potential contrast agents for MR arthrography: in vitro evaluation and practical observations. AJR 1987; 149:97–104.
8. Pettersson H, Eliasson J, Egund N, et al. Gadolinium-DTPA enhancement of soft tissue tumors in magnetic resonance imaging—preliminary clinical experience in five patients. Skeletal Radiol 1988; 17:319.
9. Markisz JA, Knowles RJR, Altchek DW, Schneider R, Whalen JP, Cahill PT. Segmental patterns of avascular necrosis of the femoral heads: early detection with MR imaging. Radiology 1987; 162:717–20.
10. Mitchell DG, Rao VM, Dalinka MK, et al. Femoral head avascular necrosis: correlation of MR imaging, radiographic staging, radionuclide imaging, and clinical findings. Radiology 1987; 162:709–15.
11. Beltran J, Herman LJ, Burk JM, et al. Femoral head avascular necrosis: MR imaging with clinical-pathological and radionuclide correlation. Radiology 1988; 166:215–20.
12. Coleman BG, Kressel HY, Dalinka MK, Scheibler ML, Burk DL, Cohen EK. Radiographically negative avascular necrosis: detection with MR imaging. Radiology 1988; 168:525–8.
13. Deutsch AL, Mink JH, Waxman AD. Occult fractures of the proximal femur: MR imaging. Radiology 1989; 170:113–6.
14. Yao L, Lee JK. Occult intraosseous fracture: detection with MR imaging. Radiology 1988; 167:749–51.
15. Berger PE, Ofstein RA, Jackson DW, Morrison DS, Silvino N, Amador R. MRI demonstration of radiographically occult fractures: what have we been missing? Radiographics 1989; 9:407–36.
16. Dunnill MS, Anderson JA, Whitehead R. Quantitative histological studies on age changes in bone. J Pathol Bacteriol 1967; 94:275–91.
17. Dooms GC, Fisher MR, Hricak H, Richardson M, Crooks LE, Genant HK. Bone marrow imaging: magnetic resonance studies related to age and sex. Radiology 1985; 155:429–32.
18. Cohen MD, Klatte EC, Bachner R, et al. Magnetic resonance imaging of bone marrow disease in children. Radiology 1984; 151:715–8.
19. Langman J. Medical embryology. 3rd ed. Baltimore: Williams & Wilkins, 1975:145–6.
20. Walker HS, Dietrich RB, Flannigan BD, Lufkin RB, Peacock WJ, Kangarloo H. Magnetic resonance imaging of the pediatric spine. Radiographics 1987; 7:1129–1152.
21. Baker LL, Goodman SB, Perkash I, Lane B, Enzmann DR. Benign versus pathologic compression fractures of vertebral bodies: assessment with conventional spin-echo, chemical-shift and STIR MR imaging. Radiology 1990; 174:495–502.
22. Vogler JB III, Murphy WA. Bone marrow imaging. Radiology 1988; 168:679–93.
23. Sugimura K, Yamasaki K, Kitagaki H, Tanaka Y, Kono M. Bone marrow diseases of the spine: differentiation with T1 and T2 relaxation times in MR imaging. Radiology 1987; 165:541–4.
24. Hajek PC, Baker LL, Goobar JE, et al. Focal fat deposition in axial bone marrow: MR characteristics. Radiology 1987; 162:245–9.
25. Ross JS, Masaryk TJ, Modic MT, Carter JR, Mapstone T, Dengel FH. Vertebral hemangiomas: MR imaging. Radiology 1987; 165:165–9.
26. Beltran J, Simon DC, Levy M, Herman L, Weis L, Mueller CF. Aneurysmal bone cysts: MR imaging at 1.5 T. Radiology 1986; 158:689–90.
27. Munk PL, Helms CA, Holt RG, et al. MR imaging of aneurysmal bone cysts. AJR 1989; 153:99–101.
28. Ramsey RG, Zacharias CE. MR imaging of the spine after radiation therapy: easily recognizable effects. AJR 1985; 144:1131–5.
29. Remedios PA, Colletti PM, Raval JK, et al. Magnetic resonance imaging of bone after radiation. Magn Reson Imaging 1988; 6:301–4.

30. Casamassina F, Ruggiero C, Caramella D, et al. Hematopoietic bone marrow recovery after radiation therapy: MRI evaluation. Blood 1989; 73: 1677–81.

31. Kaplan PA, Asleson RJ, Klassen LW, Duggan MJ. Bone marrow patterns in aplastic anemia: observations with 1.5-T MR imaging. Radiology 1987; 164:441–4.

32. Bloem JL. Transient osteoporosis of the hip: MR imaging. Radiology 1988; 167:753–5.

33. Wilson AJ, Murphy WA, Hardy DC, Totty WG. Transient osteoporosis: transient bone marrow edema? Radiology 1988; 167:757–60.

34. Beltran J, Simon DC, Katz W, Weis LD. Increased MR signal intensity in skeletal muscle adjacent to malignant tumors: pathologic correlation and clinical relevance. Radiology 1987; 162:251–5.

35. Modic MT, Pflanze W, Feiglin DH, Belhobek F. Magnetic resonance imaging of musculoskeletal infections. Radiol Clin North Am 1986; 24:247–58.

36. Scott JA, Palmer EL. Musculoskeletal infection of the extremities: evaluation with MR imaging. Radiology 1988; 168:284–5.

37. Quinn SF, Murray W, Clark RA, Cochran C. MR imaging of chronic osteomyelitis. J Comput Assist Tomogr 1988; 12:113–7.

38. Unger EC, Moldofsky PJ, Gatenby RA, et al. Diagnosis of osteomyelitis by MR imaging. AJR 1988; 150:605–18.

39. Bongartz G, Bock E, Horbach T, Requardt H. Degenerative cartilage lesions of the hip—magnetic resonance evaluation. Magn Reson Imaging 1989; 7:179–86.

40. Roberts MC, Kressel HY, Fallon MD, Zlatkin MB, Dalinka MK. Paget's disease: MR imaging findings. Radiology 1989; 173:341–5.

10

MRI Evaluation of Soft Tissue

JOHN A. MARKISZ

MRI is rapidly becoming an important technique for the evaluation of disorders of the soft tissue structures of the extremities. The inherent contrast of MRI, unobtainable with any other modality, permits identification of pathologic features and differentiation among various tissues to be made quite readily. The ability to image directly in multiple planes enables the extent of a lesion to be more precisely delineated. Significant improvements in surface coil technology over the past several years have made resolution of MRI equal to or better than that available with CT. For these reasons, MRI for musculoskeletal evaluation is now competitive with other imaging techniques.

The selection of imaging techniques is very dependent on the particular problem to be solved. Both extremities can often be imaged together using the head or body coils, if the concern is relatively vague and does not involve a known mass or specific condition. In this way, a direct comparison of normal and abnormal sides can be made, and congenital variants accounted for. If the interest lies in a palpable mass or a specific relatively small region, a surface coil

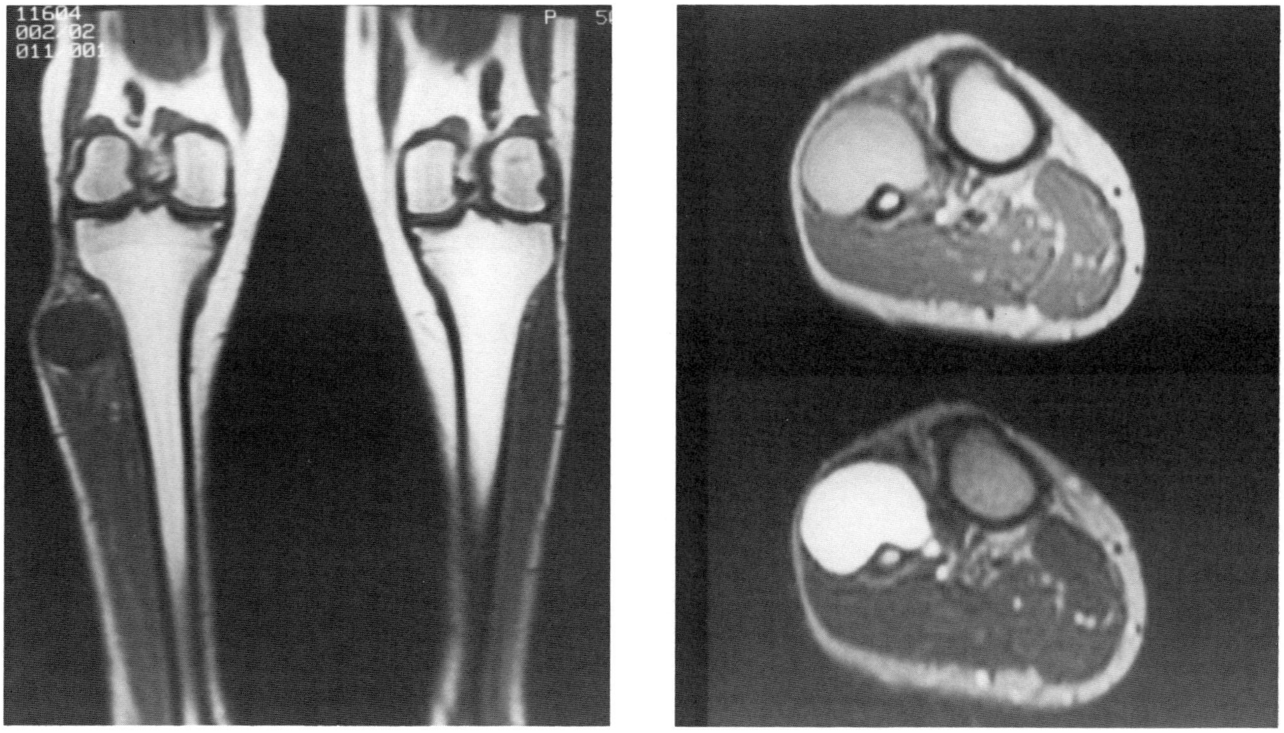

A B

FIGURE 10–1
A benign cyst, smooth, well-circumscribed, and homogeneous, is represented by low signal intensity on the coronal T1-weighted image, increasing in intensity on the axial images (proton density above, T2-weighted below). A. Coronal T1-weighted image. B. Axial images.

examination of the area in question can be performed. The selection of planes and pulse sequences depends on the region under investigation, and should always include both heavily T1-weighted and T2-weighted images, although not necessarily in the same planes. Slice thickness and interslice spacing may have to be revised, depending on the size of the abnormality expected and the extent of the region to be covered. A large mass may be adequately visualized with 5-mm sections and 5-mm spacings, whereas a small lesion may require 3-mm contiguous sections to properly evaluate the abnormality and its relationship to the surrounding structures. Fields of view can be adjusted to improve resolution, especially if the body coil is utilized.

Musculoskeletal problems that are amenable to MRI examination include neoplasms, infections, and trauma-induced conditions [1–3]. The intrinsic contrast in MRI provides better discrimination to be made between normal and abnormal tissues than can be obtained from CT, as well as adequately defining relationships to vascular and osseous structures [4]. MRI can also provide useful information as to the size and nature of residual masses after a tumor is excised or treated with chemo- or radiation therapy [5]. The major disadvantage inherent in this modality is the inability to image calcium [6].

In cases of abnormal masses, the ideal diagnostic device distinguishes between malignant and benign tumors. Although early reports promised that MRI relaxation times would be capable of doing this, experience has shown significant overlap in T1 and T2 values for malignant and benign conditions. However, a combination of

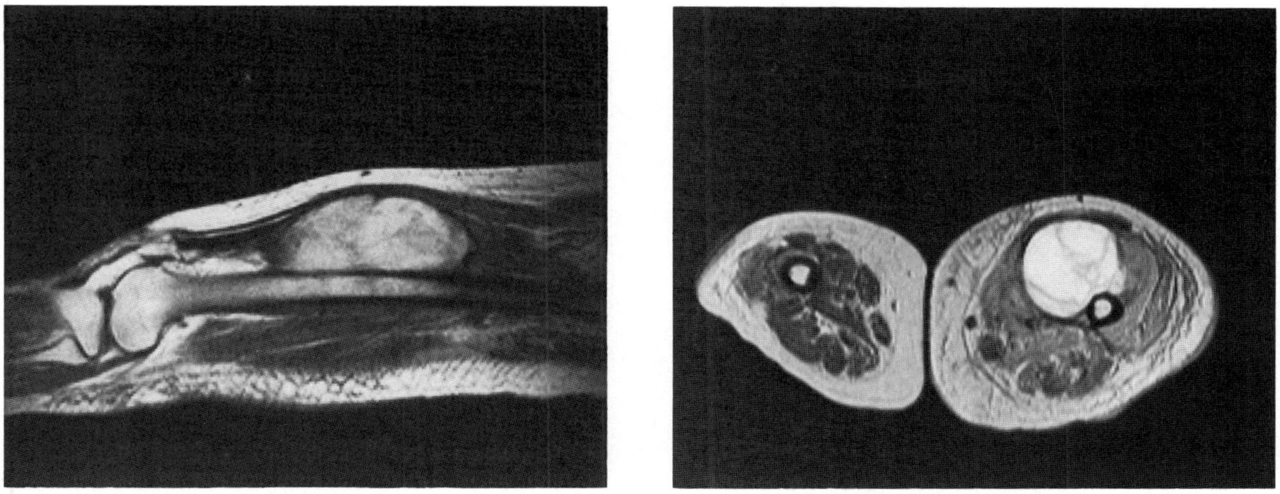

FIGURE 10–2
Smoothly contoured fibroliposarcoma, adjacent to the femur. A. *T1-weighted coronal image.* B. *T2-weighted axial images. Note the increased signal in surrounding muscles due to edema.*

anatomic analysis and changes in signal intensities often strongly suggest the type of condition being investigated. In general, smoothly contoured, well-delineated, homogeneous masses are usually indicative of a benign condition, whereas irregular, heterogeneous masses, crossing tissue planes, are most suggestive of a malignancy. Unfortunately, there is no sharp boundary separating malignant from benign conditions (Figures 10–1 and 10–2).

Changes in signal intensity on the various echo sequences very often aid in determining the type of tissue present, thus limiting the possible diagnoses, and occasionally accurately identify a particular entity. Certain materials can be characterized by their MRI appearance, thus providing (noninvasively) information essential to a diagnosis. Fat is the brightest, naturally found material on T1-weighted images, and decreases in intensity slightly on T2-weighted images. The intensity demonstrated by hemorrhage is time-dependent, and except for the acute stage, usually some part is very intense on all echo sequences, bright on the T1 sequences, becoming even brighter on the T2 images. Muscle exhibits an intermediate signal intensity on the T1-weighted images; the signal decreases on the T2 views. Fibrotic tissue is always signal-poor, becoming darker on the T2 views than on the T1 images. Ligaments and tendons are dark, signal-poor structures. Areas of calcification produce no signal on any MR images. Fluids vary in their T1-weighted appearance, depending primarily on their protein content, and increase significantly in signal intensity, becoming bright on the T2-weighted views (Figure 10–3). Fluid with a high concentration of protein, such as an abscess, can sometimes be bright on T1-weighted images as well (Figures 10–4 and 10–5) [2,7].

Infection within the soft tissues is often a diagnostic dilemma, presenting with nonspecific findings. Radiographs are most often negative, and unless air is present within an abscess, it is difficult to differentiate from surrounding soft tissue structures on CT as well [8]. If several pulse sequences are used, at least one (usually the T2-weighted images) will document the presence of an abnormality. An abscess will present as a distinct fluid collection on MRI, separate from surrounding structures, usually heterogeneous. Although the MRI findings are not specific, with a proper clinical history, a diagnosis can be made. In addition to detecting an abscess, an accurate assessment of the size, extent, and

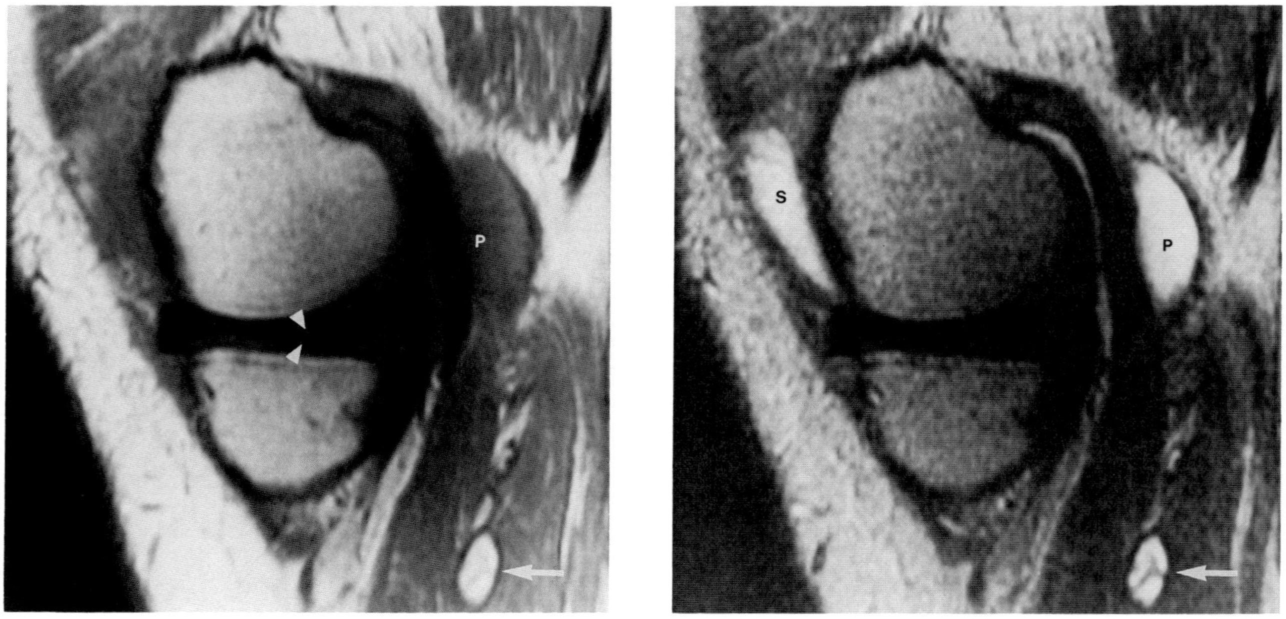

A B

FIGURE 10–3

Traumatic injury to the leg. A. T1-weighted image demonstrates complex tear of the posterior horn of the medial meniscus (arrowheads)*; abnormal signal in the anterior horn, not quite extending to a meniscal surface on this section; popliteal cyst between the medial head of the gastrocnemius and semimembranosus muscles (P); and high signal area of hemorrhage* (arrow) *in the medial head of the gastrocnemius muscle. B. On the T2-weighted images the meniscal tears are no longer visible. A suprapatellar effusion (S) is now evident as a very bright region, as is the popliteal cyst. An area of hemorrhage has increased in signal, with dark rind and septations of hemosiderin-laden macrophages visible.*

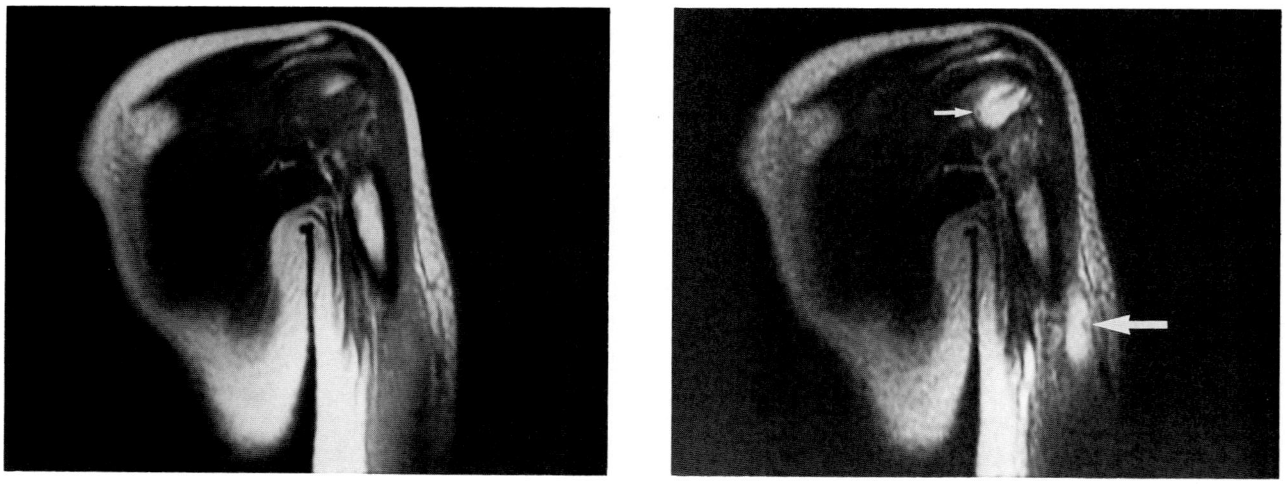

A B

FIGURE 10–4

A. Both the septic effusion of the shoulder and an abscess in the mid upper arm are not discernible on this coronal T1-weighted image. B. Both the effusion (small arrow) *and the abscess* (large arrow) *are evident on the T2-weighted image.*

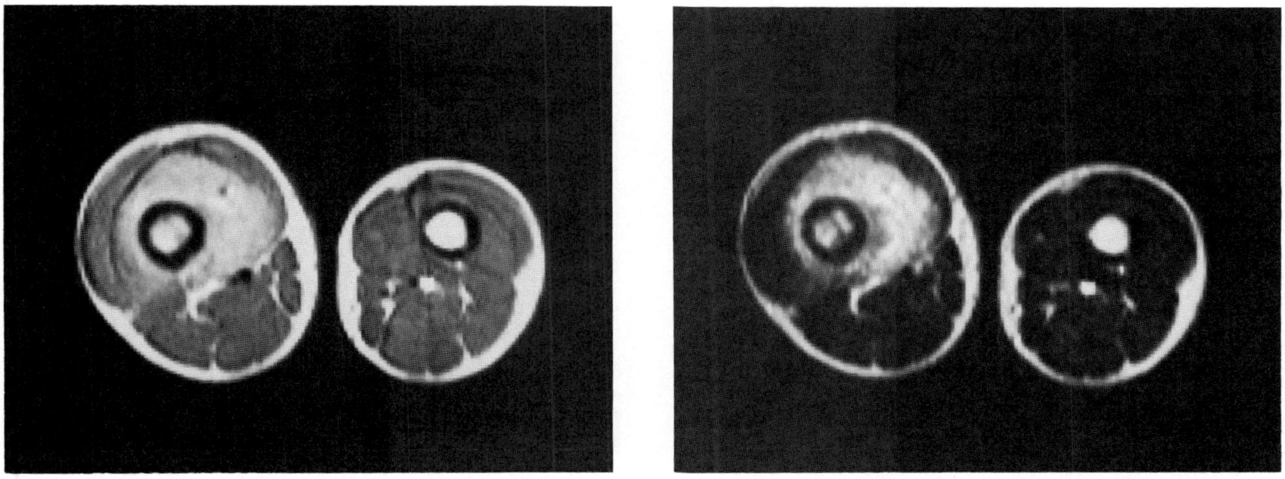

A B

FIGURE 10–5
A. The swollen right thigh contains an abscess, bright on the T1-weighted image.
B. The appearance is brighter and heterogeneous on the T2-weighted image.

involvement of other structures can be made [9]. This is especially true with the determination of bone involvement. Reactive changes may cause an increased uptake of radionuclide, leading to an overestimation of the presence of osteomyelitis. MRI, however, will reliably determine infections within the marrow space, as described in the previous chapter.

The MRI appearance of hemorrhage in the soft tissues can be quite varied (much more so than it is in the brain), and is distinctly time-dependent (Figures 10–6, 10–7, and 10–8) [10,11]. Paramagnetic materials (e.g., ferric [Fe^{+3}] complexes) contain unpaired electrons that can alter relaxation mechanisms and significantly change observed signal intensities. An acute hemorrhage may be difficult to detect, appearing low or isointense to soft tissue structures on both the T1- and T2-weighted images. Oxyhemoglobin contains iron in the ferrous (Fe^{+2}) state, which has no unpaired electrons and exerts no paramagnetic effect. As the hemorrhage ages slightly, the decreased oxygen pressure (PO_2) causes the conversion of oxyhemoglobin to deoxyhemoglobin, which although still in the ferrous state, now has four unpaired electrons and is paramagnetic. These electrons are still contained within the red blood cell, and exhibit only local magnetic susceptibility effects, resulting in a decrease in the overall T2 value and decreased MR signal [12]. Methemoglobin contains iron in the ferric (Fe^{+3}) state with five unpaired electrons and exerts both local magnetic susceptibility effects and dipole-dipole interactions, resulting in lowering both T1 and T2 values, and causing high signal intensity to be observed on both T1- and T2-weighted images [13]. A peripheral rim of low signal appears in the later stages, and decreases in signal intensity on the T2-weighted images, due to the deposition of hemosiderin by macrophages, which again shortens T2 values [11]. These effects can vary greatly, depending on the site of the bleed, whether or not it becomes walled off, how quickly it is resorbed, whether it is mixed with other fluids, and so on.

Diffuse abnormalities within the soft tissues are well documented by MRI. Subcutaneous edema can be seen as regions within the subcutaneous fat. It has significantly lower signal than fat on the T1-weighted images, and becomes bright on the T2-weighted views. Edema within muscle groups can often be diagnosed by MRI as a diffuse region of uniformly slightly increased

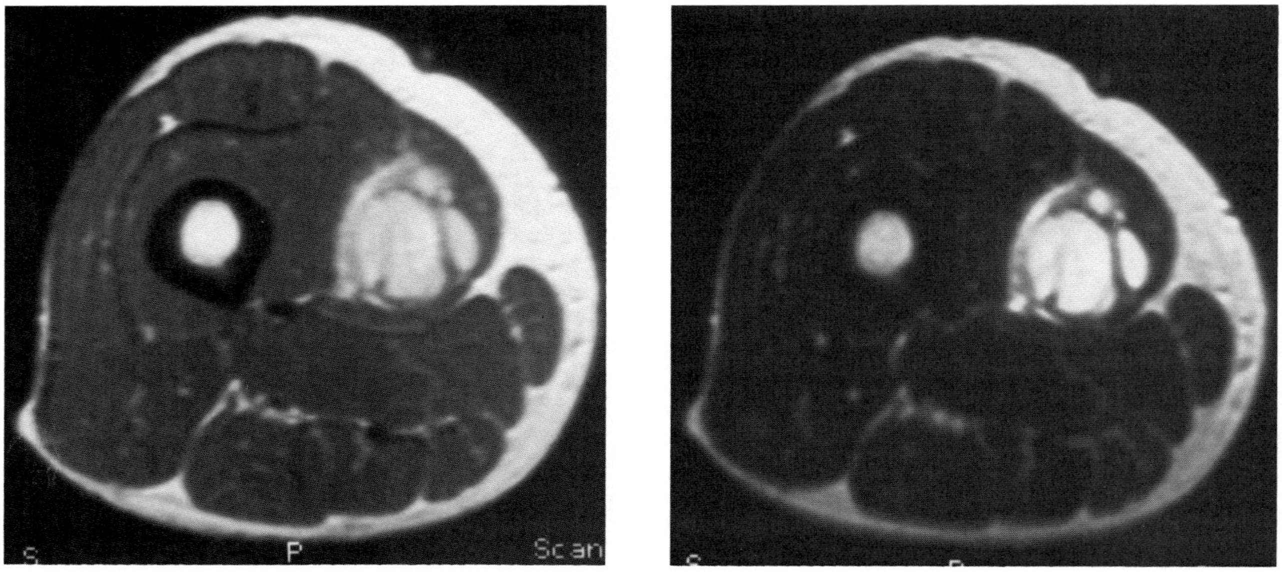

A B

FIGURE 10–6
A large resolving 3-week-old hematoma, bright on both axial images. Areas of hemosiderin deposition can be observed as dark areas on both sequences. A. T1-weighted image. B. T2-weighted image.

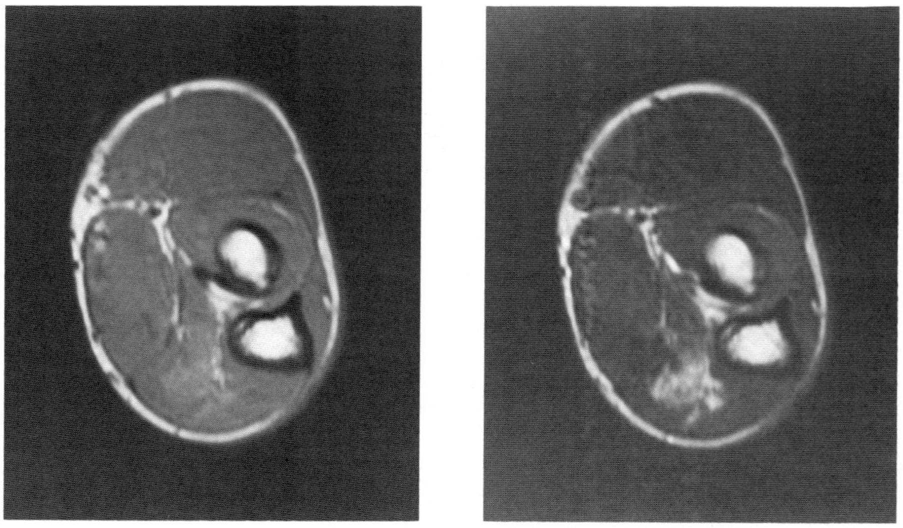

A B

FIGURE 10–7
A 48-hour-old hemorrhage in the flexor digitorum superficialis muscle is better observed as an irregular bright area within the muscle on the T2-weighted image than the T1-weighted image. A. T1-weighted image. B. T2-weighted image.

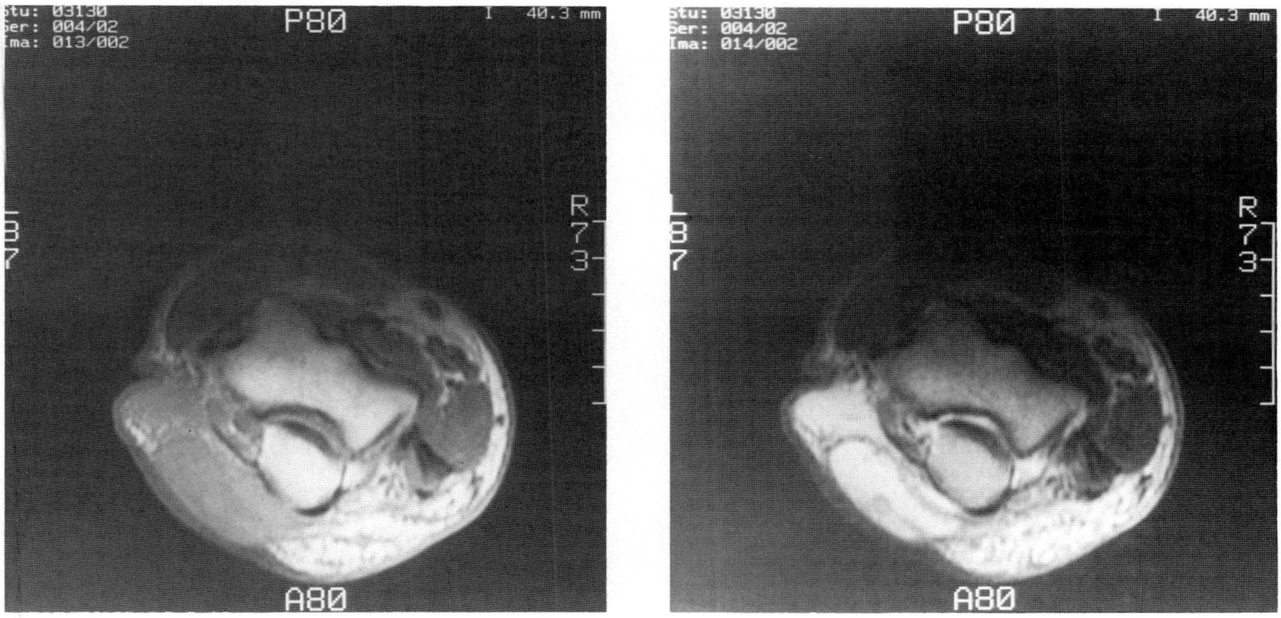

A B

FIGURE 10–8
Subcutaneous hemorrhage from a hemangioma at the elbow observed after 7 days. A. *T1-weighted image.* B. *T2-weighted image.*

signal within the muscle on the T1-weighted images, due to increased proton density, and becoming brighter on the T2-weighted images [14], due to the increased fluid within the edematous tissues. Edema within muscles is often noted around other pathologic areas (e.g., bone tumors, osteomyelitis, soft tissue tumors) and care must be taken not to confuse these findings with true pathologic features (Figure 10–9). Fatty infiltration of muscles appears as a diffuse increase in signal on all spin-echo sequences (Figure 10–10).

Both malignant and benign tumors found in the soft tissues can be demonstrated very well, and although MRI is not capable of distinguishing among all of them, several unique characteristics can sometimes aid in a diagnosis. The most common benign tumor is the lipoma, commonly found in the subcutaneous tissues and composed of fatty elements, encased by a very thin, delicate capsule. Although they present as a palpable mass, histologically these cannot be differentiated from normal adipose tissue on the MR images (Figure 10–11) [15]. Their MR appearance is, not surprisingly, identical to fat on all echo sequences [16]. The capsule is so thin that it rarely can be imaged; therefore, many palpable subcutaneous lipomas are "lost" in the normal fat, blending in with the subcutaneous adipose tissue. We have found it helpful to mark the palpable abnormality with an oil-containing capsule as an aid in interpreting the images (Figure 10–12). If a palpable subcutaneous mass is not discernible on any images within the normal adipose tissue, the diagnosis of a lipoma can be made by exclusion. Lipomas can also present intramuscularly, in which case a rim of muscular tissue is observed encapsulating the lesion (Figure 10–13). Liposarcomas can present with an MR appearance very similar to lipomas, with visible septa seen throughout the lesion (Figures 10–13 and 10–14). Histologically, these two entities are very similar [18], and we have found biopsied proven lipomas with septations that are identical in MR appearance to that of proven low-grade liposarcomas. In such fatty deposits, therefore, the possibility of a low-grade liposarcoma cannot be excluded.

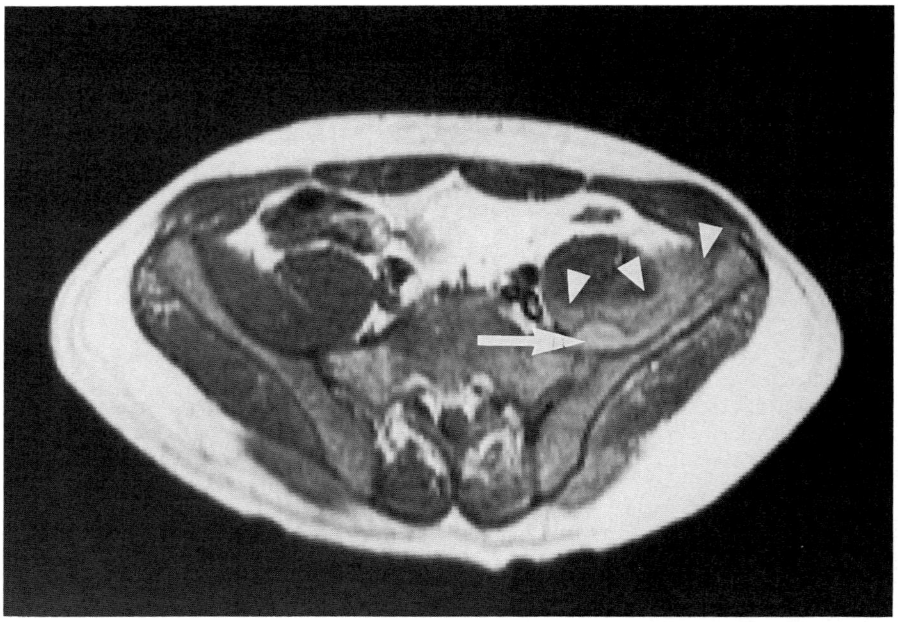

A

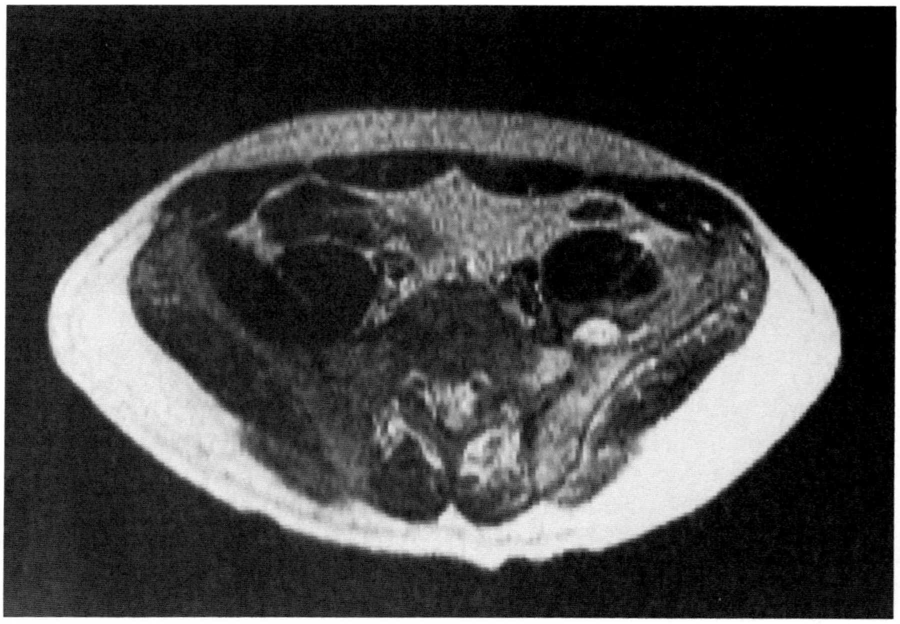

B

FIGURE 10–9
A retroiliopsoas abscess (arrow) *with edema in adjacent musculature* (arrowheads). A. *T1-weighted image.* B. *T2-weighted image.*

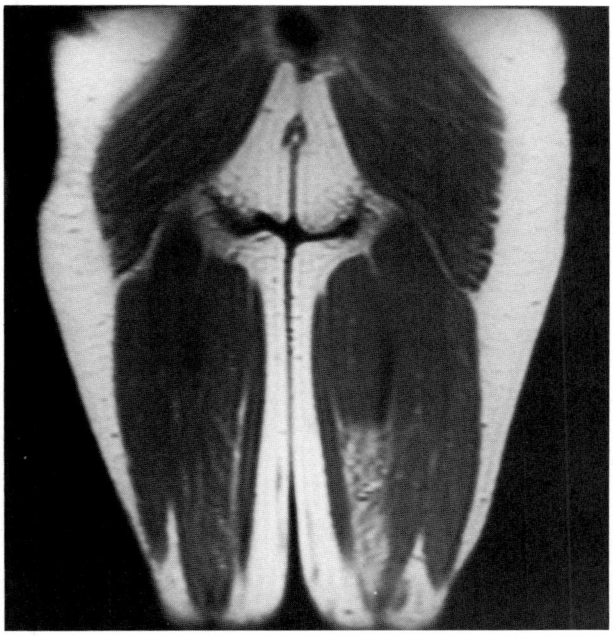

A

FIGURE 10–10
Fatty infiltration in the inferior portion of the semimembranosus muscle. A. *T1-weighted coronal image.* B. *Proton density* (left) *and T2-weighted* (right) *images.*

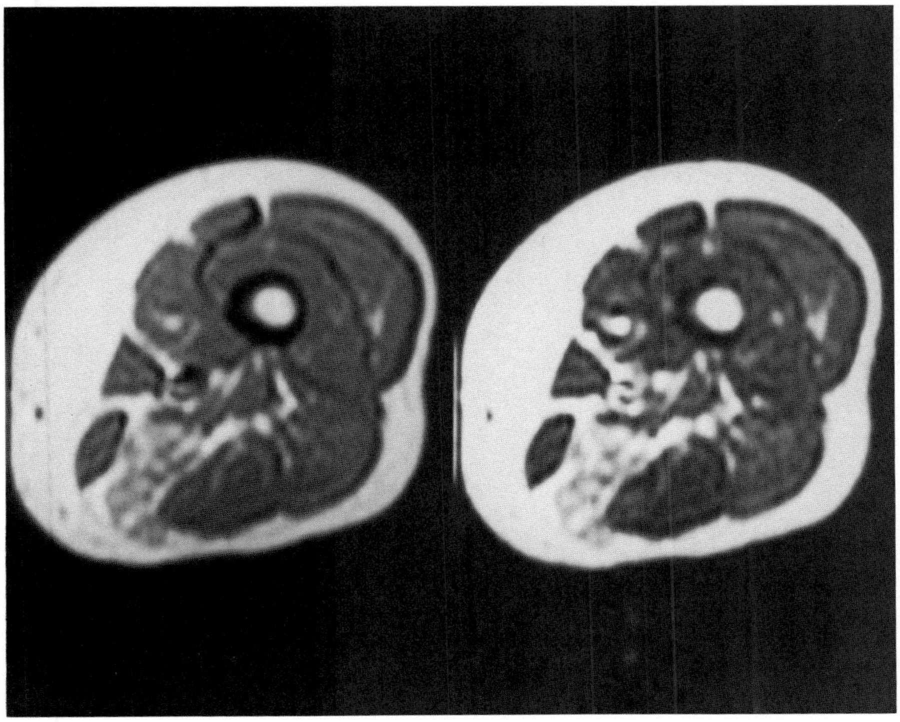

B

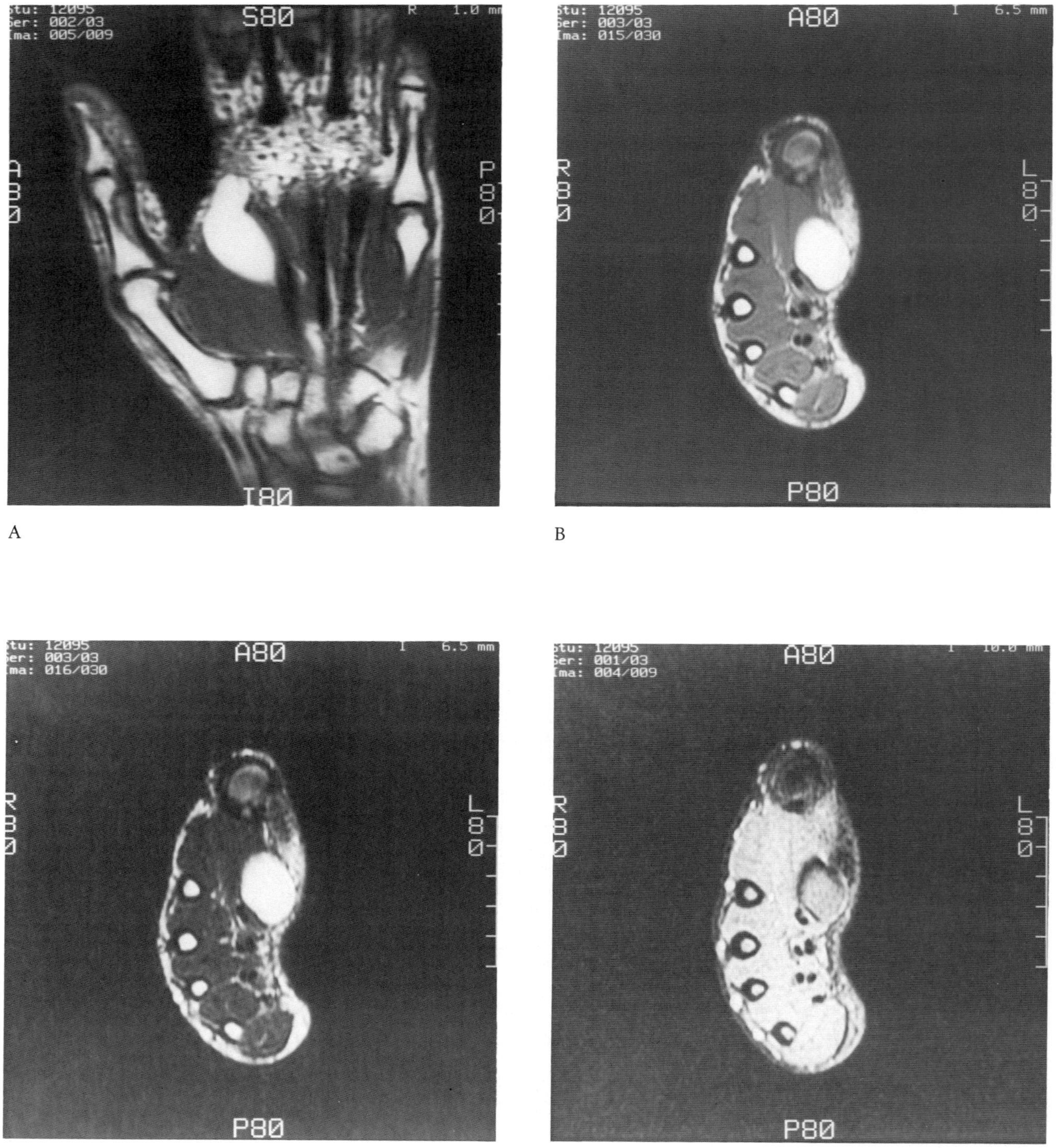

A B C D

FIGURE 10–11

A. A large lipoma in the palmar aspect of the hand, seen to be pressing against the flexor tendons of the second digit on this T1-weighted coronal image. This remains isointense with subcutaneous fat on all sequences employed. B. T1-weighted spin-echo image. C. T2-weighted spin-echo image. D. Gradient-echo image.

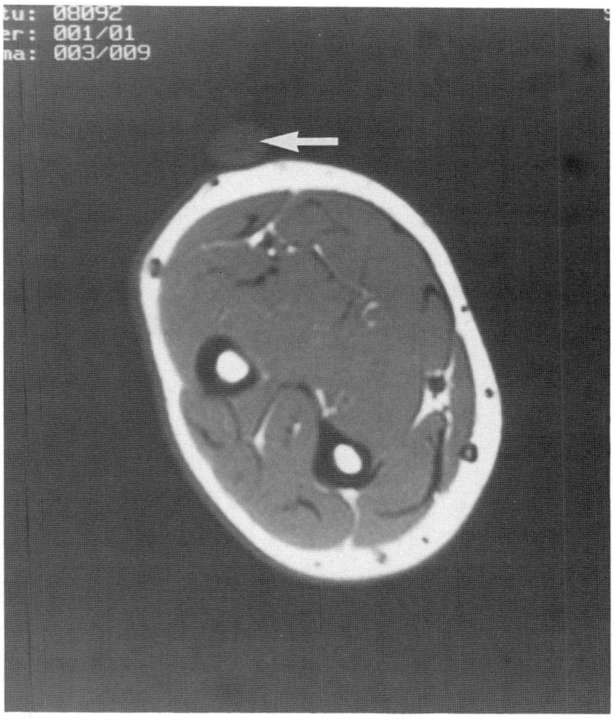

FIGURE 10–12
A palpable subcutaneous mass cannot be identified in this axial image. An oil capsule (arrow) *was used to mark the lesion, clinically identified as a lipoma.*

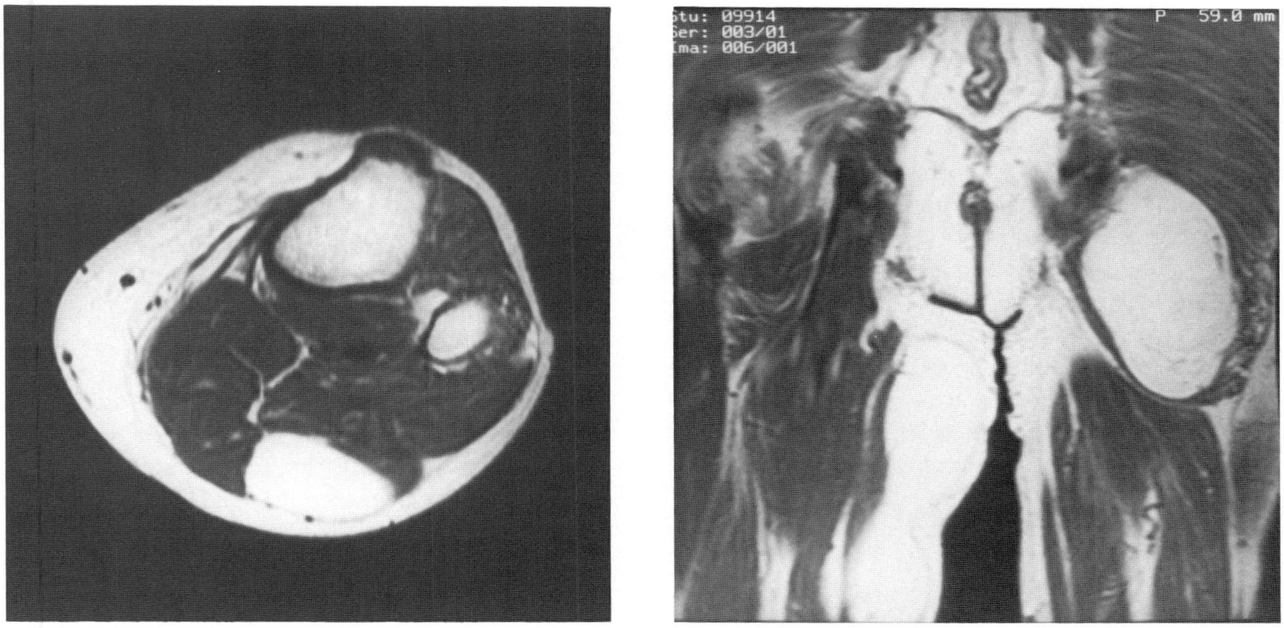

A B

FIGURE 10–13
A. *Lipoma within the gastrocnemius muscle.* B. *Large intramuscular liposarcoma (low grade).*

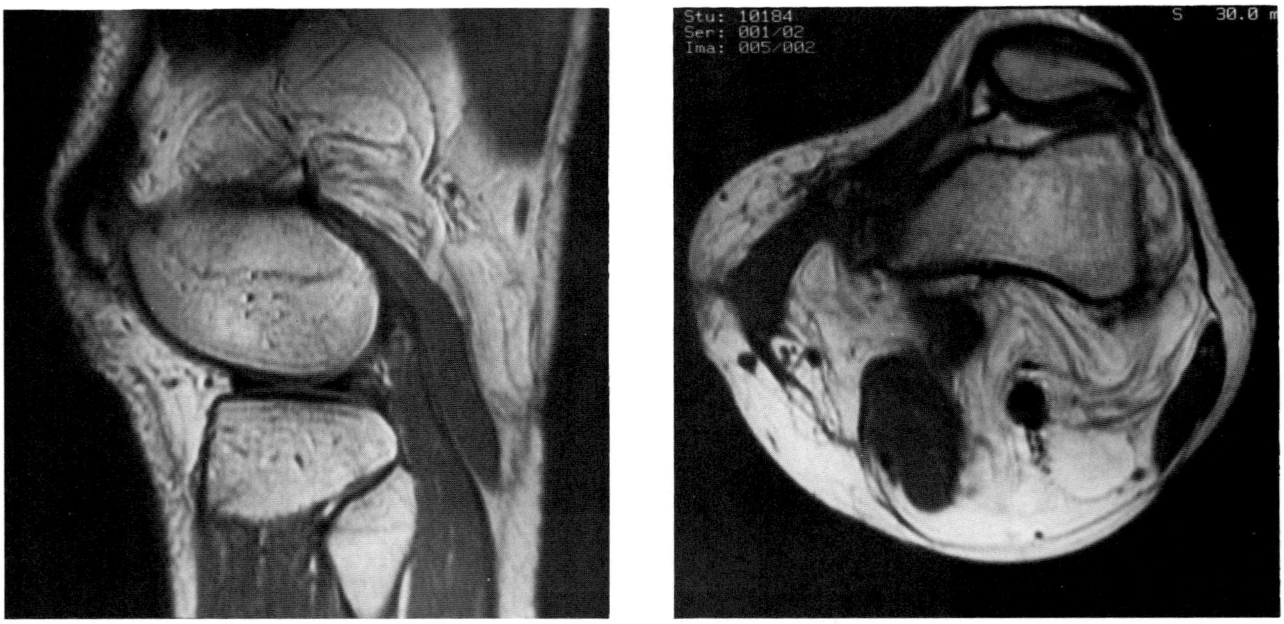

A B

FIGURE 10–14
Large intramuscular liposarcoma with multiple septations, in the popliteal fossa. A. *Sagittal view.* B. *Axial view.*

Another common type of benign palpable lesion is ordinary muscle that has herniated through the normal fascial planes, taking on an irregular, lumpy contour, which can be palpated as a "mass." This is very easily identified by MRI, which demonstrates normal muscle on all echo sequences, irregular contours, and an absence of abnormal tissues (Figure 10–15).

Hemangiomas in muscle and in the subcutaneous tissues can be diagnosed by their distinctive MRI patterns [17,18]. These lesions will be predominantly homogeneous, smoothly contoured, possibly lobulated, and slightly brighter than surrounding muscular tissues on the T1-weighted and proton density sequences [19]. On T2-weighted and gradient-echo images, these lesions become extremely bright. Multiple feeding vessels can often be visualized around the mass (Figures 10–16, 10–17, and 10–18).

Neurofibromas are seen as discrete masses along the course of a nerve, and can vary greatly in size. These appear dark on the T1 images and bright on the T2 images. Cystic masses appear as typical fluid-filled structures, becoming uniformly bright on the T2-weighted images, with the increased signal dependent on the fluid content (Figure 10–19). Ganglion cysts, synovial cysts, popliteal cysts, inflamed bursae, and cystic tumors all present with similar MRI findings (Figures 10–3 and 10–20). Anatomic location and clinical presentation aid in limiting the differential diagnosis.

Malignant tumors do not demonstrate any specific characteristics on MRI images. Calculated relaxation times, once thought the key to MR specificity, have been shown by experience to only infrequently discriminate between types of lesions and are of only very limited utility [20–22]. The masses can be identified and their extent determined; however, there is a great variation in their appearance on the scans [6,22,23]. Liposarcomas, for example, can contain 1–99% adipose tissue histologically. This allows a great variation in the appearance of the scan, as there need not be any signal resembling fat intensity (Figure 10–21). Malignant neoplasms often have

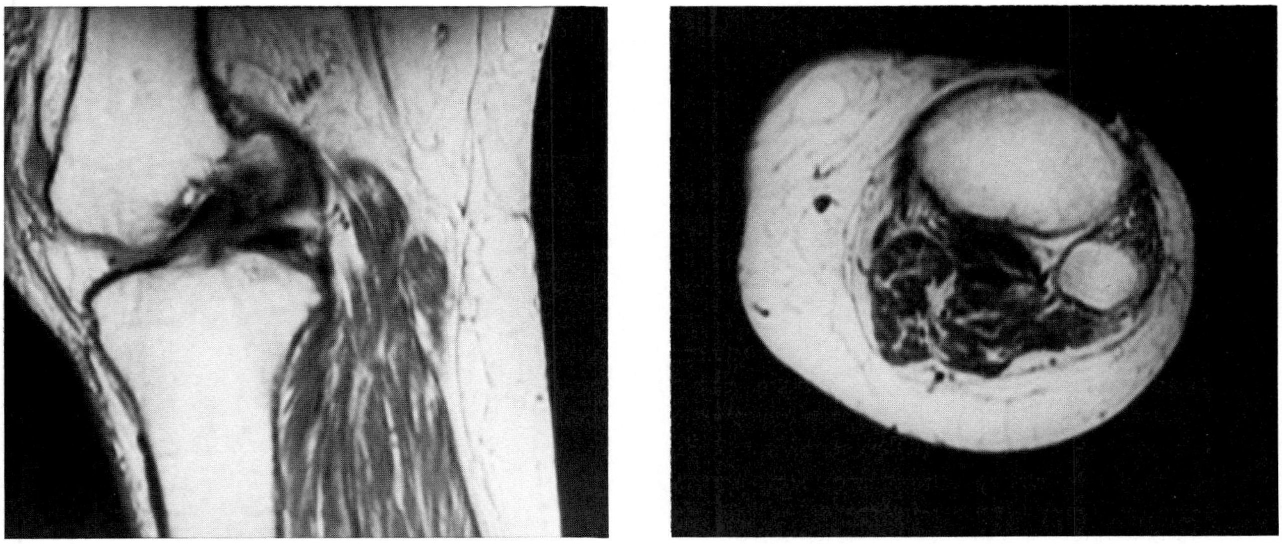

A B

FIGURE 10–15
Palpable, bulging "mass" and atrophic gastrocnemius muscle. A. Sagittal image. B. Axial T1-weighted image.

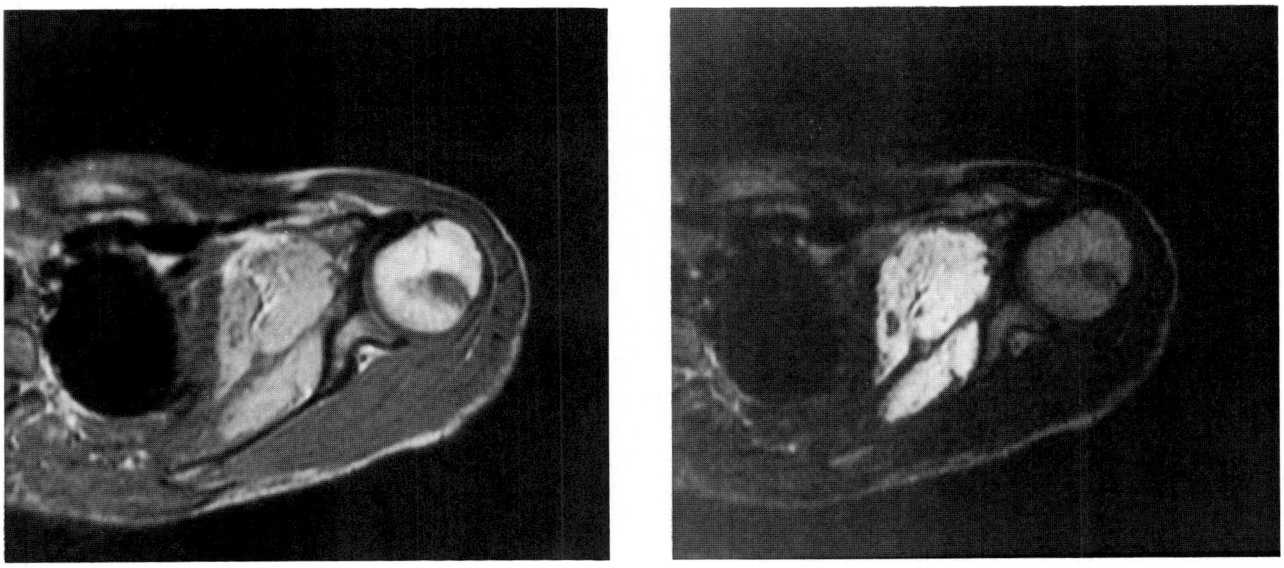

A B

FIGURE 10–16
Large hemangioma of the subscapularis muscle. A. Brighter than muscle on the proton density image. B. Becoming very bright on the T2-weighted image.

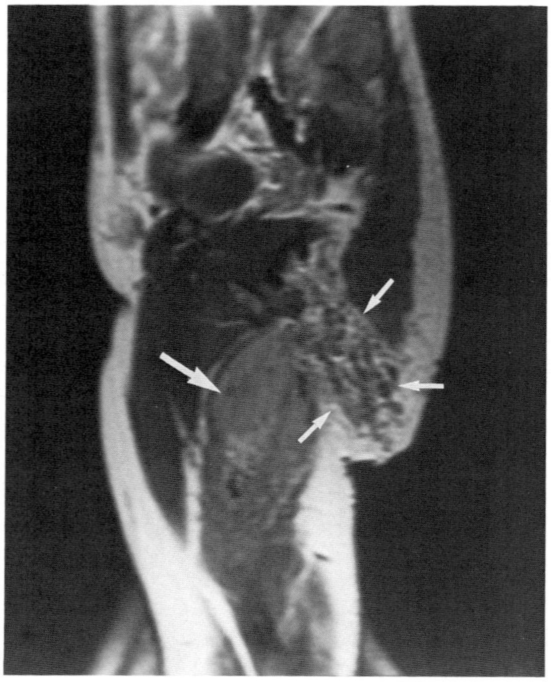

A

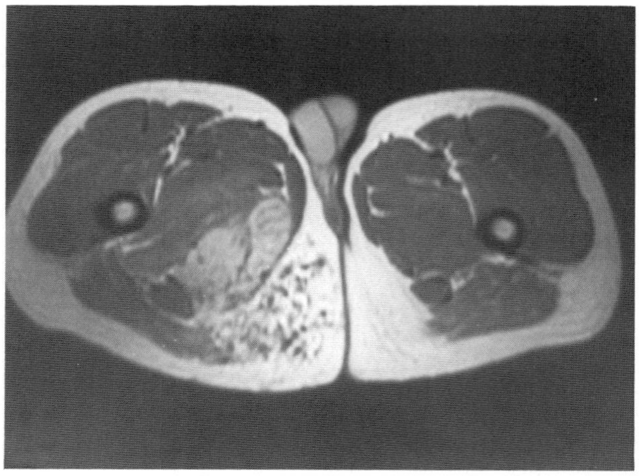

B

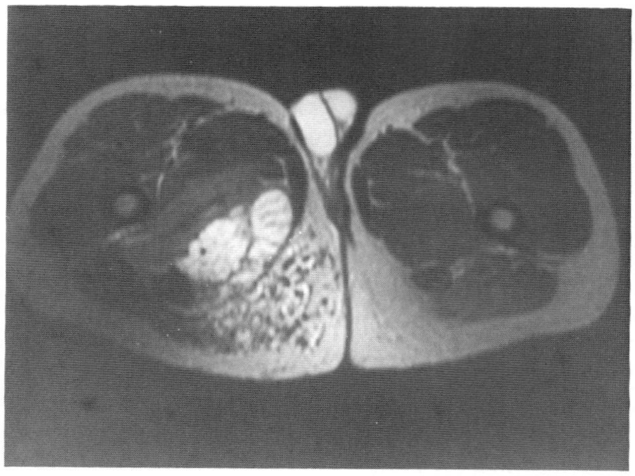

C

FIGURE 10–17

A. *Large hemangioma with subcutaneous* (small arrows) *and intramuscular* (large arrow) *components seen on T1-weighted sagittal image.* B, C. *T1-weighted and T2-weighted images, respectively, demonstrate an increase in intramuscular portion.* D, E. *After embolization therapy, T1-weighted and T2-weighted images, respectively, reveal significant decrease in size and signal intensity of the lesion.*

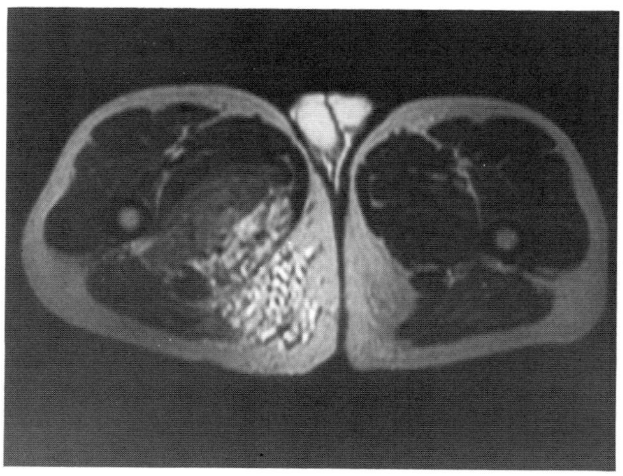

D

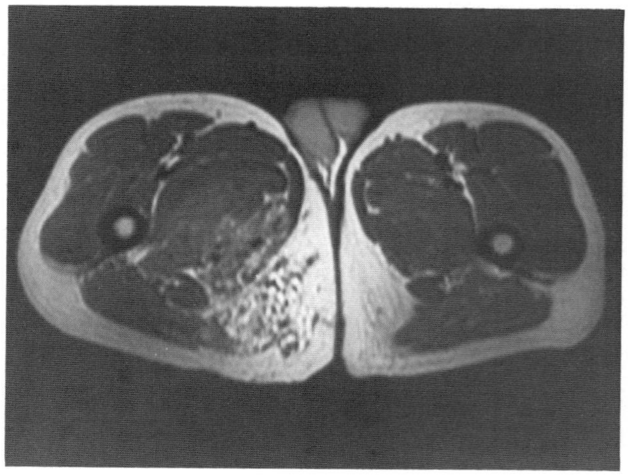

E

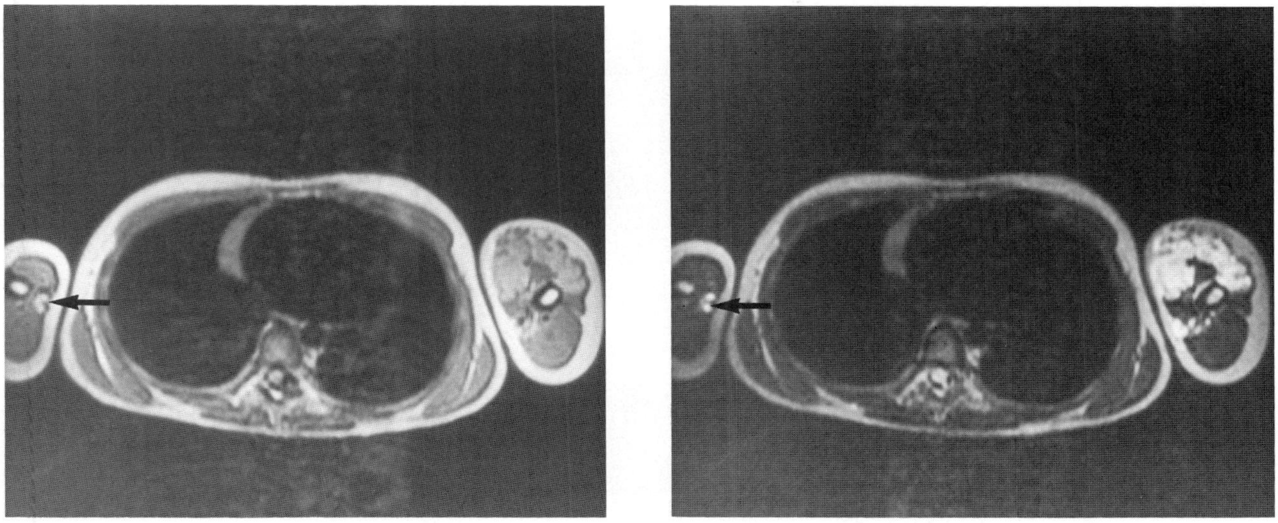

A B

FIGURE 10–18
Large, lobulated hemangioma in the left biceps brachii muscle and unsuspected small hemangioma in the right biceps (arrow). A. *T1-weighted image.* B. *T2-weighted image.*

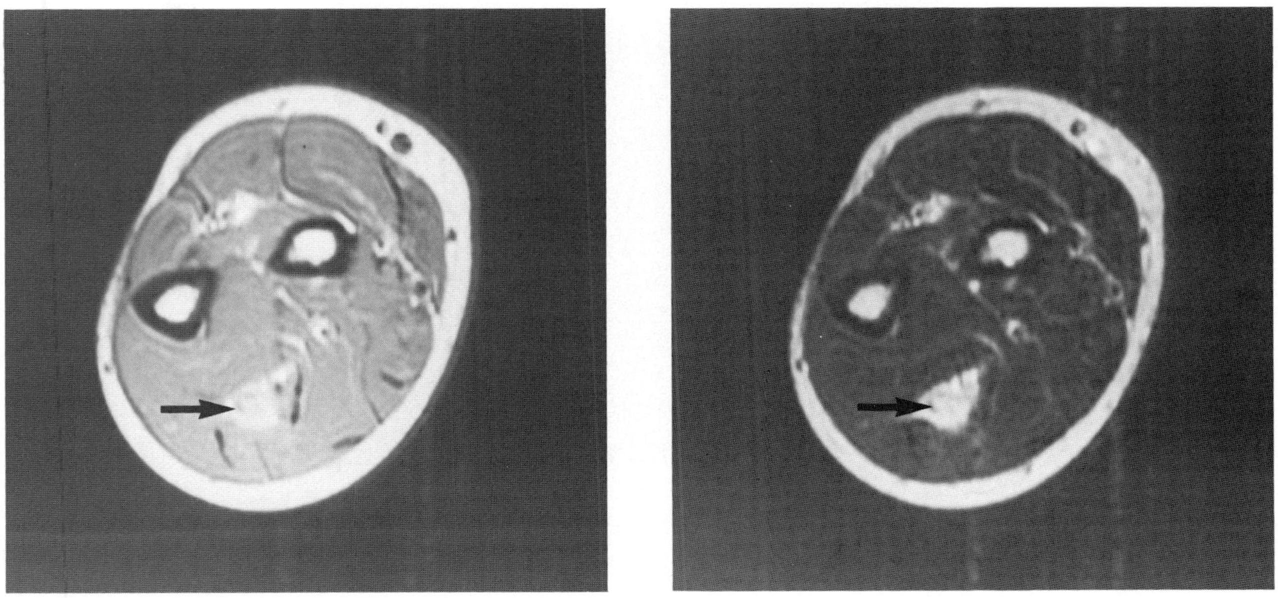

A B

FIGURE 10–19
A mass (arrow) *located along the path of the ulnar nerve, increasing in relative brightness on the T2-weighted view.* A. *T1-weighted image.* B. *T2-weighted image.*

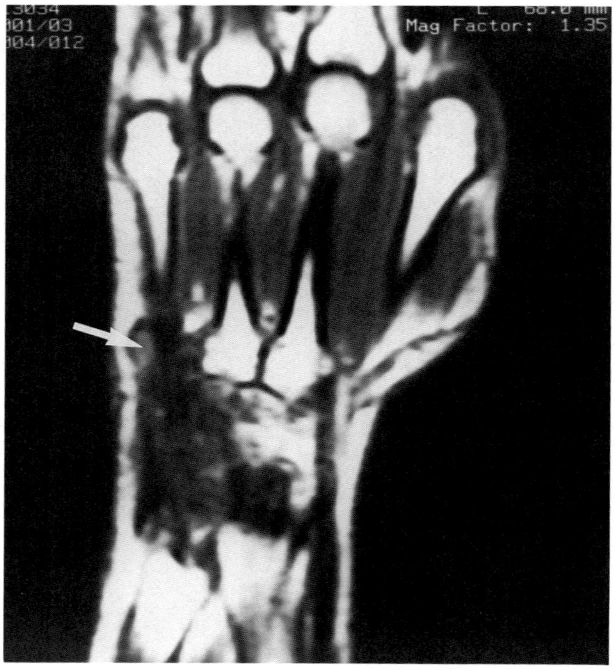

A

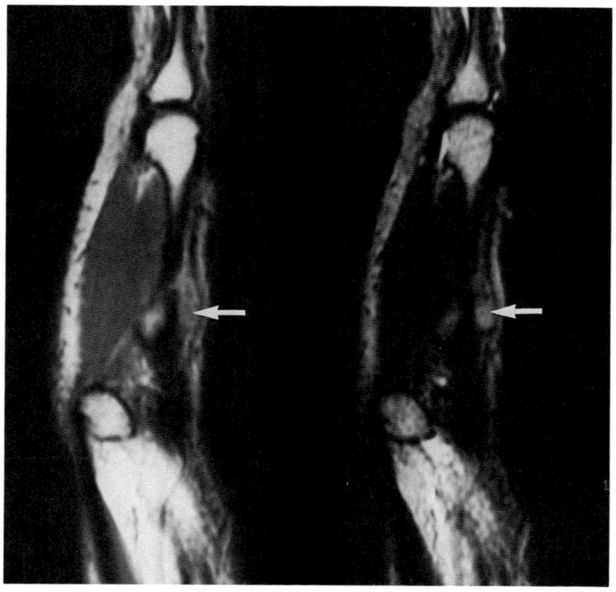

B

FIGURE 10–20
A. *Ganglion cyst off of the extensor tendon of the fifth digit on a coronal T1-weighted image* (arrow).
B. *Proton density* (left) *and T2-weighted* (right) *images in the sagittal plane of the same lesion, increasing in signal intensity.*

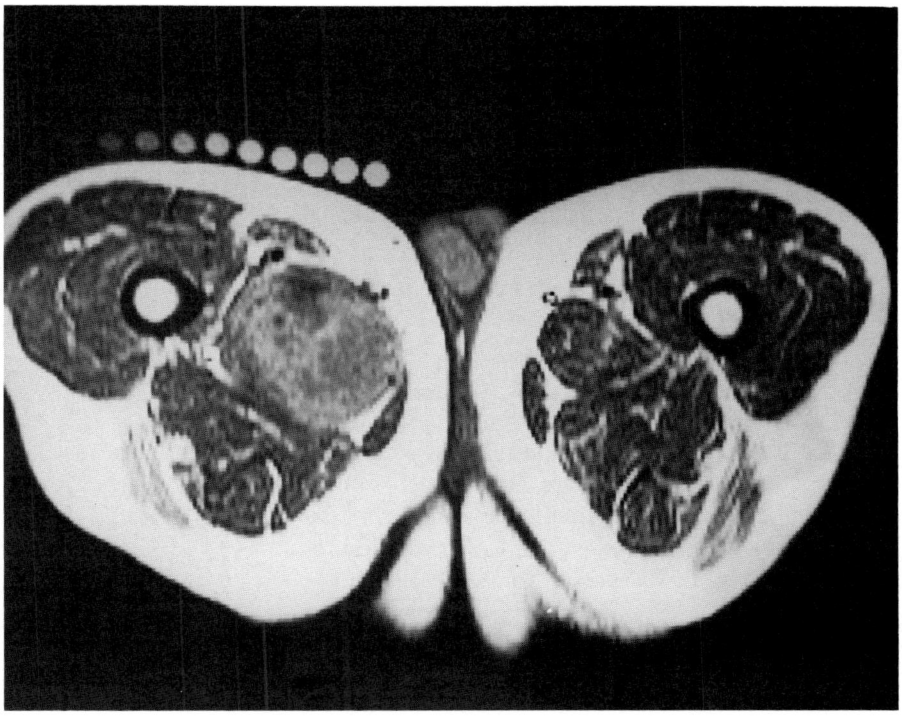

A

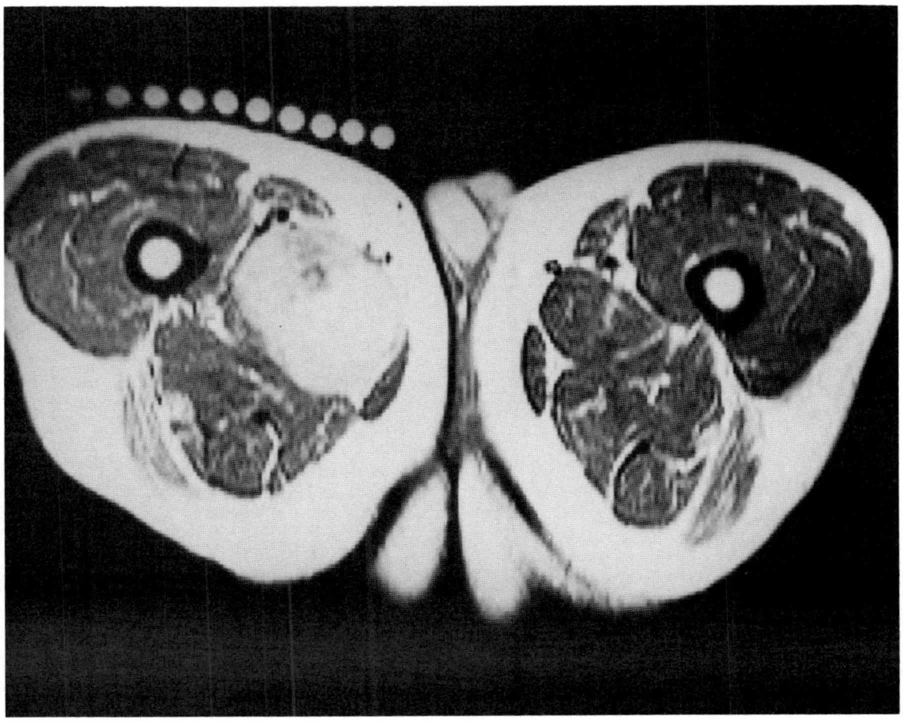

B

FIGURE 10–21
High-grade liposarcoma replacing the vastus medialis muscle, containing very little adipose tissue, is relatively signal-poor on the T1-weighted image, with the signal increasing significantly on the T2-weighted image.
A. *T1-weighted image.*
B. *T2-weighted image.*

FIGURE 10–22
Spindle-cell sarcoma involving adductor longus and magnus, gluteus maximus, biceps femoris, and semitendinosus muscles shows decreased signal intensity on the T1-weighted coronal image, increasing heterogeneously on the T2-weighted axial image. Anterior region of very bright signal represents tumor necrosis (arrow). *A. T1-weighted coronal image. B. T2-weighted axial image.*

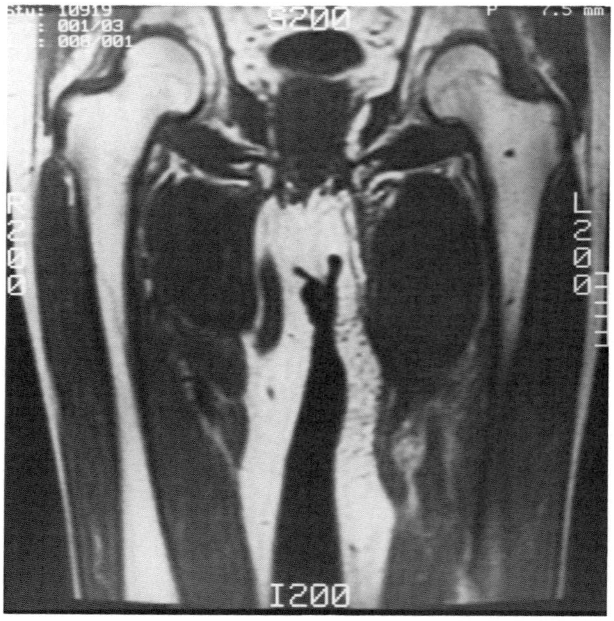

A

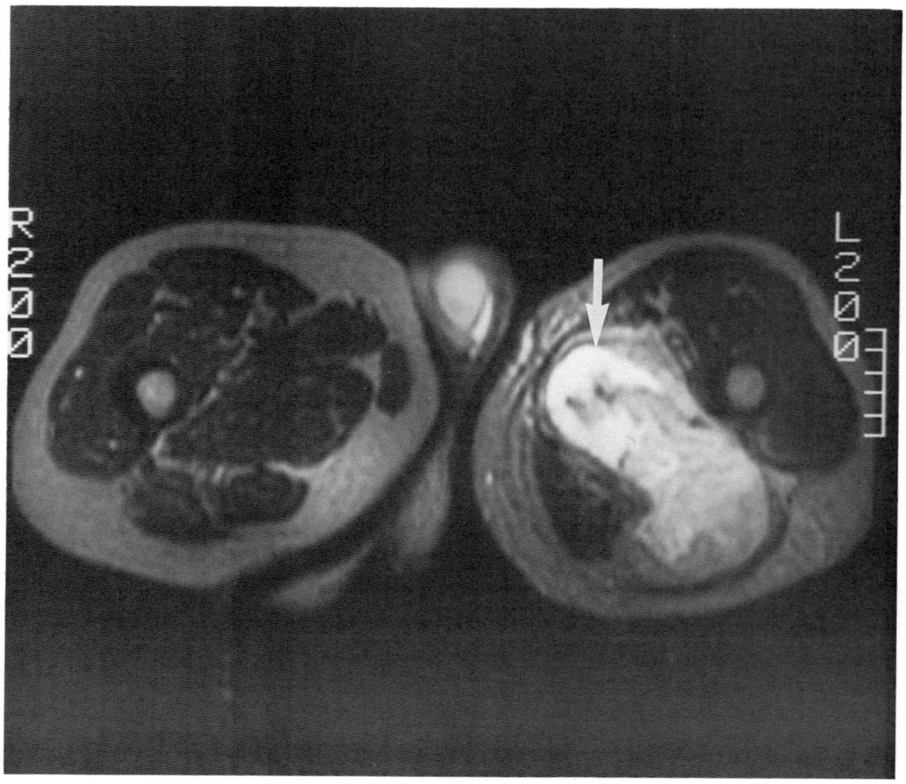

B

irregular borders crossing tissue planes and invading adjacent structures. They are more often heterogeneous on the T2-weighted images, with areas of necrosis becoming extremely bright (Figure 10–22). On the T2-weighted images, tumor is usually separable from surrounding edema, and these images are most useful to direct biopsies.

In conclusion, although MRI is a relatively new and still emerging technology, it has already significantly affected the clinical workup of soft tissue disorders of the musculoskeletal system. Additional experience and continued technologic improvements promise to increase the utility of this method, expanding applications and increasing the specificity of diagnoses.

REFERENCES

1. Haggar AM, Froelich JW. MR imaging strategies in primary and metastatic malignancy. Radiol Clin North Am 1988; 26:689–96.
2. Beltran J, Noto AM, McGhee RB, Freedy RM, McCalla MS. Infections of the musculoskeletal system: high-field-strength MR imaging. Radiology 1987; 164:449–54.
3. Unger EC, Cohen MS, Brown TR. Gradient-echo imaging of hemorrhage at 1.5 Tesla. Magn Reson Imaging 1989; 7:163–72.
4. Sartoris DJ, Resnick D. MR imaging of the musculoskeletal system: current and future status. AJR 1987; 149:457–67.
5. Vanel D, Lacombe M-J, Couanet D, Kalifa C, Spielmann M, Genin J. Musculoskeletal tumors: follow-up with MR imaging after treatment with surgery and radiation therapy. Radiology 1987; 164:243–5.
6. Pettersson H, Gillespy T, Hamlin DJ, et al. Primary musculoskeletal tumors: examination with MR imaging compared with conventional modalities. Radiology 1987; 164:237–41.
7. Scott JA, Palmer EL. Musculoskeletal infection of the extremities: evaluation with MR imaging. Radiology 1988; 168:284–5.
8. Wall SD, Fisher MR, Amparo EG, Hricak H, Higgins CB. Magnetic resonance in the evaluation of abscesses. AJR 1985; 144:1217–21.
9. Stern JJ, Stoopack PM. Magnetic resonance imaging in the diagnosis of a psoas abscess. J Comput Tomogr 1988; 12:79–80.
10. Gomon JM, Grossman RL, Ip-Yu C, Asakura T. NMR relaxation times of blood dependence on field strength, oxidation state and cell integrity. JCAT 1987; 11:685–90.
11. Rubin JI, Gomori JM, Grossman RI, Gefter WB, Kressel HY. High-field MR imaging of extracranial hemorrhage. AJR 1987; 148:813–7.
12. Gomori JM, Grossman RI, Goldberg HI, Zimmerman RA, Bilaniuk IT. Intracranial hematomas: imaging by high-field MR. Radiology 1985; 157:87–93.
13. Bradley WG, Schmidt PG. Effect of methemoglobin formation on the MR appearance of subarachnoid hemorrhage. Radiology 1985; 156:99–103.
14. Beltran J, Simon DC, Katz W, Weis LD. Increased MR signal intensity in skeletal muscle adjacent to malignant tumors: pathologic correlation and clinical relevance. Radiology 1987; 162:251–5.
15. Robbins SL, Angell M. Basic pathology. 2nd ed. Philadelphia: WB Saunders, 1976:115–6.
16. Dooms GC, Hricak H, Sollitto RA, Higgins CB. Lipomatous tumors and tumors with fatty component: MR imaging potential and comparison of MR and CT results. Radiology 1985; 157:479–83.
17. Yuh WTC, Kathol MH, Sein MA, Ehara S, Chiu L. Hemangiomas of skeletal muscle: MR findings in five patients. AJR 1987; 149:765–8.
18. Cohen EK, Kressel HY, Perosio T. MR imaging of soft-tissue hemangiomas: correlation with pathologic findings. AJR 1988; 150:1079–81.
19. Levine E, Wetzel LH, Neff JR. MR imaging and CT of extrahepatic cavernous hemangiomas. AJR 1986; 147:1299–1304.
20. Pettersson H, Slone RM, Spanier S, Gillespy J III, Fitzsimmons JR, Scott KN. Musculoskeletal tumors: T1 and T2 relaxation times. Radiology 1988; 167:783–5.
21. Sundaram M, McGuire MH, Schajowicz F. Soft-tissue masses: histologic basis for decreased signal (short T2) on T2-weighted MR images. AJR 1987; 148:1247–50.
22. Kransdorf MJ, Jelinek JS, Moser RP Jr, et al. Soft-tissue masses: diagnosis using MR imaging. AJR 1989; 153:541–7.
23. Wetzel LH, Levine E, Murphy MD. Comparison of MR imaging and CT in the evaluation of musculoskeletal masses. Radiographics 1987; 7:851–74.

11

MRI of the Spine

PETER A. KOSOVSKY

Imaging of the spinal column is difficult due to complex anatomy that requires different orientations for optimal visualization. In the past, evaluation of the spine was limited to plain films, tomography, and invasive procedures such as myelography and discography. Since the advent of CT, axial tomographic images and reformatted sagittal images now provide excellent delineation of the bony and soft tissue elements of the spine. However, for optimal visualization of the extent of epidural and intradural processes, CT examinations should be performed after the administration of intrathecal contrast [1]. MRI, unlike CT, has much greater inherent contrast resolution and utilizes direct multiplanar imaging without ionizing radiation or the need for intrathecal agents. While more experience and prospective studies are needed to define its overall role in spinal imaging, MRI, because of its noninvasive nature and lack of known side effects, is fast becoming the modality of choice for evaluating the spine [2]. The objective of this chapter is to give an *overview* of the uses of MRI of the spinal column, including its limitations and possible future applications with respect to marrow changes, disc disease, and epidural processes.

Tailoring MRI

To obtain the most information from each MRI study, multiple parameters must be appropriately tailored with consideration for optimizing each individual examination. Although different scanners of various field strengths are now available, user-selected parameters such as pulse sequences, imaging planes, surface coils, slice thickness, field of view, and number of signal averages must all be balanced to ensure an appropriate examination, patient compliance, and adequate throughput [3].

Both T1- and T2-weighted spin-echo sequences are employed for most studies. T1 spin-echo sequences have relatively short acquisition times and provide the best anatomic detail, especially in lower magnetic fields. T2 spin-echo sequences have longer acquisition times but provide myelographic effects due to increased CSF intensity.

The following techniques represent the author's preference for tailoring MRI examinations for different clinical situations, utilizing a 1.5-T Signal system (General Electric, Milwaukee, WI) and a 0.6-T Teslacon (Technicare, Solon, OH). They are given to provide a general guide, with the knowledge that parameters will vary depending on the specifics of individual MRI units and examiner preferences.

T1-weighted sagittal and coronal images at a TR of 500–750 milliseconds and a TE of 20–35 milliseconds using the standard body coil, 0.5–0.75-cm slice thickness, 2–4 signal averages, and a 256 × 128 matrix are routinely utilized. T1 weighting yields the best contrast between normal adult fatty marrow and diffuse or localized marrow abnormalities. T2-weighted images are less useful since most pathologic tissues display increased signal on these images and can become isointense with normal marrow. T2 images, however, can be useful in visualizing areas of acute bone marrow infarction and edema. New studies using short TR inversion recovery (STIR) images also show promise in imaging marrow abnormalities [4]. Although these images display increased contrast between normal and abnormal marrow, they have poorer signal-to-noise ratios than T1-weighted spin-echo sequences, resulting in limited anatomic detail.

Images utilizing the body coil, offering the largest field of view, are best for screening the vertebral column for both diffuse and focal pathologic conditions. If, however, a focal lesion is identified, a surface coil should be used for further evaluation. Although a surface coil has a limited field of view, it improves spatial resolution by increasing signal-to-noise ratios, thus allowing thinner sections and smaller voxel elements [3,5]. Both sagittal and axial surface coil images are necessary to define the epidural extent of most focal vertebral body lesions. Surface coils are utilized for all evaluations of radiculopathy. In addition, cardiac gating or refocusing pulses are used to minimize CSF pulsation artifacts.

CERVICAL SPINE

Sagittal T1-weighted images are obtained with a short TR (500–600 milliseconds) and TE (20–30 milliseconds), with a 3–5-mm slice thickness, 2–4 signal averages, a 256 × 256 matrix, and 24-cm field of view. Gradient-echo sagittal and axial images, utilizing a 10-degree flip angle, TR of 500 milliseconds, TE of 15 milliseconds, 4 signal averages, 5-mm slice thickness, and 256 × 192 matrix are also obtained. The gradient-echo images offer a myelographic effect similar to that obtained by T2-weighted images and have significantly faster acquisition times while still allowing the demonstration of disc herniation, spinal stenosis, and cord compression. It does, however, represent a trade-off of faster imaging times and better patient compliance with an increased potential for motion and magnetic susceptibility artifacts and an overall decrease in signal-to-noise ratio when compared with T2-weighted images [6–8]. T2-weighted spin-echo sequences are therefore preferred when examining the postoperative spine when metallic clips may be present or whenever cord abnormalities are suspected.

THORACIC SPINE

A T1 sagittal scout image is obtained using the body coil in order to localize the surface coil placement. A TR of 250–400 milliseconds and TE of 20–30 milliseconds, slice thickness of 0.5–1.0 cm, 2 signal averages, 256 × 128 matrix, and

48-cm field of view are routinely used. A surface coil is employed to acquire T1-weighted sagittal and axial images, and sagittal proton density and T2-weighted images. The short TR axial and sagittal images are obtained using parameters similar to those described for the cervical levels of the spine. The multislice, multiecho, T2-weighted sequences use a TR of 2000 milliseconds, TE of 40 and 80 milliseconds, 3–5-mm slice thickness, 256 × 256 matrix, 2 signal averages, and a 24-cm field of view.

LUMBAR AND SACRAL SPINE
Surface coil sequences for the lumbosacral region of the spine are analogous to those used at the thoracic level. The axial short TR images are angled through the disc spaces.

Normal Anatomy

T1-weighted sagittal and axial images provide the best delineation of spinal anatomy. The vertebral bodies, discs, ligamentous structures, apophyseal joints, neural foramina, and bony cortical margins, as well as epidural fat and nerve roots, are easily demonstrated [9] (Figures 11–1, 11–2, and 11–3). Gradient-echo and T2-weighted sagittal images provide myelographic effects that allow evaluation of spinal canal dimensions and epidural processes that result in compression of the thecal sac (Figure 11–4). T2-weighted images also provide information concerning relative disc hydration, with desiccated discs becoming hypointense to normally hydrated discs (Figure 11–5).

Unlike the lumbar foramina, the cervical neural foramina are best visualized in oblique or axial planes. This is a result of the cervical foramina being oriented 45 degrees away from the straight sagittal planes routinely utilized for visualizing the lumbar neural foramina [10]. High-resolution surface coil studies have shown that the anterior epidural venous plexus appears as a

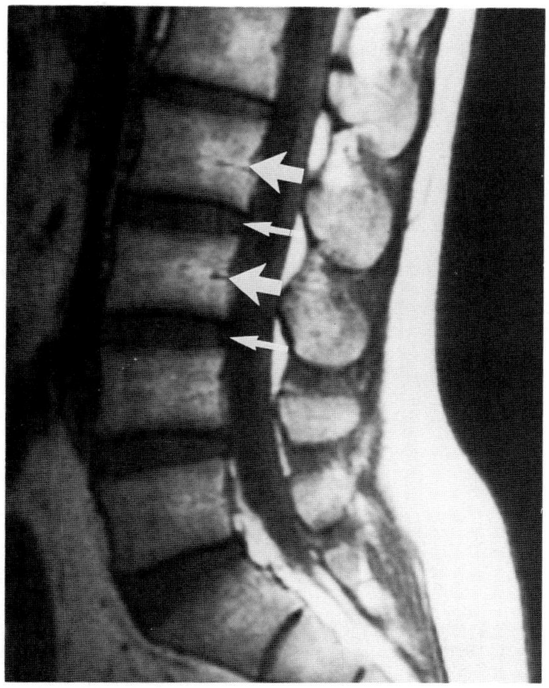

FIGURE 11–1
T1-weighted midline sagittal image of the lumbosacral section of a normal adult spine. The vertebral bodies (large arrows) *are slightly higher in signal than the intervertebral discs* (small arrows).

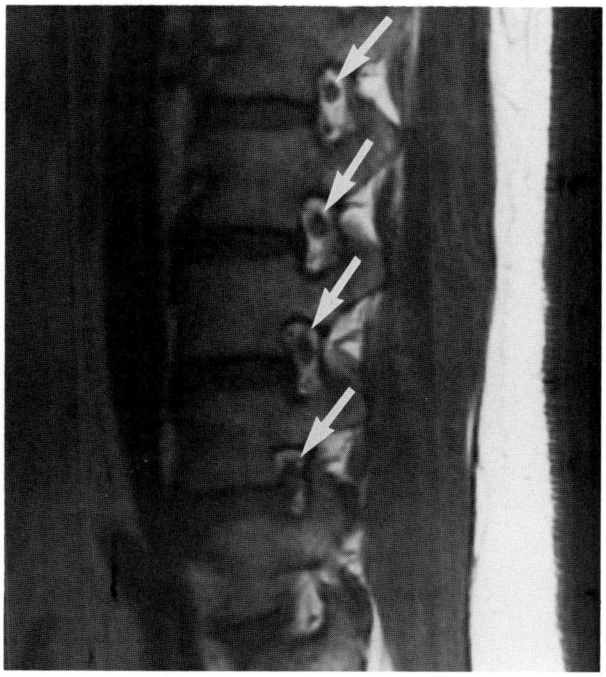

FIGURE 11–2 ▶
T1-weighted parasagittal image of the lumbosacral section of a normal spine. The nerve roots (arrows) *are shown as they traverse the neural foramina surrounded by normal bright intraforaminal fat.*

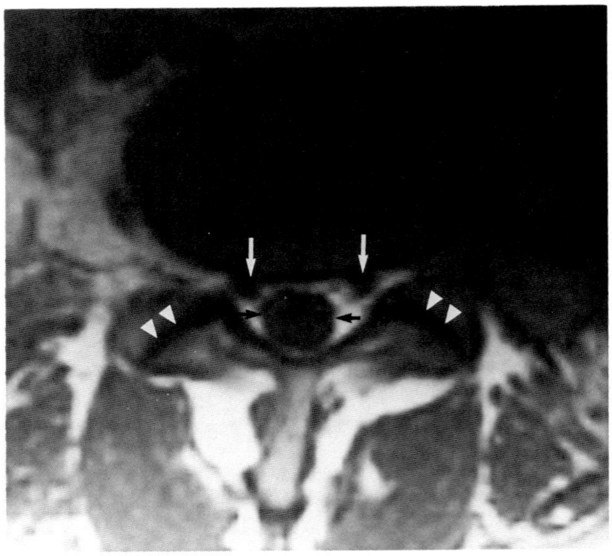

FIGURE 11–3
T1-weighted axial image of the lumbosacral section of a normal spine. The nerve roots (white arrows) *are surrounded by bright epidural fat. The thecal sac* (black arrows) *and apophyseal joints* (white arrowheads) *are well visualized.*

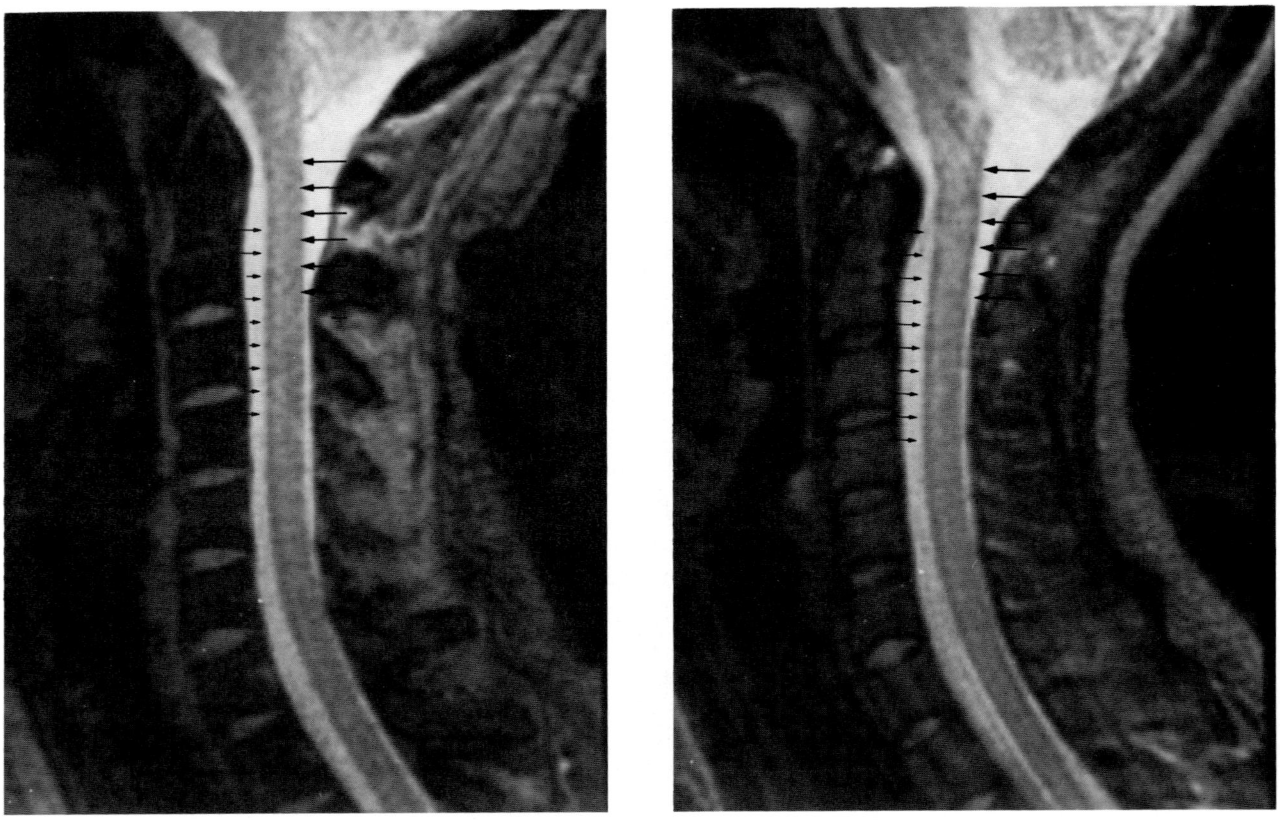

A B

FIGURE 11–4
A. *Gradient-echo midline sagittal image at the cervical level of a normal spine. The cervical cord* (small black arrows) *is surrounded by bright CSF resulting in a myelographic effect.* B. *T2-weighted midline sagittal image at the cervical level of a normal spine, producing similar myelographic effect.*

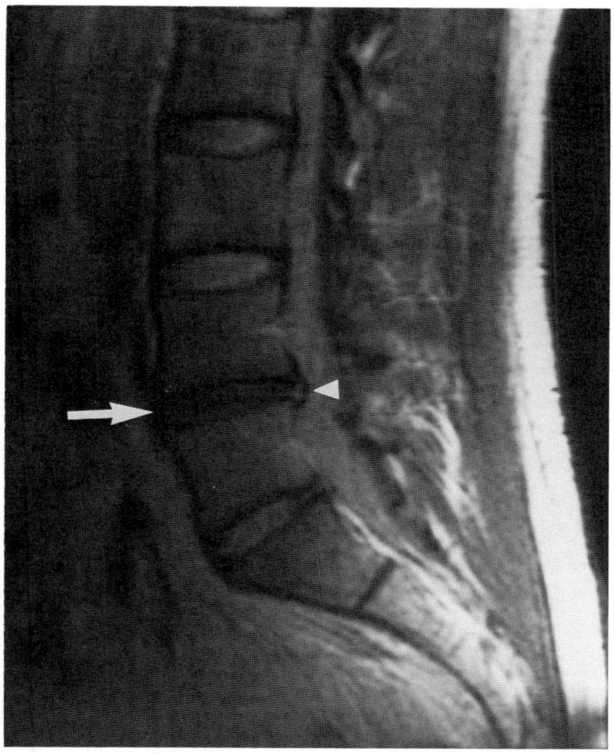

◀ *FIGURE 11–5*
T2-weighted midline sagittal image at the lumbosacral level of the spine. The relatively desiccated L4-5 disc (large arrow) *is decreased in signal with a small posterior herniation* (arrowhead).

bright signal band on sagittal and axial T1-weighted images. This high signal is most likely secondary to slow flow within the veins. Axial images can best demonstrate the dorsal and ventral nerve roots, whereas oblique 45-degree images best demonstrate the cervical nerve roots as they course through their respective foramina (Figure 11–6) [11]. Oblique images are, however, rarely utilized in routine clinical imaging.

Anatomic studies on cadaver specimens have been performed to correlate the MRI signal intensities with the various components of the intervertebral disc. Normal discs have a high-signal central region, with a peripheral area of decreased signal on T2-weighted images. The cen-

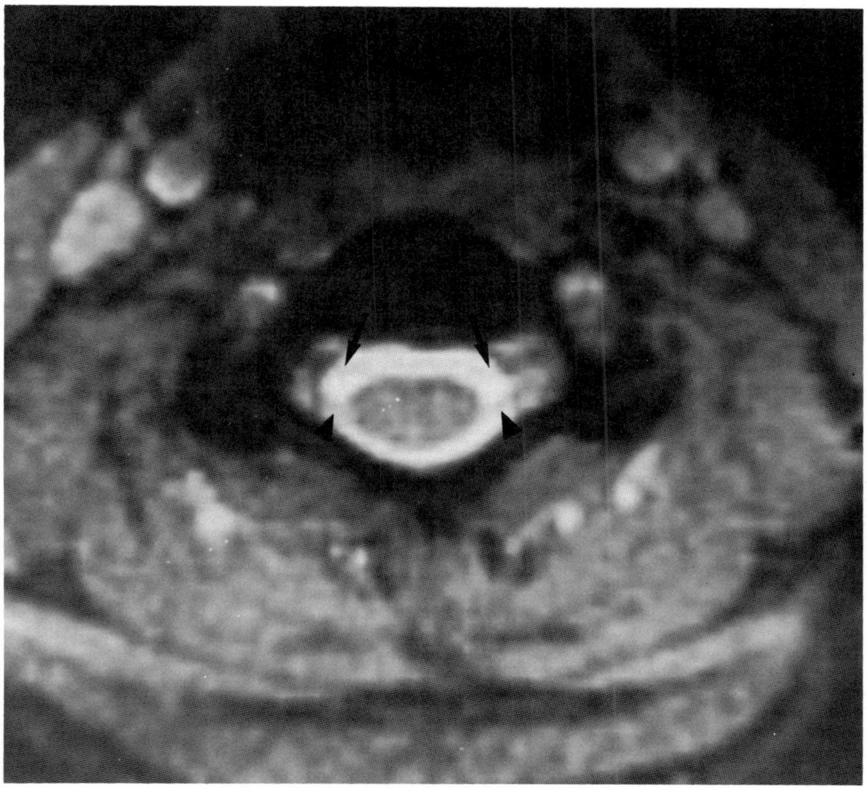

FIGURE 11–6
Gradient-echo axial image at the cervical level. The dorsal nerve roots (arrowheads) *and ventral nerve roots* (arrows) *emanate from the cervical cord surrounded by bright CSF.*

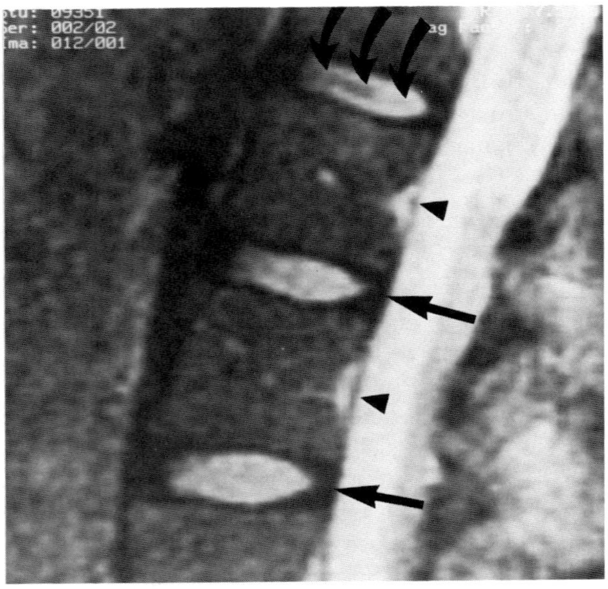

FIGURE 11–7
Magnified T2-weighted view of a disc. The high-signal central portion is transversed by the lower-signal linear fibrous band (curved arrows). *The low-signal peripheral portion* (straight arrows) *is bordered by the low-signal posterior longitudinal ligament* (arrowheads).

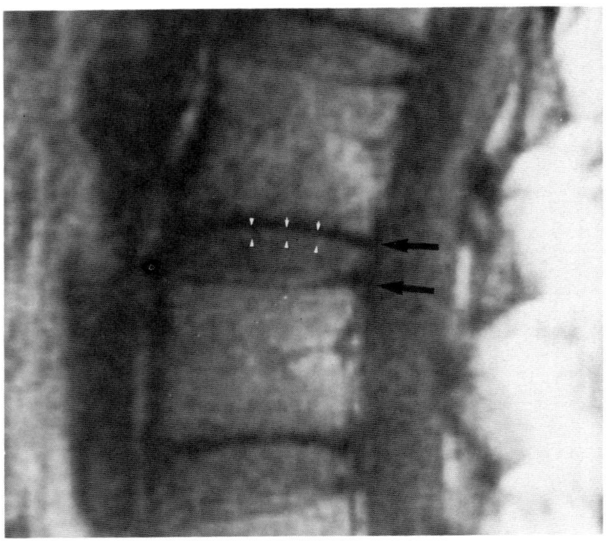

FIGURE 11–8
Magnified T1-weighted view of a disc. The low-signal central portion is surrounded by a thin band of slightly higher signal (white arrowheads) *and is bordered by the low-signal posterior longitudinal ligament* (black arrows).

tral high-signal nucleus pulposus and inner portions of the annulus fibrosis are indistinguishable from one another as are the lower-signal peripheral complex of the outer annulus and longitudinal ligaments [12]. In adults there is often a low-signal band seen on T2-weighted images traversing the central portion of lumbar discs, which give them a notched or biconvex appearance (Figure 11–7). This is thought to be due to a normal area of increased fibrous content and should not be mistaken for the truncation artifacts, commonly seen in thoracic discs, which can be eliminated by changing the phase encoding gradients [13]. The signal intensities are somewhat inverted on T1-weighted images, where the central disc is lower in signal than the slightly brighter peripheral portion, which is bounded by the low-signal band of the longitudinal ligaments (Figure 11–8).

The regions of the vertebral end-plates are bordered by linear areas of low signal on T1-weighted images. The thickness of these low-signal lines can change with different frequency encoding gradients; therefore, these low-signal lines do not always represent true delineation of the cortical bone of vertebral end-plates [12].

Normal Adult Bone Marrow

Bone marrow is composed of hematopoietically active "red marrow" and hematopoietically inactive "yellow marrow." Red marrow contains approximately 50% fat, while about 80% of yellow marrow is fat. At birth the entire marrow cavity is hematopoietically active red marrow. During the first three decades of life there is a continuous conversion of red to yellow marrow that progresses from the appendicular to the axial skeleton. The adult pattern of marrow distribution is usually achieved by age 25. Red marrow remains present primarily in the skull, vertebrae, sternum, pelvic bones, and proximal ends of the femoral shafts. With advancing age there is an increase in the fat content within the residual red marrow throughout the vertebrae, which may be secondary to osteoporosis and may lead to a heterogeneous appearance [14]. The large component of fat (with a short T1 and long T2 relaxation time) in adult marrow forms a relatively bright background for most pathologic processes

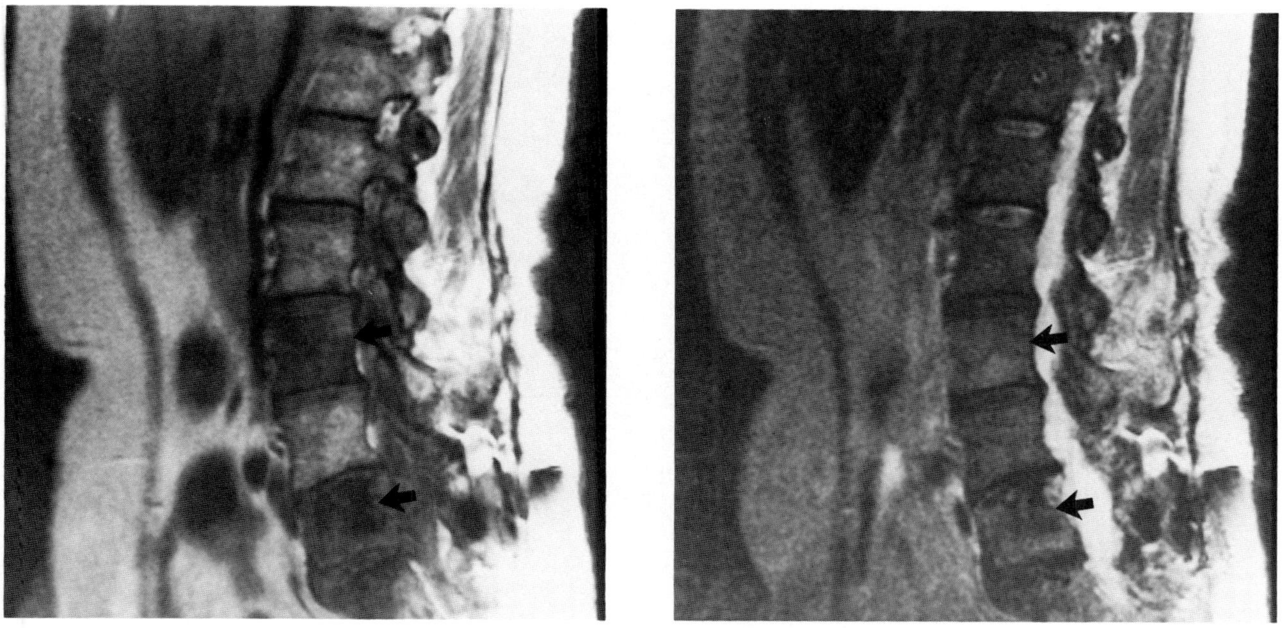

A B

FIGURE 11–9
A. *T1-weighted image. Arrows indicate low-signal areas in L-3 and L-5 vertebral bodies, secondary to infiltration by metastatic carcinoma of the breast.* B. *T2-weighted image. Both the L-3 and L-5 vertebral bodies* (arrows) *are virtually isointense with the uninvolved vertebrae.*

(which have long T1 and long T2 relaxation times). Therefore, the differences in T1 relaxation times offer the greatest contrast between normal and abnormal adult marrow. Since T2 values of normal marrow and pathologic tissue can overlap, the appearance of pathologic areas can become indistinguishable from normal marrow on T2-weighted images (Figure 11–9).

The cellular red marrow content in children makes it difficult to distinguish focal or infiltrating processes from an already low-signal background. Recent early reports of STIR imaging show promise in delineating abnormal marrow changes in children after marrow transplantation [4]. Focal or infiltrating processes in adults can also be masked by hyperplastic marrow from a concurrent anemia.

Abnormal Bone Marrow Patterns

In adults the interpretation of low-signal areas in fatty vertebral marrow must be done with caution. Areas of residual cellular red marrow, primary and secondary neoplasms, old infarction, and regions of fibrosis can yield a similar appearance. These abnormal signal areas seldom represent a problem if there is correlation with the patient's clinical history and other diagnostic studies. If these studies are not always specific, MRI is still of considerable value in determining the most appropriate regions for diagnostic biopsies and as a baseline study to be compared with future examinations. Although there are various dynamic pathophysiologic processes responsible for individual diseases, most pathologic entities present in one of the following four patterns on T1-weighted images.

DIFFUSE DECREASED SIGNAL

Hypercellular marrow processes result in diffuse decreased signal on T1-weighted images. This results from a prolongation of T1 relaxation times secondary to various parameters, such as increased edema and cellularity. Generally, in adults, abnormally low-signal marrow can be rec-

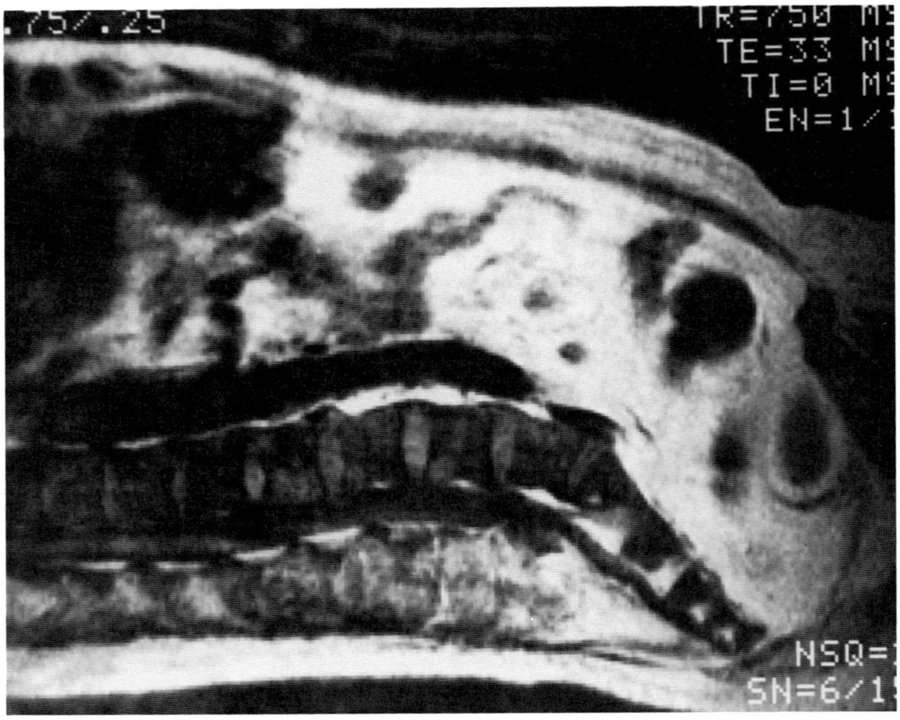

FIGURE 11–10
T1-weighted sagittal image. Diffuse low signal is seen throughout the vertebral marrow in a patient with multiple myeloma.

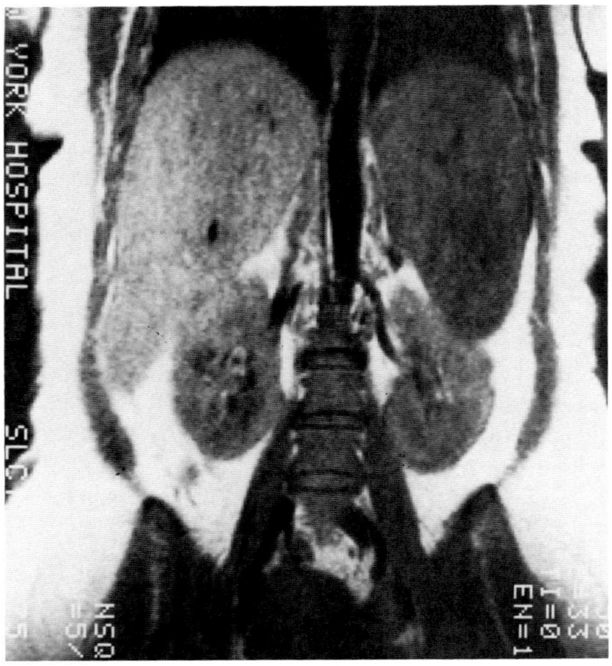

FIGURE 11–11
T1-weighted coronal image. Diffuse low signal is seen throughout the vertebral marrow cavity in a patient with primary amyloidosis.

ognized if it is of lower signal intensity than normal disc material on T1-weighted sagittal or coronal images. Since replacement of normal marrow in leukemia, lymphoma, and multiple myeloma occurs in regions of red marrow prevalence, imaging of the spinal column, sacrum, and pelvic bones is sensitive, though not specific, for these disorders (Figure 11–10). Another primary marrow disorder, myelofibrosis, can also result in a decreased signal from within the marrow cavity due to the replacement of normal marrow by fibrotic material. Other systemic disorders such as amyloidosis, Gaucher's disease, and hemochromatosis also present with low-signal-intensity marrow. Normal fatty marrow is infiltrated by plasma cells producing abnormal amyloid proteins in amyloidosis, by Gaucher's cells in Gaucher's disease, and by excess iron deposition in hemochromatosis (Figures 11–11 and 11–12) [14,15]. Occasionally, diffuse metastatic involvement of the vertebral column and even normal hyperplastic marrow can mimic these infiltrative patterns (Figure 11–13). Although many of these disorders may be clinically evident, MRI can be useful for demonstrating the

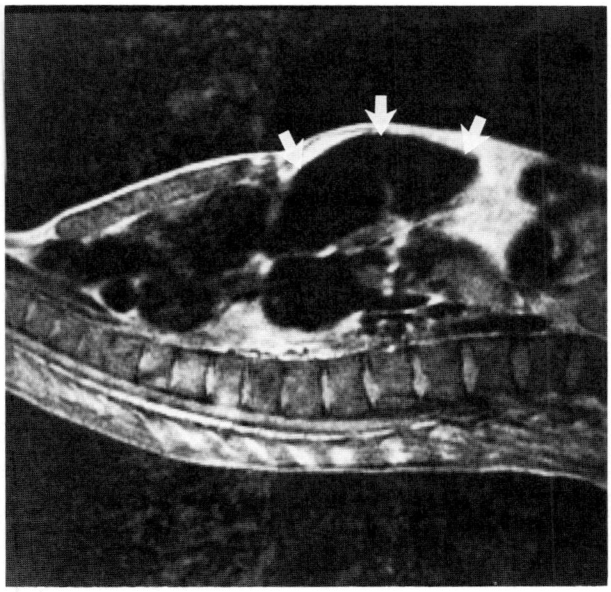

FIGURE 11–12
T1-weighted sagittal image. Diffuse low signal is seen throughout the vertebral marrow in a patient with thalassemia major and secondary hemochromatosis. The low-signal liver (white arrows) *is due to iron deposition.*

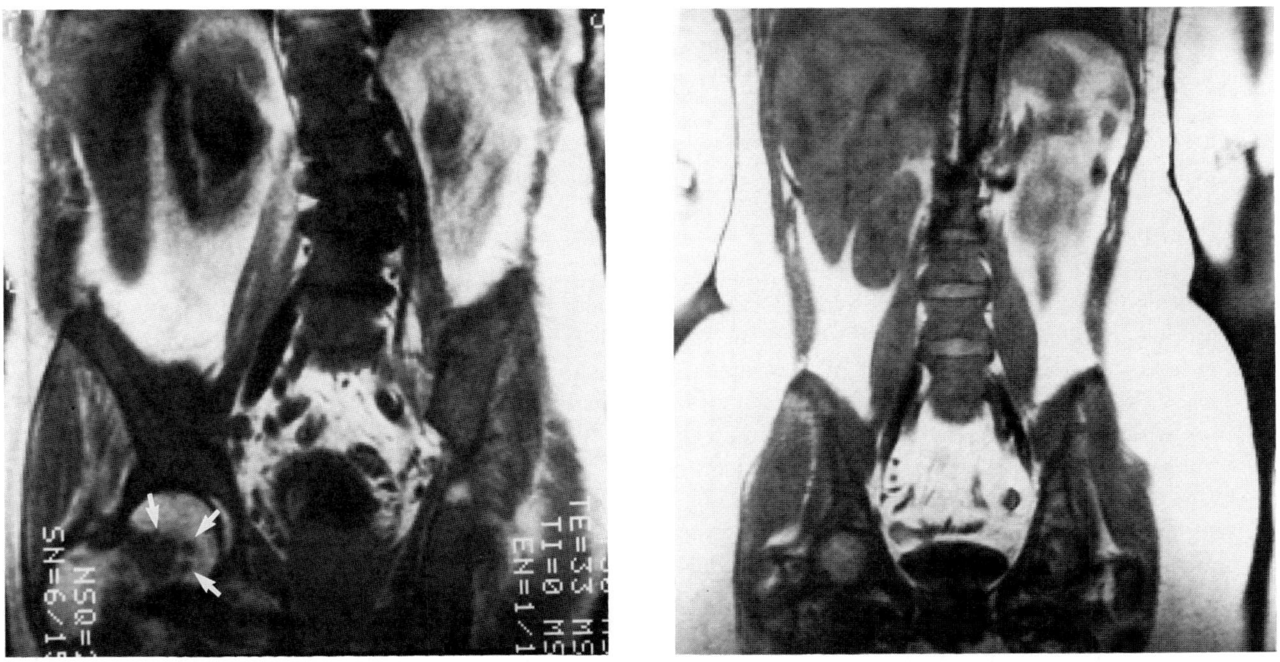

A B

FIGURE 11–13
A. *T1-weighted coronal image. Low-signal vertebral marrow is secondary to diffuse infiltration by metastatic prostate carcinoma. Focal low-signal metastases are also noted in right femoral neck* (arrows). B. *T1-weighted coronal image. Diffuse infiltration of the marrow space of the vertebrae by metastatic breast carcinoma is demonstrated.*

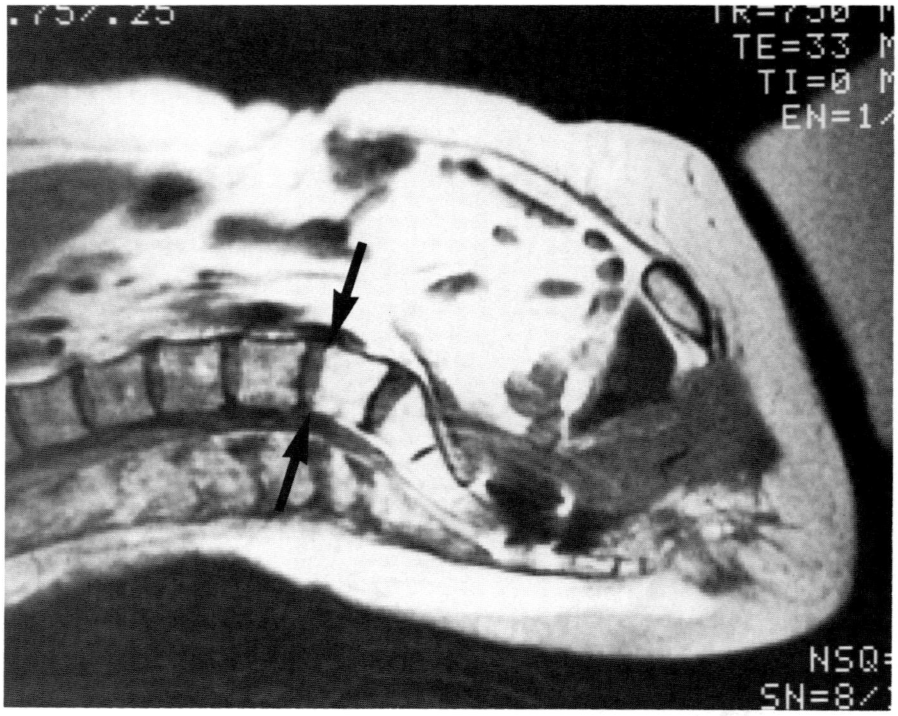

FIGURE 11–14
T1-weighted sagittal image. The edge of the radiation portal is sharply demarcated at the top of the L-5 vertebral body (arrows), *with hyperintense signal from the radiated vertebral bodies below this level.*

full extent of each disease. Although T2-weighted images are of relatively limited value, focal areas of marrow edema exhibit increased signal in areas of acute infarction, as in patients in sickle cell crisis [16].

DIFFUSE INCREASED SIGNAL

Depletion of hematopoietically active elements can be due to primary marrow hypoplasia, as in aplastic anemia, or to iatrogenic causes such as radiation or chemotherapy. All of these conditions ultimately result in the replacement of myeloid elements by fat. This results in a marrow of increased signal intensity from the T1-shortening effects of the increased fat content. Although spin-echo sequences are not sensitive enough to differentiate between hypoplasia and normal adult marrow, radiation changes do have a characteristic appearance that has been reported to occur within 2 weeks after patients undergo therapy [14,15,17]. These changes are easily recognized by marked increased signal within the sharply demarcated boundaries of the radiation portals (Figure 11–14) [17].

Occasionally after radiation therapy, patchy low-signal areas may appear within the treated area. These focal areas may represent fibrosis, tumor infiltrate, or normal red marrow, all of which are indistinguishable on T1 images [18].

FOCAL AREAS OF DECREASED SIGNAL

Focal lesions may involve all or part of a vertebral body. Probably the most common benign focal lesions of decreased signal are secondary to Schmorl's nodes (Figure 11–15). These are caused by the herniation of disc material through the vertebral end-plates into the vertebral bodies. They are usually well delineated and easily differentiated from metastases. If necessary, correlation with plain films or CT demonstrates their characteristic sclerotic margins.

Although primary marrow diseases such as lymphoma and myeloma can present early as focal lesions (Figure 11–16), most malignant focal lesions of decreased signal in the adult spine are secondary to metastatic deposits. MRI should be the study of choice for evaluating suspected metastatic disease of the spine, since it easily

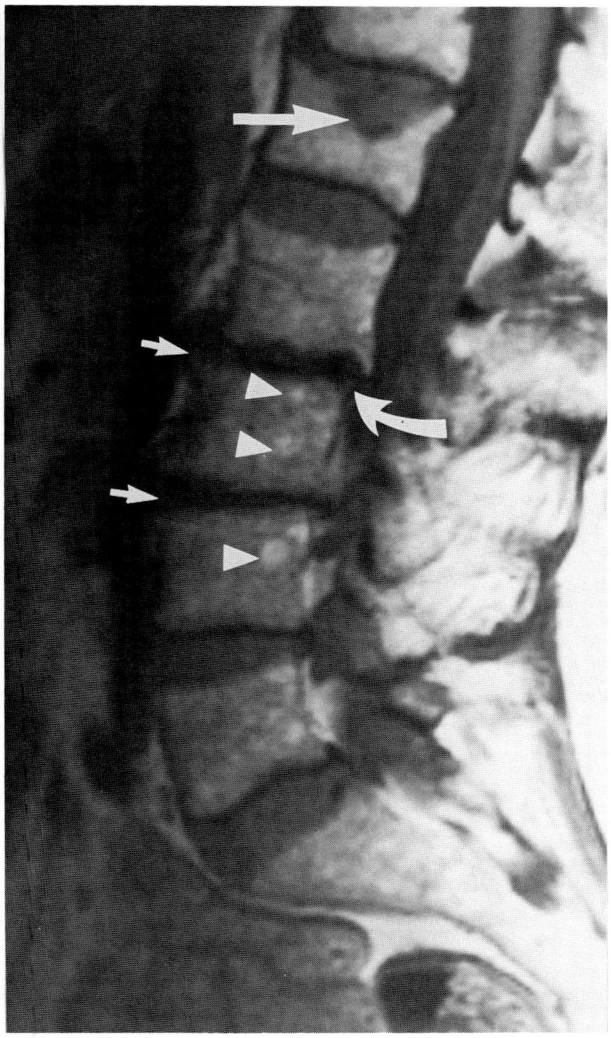

FIGURE 11–15
T1-weighted sagittal image. Schmorl's node is evident in the superior portion of the L-1 vertebral body (large arrow). *Decreased disc space* (small arrows), *retrospondylolisthesis of L-2 on L-3* (curved arrow) *and focal fatty deposits* (arrowheads) *are also demonstrated.*

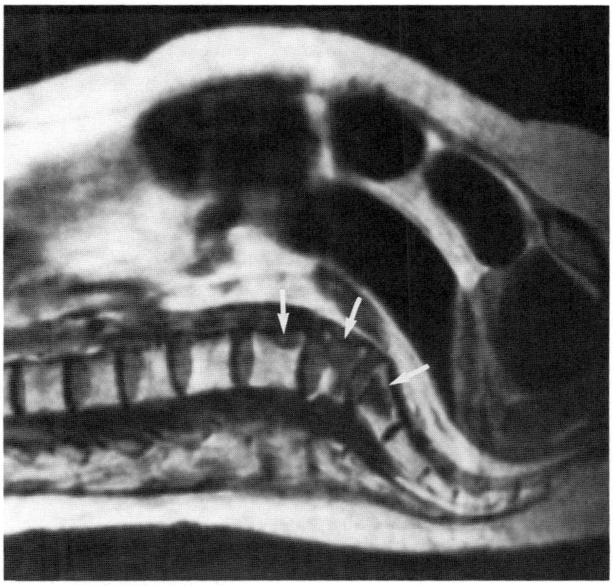

FIGURE 11–16
T1-weighted sagittal image. Focal involvement of the L-4, L-5, and S-1 vertebral bodies (arrows) *is seen in a patient with non-Hodgkin's lymphoma.*

demonstrates vertebral involvement, soft tissue paravertebral tumor extension, and cord compression. It also has the advantage of depicting abnormal areas that could go unnoticed between regions of myelographic blocks [19]. On T1-weighted images, tumor deposits may manifest themselves as regions of intermediate signal against a bright background of fatty marrow, with varying T2 signal characteristics. Sclerotic metastases will demonstrate no signal on both T1- and T2-weighted images due to a lack of mobile protons in dense bone (Figure 11–17). These signal characteristics can be mimicked by benign bone islands and regions of sclerosis secondary to prior compression fractures. MRI is more sensitive than any other imaging modality for the detection of malignant marrow infiltration. This is especially evident in multiple myeloma, where bone scans are often normal and detection of plain film changes requires significant bone destruction [20].

Primary bone tumors of the spine can also be demonstrated on MRI. However, plain films and CT are superior in characterizing tumor matrix and periosteal reaction. The role of MRI in these primary lesions is to demonstrate vascular

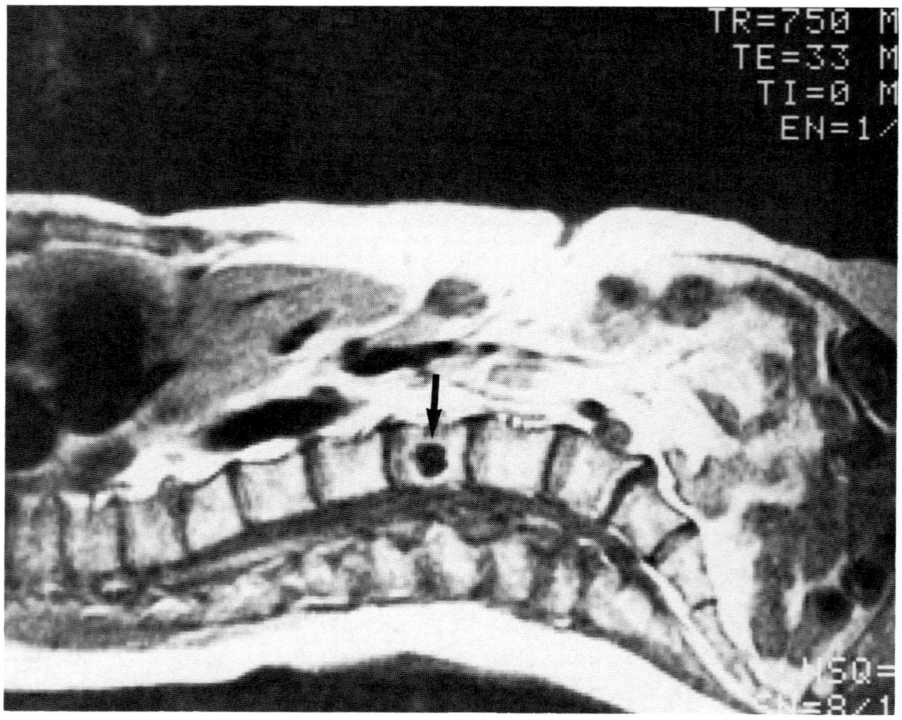

FIGURE 11–17
T1-weighted sagittal image. L-3 sclerotic metastasis from carcinoma of the breast (arrow) *is shown. This lesion remained signal-deficient on the T2-weighted images.*

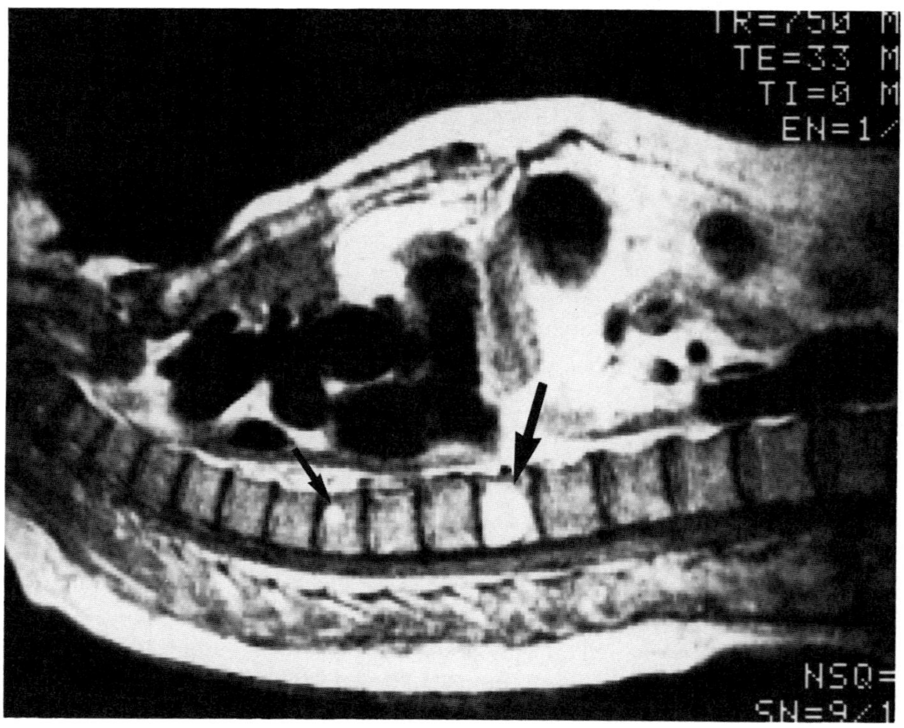

FIGURE 11–18
T1-weighted sagittal image. A large lower thoracic vertebral hemangioma (large arrow) *with a smaller hemangioma or area of focal fatty deposition is seen in an upper thoracic vertebral body* (small arrow).

encasement, soft tissue involvement, and epidural extension, without using intrathecal contrast [21–23].

FOCAL REGIONS OF INCREASED SIGNAL

While most pathologic tissues have long T1 values, vertebral hemangiomas and focal areas of fatty deposition have short T1 values, and appear as focal regions of noticeably increased signal against the normally bright adult marrow (Figure 11–18). Studies have shown that the increased signal of vertebral hemangiomas on T1-weighted images is due to the prominent amount of adipose tissue contained within the stroma surrounding the tumors' thin capillary-like vessels, and is not flow-related [24]. Since the T1 signal characteristics are due to fat, small hemangiomas are indistinguishable from regions of focal fatty replacement in normal hematopoietic marrow. These fatty deposits become more prevalent with advancing age and may be related to localized regions of stress and secondary changes of spondylosis, spondylolysis, and degenerative disc disease [25]. Although rare, intraosseous lipomas may also be considered in the differential diagnosis of high-signal lesions, although differentiating between these entities is seldom of clinical significance when they are contained within the vertebral body. Vertebral hemangiomas, however, usually increase in signal on T2-weighted images (presumably secondary to blood pooling) while fatty focal deposits or lipomas fade slightly. If necessary, further evaluation can be obtained with plain films and CT, where vertebral hemangiomas have a characteristic striated appearance and interosseous lipomas usually have a well-defined sclerotic margin [25,26].

Degenerative Disease

Degenerative changes of the spinal column include discogenic disease and hypertrophy of its bony and ligamentous structures. Bony changes such as osteophyte formation and facet hypertrophy can result in stenosis of the central canal and narrowing of the lateral recesses and neural foramina. Although MRI can demonstrate these hypertrophic changes, CT is still the method of choice for the evaluation of bony abnormalities, due to its ability to obtain high-resolution images of cortical bone. Focal disc herniations and ligamentous hypertrophy are demonstrated well by both modalities. Any of the above degenerative changes may present with the clinical symptoms of a radiculopathy, a myelopathy, or both. Therefore, MRI has become the procedure of choice in many institutions for evaluating the spine, since it is noninvasive and can simultaneously evaluate degenerative disease as well as possible intrinsic cord abnormalities.

CERVICAL SPINE

Prospective studies have shown that high-resolution surface coil examinations of cervical radiculopathy are as sensitive as postmyelographic CT in demonstrating the level of disease, but are less specific in demonstrating disease cause. This slightly lower level of specificity is due to CT's better distinction between bone and soft tissue [27,28].

Isolated cervical disc herniations are relatively rare. Similar to the more common lumbar disc herniations, they present as focal protrusions of disc material that distort the bright epidural venous plexus and epidural fat (Figure 11–19). Degenerative changes secondary to hypertrophy of the uncinate process and osteophyte formation are a more common cause of cervical radiculopathy due to narrowing of the neural foramina and nerve root compression. Spondylitic ridging is also common and is caused by disc degeneration in conjunction with osteophyte formation. This can result in compression and narrowing of the subarachnoid space by an anterior defect composed of protruding disc material sandwiched between upper and lower osteophytes [29] (Figure 11–20).

THORACIC SPINE

As in the cervical region, thoracic disc herniations are relatively uncommon. MRI has been found to be equal to postmyelography CT and superior to myelography alone in the demonstration of thoracic disc herniations [30]. MRI can reveal cord compression even before clinical symptoms are present. T2-weighted images provide the best delineation of the extent of disc

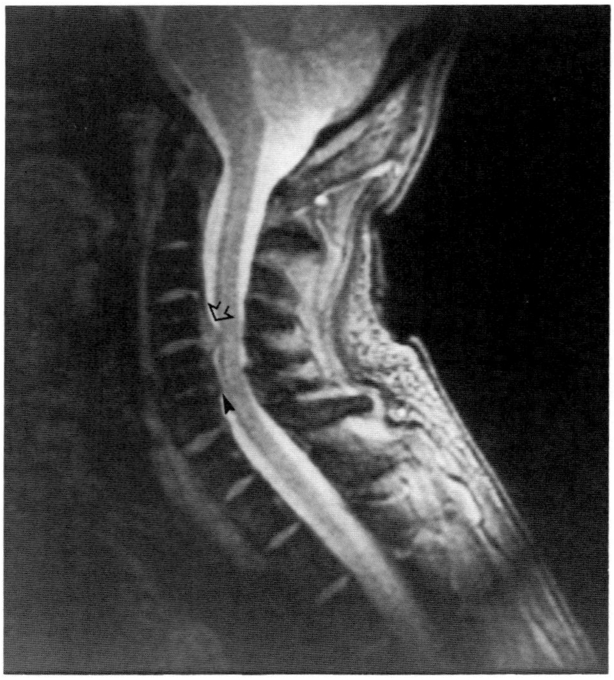

FIGURE 11–19
Gradient-echo sagittal image at the cervical level. A large C4-5 disc herniation (open arrow) *and a C5-6 bulging disc* (arrowhead) *are demonstrated.*

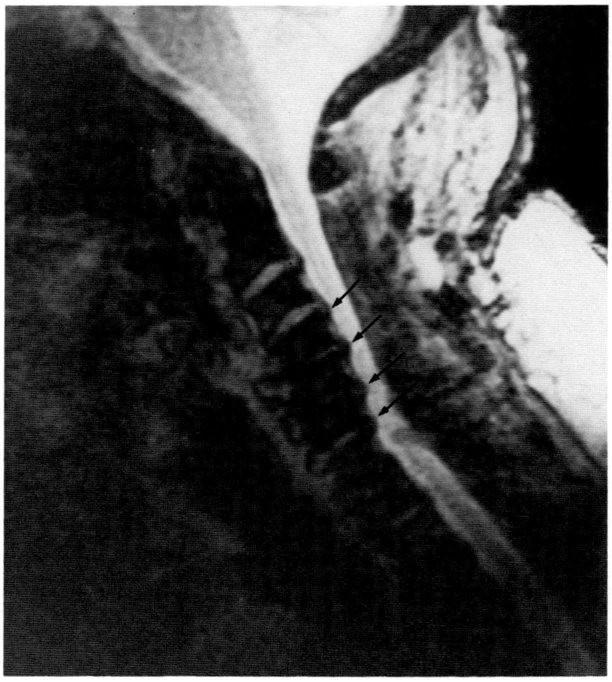

FIGURE 11–20
Gradient-echo sagittal image at the cervical level. There is diffuse spondylitic ridging (arrows), *resulting in cervical stenosis.*

protrusions because of their myelographic effects [31].

LUMBAR AND SACRAL SPINE

Degenerative disc disease is most common at the lumbosacral level since it is subjected to the most stress in bearing the body's weight [13]. Prospective studies have indicated that surface coil examinations for disc herniations and spinal stenosis at the lumbar level are as accurate as CT and superior to myelography alone. The advantage over myelography is especially evident at the L5-S1 level where there is normally a large epidural space formed by increased amounts of epidural fat. At this level, a disc herniation may not be large enough to distort a myelographic column [32].

Lateral disc herniations are usually undetected by myelography and often overlooked on CT and MRI examinations. Far lateral herniations may not be associated with an intraspinal component and tend to migrate cephalad, causing symptoms referred to the level above the involved disc space. Both MRI and CT can demonstrate lateral herniations equally well when the neural foramina and paravertebral spaces are carefully evaluated [33].

Disc herniations usually present as focal protrusions of similar intensity to that of the parent disc on T1-weighted images. On T2-weighted images, small herniations are usually hypointense (presumably due to decreased water content), but can become hyperintense to the parent disc in cases of large extrusions or sequestered fragments (Figure 11–21). Lumbar disc herniations are best visualized on axial and sagittal images. On axial images they displace adjacent epidural veins and fat, compressing the thecal sac or nerve roots (Figure 11–22). Midline sagittal images may show focal disruption of the outer annulus or the posterior longitudinal ligament, or both, on T2-weighted images. Parasagittal images are

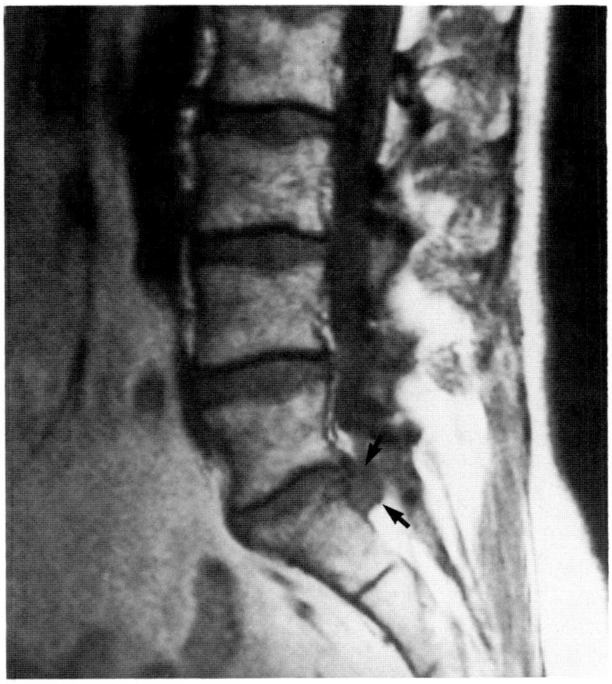

A

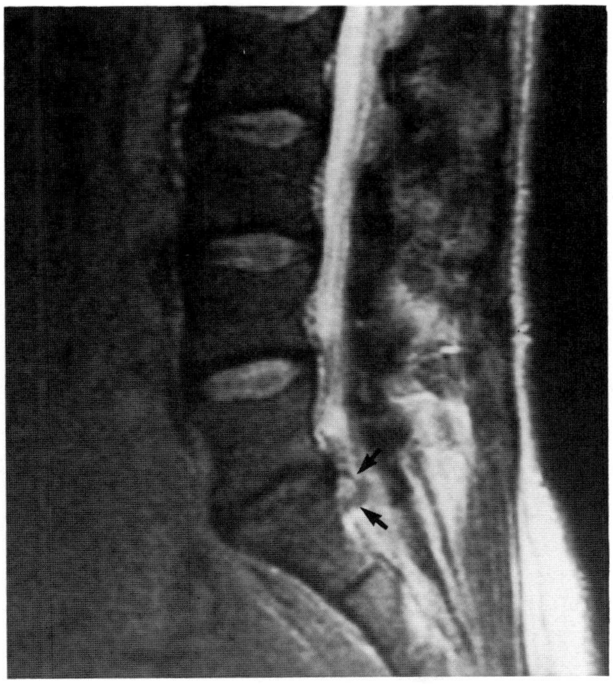

B

FIGURE 11–21
A. *T1-weighted sagittal image of the lumbosacral region. A sequestered L5-S1 disc* (arrows) *is demonstrated.*
B. *A T2-weighted image corresponding to* A *displays the high signal in the sequestered disc fragment* (arrows).

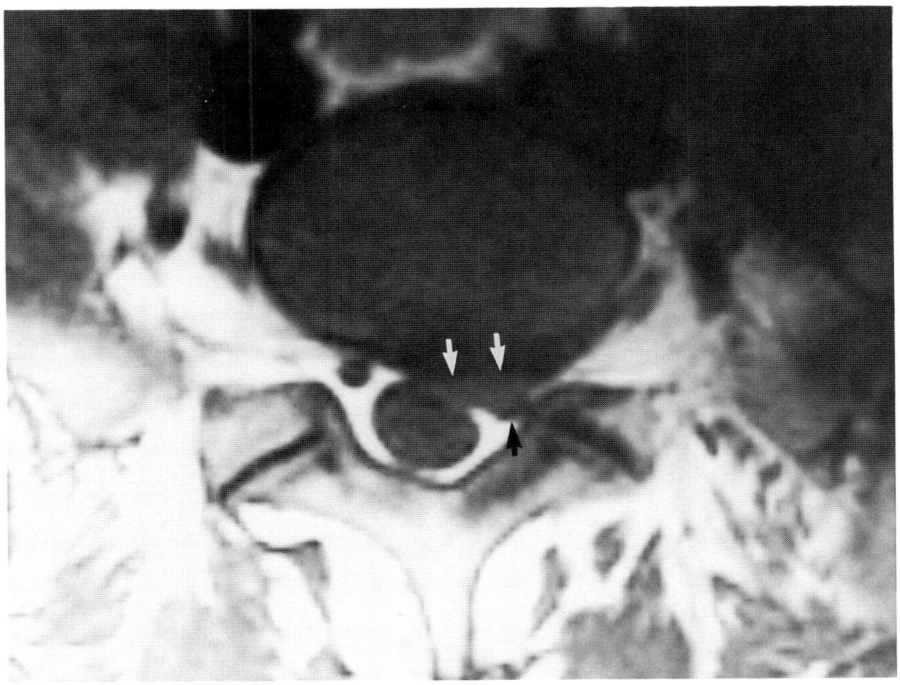

FIGURE 11–22
T1-weighted axial image. A left-sided herniated lumbar disc (white arrows) *distorts the thecal sac and compresses the left nerve root posteriorly* (black arrow).

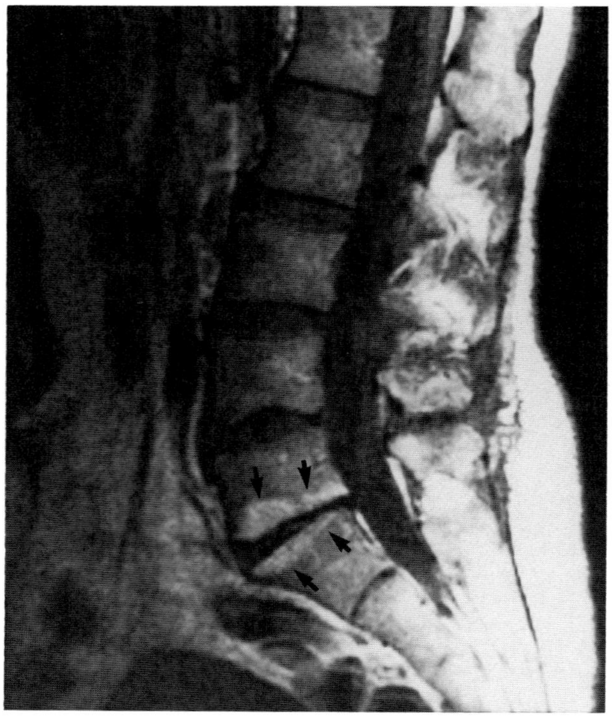

FIGURE 11–23
T1-weighted sagittal image at the lumbosacral level. Bright end-plates around the L5-S1 disc (arrows) *represent type II end-plate changes.*

also useful for depicting lateral herniations into the neural foramina, which obliterate the normal intraforaminal fat [13,34].

Three types of signal intensity changes have been reported in the marrow in the region of the vertebral end-plates adjacent to degenerated discs. Type I end-plate changes show low signal on T1-weighted images and increase in relative signal intensity on T2-weighted images. Type II end-plate changes exhibit high signal on both T1- and T2-weighted images, while type III changes demonstrate low signal on both echo sequences. Type I changes can convert to the type II variety, which have been reported to remain relatively constant with the passage of time. Type I changes probably represent early acute inflammation associated with marrow edema, type II changes are caused by yellow marrow replacement within the end-plates (Figure 11–23), while type III changes correlate with bony sclerosis of the end-plates [35,36].

HYPERTROPHIC DEGENERATIVE CHANGES

CT still remains the examination of choice for demonstrating bony hypertrophy. It is also preferred for evaluating spondylolisthesis (slippage of the vertebral bodies) and spondylolysis (pars defects). The axial CT appearance of these changes are described in detail by Teplick et al. [37]. High-resolution surface coil MRI studies have also been able to demonstrate the degenerative changes of subchondral osteosclerosis, marrow changes, and hypertrophy of the ligamentum flavum [9]. Defects in the pars can be suggested by dark signal in the pars articularis, representing sclerosis around a fracture on sagittal images. However, this finding can be mimicked by partial facetectomy, by volume averaging of degenerative spurs, and by metastases [38]. Further prospective studies are needed to evaluate the sensitivity and specificity of MRI in these degenerative changes.

POSTOPERATIVE SPINE

Persistent pain is not uncommon after back surgery. It may be caused by incomplete decompression, hemorrhage, infection, recurrent disc herniation, epidural fibrosis, arachnoiditis, or

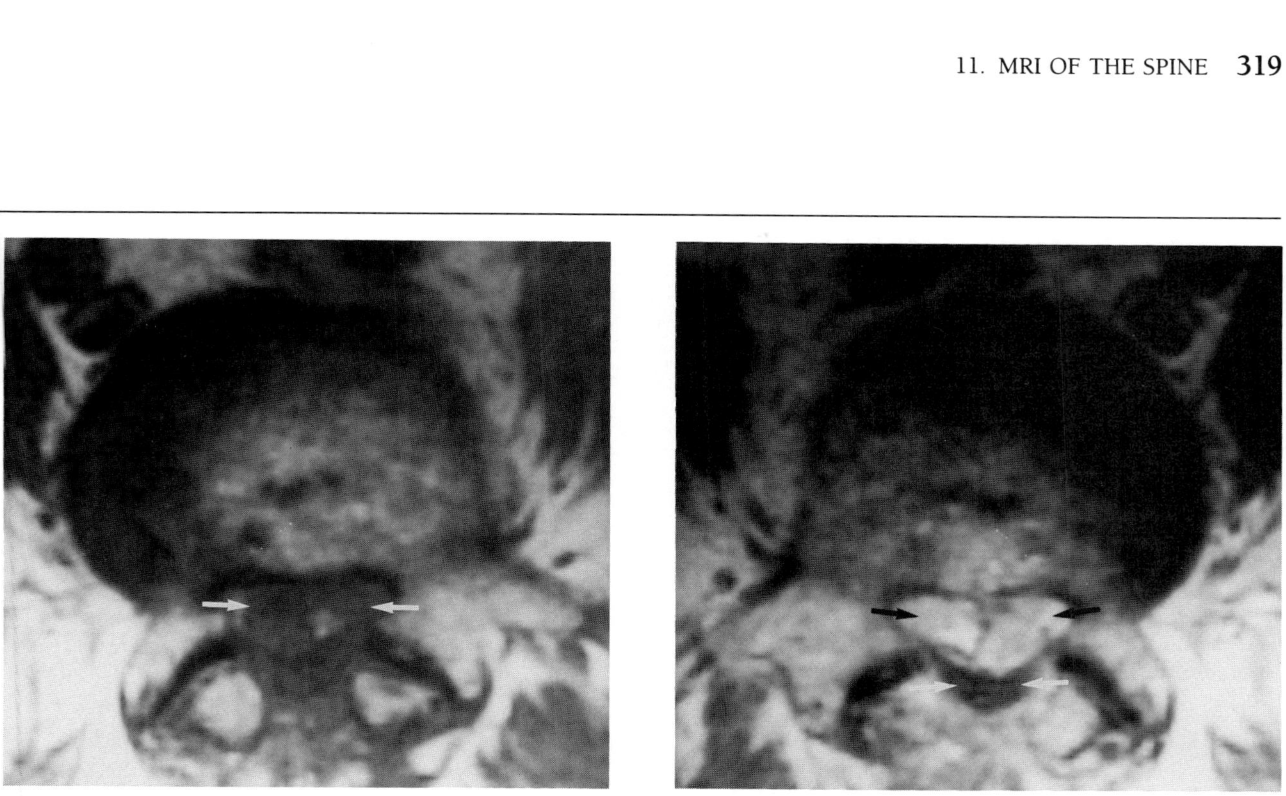

A B

FIGURE 11–24

A. *Postoperative T1-weighted axial images of the lumbosacral region. Precontrast (Gd-DTPA) study shows intermediate-signal epidural soft tissue mass* (arrows). B. *T1-weighted postcontrast study corresponding to* A *demonstrates scar enhancement* (black arrows) *with a posteriorly compressed thecal sac* (white arrows).

previously unknown lesions of the vertebral column or spinal cord unassociated with the sequelae of surgical intervention. While MRI can easily screen the spine for unknown occult lesions, chronic pain is often due to recurrent disc herniation or postoperative scarring. It is crucial to differentiate scar formation from recurrent disc herniation in the postoperative patient since reoperation on scar tissue has a poor prognosis [39]. Studies have shown that recurrent disc herniations have sharply contoured margins, while fibrosis has irregular margins [40]. Disc material remains hypointense to isointense on all spin-echo pulse sequences, when compared to the disc of origin, while scar tissue becomes hyperintense on T2-weighted images [39,40]. Perhaps the most accurate means of differentiating recurrent disc from epidural scarring is achieved by obtaining T1-weighted images before and after the intravenous administration of gadolinium diethylenetriaminepentaacetic acid (Gd-DTPA). This paramagnetic contrast agent causes heterogeneous enhancement of scar tissue on T1-weighted images obtained within 10–15 minutes after injection. During this time disc material will not enhance, but may intensify later if imaging is delayed for more than half an hour after injection. These findings are analogous to intravenous contrast media–enhanced CT studies that also demonstrate enhancement of scar tissue [41–43]. This enhancement is most likely due to abnormally leaky capillaries in the scar tissue formed by the body's attempt to heal the surgical site (Figures 11–24 and 11–25) [13].

Inflammatory Disease

Early studies have shown that MRI is useful in the evaluation of patients with rheumatoid arthritis. Perithecal pannus formation and subsequent compression of the cervicomedullary junction are well demonstrated on both T1- and T2-weighted sagittal images [44] (Figure 11–26). Other studies have found MRI to be as useful as tomography in evaluating other clinically impor-

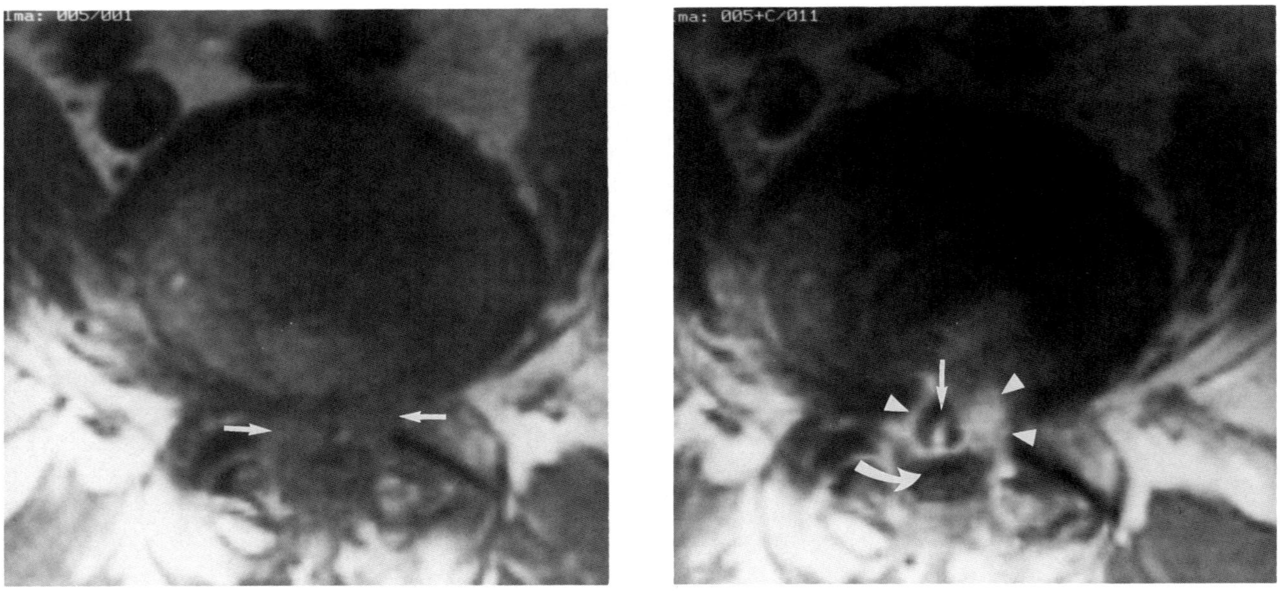

A B

FIGURE 11–25
A. Postoperative T1-weighted axial images of the lumbosacral region. Precontrast (Gd-DTPA) study demonstrates intermediate-signal soft tissue mass (arrows). *B. T1-weighted post-Gd-DTPA study demonstrates nonenhancement of a recurrent disc* (arrow) *surrounded by enhancing scar* (arrowheads) *and posteriorly displaced thecal sac* (curved arrow).

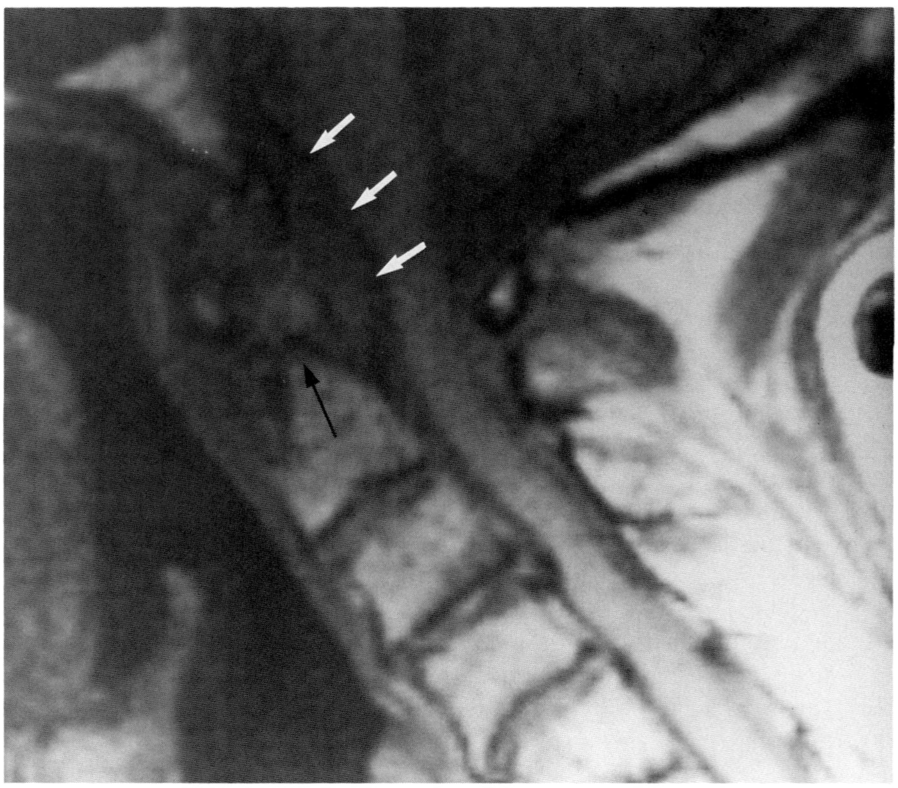

FIGURE 11–26
T1-weighted sagittal image at the cervical level. Patient with rheumatoid arthritis has a perithecal soft tissue mass (white arrows) *compressing cervicomedullary junction and erosion of the odontoid process* (black arrow).

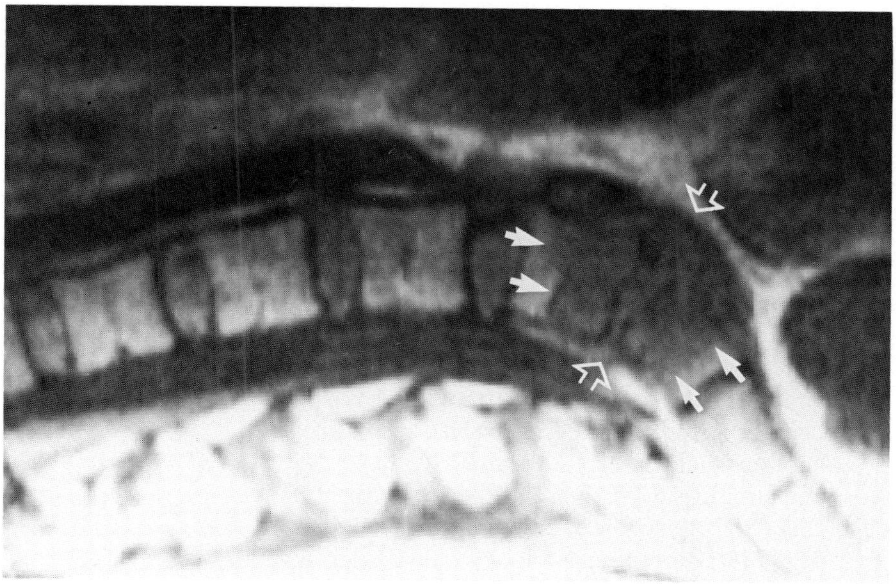

A

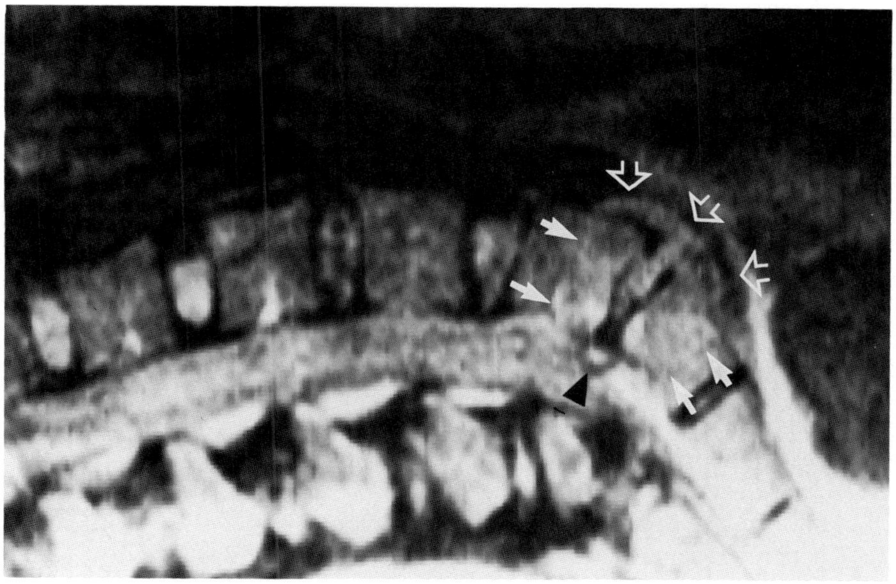

B

FIGURE 11–27

A. *T1-weighted sagittal images of the lumbosacral region. There is a confluence of low signal involving the L5-S1 disc* (open arrows) *and adjacent vertebral end-plates* (solid arrows). B. *The T2-weighted image corresponding to* A *shows an increased signal and loss of normal configuration of the L5-S1 disc, with extension of abnormal high signal into the anterior paravertebral space* (open white arrows). *Increased signal is also observed in the adjacent vertebral end-plates* (solid white arrows), *all consistent with L5-S1 vertebral osteomyelitis and discitis with anterior paravertebral abscess formation. Minimal posterior epidural extension is also demonstrated* (black arrowhead).

tant features such as erosion of the dens, subluxation, cranial settling, and atlantodental distance. MRI may become the modality of choice for evaluating cervical involvement in rheumatoid arthritis [45].

Early reports also suggest that pyogenic inflammatory disease such as epidural abscess is also best demonstrated by MRI [46]. Vertebral osteomyelitis with discitis has a characteristic appearance on MRI [47]. On T1-weighted images there is a confluence of decreased signal involving the affected portions of the vertebral bodies and discs. On T2-weighted images, these regions become hyperintense. The involved disc usually appears brighter than a normally hydrated disc, and also displays decreased height and loss of its characteristic notched configuration (Figure 11–27) [13]. This constellation of findings helps differentiate infection from metastatic disease or postchemonucleation changes that may also demonstrate increased signal in the vertebral bodies on T2-weighted images. In metastatic disease, however, the disc signal remains normal, whereas after chemonucleation the disc usually becomes hypointense in signal on T2-weighted images [13,48]. Early isolated discitis without osteomyelitis can present with abnormally increased signal in a disc on T2-weighted images without accompanying abnormal signal changes in the adjacent end-plates.

Conclusions

Although still in its infancy, MRI has already become the imaging modality of choice for the detailed evaluation of the spinal column. It is the most sensitive, noninvasive means for the evaluation of bone marrow changes, degenerative disc disease, and inflammatory changes. However, because of MRI's inability to demonstrate cortical bone, tumor matrix, and calcification as well as CT does, and MRI's physical limitations in imaging acute trauma, these modalities should serve complementary rather than competing roles in the establishment of a diagnosis, after the evaluation of plain films. MRI, in an outpatient setting, without the need for intrathecal contrast should also become the modality of choice for following the response of neoplastic and infectious diseases of the spine after therapy.

In the future, faster scanning techniques (FISP, FLASH, GRASS), new contrast agents, and better surface coil technology will help decrease motion artifacts, increase specificity, and further improve spatial resolution. Furthermore, in vivo spectroscopy holds the promise of expanding the role of MRI from merely visualizing pathologic features to increasing our knowledge of the metabolic and biochemical nature of the disease processes themselves.

REFERENCES

1. Hesselink JR. Spinal imaging: history, achievements, remaining frontiers. AJR 1988; 150:1223–9.
2. Haughton VM. MR imaging of the spine. Radiology 1988; 166:297–301.
3. Modic MT. The spine. In: Higgins CB, Hricak H, eds. Magnetic resonance imaging of the body. New York: Raven Press, 1987.
4. Porter BA, Redmond J, Dunning DM, et al. Low-field STIR imaging of spinal malignancies. Radiology 1988; 169(P):65. abstract.
5. Edelman RR, Shoukimas GM, Stark DD, et al. High-resolution surface-coil imaging of lumbar disk disease. AJR 1985; 144:1123–9.
6. Czervionke LF, Daniels DL, Wehrli FW, et al. Magnetic susceptibility artifacts in gradient-recalled echo MR imaging. AJNR 1988; 9:1149–55.
7. Enzmann DR, Rubin JB. Cervical spine: MR imaging with a partial flip angle, gradient-refocused pulse sequence. Part I: general considerations and disk disease. Radiology 1988; 166:467–72.
8. Hedberg MC, Drayer BP, Flom RA, et al. Gradient echo (GRASS) MR imaging in cervical radiculopathy. AJNR 1988; 9:145–51.
9. Grenier N, Kressel HY, Schiebler ML, Grossman RI, Dalinka MK. Normal and degenerative posterior spinal structures: MR imaging. Radiology 1987; 165:517–25.
10. Daniels DL, Hyde JS, Kneeland JB, et al. The cervical nerves and foramina: local-coil MR imaging. AJNR 1986; 7:129–33.
11. Flannigan BD, Lufkin RB, McGlade C, et al. MR imaging of the cervical spine: neurovascular anatomy. AJNR 1987; 8:27–32.
12. Pech P, Haughton VM. Lumbar intervertebral disk: correlative MR and anatomic study. Radiology 1985; 156:699–701.

13. Modic MT, Masaryk TJ, Ross JS, Carter JR. Imaging of degenerative disc disease. Radiology 1988; 168:177–86.
14. Vogler JB III, Murphy WA. Bone marrow imaging. Radiology 1988; 168:679–93.
15. Porter BA, Shields AF, Olson DO. Magnetic resonance imaging of bone marrow disorders. Radiol Clin North Am 1986; 24:69–87.
16. Rao VM, Fishman M, Mitchell DG, et al. Painful sickle cell crisis: bone marrow patterns observed with MR imaging. Radiology 1986; 161:211–5.
17. Kaplan PA, Asleson RJ, Klassen LW, Duggan MJ. Bone marrow patterns in aplastic anemia: observations with 1.5-T MR imaging. Radiology 1987; 164:441–4.
18. Jahre C, Sze G, Krol G, Zimmerman RD, Deck MDF. Temporal and signal characteristics of marrow radiation changes on MR imaging. Radiology 1988; 169(P):101. abstract.
19. Smoker WRK, Godersky JC, Knutzon RK, et al. The role of MR imaging in evaluating metastatic spinal disease. AJR 1987; 149:1241–8.
20. Daffner RH, Lupetin AR, Dash N, et al. MRI in the detection of malignant infiltration of bone marrow. AJR 1986; 146:353–8.
21. Beltran J, Noto AM, Chakeres DW, Christofordis AJ. Tumors of the osseous spine: staging with MR imaging versus CT. Radiology 1987; 162:565–9.
22. Sze G, Ulchanco LS III, Brant-Zawadzki MN, et al. Chordomas: MR imaging. Radiology 1988; 166:187–91.
23. Aisen AM, Martel W, Braunstein EM, et al. MRI and CT evaluation of primary bone and soft-tissue tumors. AJR 1986; 146:749–56.
24. Ross JS, Masaryk TJ, Modic MT, Carter JR, Mapstone T, Dengel JH. Vertebral hemangiomas: MR imaging. Radiology 1987; 165:165–9.
25. Hajek PC, Baker LL, Goodbar JE, et al. Focal fat deposition in axial bone marrow: MR characteristics. Radiology 1987; 162:245–9.
26. Laredo JD, Reizine D, Bard M, Merland J-J. Vertebral hemangiomas: radiologic evaluation. Radiology 1986; 161:183–9.
27. Modic MT, Masaryk TJ, Mulopulos GP, Bundschuh C, Han JS, Bohlman H. Cervical radiculopathy: prospective evaluation with surface coil MR imaging, CT with metrizamide and metrizamide myelography. Radiology 1986; 161:753–9.
28. Karnaze MC, Gado MH, Sartor KJ, et al. Comparison of MR and CT myelography in imaging the cervical and thoracic spine. AJR 1988; 150:397–403.
29. Haughton VM, Daniels DL, Czervionke LF, Williams AL. Cervical spine. In: Stark DD, Bradley WG, eds. Magnetic resonance imaging. St. Louis: CV Mosby, 1988; 614–31.
30. Ross JS, Perez-Reyes N, Masaryk TJ, Bohlman H, Modic MT. Thoracic disk herniation: MR imaging. Radiology 1987; 165:511–5.
31. New PFJ, Shoukimas GM. Thoracic spine and spinal cord. In: Stark DD, Bradley WG, eds. Magnetic resonance imaging. St. Louis: CV Mosby, 1988; 632–65.
32. Modic MT, Masaryk T, Boumphrey F, Goormastic M, Bell G. Lumbar herniated disk disease and canal stenosis: prospective evaluation by surface coil MR, CT and myelography. AJNR 1986; 7:709–17.
33. Osborn AG, Hood RS, Sherry RG, et al. CT/MR spectrum of far lateral and anterior lumbosacral disk herniations. AJNR 1988; 9:775–8.
34. Masaryk TJ, Ross JS, Modic MT, et al. High resolution MR imaging of sequestered lumbar intervertebral disks. AJNR 1988; 9:351–8.
35. de Roos A, Kressel HY, Spritzer C, et al. MR imaging of marrow changes adjacent to end plates in degenerative lumbar disk disease. AJR 1987; 149:531–4.
36. Modic MT, Steinberg PM, Ross JS, Masaryk TJ, Carter JR. Degenerative disk disease: assessment of changes in vertebral body marrow with MR imaging. Radiology 1988; 166:193–9.
37. Teplick JG, Laffey PA, Berman A, et al. Diagnosis and evaluation of spondylolisthesis and/or spondylolysis on axial CT. AJNR 1986; 7:479–91.
38. Johnson DW, Farnum GN, Latchaw RE, et al. MR imaging of the pars interarticularis. AJR 1989; 152:327–32.
39. Bundschuh CV, Modic MT, Ross JS, et al. Epidural fibrosis and recurrent disk herniation in the lumbar spine: MR imaging assessment. AJR 1988; 150:923–32.
40. Hockhauser L, Kieffer SA, Cacayorin ED, et al. Recurrent postdiskectomy low back pain: MR-surgical correlation. AJNR 1988; 9:769–74.
41. Hueftel MG, Modic MT, Ross JS, et al. Lumbar spine: postoperative MR imaging with Gd-DTPA. Radiology 1988; 167:817–24.
42. Teplick JG, Haskin ME. Computed tomography of the postoperative lumbar spine. AJR 1983; 141:865–84.

43. Teplick JG, Haskin ME. Intravenous contrast-enhanced CT of the postoperative lumbar spine: improved identification of recurrent disk herniation, scar, arachnoiditis and diskitis. AJNR 1984; 5:373–83.
44. Aisen AM, Martel W, Ellis JH, McCune WJ. Cervical spine involvement in rheumatoid arthritis: MR imaging. AJNR 1988; 9:565–71.
45. Bundschuh C, Modic MT, Kearney F, et al. Rheumatoid arthritis of the cervical spine: surface-coil MR imaging. AJNR 1988; 9:565–71.
46. Angtuaco EJC, McConnell JR, Chadduck WM, et al. MR imaging of spinal epidural sepsis. AJR 1987; 149:1249–53.
47. Modic MT, Feiglin DH, Piraino DW, et al. Vertebral osteomyelitis: assessment using MR. Radiology 1985; 157:157–66.
48. Huckman MS, Clark JW, McNeill TW, et al. Chemonucleation and changes observed on lumbar MR scan: preliminary report. AJNR 1987; 8:1–4.

12

MRI of the Joints

JOHN A. MARKISZ

Imaging of the joints using MR techniques is becoming an increasingly important and versatile method of evaluating a wide variety of clinical complaints. These disorders can essentially be divided into two types: the first in which the specific mechanical components of the joint itself are scrutinized (e.g., tendon, ligamentous, meniscal, and articular problems), and the second examining more general musculoskeletal problems that secondarily affect the operation of the joint (e.g., tumor, infection, trauma, and bone infarction). Specialized surface coils (Chapter 8) are necessary to obtain the submillimeter resolution required for a proper evaluation of the tendons, ligaments, menisci, and other small, intricate structures necessary for normal articular motion [1–4]. Although these surface coil techniques may be appropriate for evaluating nonarticular abnormalities as well, the rather limited useful field of view inherent in these methods does not afford an adequate evaluation of tissues somewhat removed from the joint spaces, especially around the hip and the shoulder [5]. In almost all cases involving these areas, resolution obtained utilizing the body coil is usually sufficient to permit an accurate diagnosis, especially when acquired with off-center imaging and a decreased field of view. The wrist, hand, ankle, and foot are compact enough regions that surface coil imaging usually provides complete coverage of all of the structures of interest. For cases in which an osseous, infectious, or soft tissue abnormality is suspected, imaging protocols are similar to those described in Chapters 9 and 10 for bone and soft tissue diseases.

Many problems involving joints are similar and can be treated in a comparable manner. All depend on a knowledge of anatomy (Chapter 2) in order to choose the proper imaging planes and sequence parameters. Protocols for surface coil evaluation of joints vary with the individual joint, the type of surface coil used, and the particular problem being investigated. Evaluations involving the joint space and related structures usually employ thin (3-mm), contiguous, or closely spaced sections, and both T1- and T2-weighted sequences are acquired. For proper evaluation of the extremities, the appropriate surface coil should be selected. A coil only slightly larger than the area of interest is optimal. For the thick joints, a wraparound or enveloping surface coil eliminates the problem of severely decreased signal from regions relatively remote from the coil. Many specialized coils have been developed in various form-fitting shapes for the knee, shoulder, wrist, cervical, temporomandibular (TMJ), and even digital and sternoclavicular joints.

When evaluating a surface coil image of a joint, it is important to realize some of the limitations, imposed by the imaging modality itself, that may become an impediment to accurate evaluation if these constraints are not recognized. One such restriction involves assessing the thickness of the cortical bone and articular cartilage, both of which are visualized quite well with surface coils. Chemical shift artifact (Chapter 8) can seriously alter the appearance of these structures [6]. In Figure 12–1, the right and left images demonstrate the same anatomic sagittal section of a knee. The only difference between the two images is that the phase and frequency directions were exchanged prior to data acquisition. This effectively rotated the direction of the chemical shift by 90 degrees. The apparent thicknesses of the articular cartilage and cortical bone are quite dissimilar on the two different images, even though they represent exactly the same section of anatomy. Quantitative measurements of small or thin structures may not be accurate due to this alteration in chemical shift in the depiction of thin structures. Qualitative evaluation, in terms of irregularities, erosions, and interruption, however, are still valid [7].

Tissue Abnormalities

Trauma or unexplained pain are the more common reasons for evaluating a joint by MRI. The imaging planes chosen depend on the anatomy of the structure to be evaluated. Since routine x-rays usually have already been obtained, visible fractures have been ruled out, and tendon, ligamentous, or soft tissue damage is being considered. Stress fractures, occult fractures, and bone contusions are not always detected on plain films, and should always be considered, even if x-rays demonstrate normal findings or bone injury is not suspected (Figure 12–2) [8–10].

A complete rupture of a tendon or ligament can be diagnosed either by observing a discontinuity in the structure along its longitudinal axis, or more often, by not observing the structure in any one of a series of sequential images obtained perpendicular to its path (Figure 12–3) [11]. Care must be taken not to confuse imaging of a relaxed tortuous structure with a tear. A slack tendon or ligament often bends out of the imaging plane and appears to be severed, although it actually is intact. Since it is extremely difficult to obtain an imaging plane that always includes the entire structure of interest, consecutive sections should be examined to evaluate continuity, if a tear is suspected.

Images obtained perpendicular to the path of a tendon or ligament are particularly useful in determining patency. Tendons and ligaments normally have no MR signal, and are easily visible on both T1- and T2-weighted images as regions of signal void. Observation of one axial section through the normal tendon course that does not demonstrate the presence of the tendon is sufficient to diagnose a complete rupture (Fig-

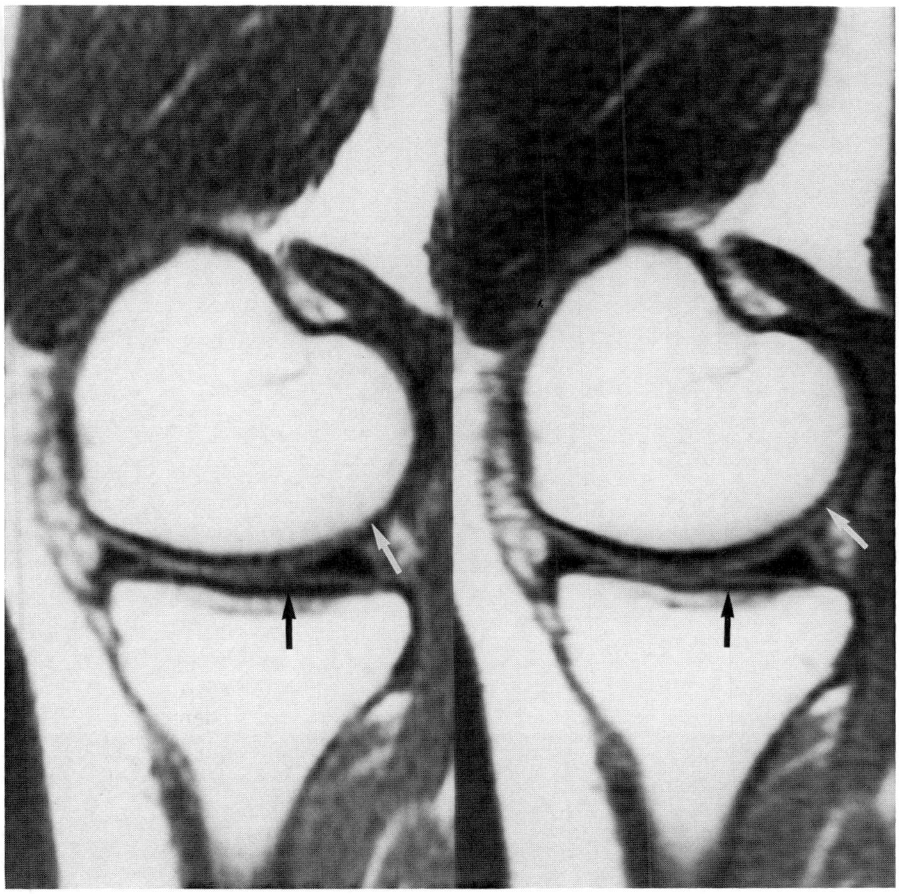

A B

FIGURE 12–1
Precisely the same anatomic section, with phase and frequency encoding directions exchanged, is shown on the left (A) *and the right* (B). *Note the apparent differences in thickness of articular cartilage* (white arrows) *and cortical bone* (black arrows) *in the two images.*

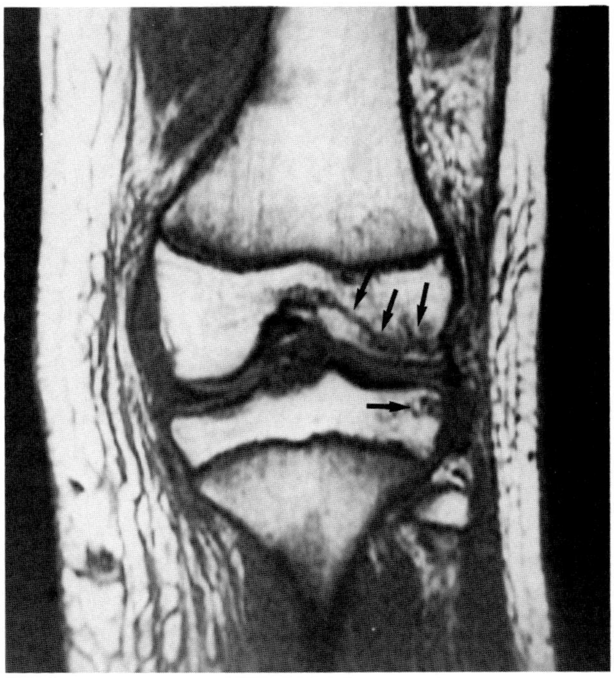

FIGURE 12–2
Occult fractures (arrows) *in the lateral femoral condyle and lateral tibial plateau are seen in a 12-year-old involved in an auto accident. Note the thick, open epiphyseal plates* (open arrows).

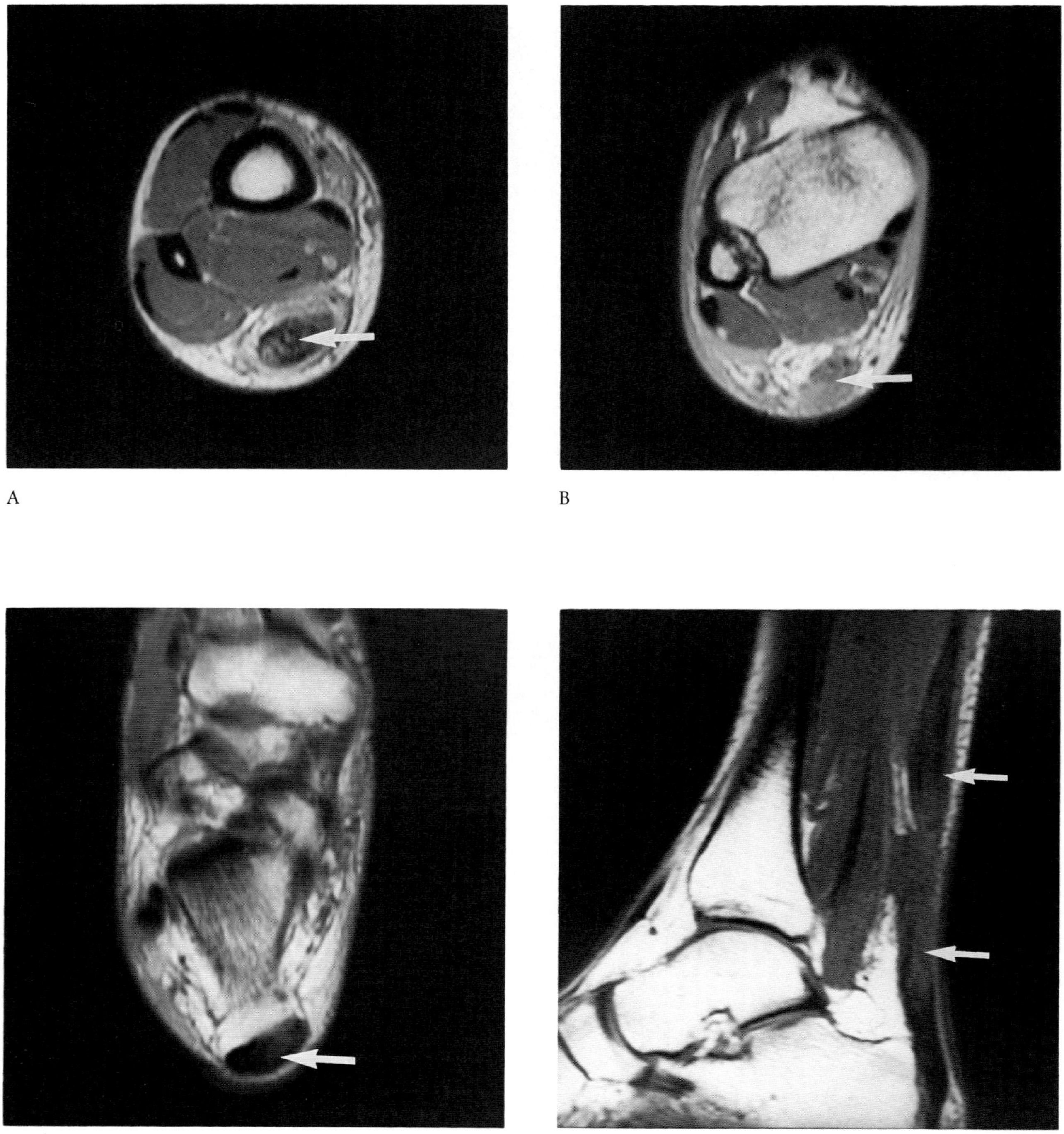

A B C D

FIGURE 12–3

A. *At the distal level of the tibia and fibula, the Achilles tendon* (arrow) *is lacerated.* B. *At the level of the malleoli, there is no evidence of a tendon. An inflammatory mass* (arrow) *is seen in the area normally occupied by the tendon.* C. *A normal-appearing attachment at the calcaneus is seen* (arrow). D. *This sagittal view through the plane of the Achilles tendon demonstrates the separation* (arrow).

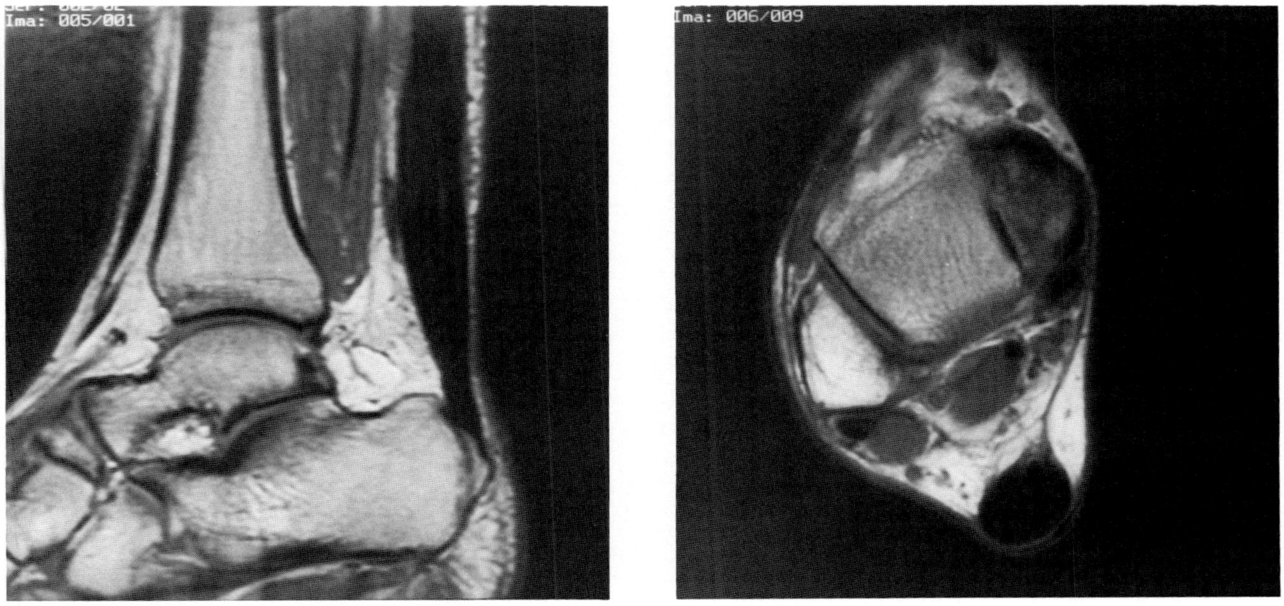

A B

FIGURE 12–4
An abnormally thickened Achilles tendon (A) *is secondary to chronic trauma.* A. *Sagittal image.* B. *Axial image.*

ure 12–3). Chronic tears are visualized as a hypertrophied tendon on the axial images (Figure 12–4). Any signal within a ligament or tendon on T1- or T2-weighted images is abnormal and suggestive of a partial tear or of inflammation. A partial tear manifests itself as a sudden irregularity in contour or area of sudden narrowing. Inflammation can be observed as increased signal within the tendon on either T1- or T2-weighted images. We have found gradient-echo images especially sensitive to inflammatory changes within tendons (Figure 12–5). Fluid within the sheath of an intact tendon, consistent with tenosynovitis, can be observed best on T2-weighted images as a very bright region surrounding the signal-poor tendon (Figure 12–6).

The MR appearance of fluid varies on T1-weighted images, depending on the composition of the solution. Liquids usually are low-signal regions on these images, with the observed signal increasing with increasing protein content. Joint effusions present as regions of moderate signal within the joint space on T1-weighted images,

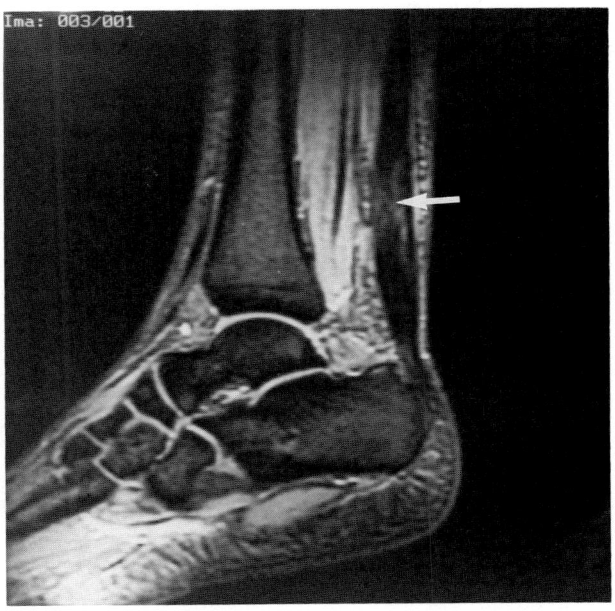

FIGURE 12–5
Increased signal is seen on this gradient-echo image of a patient with chronic inflammatory changes (arrow) *in the Achilles tendon.*

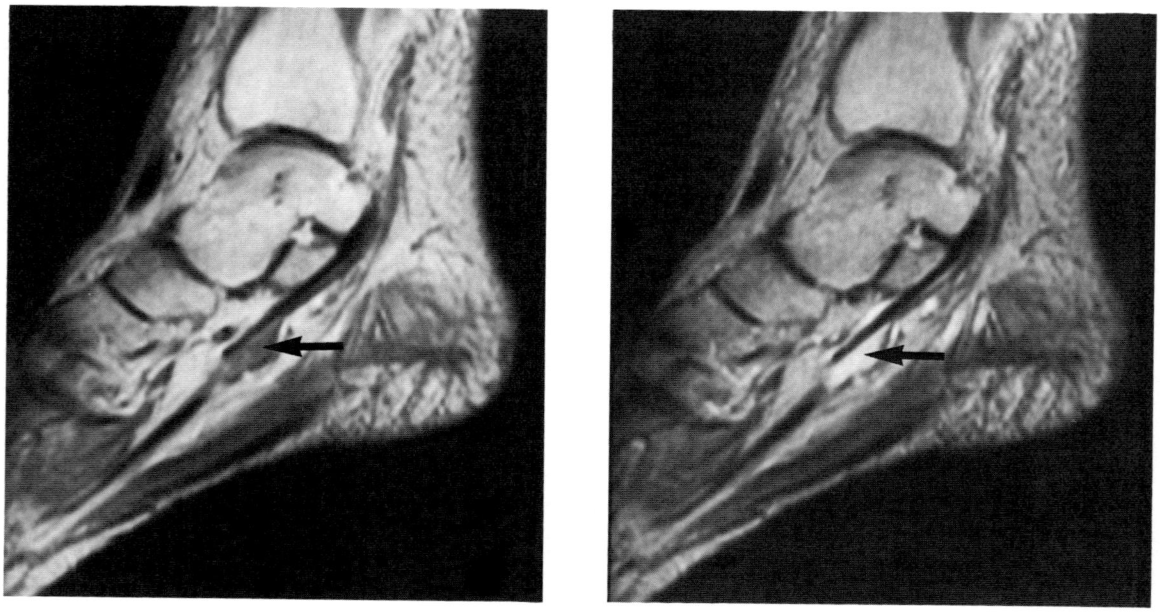

A B

FIGURE 12–6
A. *A T1-weighted image demonstrates abnormal signal* (arrow) *in the vicinity of the dark tibialis posterior tendon.* B. *On the T2-weighted image, the fluid within the tendon sheath becomes very bright. Note that the tendon cannot be imaged completely in a single plane. The sudden appearance and disappearance are due to its tortuous course.*

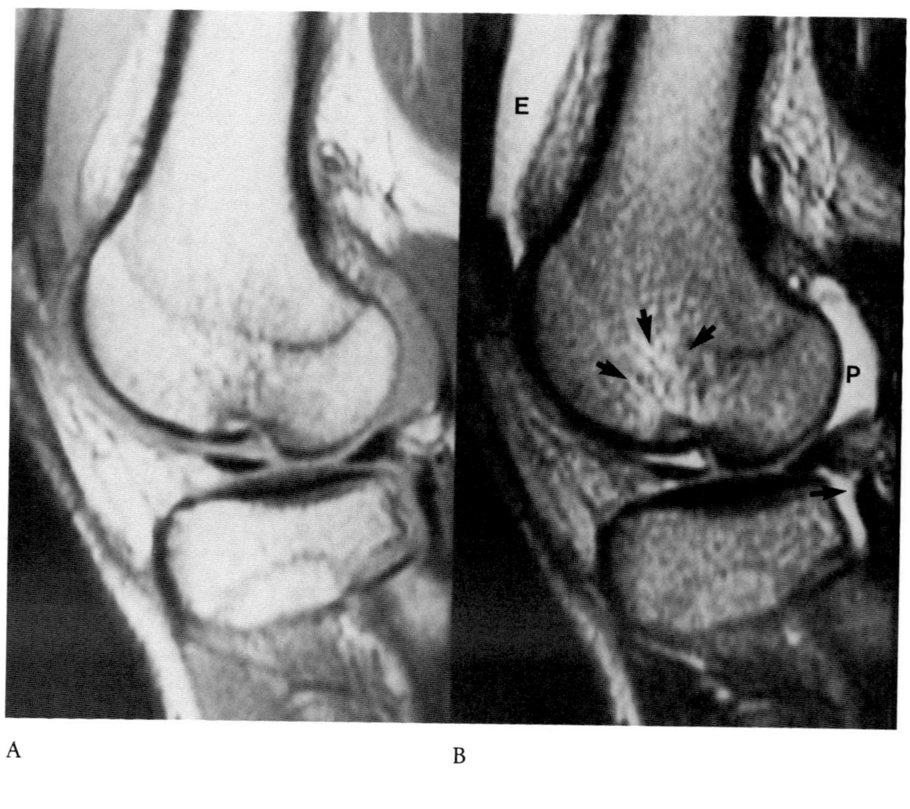

A B

FIGURE 12–7
The T1-weighted image (A) *and T2-weighted image* (B) *demonstrate a fracture of the lateral femoral condyle, with edema within the marrow cavity* (small arrows), *a posterior joint effusion (P), a suprapatellar effusion (E), and fluid within the tendon sheath of the popliteus tendon* (black arrow) *all increasing in signal intensity on the T2-weighted image.*

and become quite bright on the T2-weighted views (Figure 12–7). The normal joint space contains only a very small amount of synovial fluid, with the knee, the body's largest joint, containing less than 0.5 ml under normal conditions. It is, therefore, generally abnormal to observe fluid within a joint space on a T2-weighted image. We have, however, noted small regions of fluid within the hip joints of nonsymptomatic individuals, when the scan is done at high field strength (1.5 T). Inflamed bursae exhibit a similar MR appearance, representing a loculated region of high T2 signal outside of the joint space. The presence and anatomic location of bursae around the joints can be highly variable. Synovial cysts similarly are very bright on the T2 views (Figure 12–8). Septic effusions in joints are similar in appearance to sterile effusions, increasing strongly in signal intensity on the T2-weighted images, but sometimes they demonstrate an increased signal on T1-weighted images as well, due to the high protein content of the fluid (Figures 12–9 and 12–10).

Cortical bone can be recognized as a signal-free region on MR images; however, cortical dis-

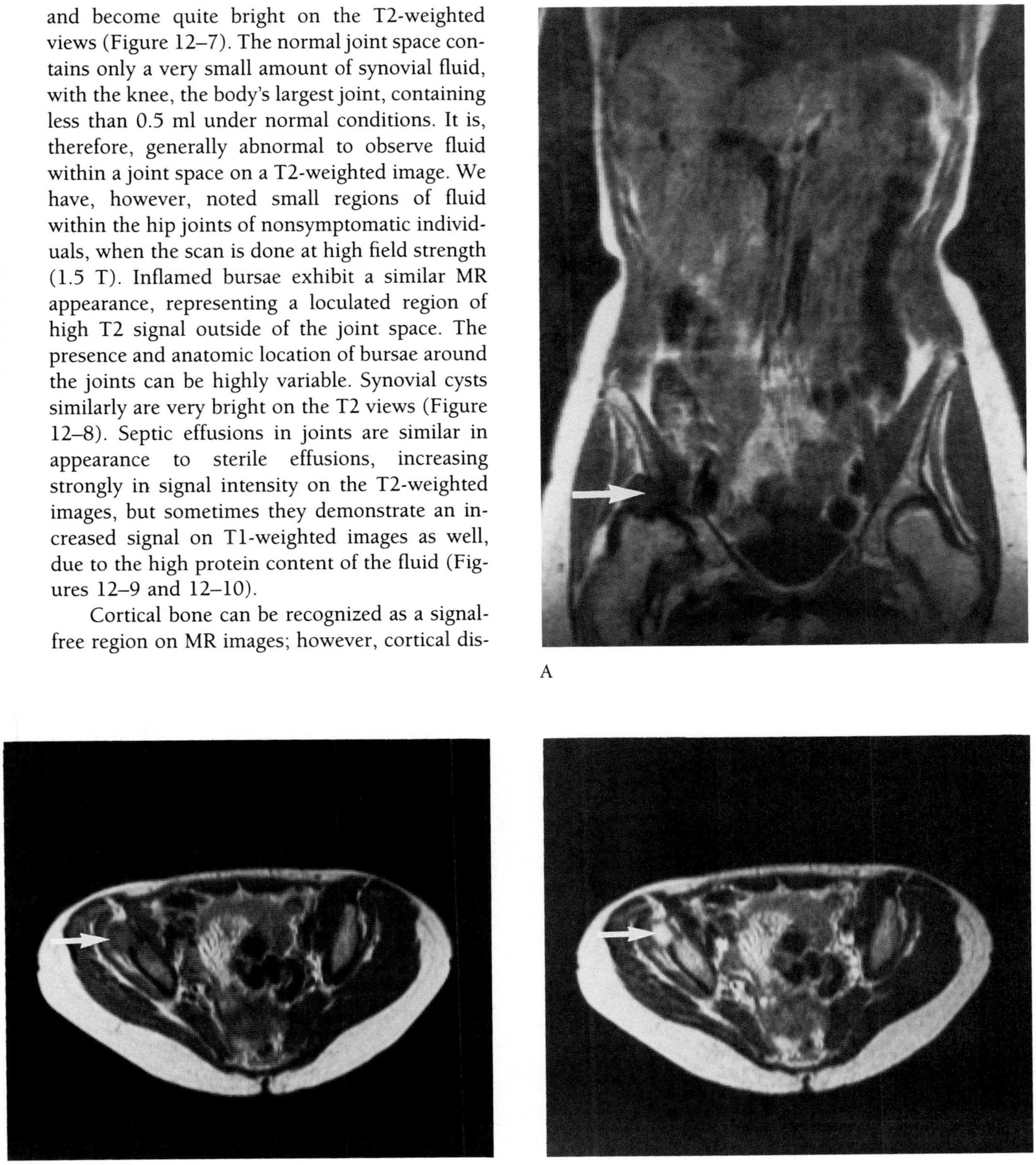

A

B

C

FIGURE 12–8

A. *A coronal view shows a synovial cyst* (arrow) *off of the superior lateral aspect of the right hip.* B, C. *T1-weighted and T2-weighted images, respectively, of the cyst illustrate the increased signal in the cyst* (C).

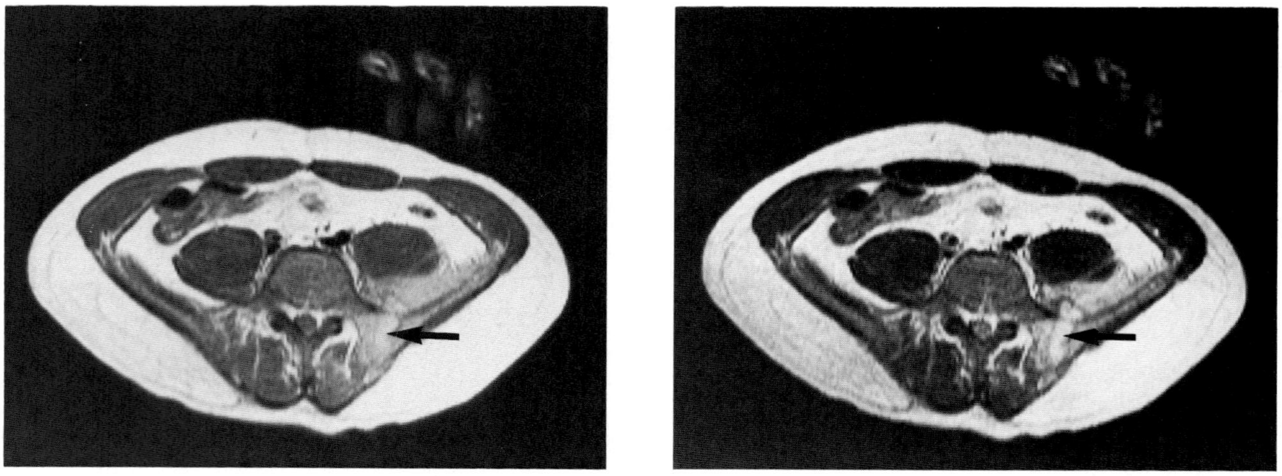

A B

FIGURE 12–9
Septic effusion of the left sacroiliac joint (arrow) *increases in signal intensity from moderately intense on the T1-weighted image to bright on the T2-weighted image.* A. *T1-weighted image.* B. *T2-weighted image.*

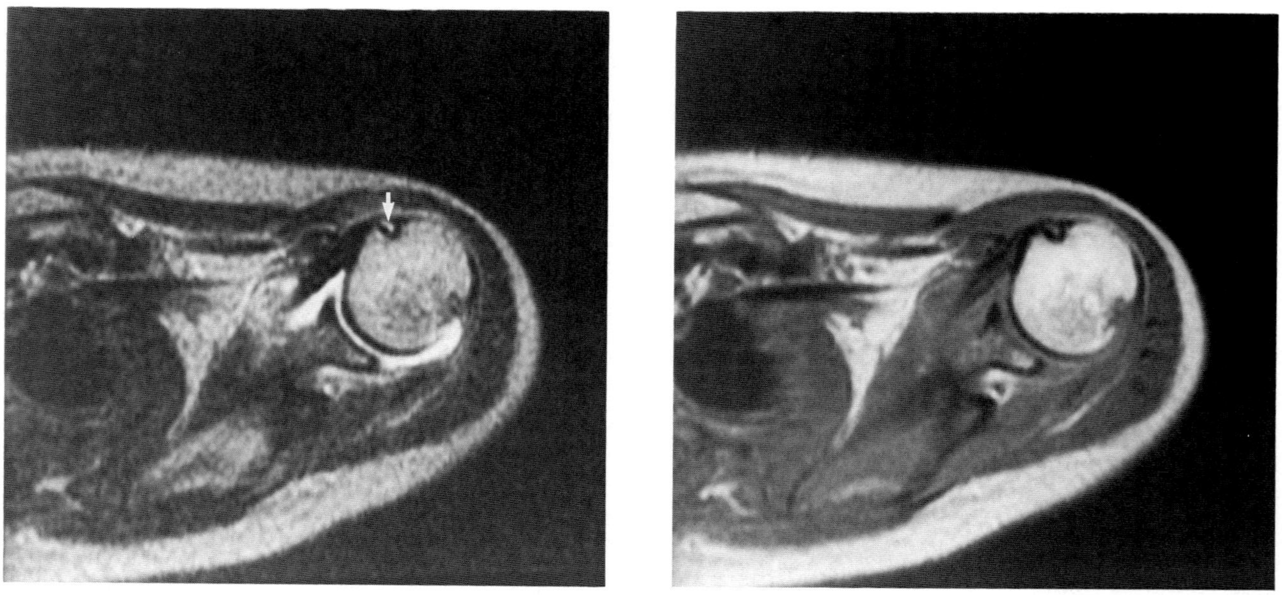

A B

FIGURE 12–10
Septic effusion within the joint space of the left shoulder is essentially isointense to muscle on the T1-weighted image (compare to Figure 12–9). B. *On the T2-weighted image, the effusion becomes bright and a small amount of fluid is observed within the biceps tendon sheath in the bicipital groove* (arrow).

orders are usually identified more easily and earlier by plain film, CT, or scintigraphy than they are by MRI. Joint space narrowing, osteophyte formation, erosion of cortical bone, and other changes due to osteoarthritis or any erosive process *can* be observed by MRI, although these changes are observed earlier and in more detail on plain films. The value of MRI for the evaluation of such disorders lies in its ability to detect and document presence of inflamed synovium, erosion of articular cartilage, small effusions, and soft tissue edema, all of which may be missed on routine radiographs. Osteochondral bodies can often be well seen on T2-weighted images silhouetted by the bright effusion [12]. The initial erosive changes and pannus formation that occur in rheumatoid arthritis, however, may actually be appreciated first by MRI, since the earliest manifestations of this disease can be difficult to detect radiographically (Figure 12–11) [13]. Juvenile rheumatoid arthritis has been a particular diagnostic problem when it is evaluated by radiographs in the early stages, and preliminary indications are that MRI may be the imaging modality of choice (Figures 12–12 and 12–13) [14,15].

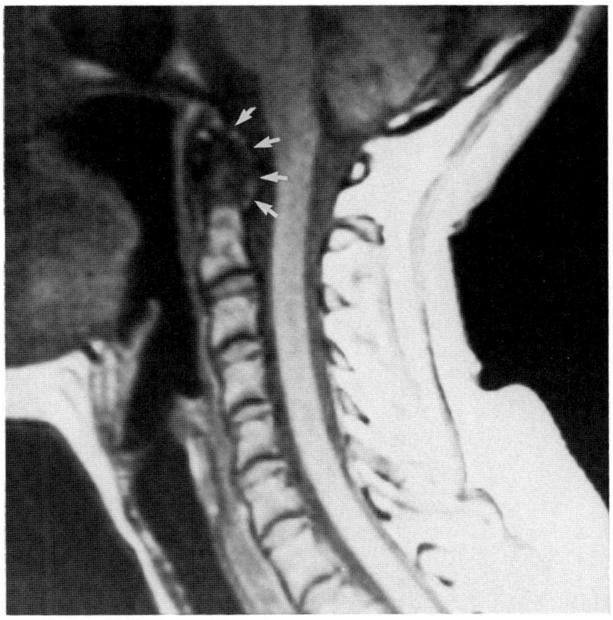

FIGURE 12–11
Erosion of the odontoid process (arrows) *is well demonstrated in this patient with rheumatoid arthritis.*

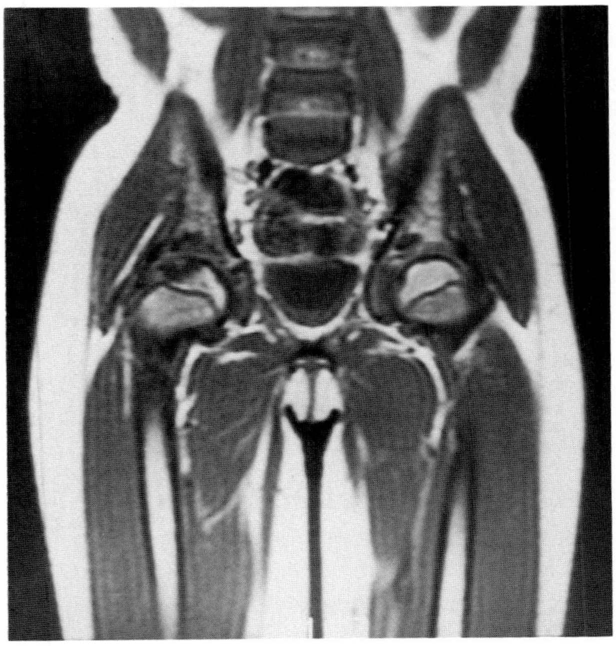

FIGURE 12–12
Erosive changes and pannus formation in juvenile rheumatoid arthritis of the right hip in this 8-year-old girl.

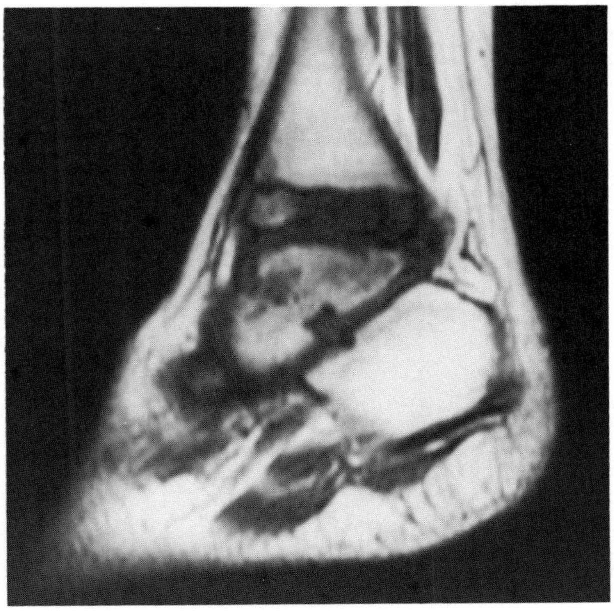

FIGURE 12–13
The normally bright epiphysis at the distal end of the tibia has been replaced by the erosive changes of juvenile rheumatoid arthritis.

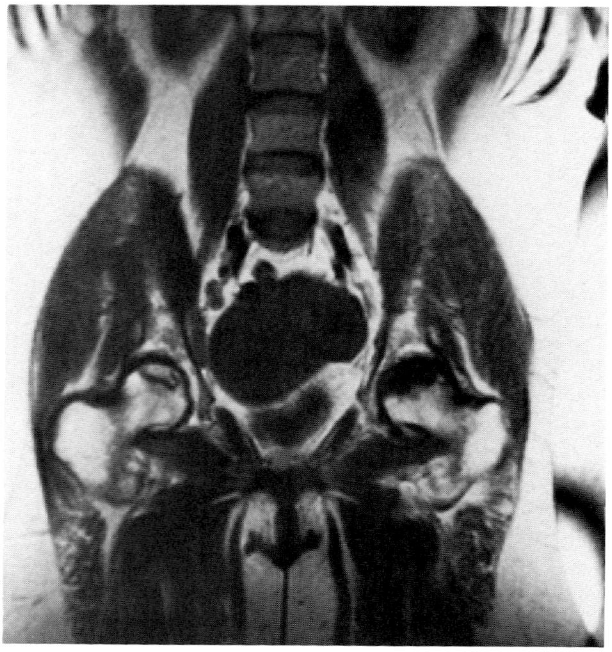

FIGURE 12–14
Bilateral osteonecrosis of the femoral heads illustrates the "ring" pattern (left) *and homogeneous region of decreased signal* (right).

Ischemic necrosis of bone, arising from a variety of causes, is a disabling condition that can be detected in very early stages by MRI, at which point therapeutic measures may be instituted [16–22]. Weight-bearing surfaces of the femoral condyles, femoral heads, and tibial plateaus, along with the humeral head and scaphoid, are very susceptible to this process. The disease may present either as a ring of decreased signal around an area of fat intensity, or as an area of homogeneously decreased signal, on the T1 images (Figures 12–14 and 12–15). If the event is recent, areas of surrounding edema may be apparent on the T2-weighted images, which may also demonstrate subchondral cyst formation. The abnormality often remains signal-poor on the delayed echo views, due to fibrosis and reactive bone formation.

Soft tissue abnormalities are readily evaluated by surface coil examination of the joints. The principles and patterns of benign and malignant neoplasms, infections, and trauma are identical to those described in Chapter 10. Synovial sarcomas are a relatively rare malignancy, containing cystic and hemorrhagic areas [23]. These often-recurrent lesions present as slow growing masses that become bright on the T2-weighted images (Figure 12–16). These can be difficult to differentiate from synovial cysts (see Figure 12–8), pigmented villonodular synovitis (PVNS), and monophasic synoviomas. PVNS arises from the joints, most often from the knee. It is a locally aggressive lesion in which hyperplastic synovium contains lipid- and hemosiderin-filled macrophages. The MR features include a heterogeneous mass, with characteristics of hemosiderin (dark on both T1- and T2-weighted images) and fat (bright on T1, fading slightly on T2) within it, usually along with cystic components (Figure 12–17) [24,25].

The Temporomandibular Joint

Pain secondary to internal derangement of the TMJ has been estimated to occur in as much as 28% of the general population, usually attributed to chronic malocclusive problems, bruxism, trauma (including dental), and stress [26,27]. The normal functional anatomy of the TMJ is

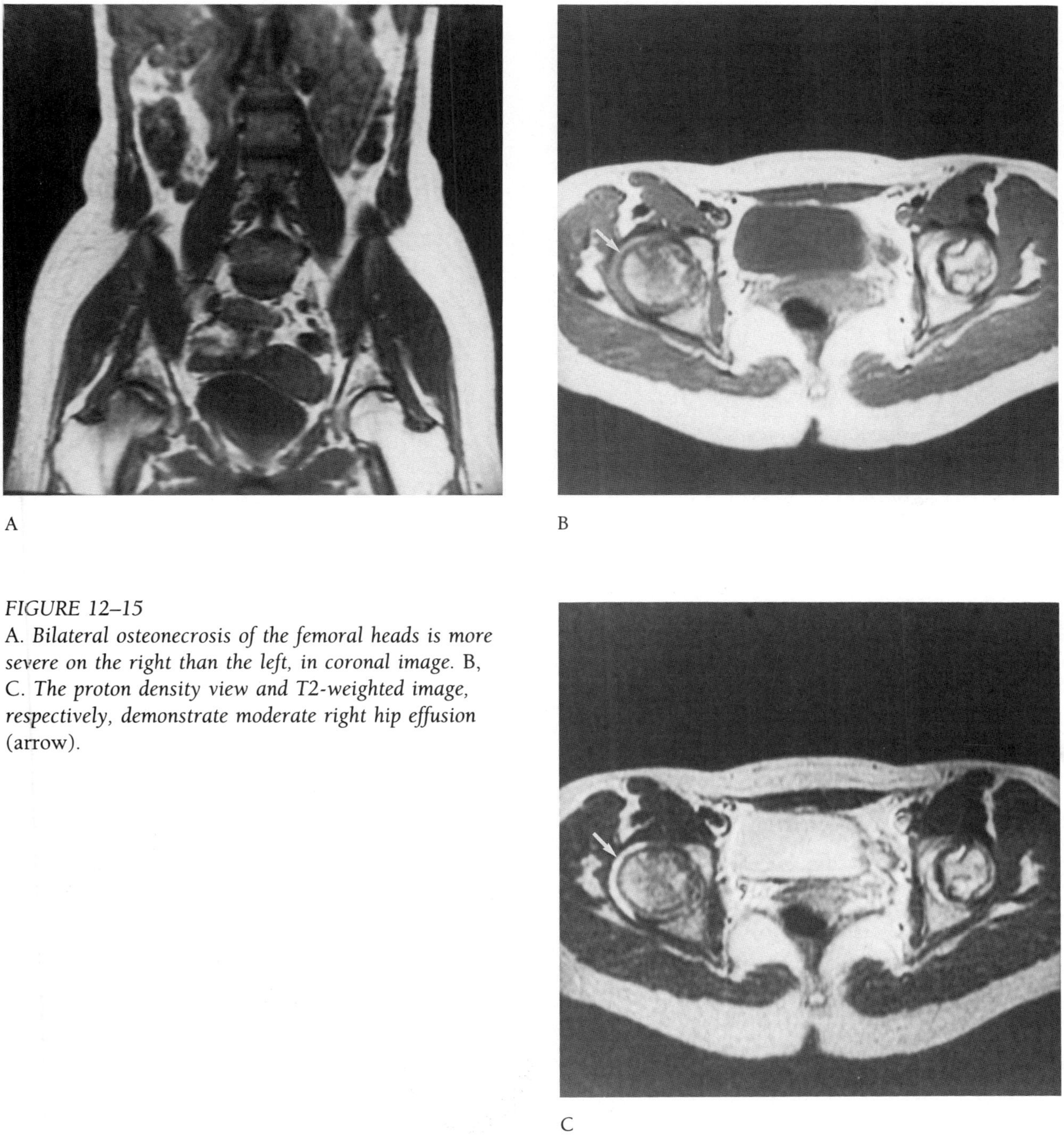

A B C

FIGURE 12–15
A. Bilateral osteonecrosis of the femoral heads is more severe on the right than the left, in coronal image. B, C. *The proton density view and T2-weighted image, respectively, demonstrate moderate right hip effusion* (arrow).

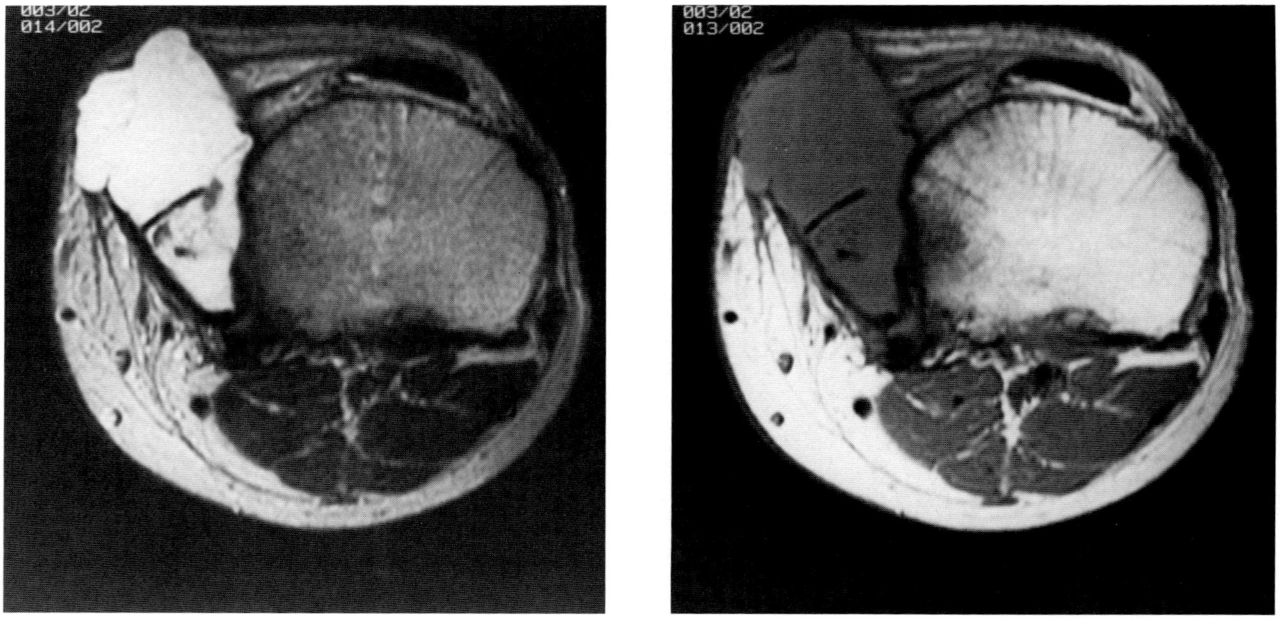

A B

FIGURE 12–16
Recurrent synovial sarcoma at the knee. A. T1-weighted image. B. T2-weighted image.

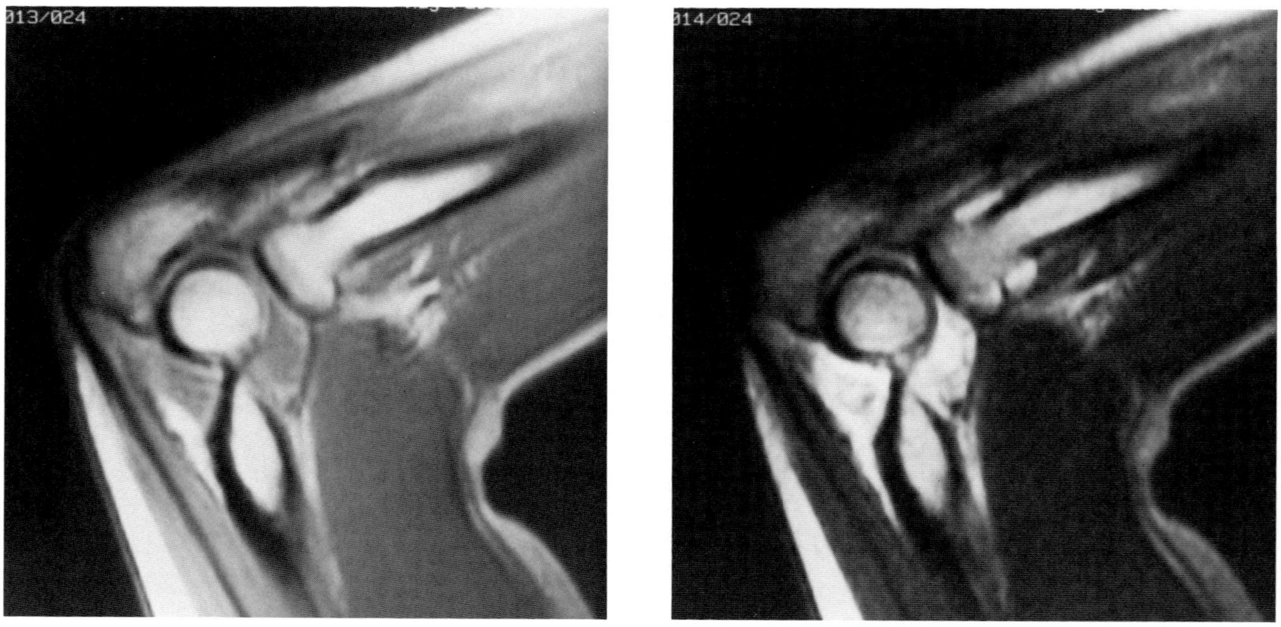

A B

FIGURE 12–17
Pigmented villonodular synovitis involving the elbow. A. Sagittal T1-weighted image. B. T2-weighted sections.

presented in Chapter 2. The mandibular condyle and articular eminence of the temporal bone are bright structures, due to the fat in the marrow cavity. The articular disc (fibrocartilage) and cortical bone are low-signal structures. Muscle is intermediate in signal intensity, somewhat lower than brain tissue. In a normal individual, the concave inferior surface of the intermediate zone of the articular disc always remains in contact with the articular surface of the mandibular condyle. In the closed-mouth position, the condyle is centered within the mandibular fossa, with the posterior band of the disc above it (Figure 12–18). In the opened-mouth position, the condyle is inferior to the articular eminence (Figure 12–19). The two bony structures are separated by an almost horizontal disc.

Internal derangements of the TMJ most often involve a displacement of the articular disc [28]. Most commonly the disc is displaced anteriorly, and opening of the jaw can occur both with and without reduction (Figures 12–20 and 12–21). Anterior displacement with reduction is frequently accompanied by an audible click as the jaw is opened. In the closed-mouth position, the disc is displaced anteriorly, with the articular surface of the condyle in contact with the posterior band of the bilaminar band. As the condyle rotates anteriorly, it moves under the posterior zone to a more normal position under the intermediate zone, producing the "click." In cases of anterior displacement without reduction, the condyle is unable to translate under the disc to assume a more normal position. Rather than rotate over the condyle, the disc moves ahead of it, acting as an obstruction to complete opening and causing pain. The disc can also be displaced laterally and medially and, rarely, posteriorly. Other common problems include deformities or fragmentation of the disc and arthritic changes of the mandibular condyle.

Since the vast majority of problems involve internal derangements, T1-weighted images of the TMJ are most important in order to best define the anatomy of the region [29,30]. Images must obviously be obtained with the mouth opened and closed in order to determine the status and relative position of the articular disc, as well as the bony structures. Two 3-inch coupled surface coils can be used to obtain inter-

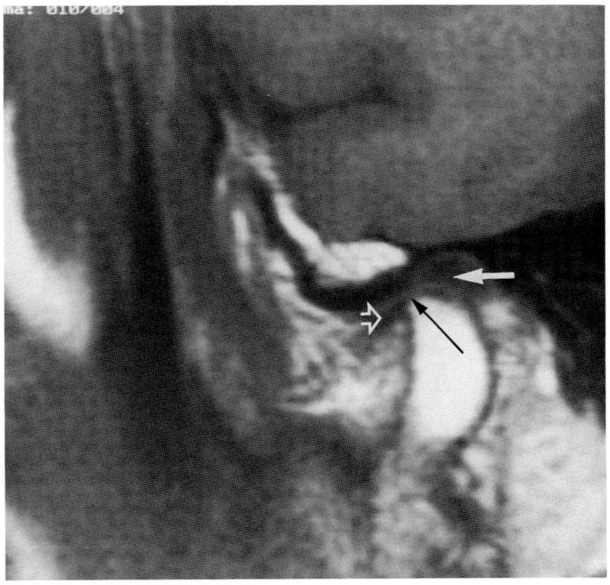

FIGURE 12–18
Normal closed-mouth view of the temporomandibular joint. Note that the intermediate zone (black arrow) *is in contact with the mandibular condyle. The posterior band* (white arrow) *is directly superior to the condyle, while the anterior band* (open arrow) *lies anteriorly.*

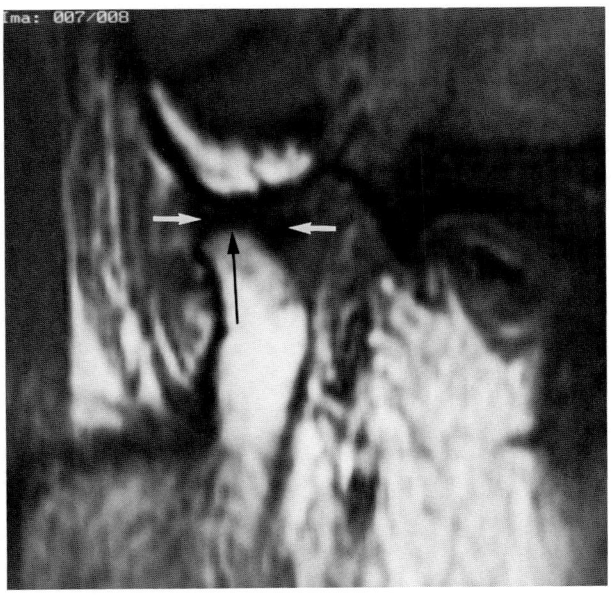

FIGURE 12–19
In the normal open-mouth view, the thin intermediate zone (black arrow) *remains in contact with the mandibular condyle and lies directly superior to it. The disc* (white arrows) *is almost horizontal.*

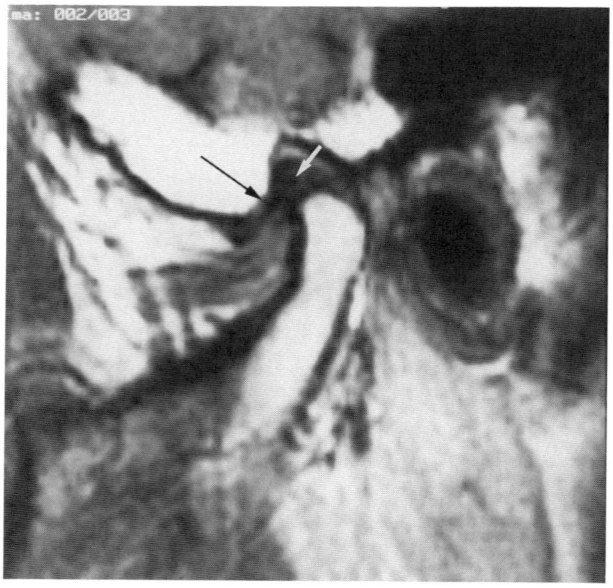

A

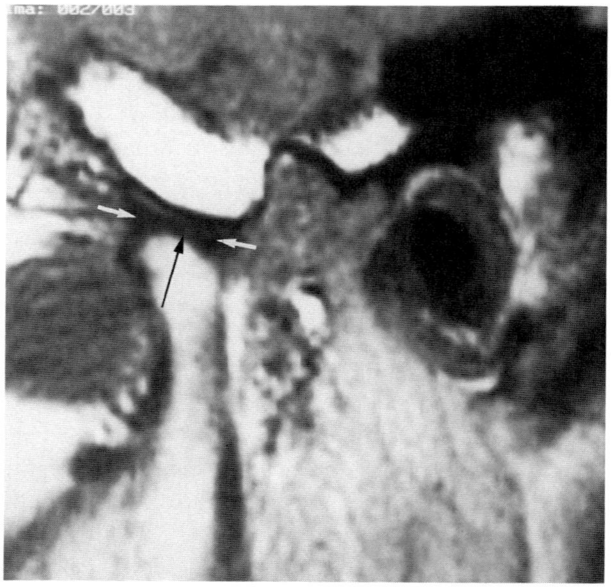

B

FIGURE 12–20
Anterior displacement with reduction. A. The closed-mouth view demonstrates an anteriorly displaced disc, with the mandibular condyle in contact with the posterior band (white arrow) *rather than the intermediate zone* (black arrow). B. *On opening, the condyle slips under the disc and is in proper position, in contact with the intermediate zone* (black arrow). *The disc* (white arrows) *is almost horizontal.*

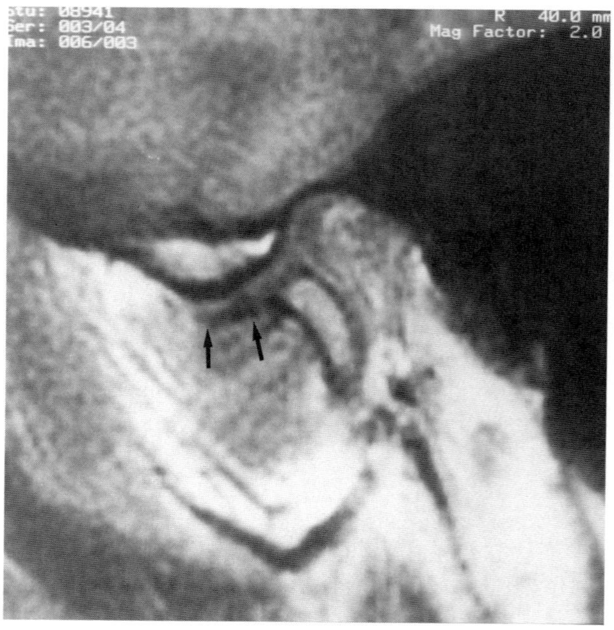

FIGURE 12–21
Anterior displacement without reduction. The articular disc (arrows) *lies anterior to the condyle acting as an impediment to further anterior motion. Note the erosion of the condyle secondary to arthritis.*

leaved 3.0-mm images of both TMJs simultaneously [31,32]. It is difficult for a patient, when lying supine with the mouth open, to keep from swallowing and moving the jaw due to the normal accumulation of saliva. For this reason we use four excitations for the closed-mouth view and only two for the open-mouth sequence, along with a 12-cm rectangular field of view, 256 × 256 matrix, TR of 500 milliseconds, and a TE of 20 milliseconds. A sterile tongue depressor or barrel of a syringe is used to keep the mouth open and immobile. Commercial devices are also available. Acquisition times are 5 minutes, 9 seconds, and 2 minutes, 46 seconds, respectively, for closed- and open-mouth views and together with a 56-second scout view, the examination on both TMJs can be performed relatively quickly and easily. In cases where additional pathologic conditions are suspected (e.g., inflammation, fluid, tumor, etc.), T2-weighted images should be obtained. Three-dimensional gradient-echo imaging of the TMJ has been proposed and does show considerable promise [33,34].

The Shoulder

Unexplained shoulder pain is a common clinical problem that can be attributed to numerous and varied causes. While many different modalities are currently being used to evaluate complaints of shoulder pain, none of them are entirely satisfactory. MRI can now offer noninvasive images of diagnostic quality to evaluate disorders of the articular structures of the shoulder, especially the rotator cuff. For this reason, imaging of the shoulder using surface coil techniques is rapidly gaining acceptance as the examination of choice for evaluating shoulder pain.

Types of surface coils used for shoulder imaging vary significantly and include: single, flat 5-inch circular coils placed posteriorly; dual-coupled, flat 5-inch circular coils, with one placed anteriorly and one posteriorly; and molded single-loop, partially enveloping coils that fit over and around the shoulder. The advantage of the latter two designs is a marked increase in signal-to-noise (S/N) from anterior sections, which lie relatively distant from a single, posteriorly positioned coil.

In addition to observing the usual articular structures, an evaluation of the shoulder must include an analysis of the integrity of the rotator cuff. The imaging planes best suited for MRI evaluation are the axial, paracoronal (along the course of the supraspinatus muscle and tendon), and parasagittal (perpendicular to the direction of the supraspinatus muscle). The anatomy is well defined by T1-weighted images obtained in each plane. The axial images afford excellent visualization of the glenoid labrum, the insertion of the subscapularis tendon, and the biceps tendon. The paracoronal images permit complete visualization of the supraspinatus tendon and muscle, the subacromial and subdeltoid bursae, the acromioclavicular joint, and the superior and inferior aspects of the glenoid labrum. The parasagittal views allow observation of the rotator cuff tendons from anterior to posterior and their relationship to the acromioclavicular joint. T2-weighted images highlight pathologic features; indicate areas of increased signal in neoplasms, fluid collections, and edema; and are most useful if obtained in the coronal oblique plane. All images at our institution are obtained using a 5.0-mm slice thickness with a 1.0-mm interslice spacing, and a 12- to 16-cm field of view.

Damage to the rotator cuff usually occurs as a chronic process, sometimes intensified by an acute injury [35–37]. Rotator cuff injuries have been classified as a continuum of disease states [38]. The vast majority of lesions are thought to result from impingement of bony abnormalities of the acromion or acromioclavicular joint, or of the coracoacromial ligament on the supraspinatus tendon. In the earliest stages, typically observed in athletic young adults with a history of acute misuse of the joint, reversible edema and hemorrhage are present within the supraspinatus tendon. This can progress to fibrosis and tendinitis, and then to partial or complete thickness tears, producing chronic symptoms in older patients. Distinct rupture with visualization of the separated ends of the tendon (stage III) are usually found in patients over 40 years old. Spontaneous rupture of a tendon is rare. Frank tears rarely occur acutely, but rather are thought to require pre-existing damage to at least 30% of the tendon [37]. Complaints arising from an

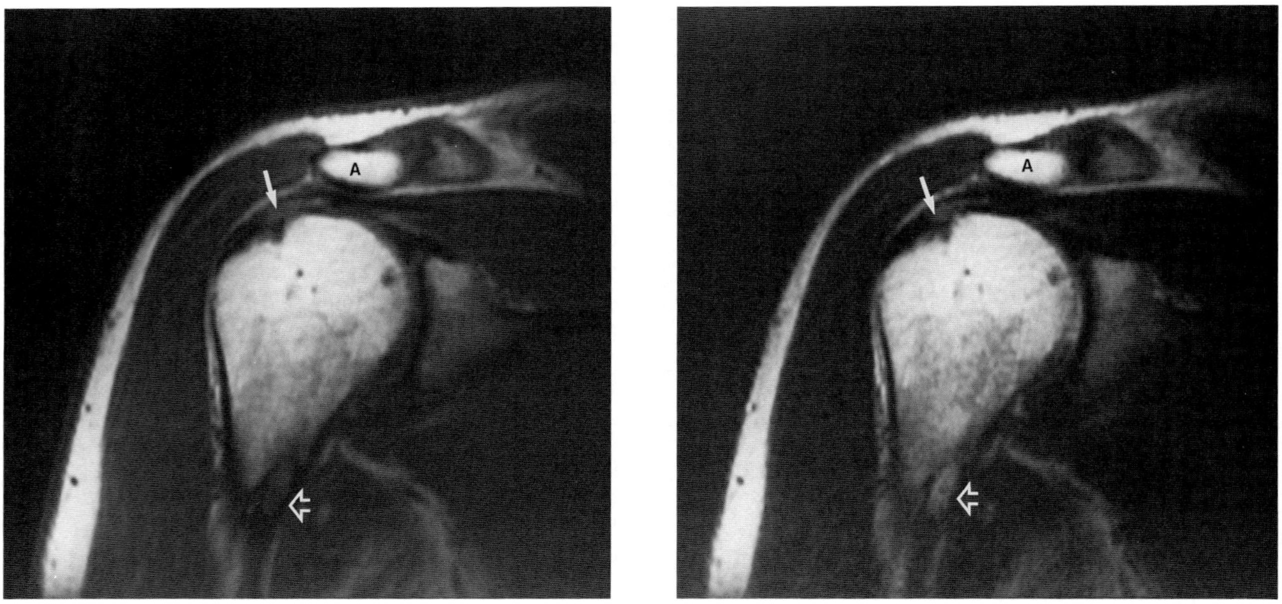

A B

FIGURE 12–22
The high-signal abnormality within the supraspinatus tendon on the T1-weighted image does not increase significantly on the T2-weighted image, consistent with tendinitis. Note impingement on supraspinatus tendon by the acromion (A) *and bone cyst* (open arrow). A. *T1-weighted image.* B. *T2-weighted image.*

acute event are thought to be due to further damage to an already degenerating or damaged tendon.

The MRI appearance of a normal tendon is that of a signal-poor structure—any signal within it is abnormal. Tendinitis rarely demonstrates increased signal within the tendons on the T1-weighted images (Figure 12–22). Acutely inflamed tendons may exhibit increased T2 signal, although in the majority of patients with tendinitis, the signal will not increase significantly. Partial thickness tears can be observed as irregular high-signal areas within the tendon that fill with joint fluid, particularly evident as very bright regions on the T2-weighted images. Full thickness tears are similar in appearance, except that the high signal abnormality extends across the entire thickness of the tendon (Figure 12–23). In a complete rupture, the discontinuity of the tendon is evident, and the supraspinatus muscle retracted (Figure 12–24). The lack of visualization of the subacromial and subdeltoid fat planes is helpful indirect evidence of the presence of a significant tear. Tears of the glenoid labrum are seen as high-signal regions within the normally signal-poor labrum, and are well visualized on gradient-echo images (Figure 12–25).

Classification of changes accompanying rotator cuff abnormalities due to impingement syndrome have been advanced by several different investigators [39–41]. These progress numerically in degree of severity, from a normally contoured tendon with some signal within it (tendinitis), to areas of thinning and irregularity (tendon degeneration), to partial tears (fluid interposed in tendon on T2 image, but some continuity retained), to a complete discontinuity in the tendon. Sensitivities and specificities of as high as 91% and 94%, respectively, have been found for the MR evaluation of rotator cuff tears [40,42]. Kneeland et al. [43] were able to demonstrate 20 of 22 proven tears by MRI in their series of 25 patients.

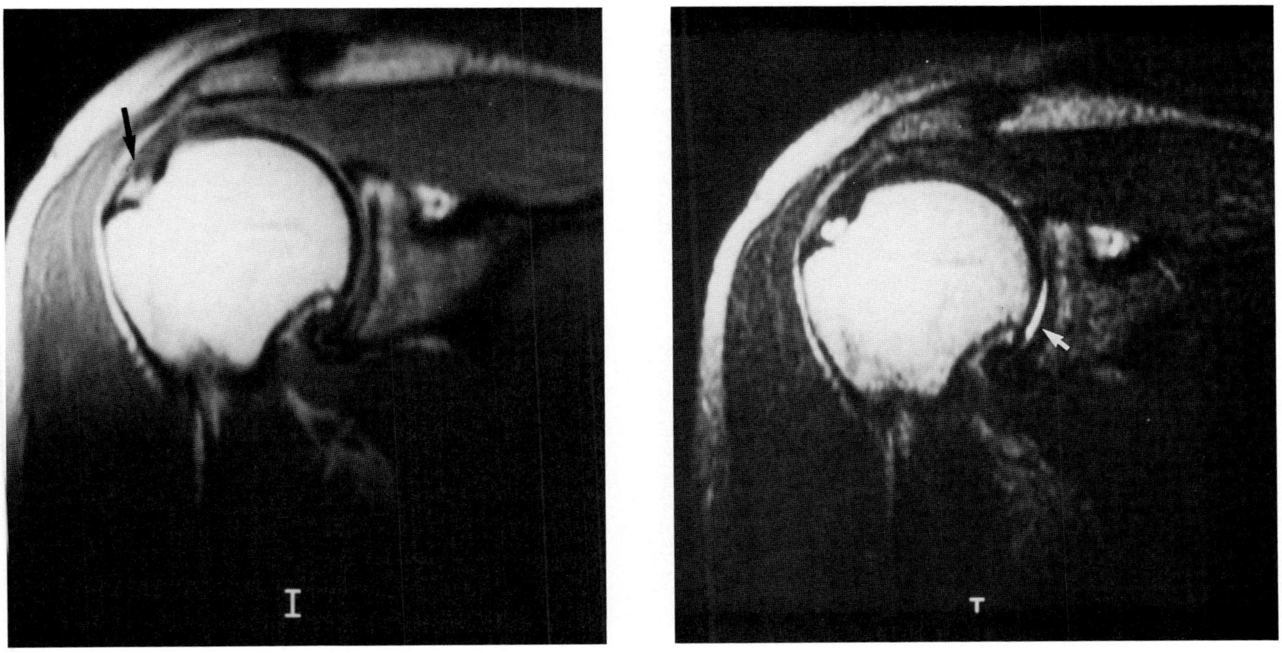

A B

FIGURE 12–23
Full thickness tear of the supraspinatus tendon with increased signal on the T1-weighted image, increasing in signal intensity on the T2-weighted image. Note the small effusion (arrow). A. *T1-weighted image.* B. *T2-weighted image. (Courtesy of Dr. J. Bruce Kneeland, Hospital of the University of Pennsylvania.)*

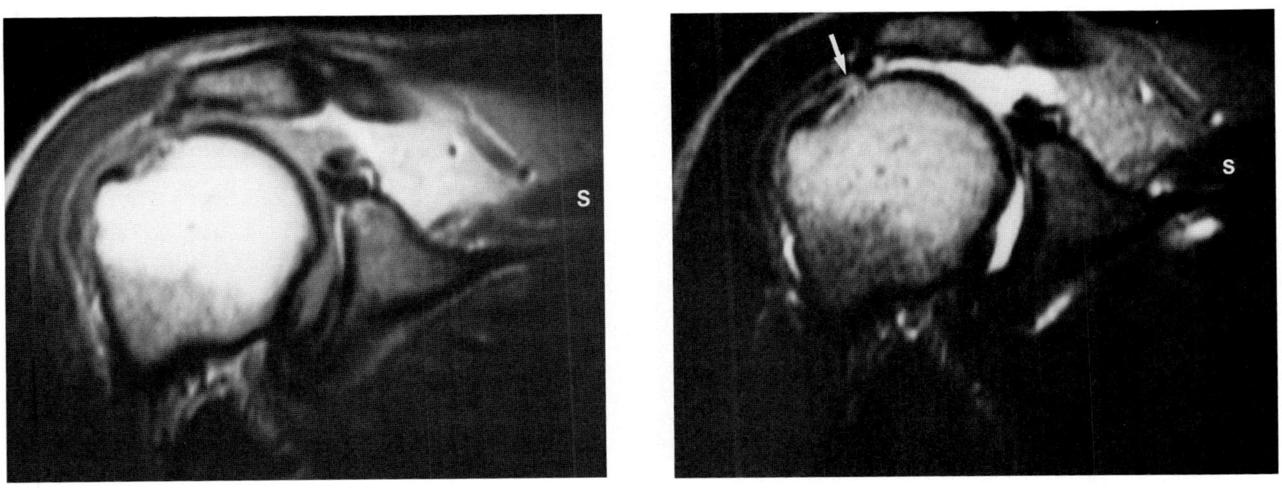

A B

FIGURE 12–24
Complete rupture of the supraspinatus tendon (arrow) *with retraction of the supraspinatus muscle (S).* A. *A T1-weighted image (A) demonstrates medium intensity effusion within joint space. B. The effusion appears bright on the T2-weighted view. (Courtesy of Middletown Radiology Groups, Middletown, NY.)*

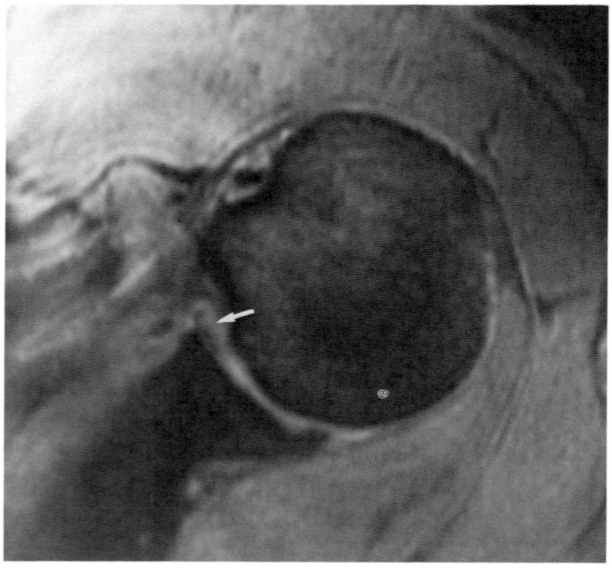

FIGURE 12–25
Tear of the anterior glenoid labrum (arrow) *is seen on this gradient-echo image.*

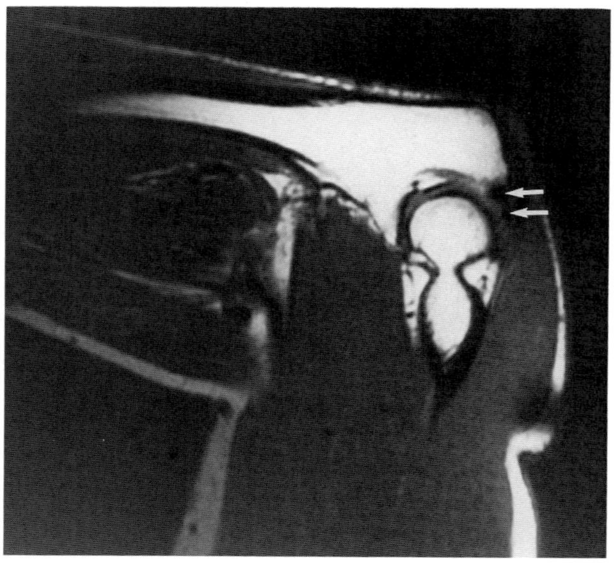

FIGURE 12–26
Osteochondral bodies (arrows) *are seen within the joint space of the elbow.*

The Elbow

MR images of the normal elbow are presented in Chapter 2. It is a relatively simple joint, without any distinctive features [44,45]. The indications for an MRI examination of the elbow are essentially the same as for any joint—high-resolution, noninvasive assessment of either unexplained pain or evaluation of a soft tissue mass. Very few specific conditions are unique to the elbow, and evaluation of this joint is similar to most of the others (Figure 12–26).

The Wrist

The distinctive anatomy of the wrist, compared to the other joints, results in unique problems that can be evaluated quite well by MRI [46,47]. The normally flat and relatively thin contour of the wrist allows excellent visualization of the entire wrist using a small (3- to 5-inch), flat surface coil [48]. The carpal tunnel is a structure found only in the wrist. All of the flexor tendons and the median nerve are enclosed within the tunnel, formed by the carpal bones and the flexor retinaculum [49]. Carpal tunnel syndrome is a painful disorder, most often attributed to compression of or damage to the median nerve as it courses through the tunnel [50]. Multiple, highly varied causes of this compression can readily be detected by MRI, the more common being a soft tissue mass; fluid secondary to synovitis; muscular hypertrophy; or abnormal decrease in size of the tunnel secondary to trauma, osteophyte formation, amyloid deposits, or even congenital variations [51,52]. It may presage the onset of rheumatoid arthritis [53]. Extensor tendon abnormalities and pain in an ulnar nerve distribution can also be easily evaluated by MRI (Figures 12–27, 12–28, and 12–29). The imaging planes chosen would depend on the particular problem at hand. Inspection of the triangular articular disc is best made by obtaining contiguous coronal T1 images through the region. The normal disc (as with the articular disc of the TMJs and the menisci of the knee) should be a triangular region of signal void (Figure 12–30). Abnormalities secondary to trauma or degenerative change can be observed as regions of increased signal within the fibrocartilaginous disc or as changes in the normally triangular contour [54].

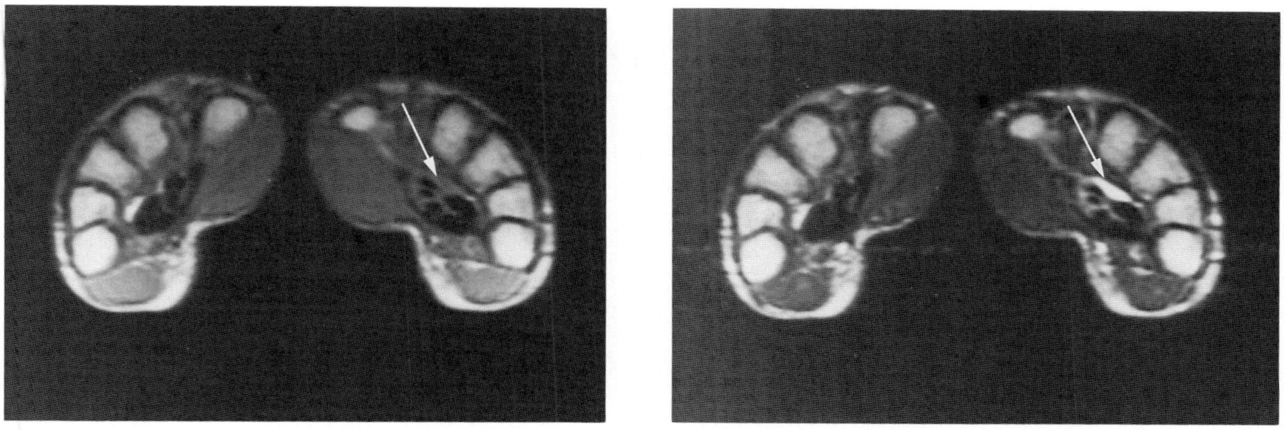

A B

FIGURE 12–27
A ganglion cyst in the dorsal aspect of the carpal tunnel of the left wrist (arrow). A. *T1-weighted image.* B. *T2-weighted image.*

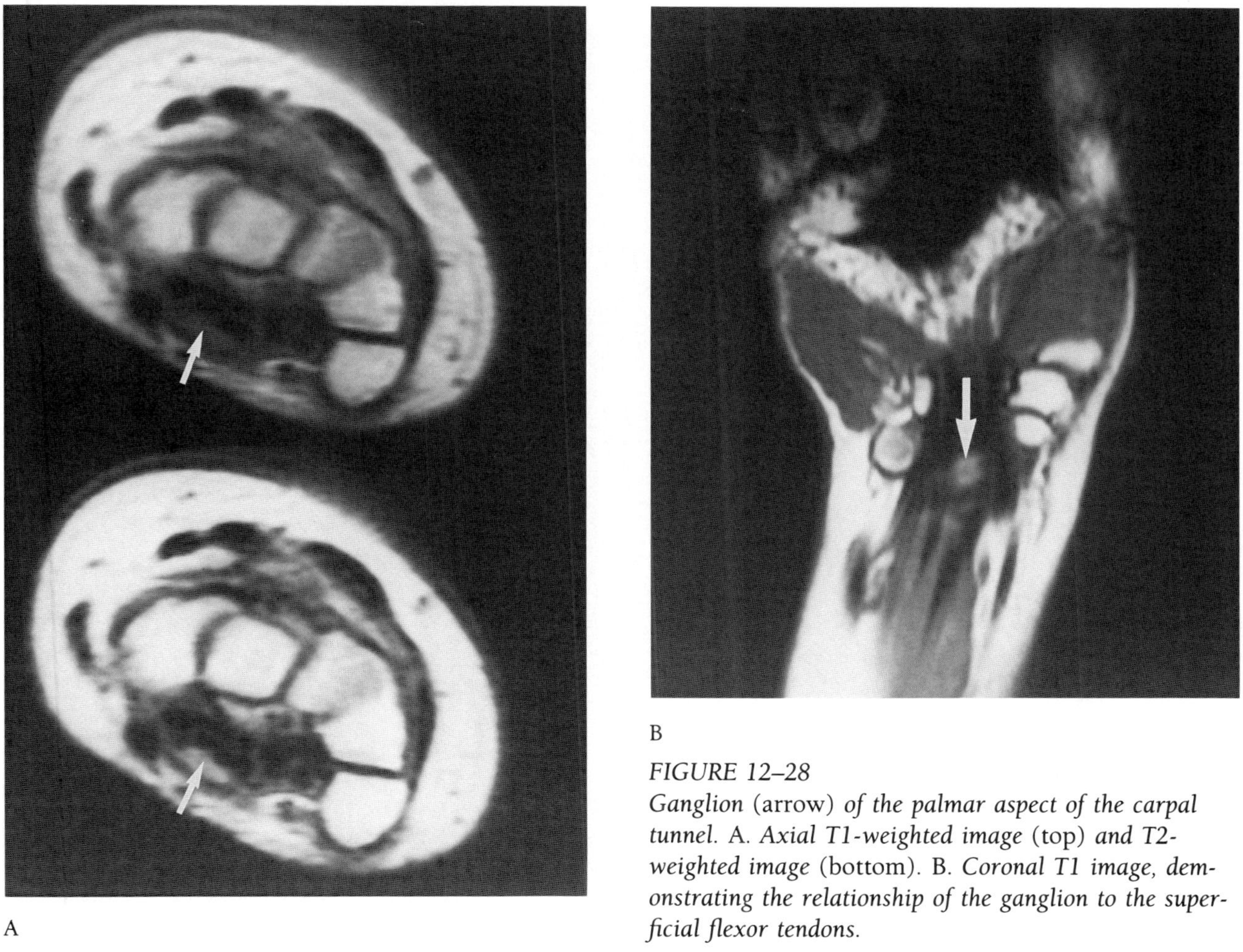

A

B

FIGURE 12–28
Ganglion (arrow) *of the palmar aspect of the carpal tunnel.* A. *Axial T1-weighted image* (top) *and T2-weighted image* (bottom). B. *Coronal T1 image, demonstrating the relationship of the ganglion to the superficial flexor tendons.*

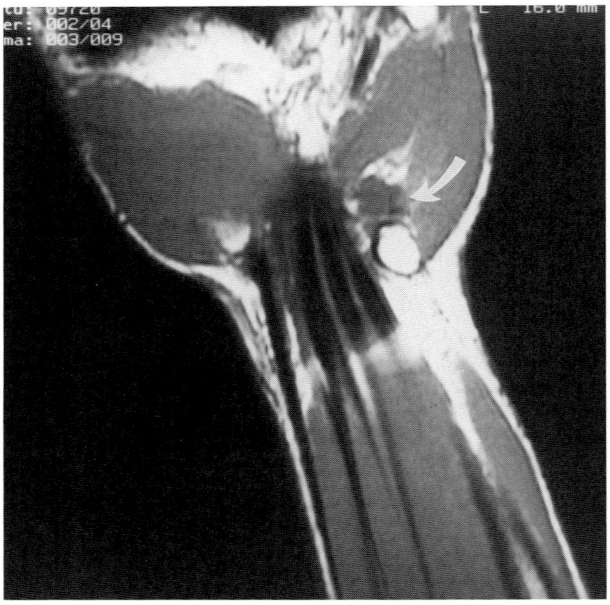

A

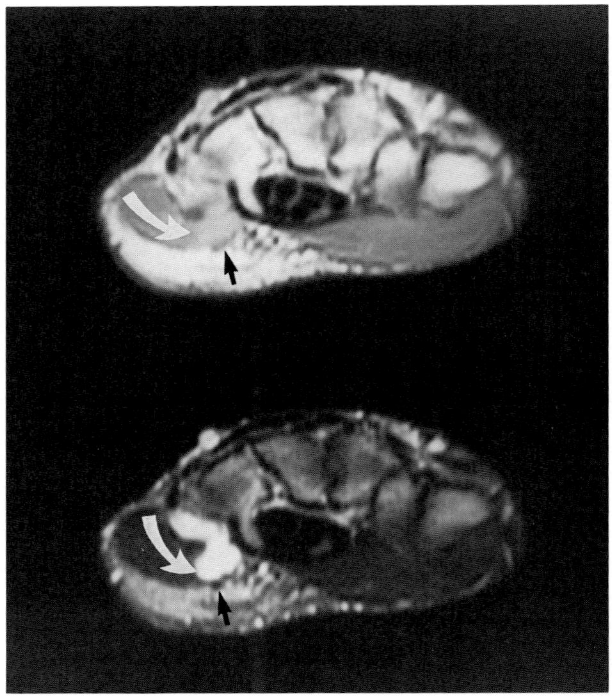

B

FIGURE 12–29
Ganglion (curved arrow) *outside of the carpal tunnel compressing the ulnar nerve* (straight arrow). A. *Proton density image* (top) *and T2-weighted image* (bottom). B. *Coronal T1-weighted image.*

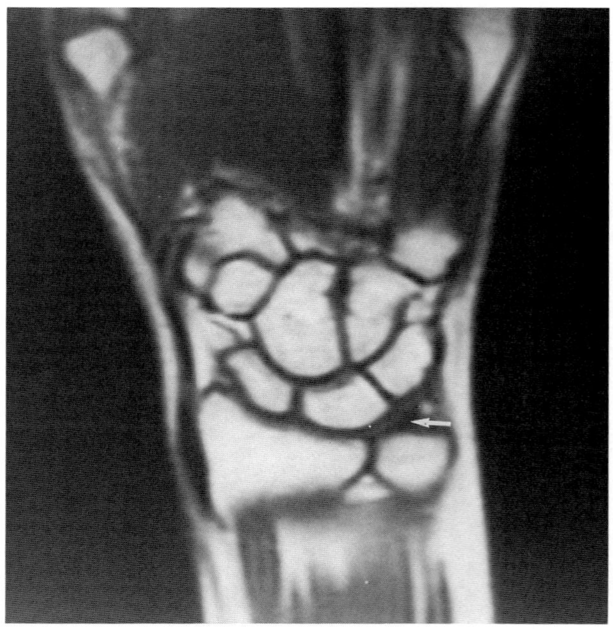

FIGURE 12–30
Normal wrist demonstrating triangular articular disc (arrow).

The Hip

The hip is the only major joint that is not routinely imaged using surface coil techniques. Our routine imaging sequence employs the body coil, obtaining T1-weighted coronal images (TR = 500 milliseconds, TE = 20 milliseconds, 5-mm sections with a 1.5-mm interslice separation) and proton density and T2-weighted axial images (TR = 1500 milliseconds, TE = 20 and 80 milliseconds, with the same slice parameters) using a large enough field of view to visualize both hips (usually 40–48 cm). If, as is usually the case, a pathologic condition is suspected in one particular hip, high-resolution T1 images of that hip (TR = 500 milliseconds, TE = 20 milliseconds) with a small field of view (20–24 cm) are also obtained. The coronal images demonstrate the anatomy very well and allow a comparison of the two hips. The presence of an infiltrative process in the femurs can best be demonstrated on the coronal views, while the axial images permit a better evaluation of joint effusions and other pathologic conditions. Joint space abnormalities and degenerative changes, including subchondral sclerosis and small irregularities secondary to osteophyte formation, can be seen on the high-resolution views, but are usually earlier and more easily identified by routine x-rays.

The most common request for imaging of the hip is to determine the cause of unexplained pain, particularly to rule out osteonecrosis. This is especially true for high-risk patients (e.g., those receiving steroid therapy) for whom MRI examination has been shown to be more sensitive than bone scintigraphy or roentgenograms (Figure 12–31). As indicated previously, T1-

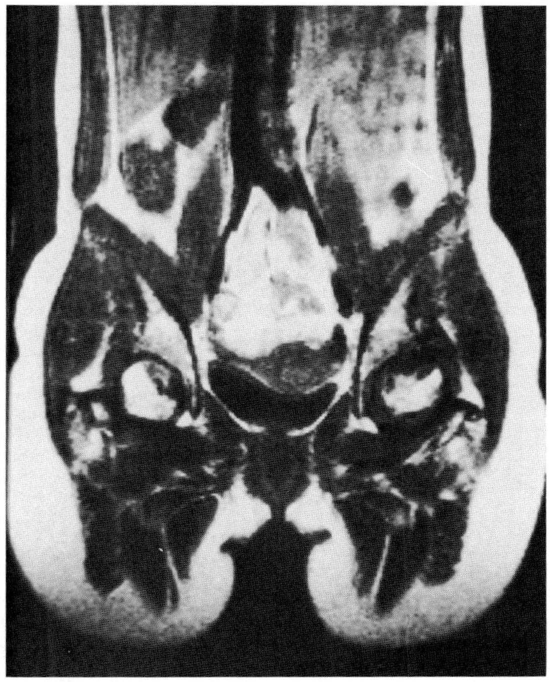

A

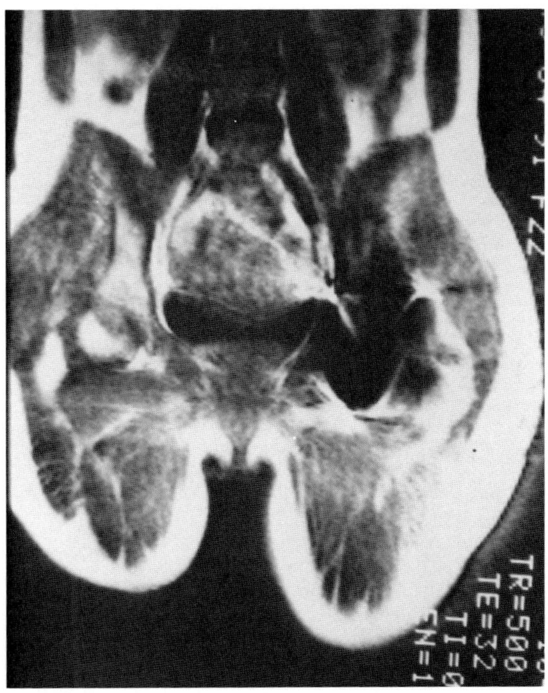

B

FIGURE 12–31 ▶
A 21-year-old woman who receives high-dose steroid therapy for colitis, with known avascular necrosis on the left and new pain on the right. A. *T1-weighted coronal images demonstrate bilateral avascular necrosis, more advanced on the left than the right.* B. *The same patient 12 months later, 11 months after a right core decompression. The coronal T1 image indicates essentially an unchanged pattern in the right femoral head. The left-sided artifact is due to total hip replacement.*

FIGURE 12–32
Legg-Calvé-Perthes Disease in an 8-year-old. A. *T1-weighted coronal images demonstrate abnormally decreased signal from the left femoral head.* B, C. *The same abnormality in T1-weighted and T2-weighted images, respectively, demonstrate a marked signal decrease in the affected femoral head on all pulse sequences. (Courtesy of Middletown Radiology Group, Middletown, NY.)*

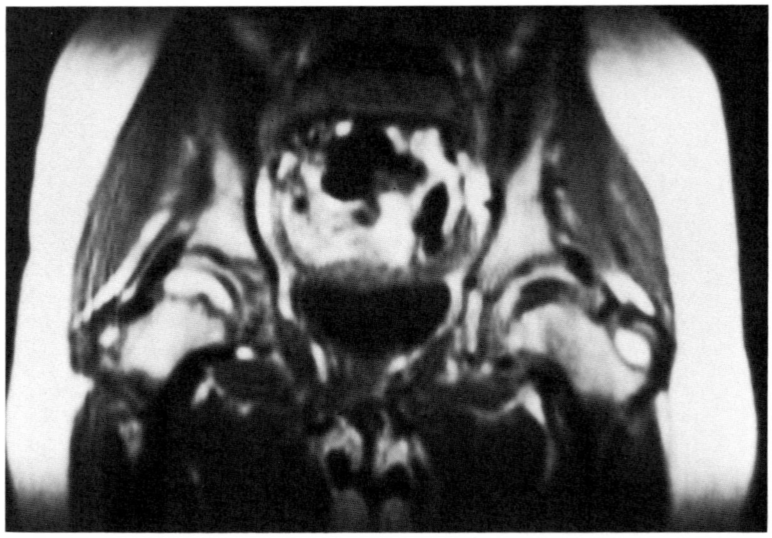

A

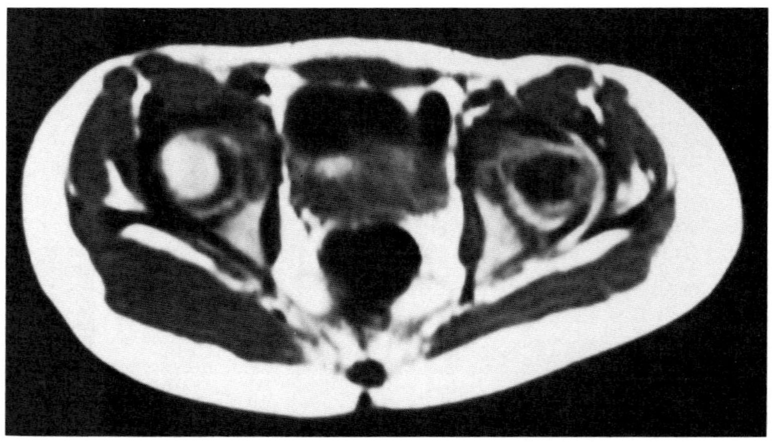

B

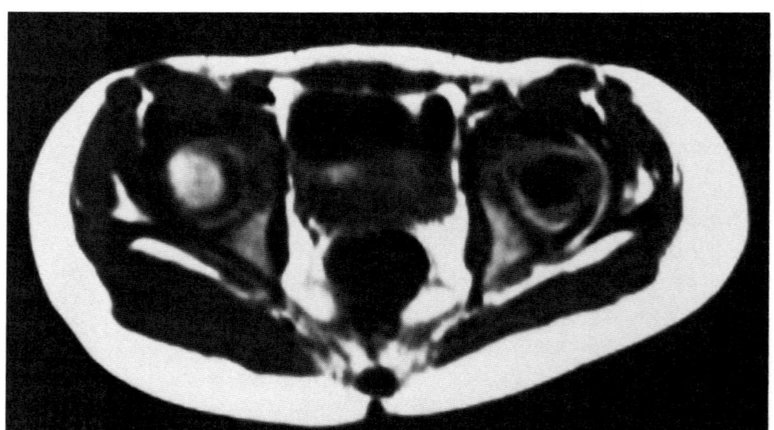

C

weighted images may demonstrate a region of decreased marrow signal, just below the cortex of the weight-bearing surface of the femoral heads, or may exhibit a ring of decreased signal (see Figures 12–14 and 12–15).

The ability of MRI to image osseous, cartilaginous, and soft tissue structures has made it a useful modality for the evaluation of the pediatric hip (Figure 12–32) [55]. Early diagnosis is essential for proper maturation when evaluating congenital hip dislocations [56]. While radiographs only clearly demonstrate calcified structures, and ossification of the femoral head is often delayed, MRI visualizes all of the pertinent structures and permits an early evaluation of this potentially disabling condition [57].

The Ankle

The primary advantage of MRI for the evaluation of the ankle lies in its ability to visualize the clinically important tendons: the peroneus longus and brevis, the tibialis posterior, and the Achilles tendons [58,59]. The Achilles tendon is the thickest in the body, originating from the most inferior portion of the gastrocnemius-soleus muscle complex and inserting into the midportion of the posterior aspect of the calcaneus, just above the calcaneal tuberosity. It can normally be visualized in its entirety on several sagittal sections (Figure 12–33). Chronic inflammation causes an abnormal thickening in the central portion of the tendon, which may be appreciated on the sagittal sections (see Figures 12–4A and 12–5). This chronic change can better be observed on axial sections (see Figure 12–4B), as a loss of the normal concavity of the anterior portion of the tendon [60].

The peroneal, posterior tibial, and flexor tendons can best be evaluated for continuity and hypertrophic changes on axial T1-weighted images [61]. T2-weighted axial images demonstrate inflammatory changes within the tendons and fluid within the tendon sheaths (Figure 12–34). The continuity of these tendons can be followed on sequential T1 axial images to below the level of the malleoli. Further distally, these tendons routinely have multiple insertions, making the determination of a tear at only one of these points of attachment extremely difficult.

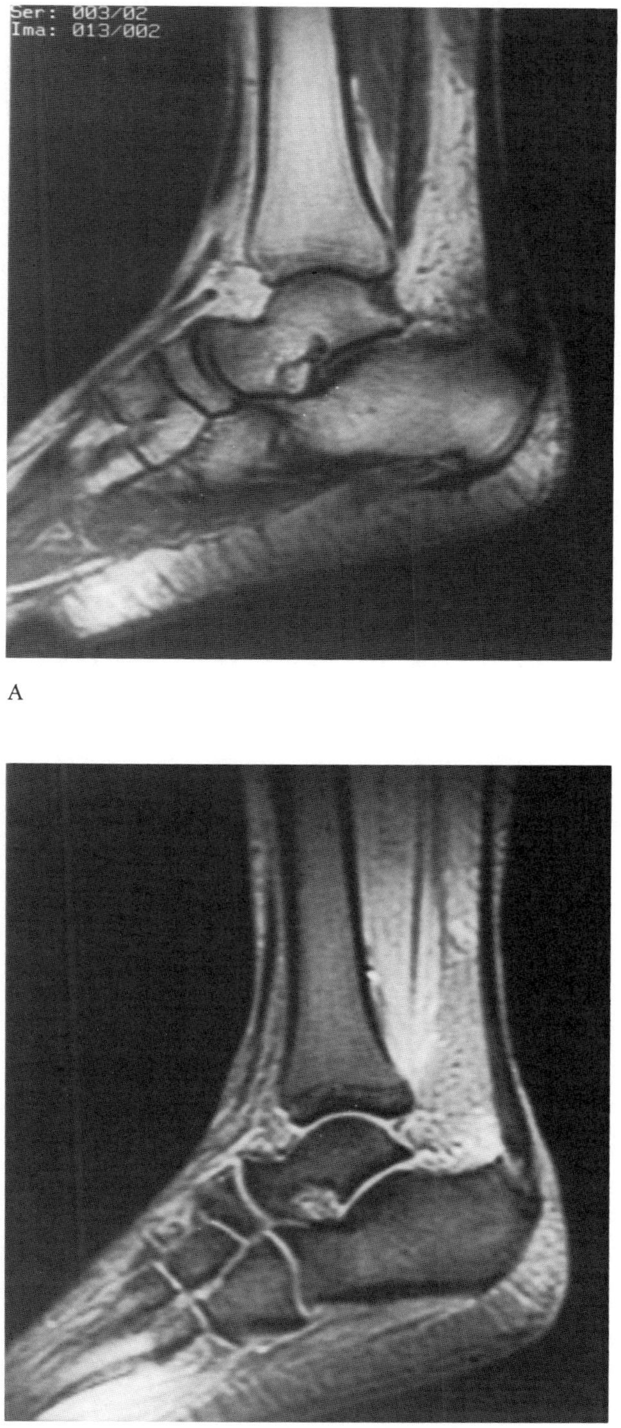

A

B

FIGURE 12–33
Partial avulsion of the Achilles tendon at its calcaneal insertion. A. Spin-echo T1-weighted image. B. Gradient-echo image.

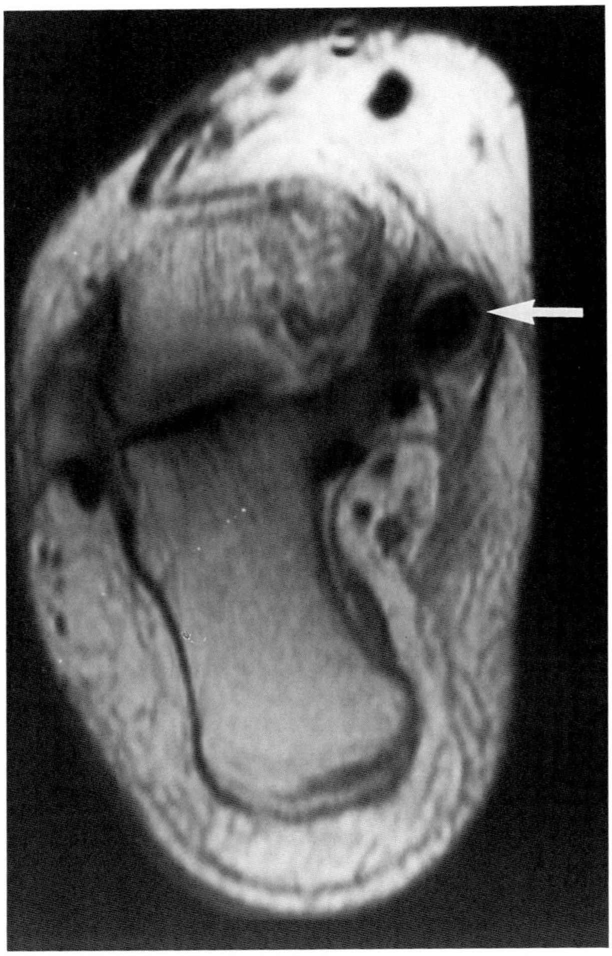

A

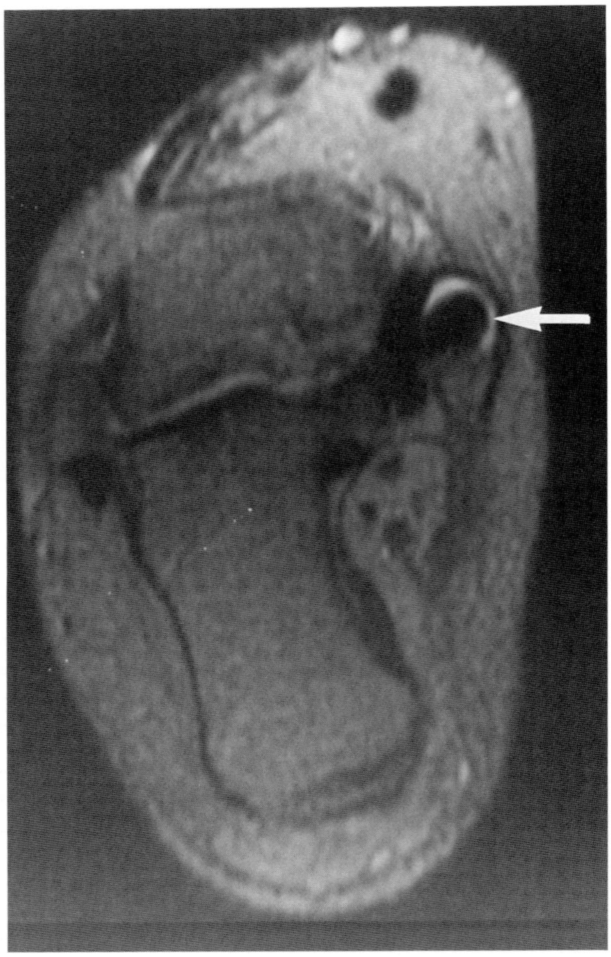

B

FIGURE 12–34
Fluid within the tendon sheath (arrow) *of a hypertrophic tibialis posterior tendon, consistent with tenosynovitis.* A. *T1-weighted image.* B. *T2-weighted image.*

Conclusions

Imaging of the joints using MRI surface coil techniques has expanded the diagnosis of joint abnormalities into a new dimension. Routine visualization of structures incapable of being imaged by other methods now allows definitive diagnoses to be made in an accurate, noninvasive manner. Although imaging strategies are complex and vary considerably for the different joints, as well as for the clinical question, information obtained is unique. New advances in MR technology and additional clinical experience promise to further increase the utility of MRI in joint evaluation.

REFERENCES

1. Axel L. Surface coil magnetic resonance imaging. J Comput Assist Tomogr 1984; 8:381–4.
2. Beltran J, Noto AM, Mosure JC, Weiss KL, Zuelzer WA, Christofordis AJ. The knee: surface-coil MR imaging at 1.5 T. Radiology 1986; 159:747–51.

3. Edelman RR, Shoukimas GM, Stark DD, et al. High-resolution surface-coil imaging of lumbar disk disease. AJR 1985; 144:1123–9.
4. Middleton WD, Macrander S, Lawson TL, et al. High resolution surface coil magnetic resonance imaging of the joints: Anatomic correlation. Radiographics 1987; 7:645–83.
5. Kneeland JB. Practical considerations in the use of surface coils for MR imaging. Appl Radiol 1988; 54–6.
6. Dick DW, Mitchell DG, Burk DL Jr, et al. Effect of chemical shift misrepresentation on cortical bone thickness on MR imaging. AJR 1988; 151:537–8.
7. Bongartz G, Bock E, Horbach T, Requardt H. Degenerative cartilage lesions of the hip—magnetic resonance evaluation. Mag Res Imag 1989; 7:179–86.
8. Deutsch AL, Mink JH, Waxman AD. Occult fractures of the proximal femur: MR imaging. Radiology 1989; 170:113–6.
9. Yao L, Lee JK. Occult intraosseous fracture: detection with MR imaging. Radiology 1988; 167:749–51.
10. Berger PE, Ofstein RA, Jackson DW, Morrison DS, Silvino N, Amador R. MRI demonstration of radiographically occult fractures: what have we been missing? Radiographics 1989; 9:407–36.
11. Beltran J, Noto AM, Herman LJ, Lubbers LM. Tendons: High-field-strength surface coil MR imaging. Radiology 1987; 162:735–40.
12. Yulish BS, Muloputos GP, Goodfellow DB, Bryan PJ, Modic MT, Dollinger BM. MR imaging of osteochondral lesions of talus. JCAT 1987; 11:296–301.
13. Beltran J, Caudill JL, Herman LA, et al. Rheumatoid arthritis: MR imaging manifestations. Radiology 1987; 165:153–7.
14. Semac, MO Jr, Deutsch D, Bernstein BH, et al. MR imaging in juvenile rheumatoid arthritis. AJR 1988; 150:873–8.
15. Yulish BS, Lieberman JM, Newman AJ, Bryan PJ, Mulopulos GP, Modic MT. Juvenile rheumatoid arthritis: assessment with MR imaging. Radiology 1987; 165:149–52.
16. Markisz JA, Knowles RJR, Altchek DW, Schneider R, Whalen JP, Cahill PT. Segmental patterns of avascular necrosis of the femoral heads: early detection with MR imaging. Radiology 1987; 162:717–20.
17. Mitchell DG, Kressel HY. MR imaging of early avascular necrosis. Radiology 1988; 169:281–2.
18. Genez BM, Wilson MR, Houk RW, et al. Early osteonecrosis of the femoral head: Detection in high-risk patients with MR imaging. Radiology 1988; 168:521–4.
19. Glickstein MF, Burk, DL Jr, Schiebler ML, et al. Avascular necrosis versus other diseases of the hip: Sensitivity of MR imaging. Radiology 1988; 169:213–5.
20. Mitchell DG, Rao VM, Dalinka MK, et al. Femoral head avascular necrosis: Correlation of MR imaging, radiographic staging, radionuclide imaging, and clinical findings. Radiology 1987; 162:709–15.
21. Beltran J, Herman LJ, Burk JM, et al. Femoral head avascular necrosis: MR imaging with clinical-pathological and radionuclide correlation. Radiology 1988; 166:215–20.
22. Coleman BC, Kressel HY, Dalinka MK, Schiebler ML, Burk DL, Cohen EK. Radiographically negative avascular necrosis: Detection with MR imaging. Radiology 1988; 168:525–8.
23. Bullough PG, Vigorita VJ. Atlas of orthopaedic pathology with clinical and radiologic correlations. London: Butterworths, 1984: 13.14–13.15.
24. Kottal RA, Vogler JB III, Matamoros A, Alexander AH, Cookson JL. Pigmented villonodular synovitis: a report of MR imaging in two cases. Radiology 1987; 163:551–3.
25. Jelinek JS, Kransdorf MJ, Utz JA, et al. Imaging of pigmented villonodular synovitis with emphasis on MR imaging. AJR 1989; 152:337–42.
26. Guralnick W, Kaban LB, Merrill RG. Temporomandibular joint afflictions. N Engl J Med 1978; 229:123–9.
27. Solberg WK, Woo MW, Houston JB. Prevalence of mandibular dysfunction in young adults. J Am Dent Assoc 1979; 98:25–34.
28. Helms CA, Kaban LB, McNeill C, Dodson T. Temporomandibular joint: morphology and signal intensity characteristics of the disk at MR imaging. Radiology 1989; 172:817–20.
29. Walter E, Huls A, Schmeizle R, Klose U, Kuper K, Kalender WA. CT and MR imaging of the temporomandibular joint. Radiographics 1988; 8:327–48.
30. Schellhas KP, Wilkes CH, Omlie MR, et al. Diagnosis of temporomandibular joint disease: Two-compartment arthrography and MR. AJR 1988; 151:341–50.

31. Wright SM, Wright RM. Bilateral MR imaging with switched mutually coupled receiver coils. Radiology 1989; 170:249–55.
32. Shellock FG, Pressman BD. Dual-surface-coil MR imaging of bilateral temporomandibular joints: Improvements in the imaging protocol. AJNR 1989; 10:595–8.
33. Conway WF, Hayes CW, Campbell RL, Laskin DM. Temporomandibular joint motion: efficacy of fast low-angle shot MR imaging. Radiology 1989; 172:821–6.
34. Schellhas KP, Wilkes CH. Temporomandibular joint inflammation: Comparison of MR fast scanning with T1- and T2-weighted imaging techniques. AJNR 1989; 10:589–94.
35. Nixon JE, DiStefano V. Ruptures of the rotator cuff. Orthop Clin North Am 1975; 6:423–47.
36. Neviaser RJ. Tears of the rotator cuff. Orthop Clin North Am 1980; 11:295–303.
37. Colfield RH. Rotator cuff disease of the shoulder. J Bone Joint Surg [Am] 1985; 67:974–9.
38. Neer CS III. Impingement lesions. Clin Orthop 1983; 173:70–7.
39. Seeger LL, Gold RH, Bassett LW, Ellman H. Shoulder impingement syndrome: MR findings in 53 shoulders. AJR 1988; 50:343–7.
40. Zlatkin MB, Ianotti JP, Roberts MC, et al. Rotator cuff tears: diagnostic performance of high resolution MR imaging. Radiology 1989; 172:223–30.
41. Zlatkin MB, Reicher MA, Kellerhouse LE, McDade W, Vetter L, Resnick D. The painful shoulder: MR imaging of the glenohumeral joint. J Comput Assist Tomogr 1988; 12:995–1001.
42. Evancho AM, Stiles RG, Fajman WA. MR imaging diagnosis of rotator cuff tears. AJR 1988; 151:751–4.
43. Kneeland JB, Middleton WD, Carrera GF, et al. MR imaging of the shoulder: diagnosis of rotator cuff tears. AJR 1987; 149:333–7.
44. Middleton WD, Macrander S, Kneeland JB, Froncisz W, Jesmanowicz A, Hyde JS. MR imaging of the normal elbow: anatomic correlation. AJR 1987; 149:543–7.
45. Bunnell DH, Fisher DA, Bassett LW, Gold RH, Ellman H. Elbow joint: normal anatomy on MR images. Radiology 1987; 165:527–31.
46. Koenig H, Lucas D, Meissner R. The wrist: a preliminary report on high-resolution MR imaging. Radiology 1986; 160:463–7.
47. Zlatkin MB, Chao PC, Osterman AL, Schnall MD, Dalinka MK, Kressel HY. Chronic wrist pain: evaluation with high-resolution MR imaging. Radiology 1989; 173:723–9.
48. Baker LL, Hajek PC, Bjorkengren A, et al. High resolution MRI of the wrist: normal anatomy. Skeletal Radiol 1987; 16:128–32.
49. Zeiss J, Skie M, Ebraheim NA, et al. Anatomic relations between the median nerve and flexor tendons in the carpal tunnel: MR evaluation in normal volunteers. AJR 1989; 153:533–6.
50. Middleton WD, Kneeland JB, Kellman GM, et al. MR imaging of the carpal tunnel: normal anatomy and preliminary findings in the carpal tunnel syndrome. AJR 1987; 148:307–16.
51. Mesgarzadeh M, Schneck CD, Bonakdarpour A. Carpal tunnel: Part I: normal anatomy. Radiology 1989; 171:743–8.
52. Mesgarzadeh M, Schneck CD, Bonakdarpour A, Mitra A, Conway D. Carpal tunnel: Part II: carpal tunnel syndrome. Radiology 1989; 171:749–54.
53. Bullough PG, Vigorita VJ. Atlas of orthopaedic pathology with clinical and radiologic correlations. London: Butterworths, 1984.
54. Golimbu CN, Firooznia H, Melone CP Jr, Rafii M, Weinreb J, Leber C. Tears of the triangular fibrocartilage of the wrist: MR imaging. Radiology 1989; 173:731–3.
55. Johnson ND, Wood BP, Noh KS, et al. MR imaging anatomy of the infant hip. AJR 1989; 153:127–33.
56. Barlow JT. Early diagnosis and treatment of congenital dislocation of the hip. J Bone Joint Surg [Br] 1962; 44:292–301.
57. Johnson ND, Wood BP, Jackman KV. Complex infantile and congenital hip dislocation: assessment with MR imaging. Radiology 1988; 168:151–6.
58. Kneeland JB, Macrander S, Middleton WD, Cates JP, Jesmanowicz A, Hyde JS. MR imaging of the normal ankle: correlation with anatomic sections. AJR 1988; 151:117–23.
59. Noto AM, Cheung Y, Rosenberg ZS, Norman A, Leeds NE. MR imaging of the ankle: normal variants. Radiology 1989; 170:121–4.
60. Quinn SF, Murray WT, Clark RA, Cochran CF. Achilles tendon: MR imaging at 1.5T. Radiology 1987; 164:767–70.
61. Rosenberg ZS, Cheung Y, Jahss MH, Noto AM, Norman A, Leeds NE. Rupture of the posterior tibial tendon: CT and MR imaging with surgical correlation. Radiology 1988; 169:229–35.

13

MRI of the Knee

JOHN A. MARKISZ

MRI of the knee is discussed separately from MRI of the other joints of the body, not because the knee is larger than the rest, but because it is the one area in which the diagnostic capability of MRI has developed most rapidly over the last several years. Even though it is a relatively new procedure, MRI of the knee is already beginning to exert a major influence on the diagnostic workup of knee pain. The development of specialized surface coils permits excellent high-resolution images to be obtained within a reasonable period of time, allowing an accurate, noninvasive evaluation of a wide variety of disorders (including meniscal and ligamentous damage, effusions, popliteal cysts, synovial and articular cartilage abnormalities, soft tissue tumors, and abnormalities within the marrow cavity) [1–5]. The structure and integrity of the intricate and delicate components of the knee are so well suited to surface coil MRI that it is rapidly becoming the modality of choice for the evaluation of knee disorders (Figure 13–1) [6].

MRI of the knee is most commonly used to noninvasively evaluate possible meniscal tears and/or ligamentous damage, for patients who

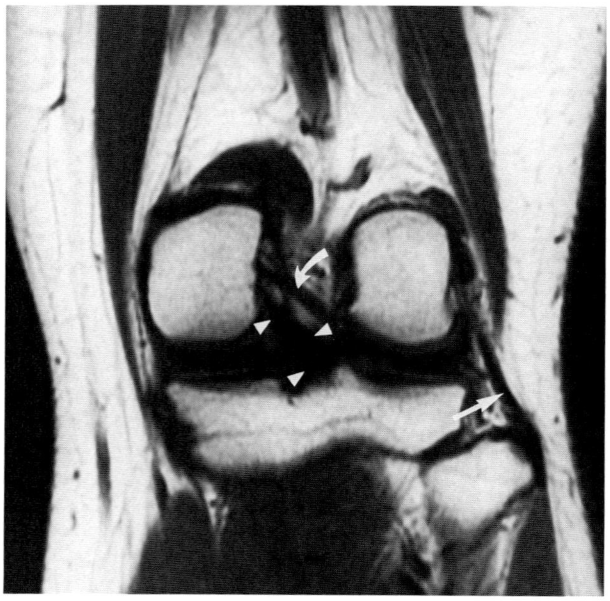

A

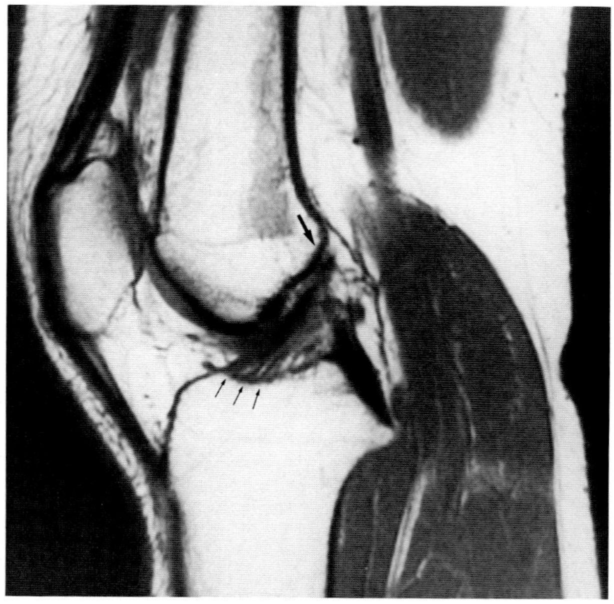

B

FIGURE 13–1

Surface coil evaluation of the knee allows excellent visualization of joint components. A. *Although all knee components are not usually observed together on the same image, this coronal view depicts the course of the posterior cruciate ligament* (arrowheads), *fibular collateral ligament* (arrow), *and ligament of Wrisberg* (curved arrow). B. *A sagittal image demonstrates the femoral* (large arrow) *and tibial* (small arrows) *attachments of the normal anterior cruciate ligament. The patchy region of increased signal above and anterior to the large arrow is active hematopoietic tissue within the marrow; this is normal for a 27-year-old woman.*

have sustained varying types of trauma, or for patients who have unexplained knee pain. Proper visualization of these small detailed structures requires the submillimeter resolution available only with the use of surface coils. The size and shape of surface coils can vary with different manufacturers. Using a wraparound surface coil, we have found the following spin-echo sequences most useful:

1. Single-echo axial, TR of 650 milliseconds, TE of 20 milliseconds, 5-mm slices with 2.5-mm interslice intervals (15 slices, imaging time of 2 minutes, 55 seconds)
2. Dual-echo sagittal, TR of 1500 milliseconds, TE of 20 and 80 milliseconds, 3-mm slices, interleaved (30 slices, 60 images, imaging time of 13 minutes, 26 seconds)
3. Single-echo coronal, TR of 600 milliseconds, TE of 20 milliseconds, 3-mm slices with a 1.0-mm interslice interval (12 slices, imaging time of 2 minutes, 15 seconds)

All are obtained using a 256 × 192 matrix and a field of view of 16 cm, and two data acquisitions. Using these sequences permits the evaluation of almost any type of disorder within the knee joint or surrounding osseous or soft tissue structures in a reasonable period of time. If a patient is allowed to lie comfortably on his or her back, with the knee extended in the surface coil, the knee is naturally rotated externally about 15–20 degrees (Figure 13–2), permitting all of the internal joint structures to be observed. This also permits optimum visualization of the cruciate ligaments, especially the anterior cruciate, with normal, nonobliqued sagittal images. Maintaining the knee in full extension allows the anterior cruciate ligament to remain taut, the

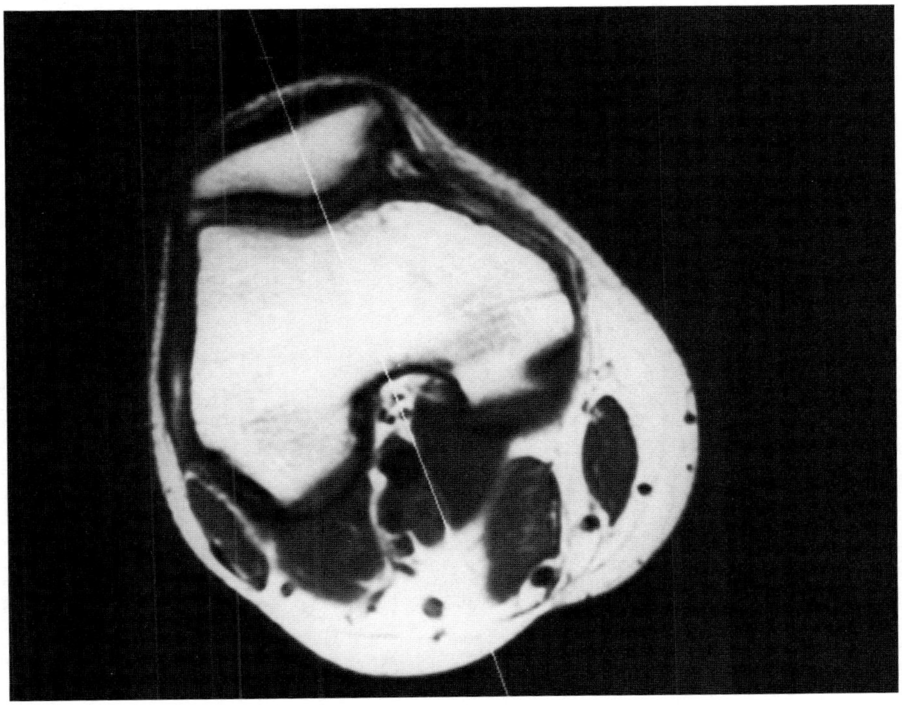

FIGURE 13–2
For the supine patient, a normally relaxed knee usually automatically undergoes external rotation of 10–20 degrees, as observed in this normal axial image. This usually permits visualization of the entire anterior cruciate ligament on a single sagittal image, as in Figure 13–1B.

position in which it is most easily visualized (see Figure 13–1B).

The normal cross-sectional anatomy of the knee is presented in Chapter 2. The usual MRI signal characteristics of the various tissues are identical to what was described in previous chapters. On T1-weighted images, fat within the marrow cavity is bright (slightly less than subcutaneous fat due to the averaging with trabecular bone). Cortical bone, areas of calcification, fibrous tissue, and ligaments, which have no mobile protons to produce a signal, normally appear very dark.

The Menisci

A normal meniscus appears triangular in shape and produces no signal on any of the echo sequences used (Figure 13–3). The menisci are best evaluated using the relatively T1-weighted images in the sagittal plane.

Any signal within the triangular meniscal compartment (as viewed on the sagittal or coronal images) is considered abnormal, and may be due either to a frank tear or to degenerative changes (Figure 13–4) [7–9]. Trauma to or degeneration of the meniscus exposes the macromolecules within it, allowing them to adsorb water molecules. There is a decrease in both T1 and T2 values, resulting in signal emanating from the surface of the exposed macromolecules. The abnormal signal can be better observed on the T1 or proton density (rather than T2) images (Figure 13–5). The views of the menisci can be magnified and photographed using high contrast and narrow window settings to accentuate abnormal changes within the meniscus (see Figures 13–3B and 13–5 through 13–10).

Using arthroscopy as the "gold standard," several studies have found that MRI detection of meniscal tears compares quite favorably with arthrography, with sensitivities, specificities, and accuracies of greater than 90% reported by different observers [6,7,10,11].

We use the grading system proposed by Mink et al. [10,12], whereby signals within the menisci are classified according to severity and clinical relevance. Signal that is contained entirely within the meniscus and does not reach the edge of the meniscal capsule or an articular

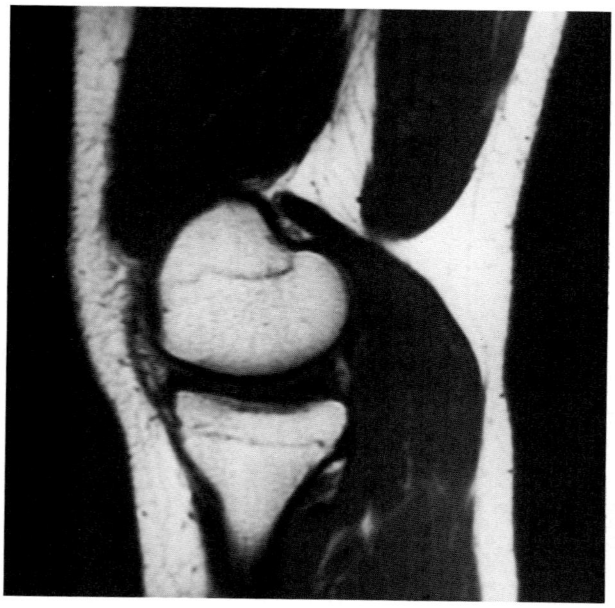

A

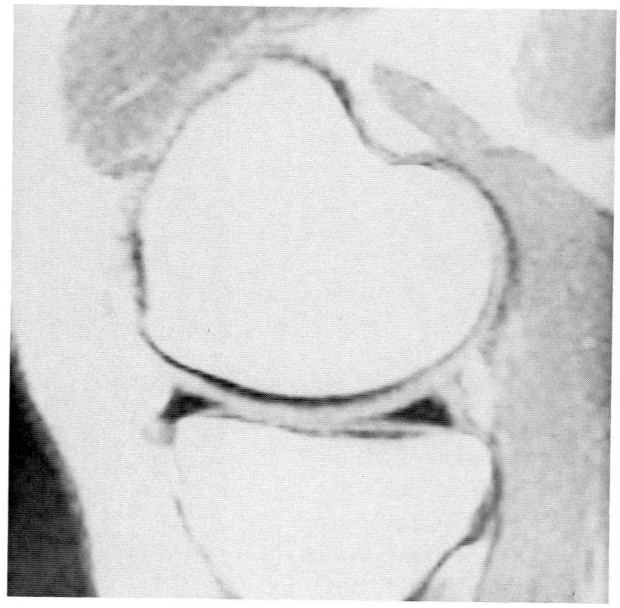

B

FIGURE 13–3
A. The anterior and posterior horns of this medial meniscus are typically normal, with no evidence of signal from within them. B. A high-contrast view confirms the absence of abnormal meniscal signal.

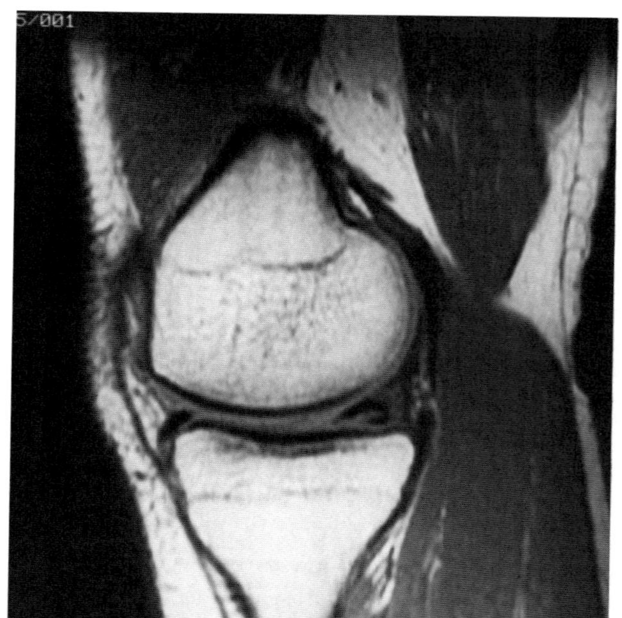

FIGURE 13–4
A large tear can sometimes be observed without the need for enhancement techniques, as shown by this tear of the posterior horn of the medial meniscus.

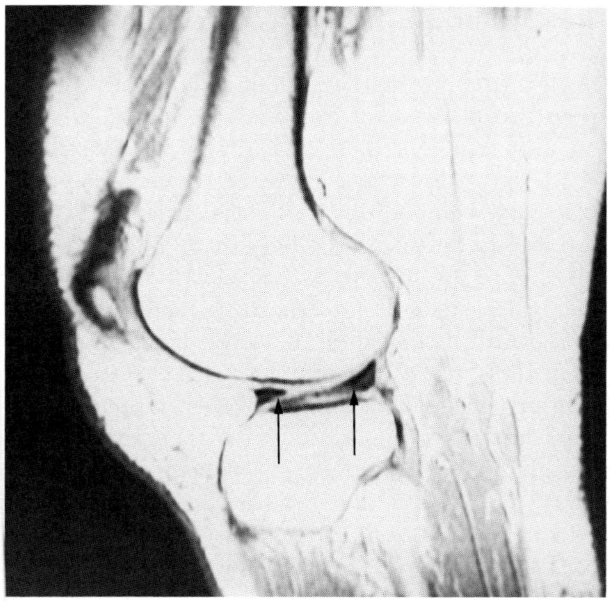

A

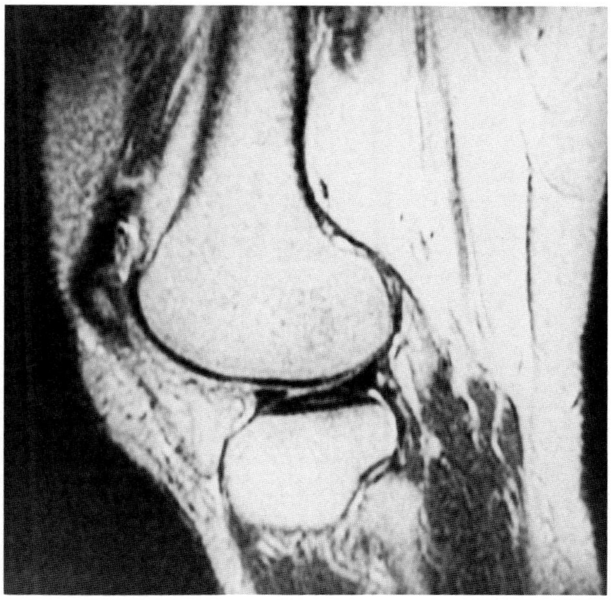

B

FIGURE 13–5
A. *Abnormal signal* (arrows) *is observed in both the anterior and posterior horns of the medial meniscus on this proton density image (TR = 1500 milliseconds, TE = 20 milliseconds), consistent with meniscal tears found at arthroscopy.* B. *On a relatively T2-weighted image (TR = 1500 milliseconds, TE = 20 milliseconds), the abnormalities are not visible.*

surface is classified as grade I signal (Figure 13–6). Grade II signal extends from the capsular surface, but does not reach an articular surface (Figure 13–7). Menisci falling into grades I and II are primarily degenerative in nature, and rarely clinically significant, although these have been described as possible precursors to clinically significant tears. MRI is an extremely sensitive way of detecting meniscal abnormalities, with some grade I and II findings being consistent over time, and some progressing to grade III. Horizontal grade II signals often follow the course of the intrameniscal blood supply, which may actually be the source of the signal observed in some cases.

Signal within a meniscus that extends to an articular surface is considered a grade III abnormality (see Figure 13–4). This classification can be further subdivided into A and B segments. Grade III A involves intrameniscal signal that appears linear or curvilinear (Figure 13–8). The thicker the signal, the more severe the tear.

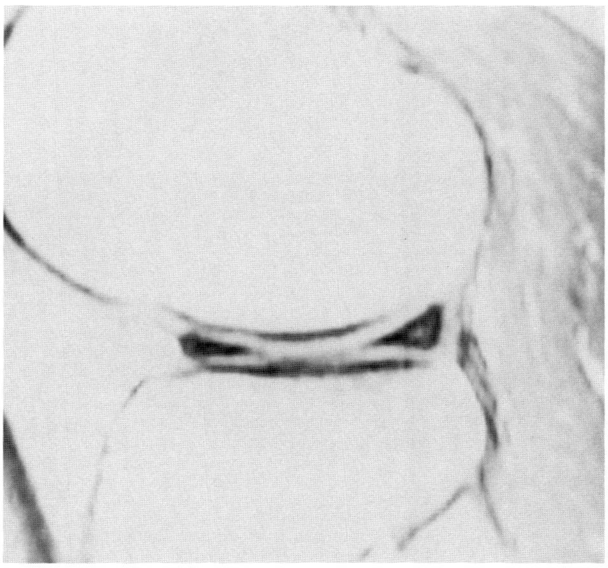

FIGURE 13–6
Grade I signal is confined totally within the signal-poor meniscus, and does not extend to an articular or capsular surface.

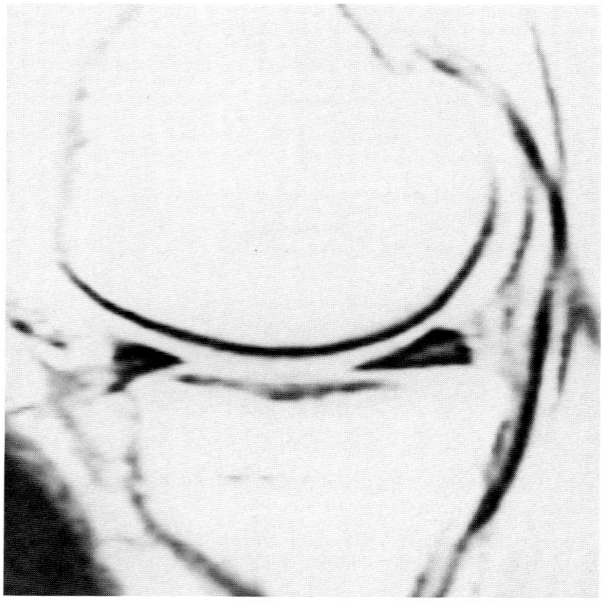

FIGURE 13–7
Grade II signal extends horizontally from the edges of the capsule without reaching an articular surface in both the anterior and posterior horns of this medial meniscus.

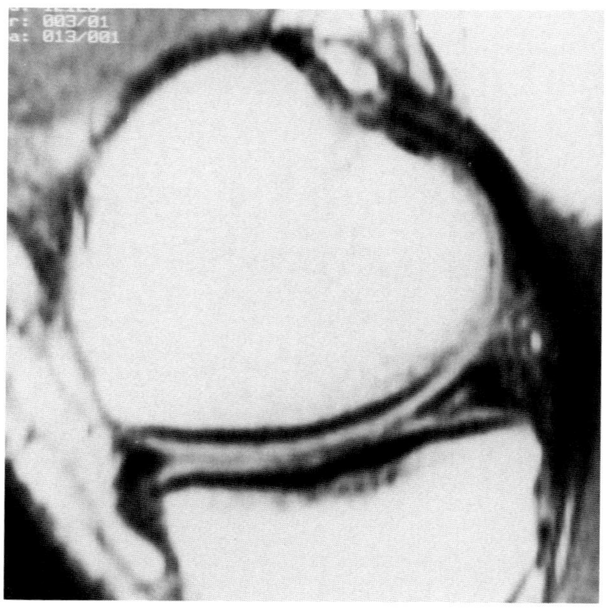

FIGURE 13–8
Grade III A signal is linear or curvilinear and extends to an articular surface.

Small, thin grade III A signal can sometimes be observed in knees, or menisci, that are only minimally symptomatic. These abnormalities are observed to be constant over several months. On occasion, some of these tears may not be visualized on arthroscopy, especially if they communicate with the tibial articular surface. This can lead to criticisms of false-positive MRI findings, since arthroscopy is currently considered the "gold standard" for evaluating meniscal injuries. Arthroscopy itself, however, has demonstrated less than a 70% accuracy when evaluating inferior (tibial) surface tears [13,14]. It appears that this category may illustrate the excellent sensitivity of MRI in detecting meniscal tears. It would appear that the exquisite sensitivity of MRI in the detection of meniscal abnormalities will eventually lead to its acceptance as the new "gold standard."

A grade III B abnormality appears as an irregular, complex signal and indicates a significant tear (Figure 13–9). The intrameniscal signal should always be observed on at least two consecutive 3-mm sections, to minimize confusion with partial volume (or averaging) artifact. Employing interleaved slicing techniques ensures that the entire meniscus is imaged and lessens the possibility that an abnormality may be missed.

The contour and expanse of the meniscus should also be evaluated. The triangular shape of the meniscus should always be preserved. Blunting of the free edge or a large tear obliterating the surface can sometimes produce a quasi–normal-appearing meniscus, with the abnormality visible only on closer inspection (Figure 13–10). The meniscus should also almost completely cover the articular surface of the tibia [15]. Failure to observe this is indicative of meniscocapsular separation (Figure 13–11). This type of tear more often involves the more secured, less mobile medial meniscus [16], but can also be observed in the lateral meniscus (Figure 13–12). Fluid between the posterior horn and the joint capsule is positive evidence of a meniscocapsular separation [17]. This area is difficult to evaluate arthroscopically, and is not often observed without MRI. Time and rest usually heal this well-vascularized region.

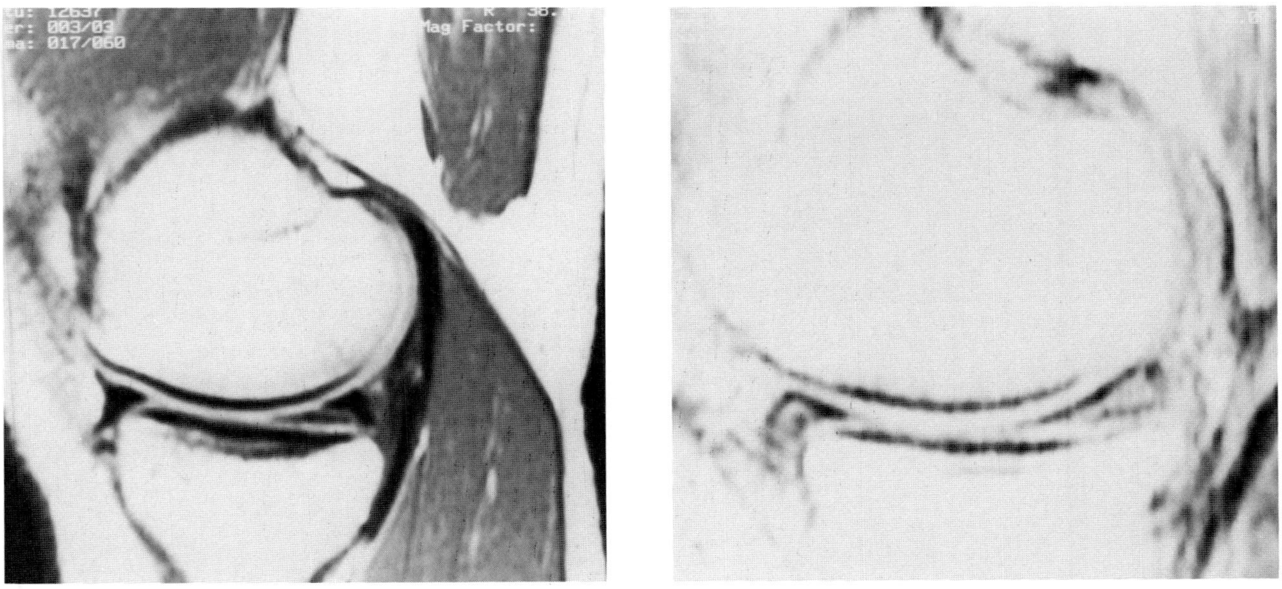

A B

FIGURE 13–9

A. A grade III B curvilinear signal within the posterior horn is observed to communicate with the articular surfaces of both the femur and the tibia. B. Complex grade III B signal extends to the tibial articular surface.

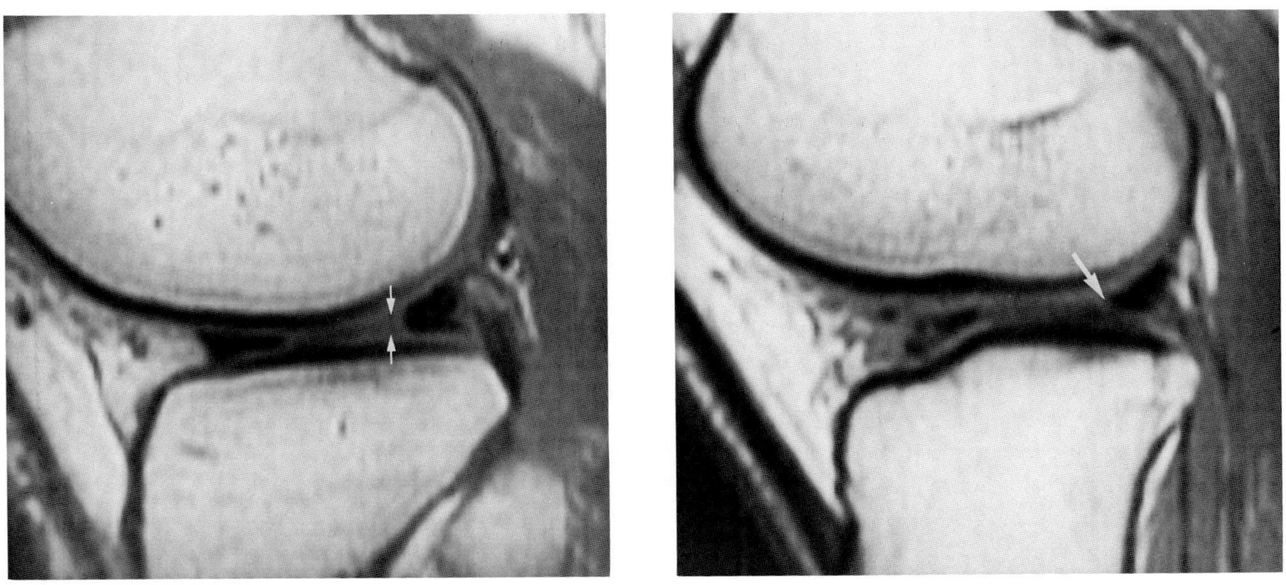

A B

FIGURE 13–10

A. The normally sharp apex of the meniscal triangle is blunted (arrows), *consistent with a tear of fraying of the free edge. B. A complex tear of the anterior horn of the lateral meniscus is shown. Although the posterior horn appears triangular in shape, the apex is not pointing through the joint, but rather up toward the femur. Closer inspection reveals a tear* (arrow) *completely through the posterior horn.*

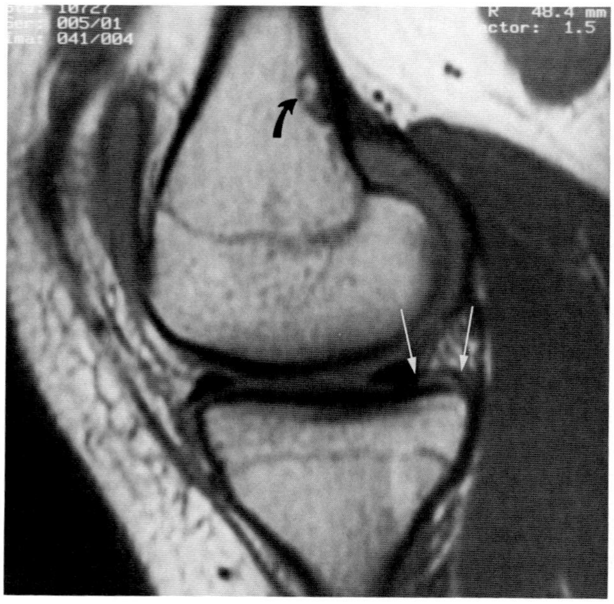

FIGURE 13–11
A large area of the posterior tibial articular surface is left uncovered by the stationary medial meniscus (arrows), *consistent with a medial capsular separation. The curved arrow indicates the ring sign of a small bone infarct.*

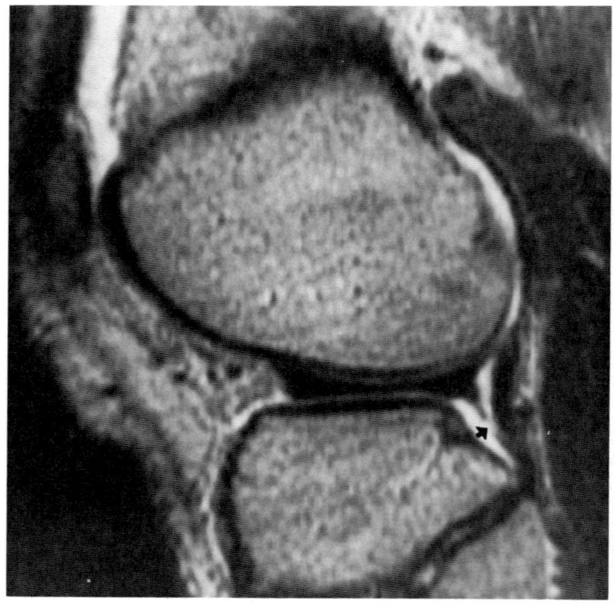

FIGURE 13–12
In this T2-weighted image, increased signal posterior to the meniscus due to fluid allows visualization of a tear (arrow) *of the inferior fascicle of the popliteal tendon.*

The same set of criteria apply when evaluating the meniscus of a patient after partial meniscectomy, except that after surgery the configuration is no longer necessarily triangular [18]. The signal void within the remaining portion of the resected meniscus should remain, and abnormalities as described above can also represent meniscal tears in the postoperative patient (Figure 13–13). After a total meniscectomy, small areas of decreased signal within the joint space may represent an incompletely resected tear or residual fragments (Figure 13–14).

A lateral discoid meniscus is estimated to occur in 1–5% of patients, from surgical and arthrographic series. The discoid shape, as opposed to the normal C shape, results in an incomplete motion of the meniscus on the tibial surface, which produces a limitation on normal mobility. The hypertrophied meniscus can best be identified by the failure to observe a tapered meniscus on sagittal views, with totally separate anterior and posterior horns (Figure 13–15) [19]. There is an increased incidence of meniscal tears observed with a discoid meniscus, as well as abnormalities in the fibular head, femoral condyles, and lateral malleolus. A medial discoid meniscus is rarely found.

Care must be taken not to mistake normal structures for meniscal tears [16]. The transverse ligament joins the anterior horns of the medial and lateral menisci. It is surrounded by fat and its lateral insertion can often simulate a grade III signal (Figure 13–16). There is a wide individual variation in thickness, and congenital absence is not uncommon. The popliteal tendon extends to the joint capsule and attaches to the posterior lateral meniscus at two different points, with the superior and inferior fascicles. Similarly, this can also easily be confused with a large tear. It is important to realize where these structures insert. If there is any doubt about this type of abnormality, it can usually be clarified by ob-

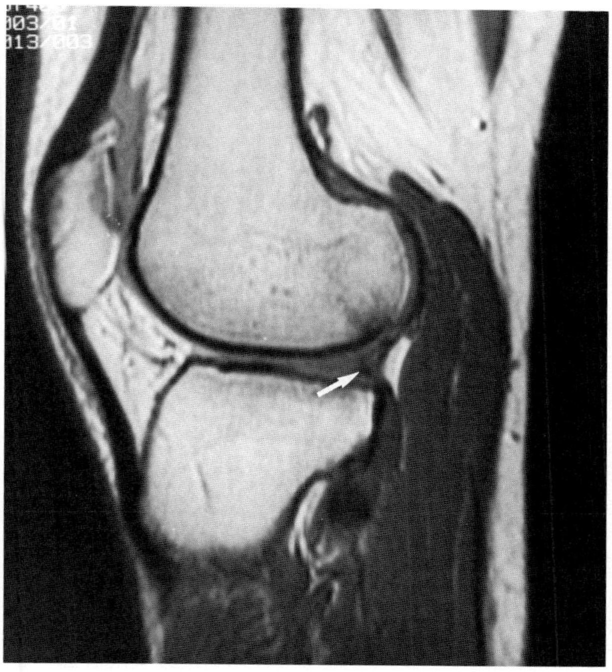

FIGURE 13–13
A complete meniscectomy. The small area of decreased signal (arrow) *is the popliteal tendon, not residual meniscus.*

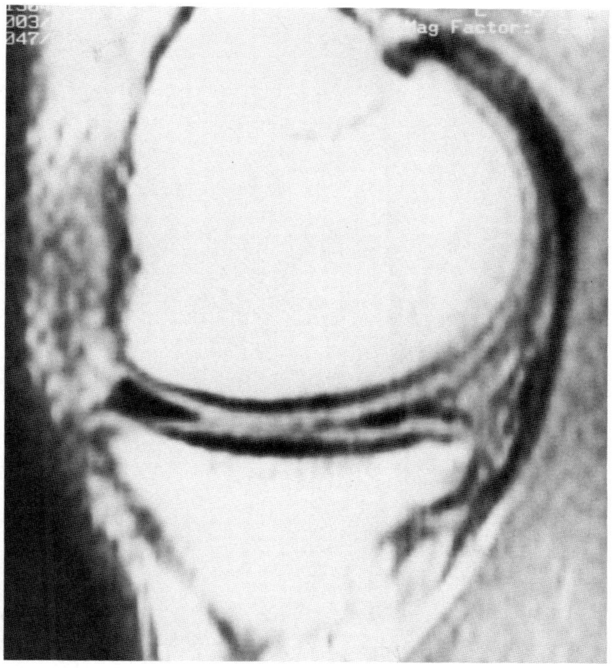

FIGURE 13–14
Arthroscopic surgery has decreased the height of the posterior horn of the medial meniscus. Signal within the meniscus is abnormal and consistent with either new or incompletely resected tears.

serving sequential sagittal sections (Figure 13–17).

The use of two-dimensional and three-dimensional gradient-echo images has been proposed as a preferred method to evaluate the menisci, since meniscal tears can be readily demonstrated and imaging time can be significantly decreased, depending on the particular imaging system involved [20–23] (Figure 13–18). Our experience at 0.5 T, 0.6 T, and 1.5 T has been that significant findings such as effusions, inflamed ligaments and tendons, and popliteal cysts can often be overlooked on such studies. The lack of a true T2 image may prevent the evaluation of unsuspected pathologic conditions; a T2 image should always be obtained. Gradient-echo images are also much more susceptible to magnetic distortion by surgical clips and other surgically implanted objects (Figure 13–19). Since quality T1- and T2-weighted spin-echo images covering the entire joint and surrounding tissues can be obtained in less than 25 minutes at 1.5 T (room utilization time), the need for replacement by gradient-echo images appears questionable. In addition, although it can be contested that meniscal tears are more easily observed on the gradient-echo images, in our experience a significant number of false-positive studies have been found [24]. Additional controlled studies are currently needed to properly evaluate the utility of this type of image (Figure 13–20).

The Ligaments

POSTERIOR CRUCIATE LIGAMENT

The posterior cruciate ligament is seen as a thick black band extending from the lateral side of the medial femoral condyle to the posterior intercondylar region of the tibia (Figure 13–21). The ligament guards against hyperextension, or abnormal posterior displacement of the tibia, rela-

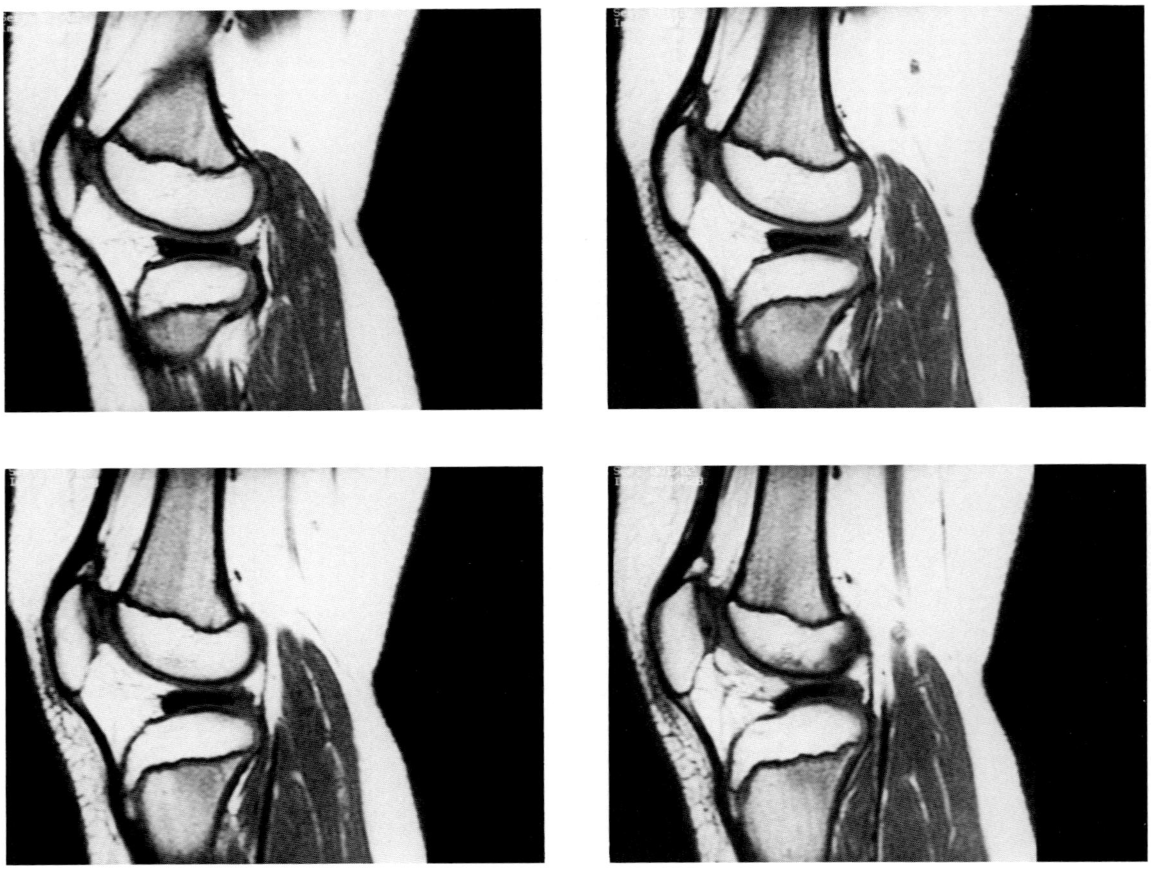

A

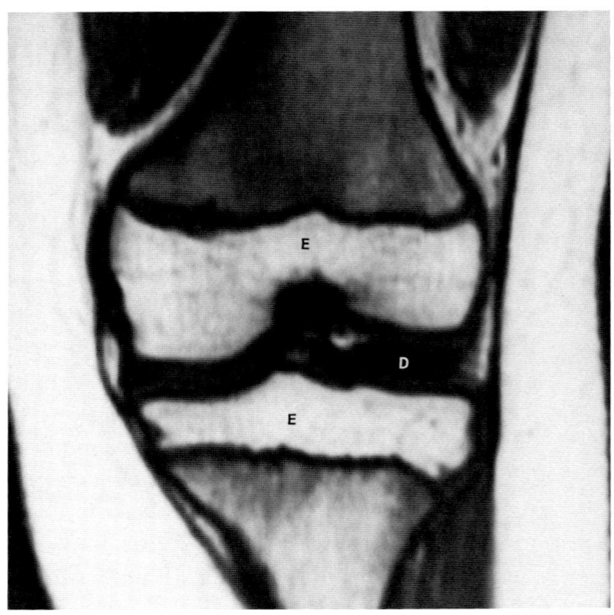

B

FIGURE 13–15
A. A discoid meniscus, as observed an 8-year-old girl, viewed in consecutive sagittal sections, demonstrates a thickened structure with no separation into distinct anterior and posterior horns. The increased signal posteriorly corresponds to a large meniscal tear. B. A coronal view illustrates just how thick this abnormal disc (D) is relative to the normal medial meniscus. The normally increased fat content of the epiphyses (E) in children is sometimes also observed in adults.

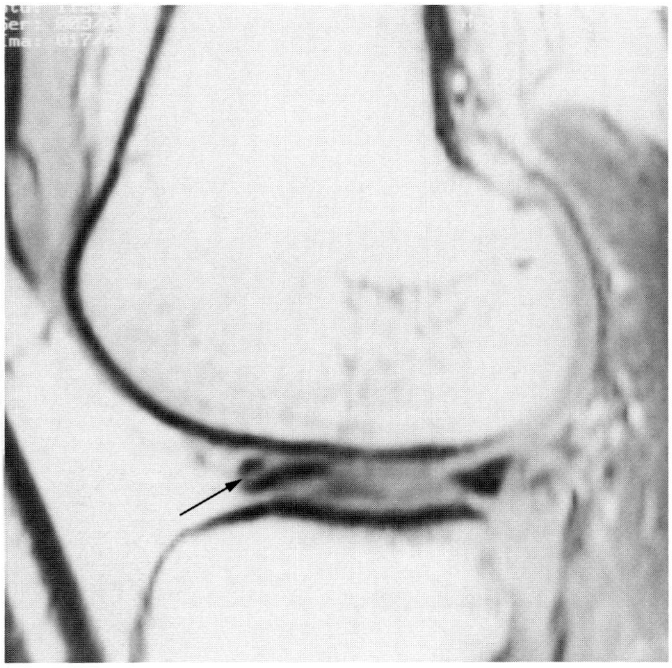

A

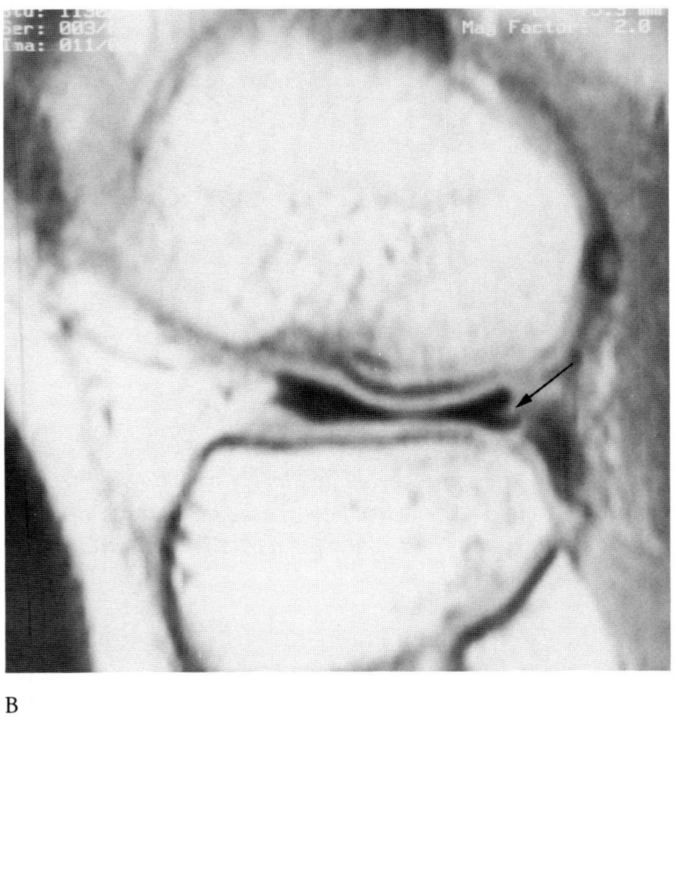

B

FIGURE 13–16

A. *The apparent abnormality* (arrow) *in the anterior horn of the lateral meniscus is due to the attachment of the transverse ligament.* B. *The noticeable irregularity in the posterior horn of the lateral meniscus* (arrow) *is due to the attachment of the popliteal tendon.*

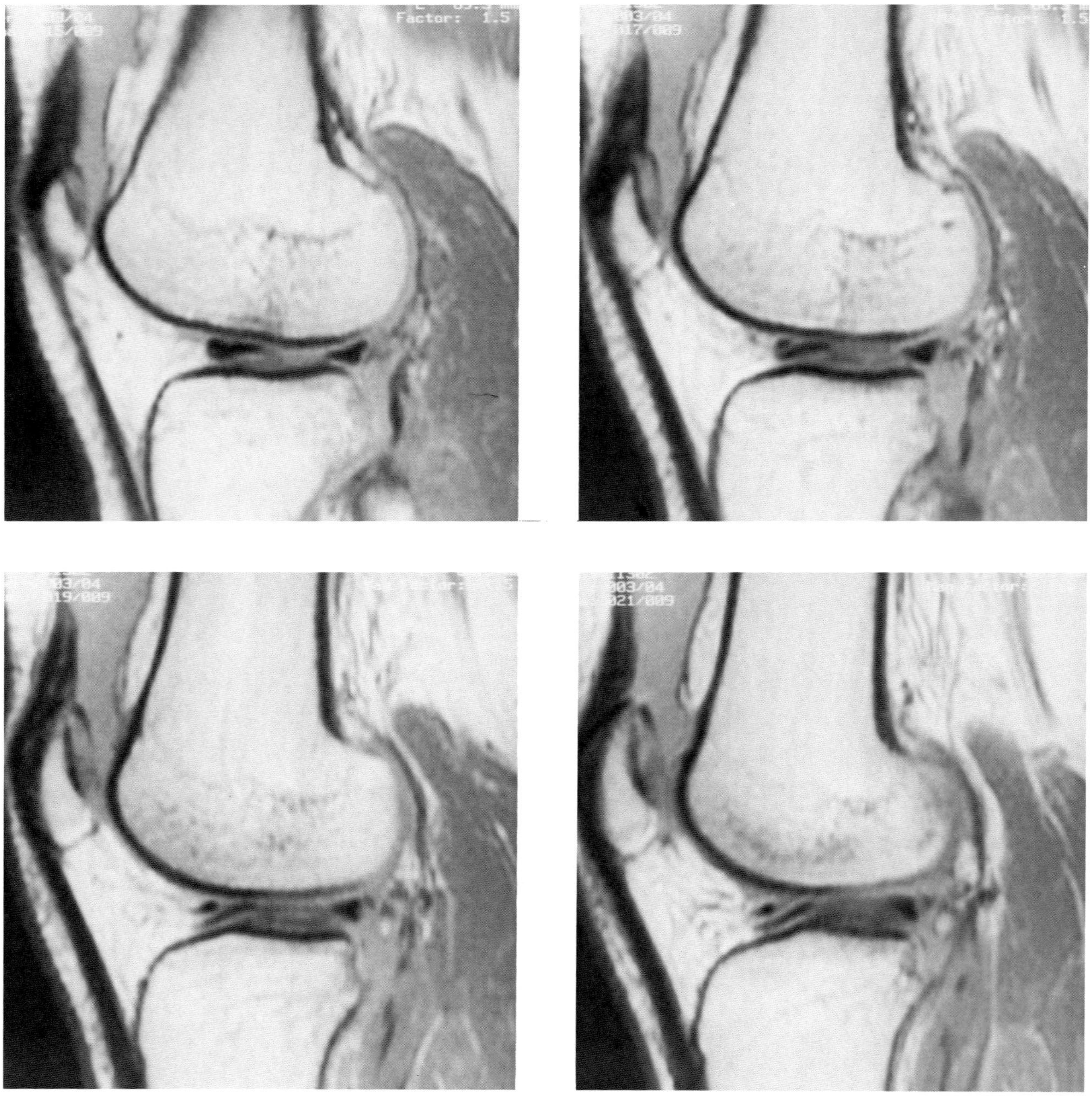

FIGURE 13–17

A determination of a true tear versus a ligament or tendon insertion often can be evaluated better by following the abnormality sequentially on consecutive sections. Four consecutive 3-mm sections demonstrate the appearance of the transverse ligament as it attaches to the anterior horn of the lateral meniscus and continues medially toward the medial meniscus.

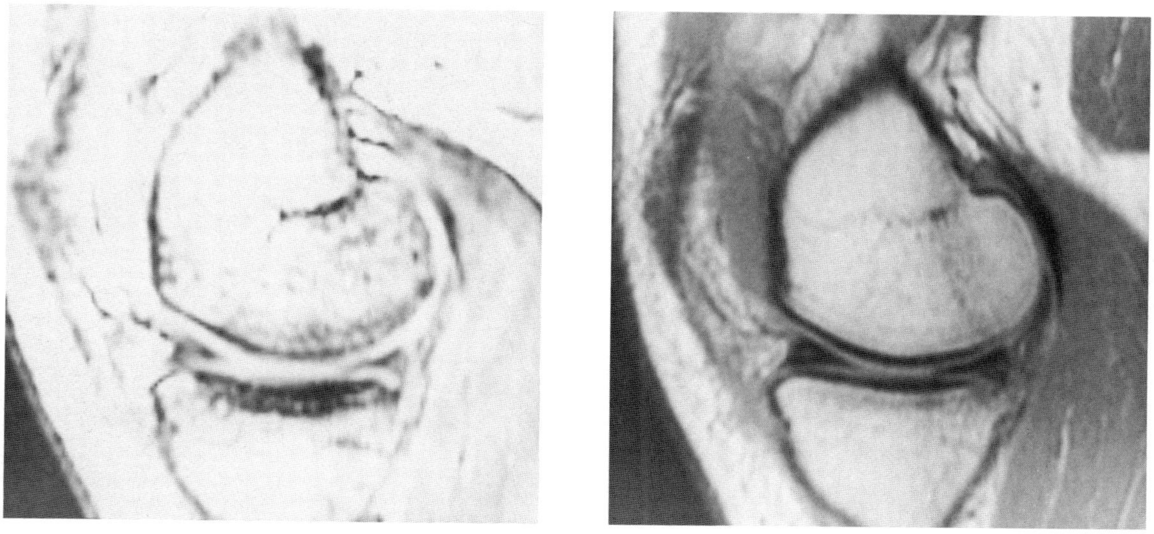

A B

FIGURE 13–18
Gradient-echo and spin-echo sagittal images of the same anatomic section demonstrate superior visualization of a grade III A meniscal tear in the posterior horn of the medial meniscus on the gradient-echo image. A. Gradient-echo sagittal image. B. Spin-echo sagittal image.

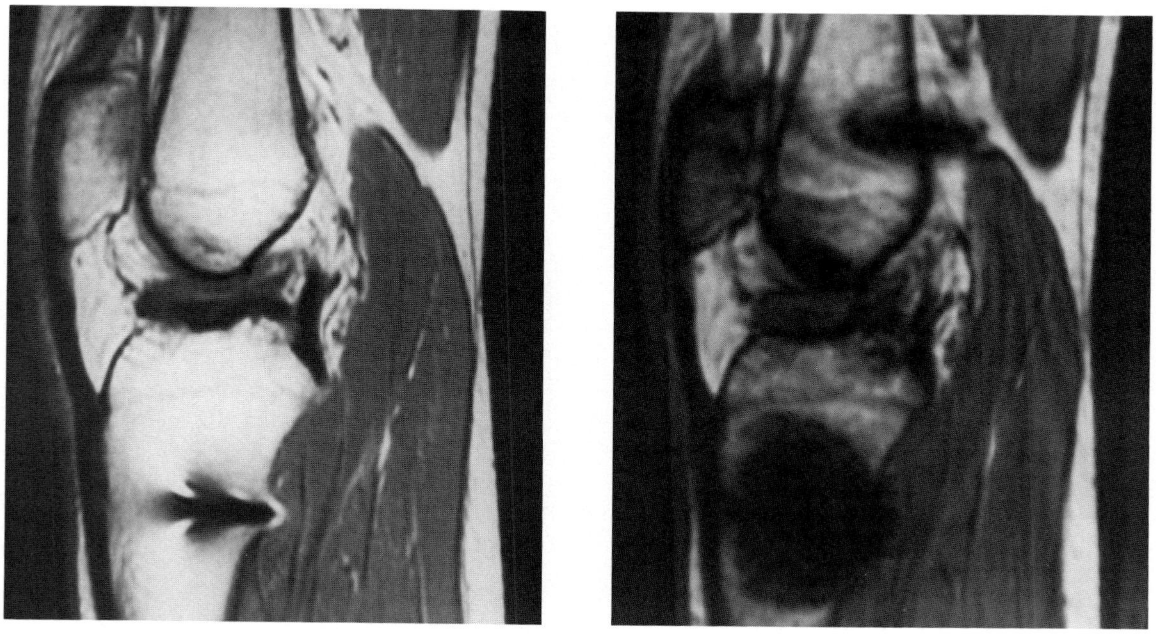

A B

FIGURE 13–19
A. Spin-echo T1-weighted image in a patient who had undergone an anterior cruciate ligament repair. Note the magnetic distortion caused by the surgical pin inserted into the tibial shaft. B. Gradient-echo image of the same anatomic section seen in A. Note the increased magnetic distortion in the tibia. Magnetic distortion is also evident from the femoral staple, although it is not observed on the spin-echo image.

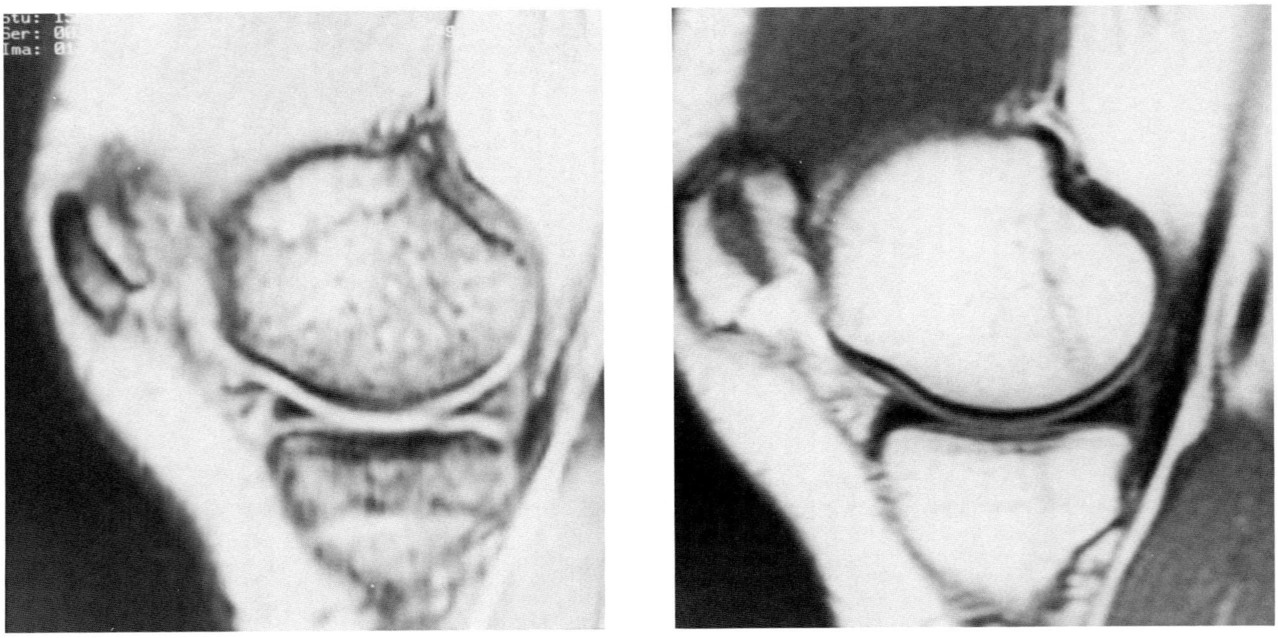

A B

FIGURE 13–20
Significant signal is found within both horns of the medial meniscus on a gradient-echo image in an asymptomatic patient who had no evidence of a tear on a corresponding spin-echo study. A. *Gradient-echo image.* B. *Spin-echo image.*

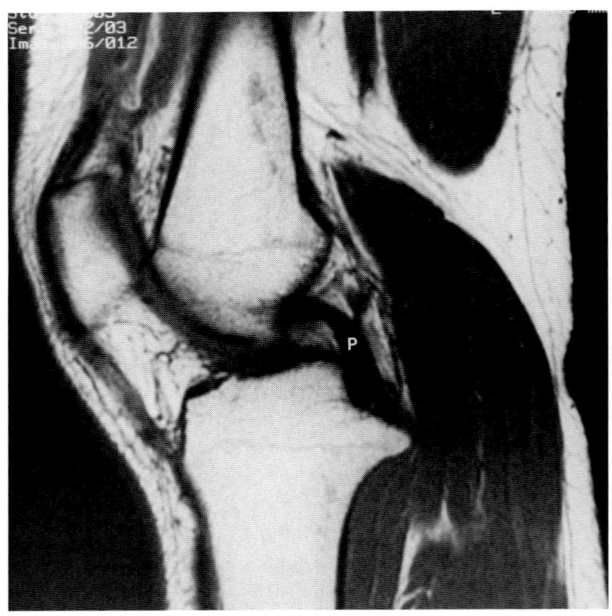

FIGURE 13–21
The normal posterior cruciate ligament (P) appears as a thick, signal-poor structure, and can almost always be completely visualized on at least one sagittal section.

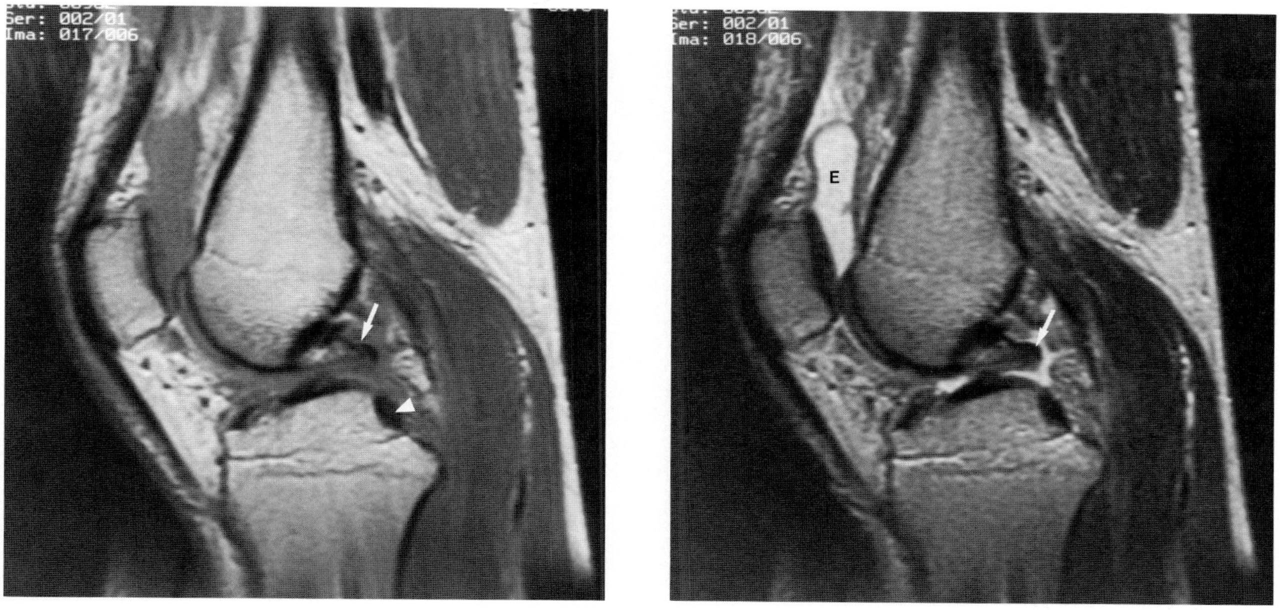

A B

FIGURE 13–22
A. A thin strand of ruptured posterior cruciate ligament (arrow) *is observed from the femoral attachment. An isolated tibial attachment* (arrowhead) *is also visible on this section.* A. *T2-weighted image demonstrates fluid in the region of the break and a suprapatellar effusion (E).*

tive to the femur [25]. It is usually observed intact on one section, although improper positioning of the knee may necessitate observing the course of the ligament over several sections.

Complete rupture of the posterior cruciate ligament can be diagnosed by a failure to visualize it intact on a single section (Figure 13–22) or on serial sections. An obvious irregularity or region of increased signal on either the T1- or T2-weighted images is suggestive of a partial tear of the ligament (Figure 13–23). The posterior cruciate ligament is a substantial structure and may be observed intact after severe trauma but secured to an avulsed piece of tibia or femur (Figure 13–24), most often detaching a piece of its tibial insertion [26]. Irregularities in cortical margins should be examined.

ANTERIOR CRUCIATE LIGAMENT

The most difficult structure to visualize in its entirety on a single section is the anterior cruciate ligament. The anterior cruciate ligament appears as a series of dark bands or fiber bundles arising from the anterior tibial plateau to the medial aspect of the lateral femoral condyle in a fan-like fashion (Figure 13–25) [27]. Due to the angulation of the ligament and the relatively thin fibers that it is composed of, true sagittal images are not able to demonstrate it properly if the knee is positioned in a true anterior-posterior direction. Obliquing the sagittal images 20 degrees laterally has been proposed as one method of compensating for the problem [28]. We have found that by permitting the patient to be relaxed and lie comfortably, the knee will routinely be placed in 15–20 degrees of external rotation (see Figure 13–2), allowing routine sagittal images to be obtained and eliminating the necessity of further manipulation. In this way, well over 95% of normal anterior cruciate ligaments can be visualized from origin to insertion on at least one sagittal section.

A complete rupture can be diagnosed as a discontinuity in the ligament with at least one

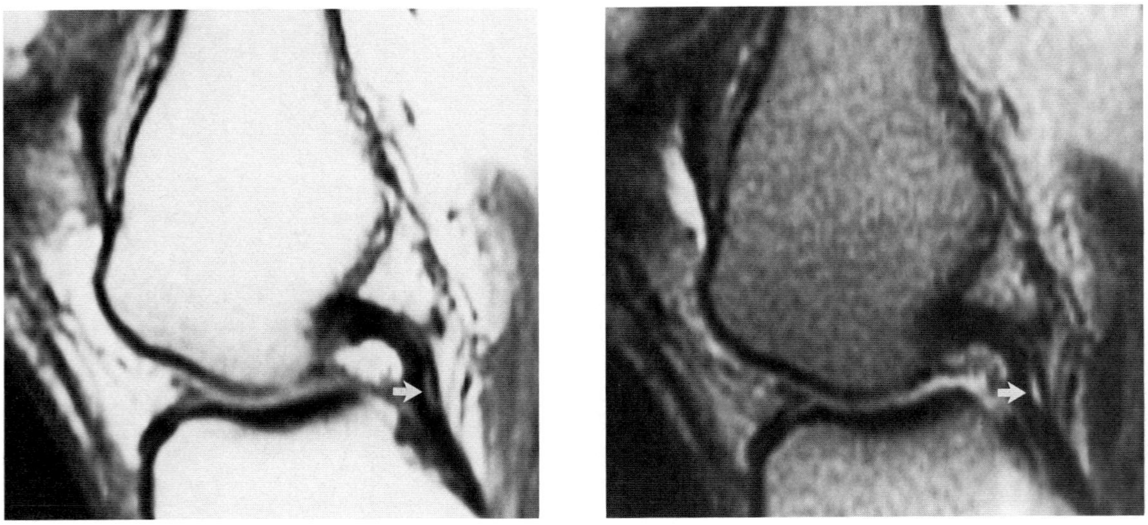

A B

FIGURE 13–23
Increased signal on both T1- and T2-weighted images is found within the normally signal-poor posterior cruciate tendon (arrows), *suggestive of a partial tear and inflammatory changes. A. T1-weighted image. B. T2-weighted image.*

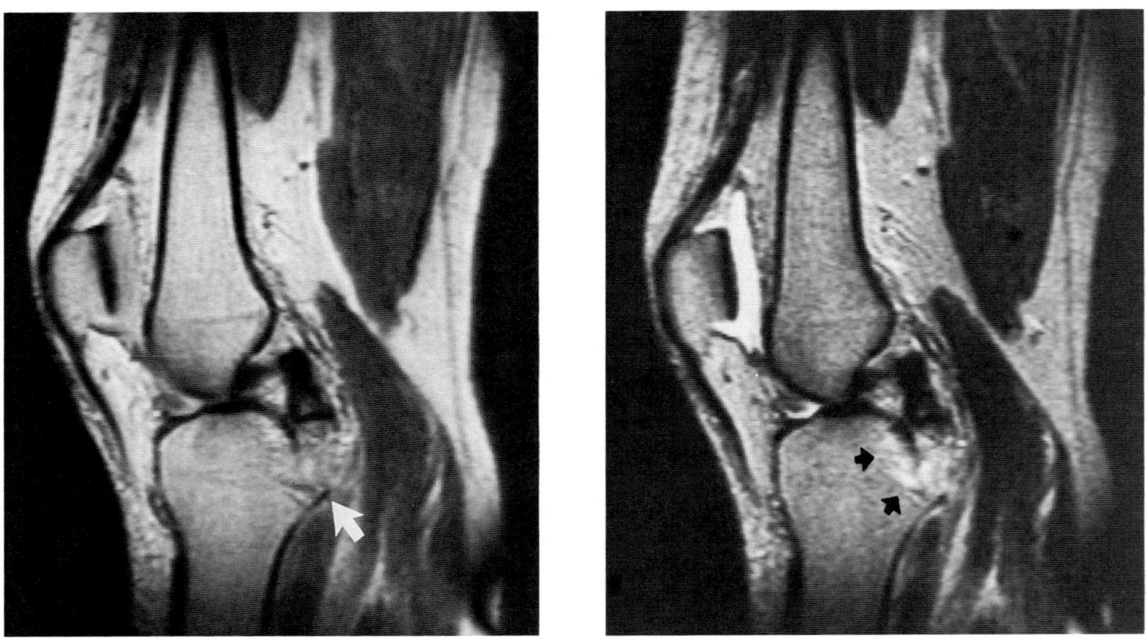

A B

FIGURE 13–24
A. *Although the posterior cruciate ligament is intact, there has been an avulsion of the tibia at the insertion of the ligament. The cortical margin at the posterior end of the tibia has been disrupted* (arrow). B. *The T2-weighted images demonstrate an area of increased signal within the marrow cavity, secondary to posttraumatic edema* (arrows).

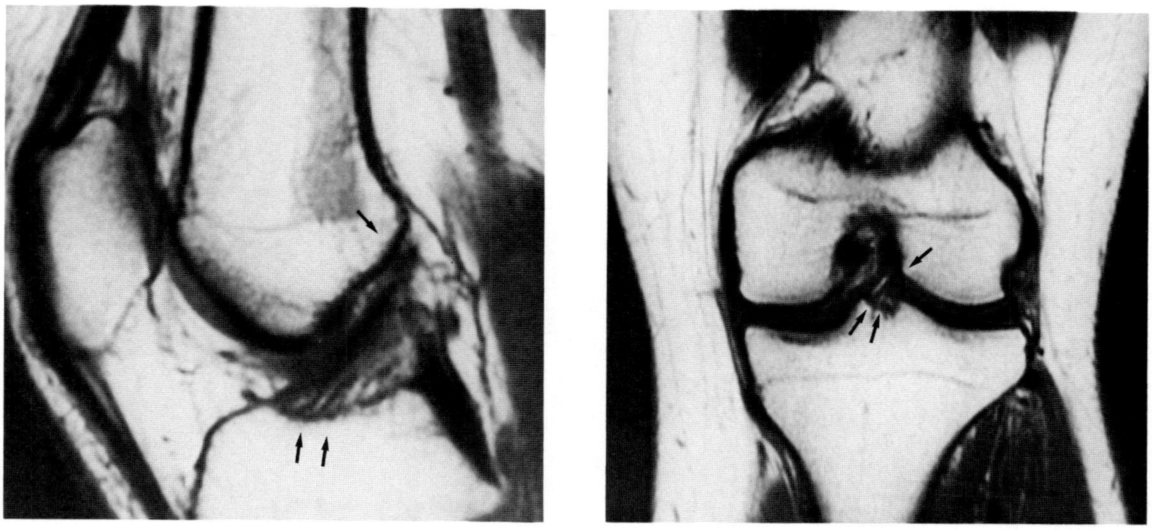

A B

FIGURE 13–25
The normal anterior cruciate ligament (arrows) *can usually be seen on a single sagittal section, and occasionally on a coronal image, extending from the lateral femoral condyle to insert on the medial aspect of the tibial plateau, just posterior to the tibial spines. A. Sagittal section. B. Coronal image.*

severed end clearly seen, usually oriented improperly (Figure 13–26) [9,29]. In an acute tear, fluid, edema, and hemorrhage will usually fill the region normally occupied by the ligament, and the T1 or proton density image may not be diagnostic. A medium-signal-intensity region due to this inflammatory mass may mimic an intact ligament. The T2-weighted image highlights the area and enables a determination of complete or partial tear to be made, with the dark ligament or fragment usually visible (Figures 13–26 and 13–27). Although failure to observe an intact anterior cruciate ligament is not conclusive proof of a complete rupture, two secondary signs can be important in helping to arrive at a diagnosis. The ligament normally prevents the femur from slipping posteriorly, relative to the tibia [30]. Severing this band permits this aberrant movement to occur and, in a knee imaged in an extended position, the posterior aspects of the femoral condyles are observed to be posterior to the tibial plateaus, rather than aligning with them as is normally observed with an intact anterior cruciate ligament (Figure 13–28). Because of this abnormal posterior location, the posterior cruciate ligament becomes more relaxed than normal, and is observed to buckle (Figure 13–29). This last sign should be used cautiously (and never by itself) since it is highly variable with individuals and with the degree of extension of the knee.

A partial rupture or incomplete tear presents with a mass of intermediate signal on T1-weighted images, increasing in signal intensity on T2 views, due to the hemorrhage and edema. In this case, normal attachment of at least one strand can usually be confirmed on the T2-weighted image (Figures 13–30 and 13–31). An incomplete tear can often simulate a normal structure. The T2-weighted images show areas of increased signal within and around the anterior cruciate ligament, indicating a partial rupture with a corresponding area of inflammation.

MRI has also been used to evaluate surgical repair of disrupted anterior cruciate ligaments (Figures 13–32 and 13–33) [31]. Initial results are not encouraging, however, with only about one-fourth of the replacement ligaments well vi-

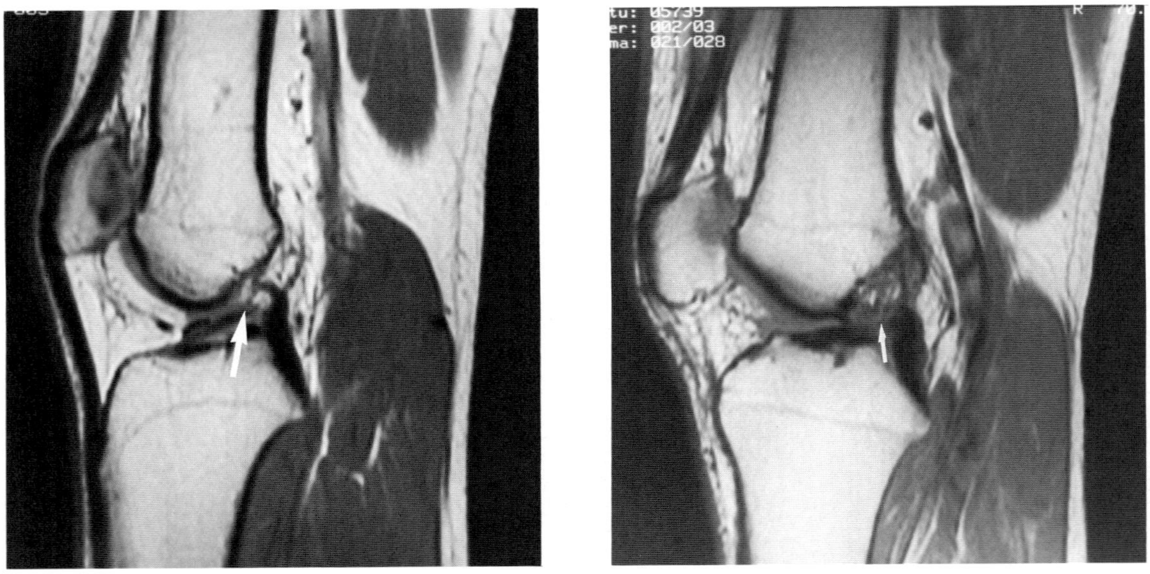

A B

FIGURE 13–26

A, B. *Ruptured anterior cruciate ligaments in two different patients. The severed ends of the ligaments* (arrows) *can be clearly seen, and are oriented posteriorly, rather than up toward the lateral femoral condyle.*

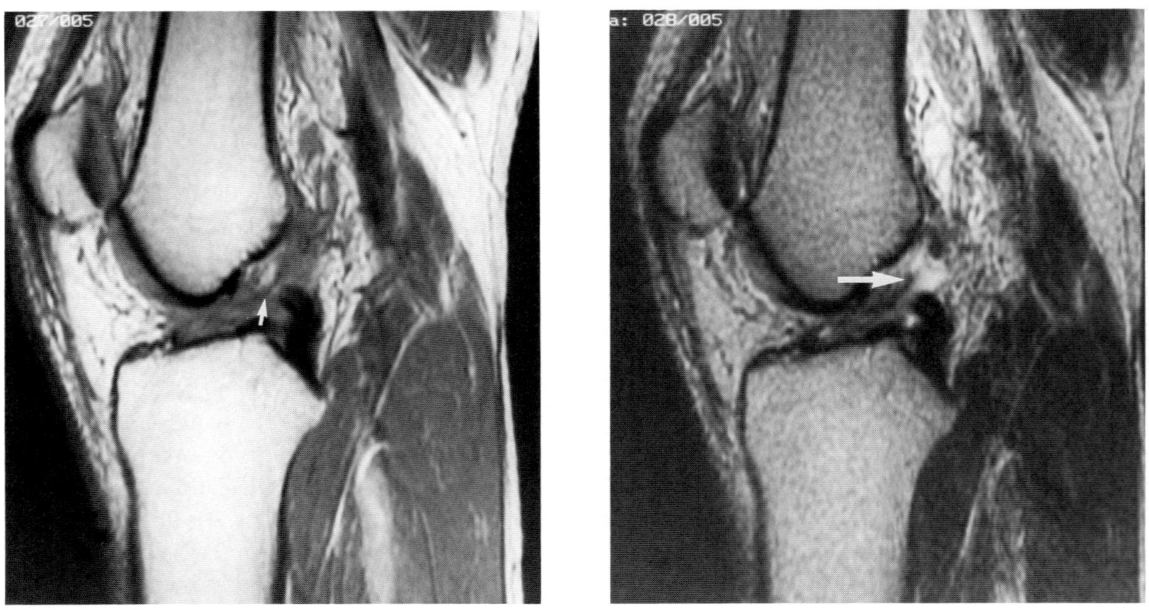

A B

FIGURE 13–27

A. *There is a region of intermediate signal intensity oriented in the direction of the anterior cruciate ligament* (arrow)*, which may be mistaken as an intact structure.* B. *A T2-weighted image demonstrates a large inflammatory mass, with the severed ligament* (arrow) *clearly outlined.*

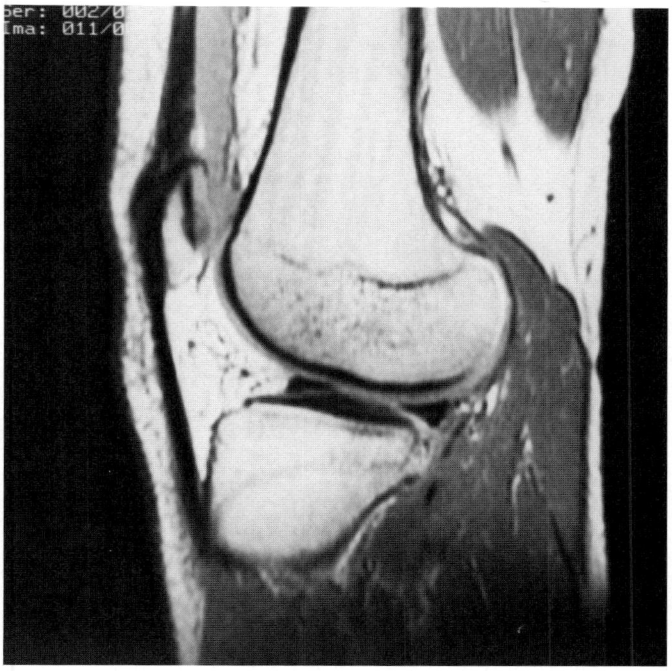

FIGURE 13–28
When the knee is fully extended, the posterior aspects of the femoral condyles should always align with the tibial plateaus. If there is a complete rupture of the anterior cruciate ligament, the femoral condyles are positioned more posteriorly than usual. This is an excellent secondary sign in cases of questionable anterior cruciate ligament tears.

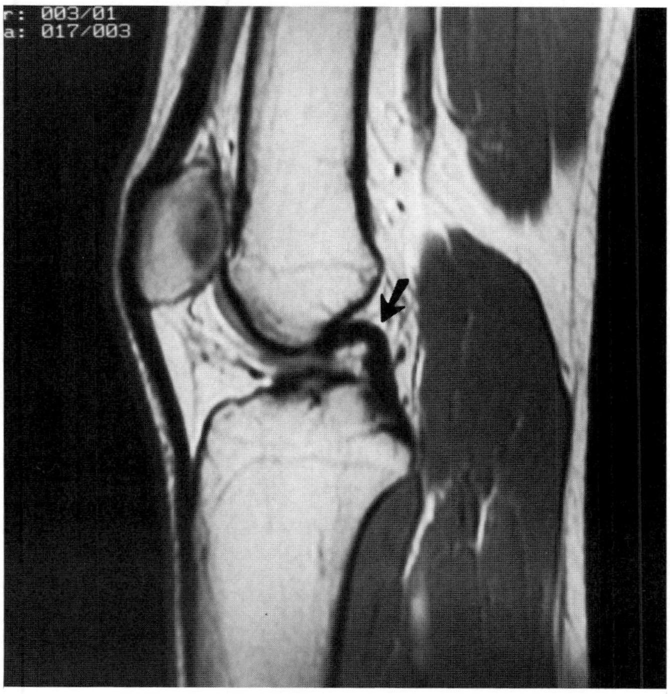

FIGURE 13–29
Buckling of the posterior cruciate ligament (arrow) *is often observed when the anterior cruciate ligament is ruptured, due to the posterior movement of the femur. This sign must be used cautiously, however, since a hyper-extended or normally lax ligament presents similarly.*

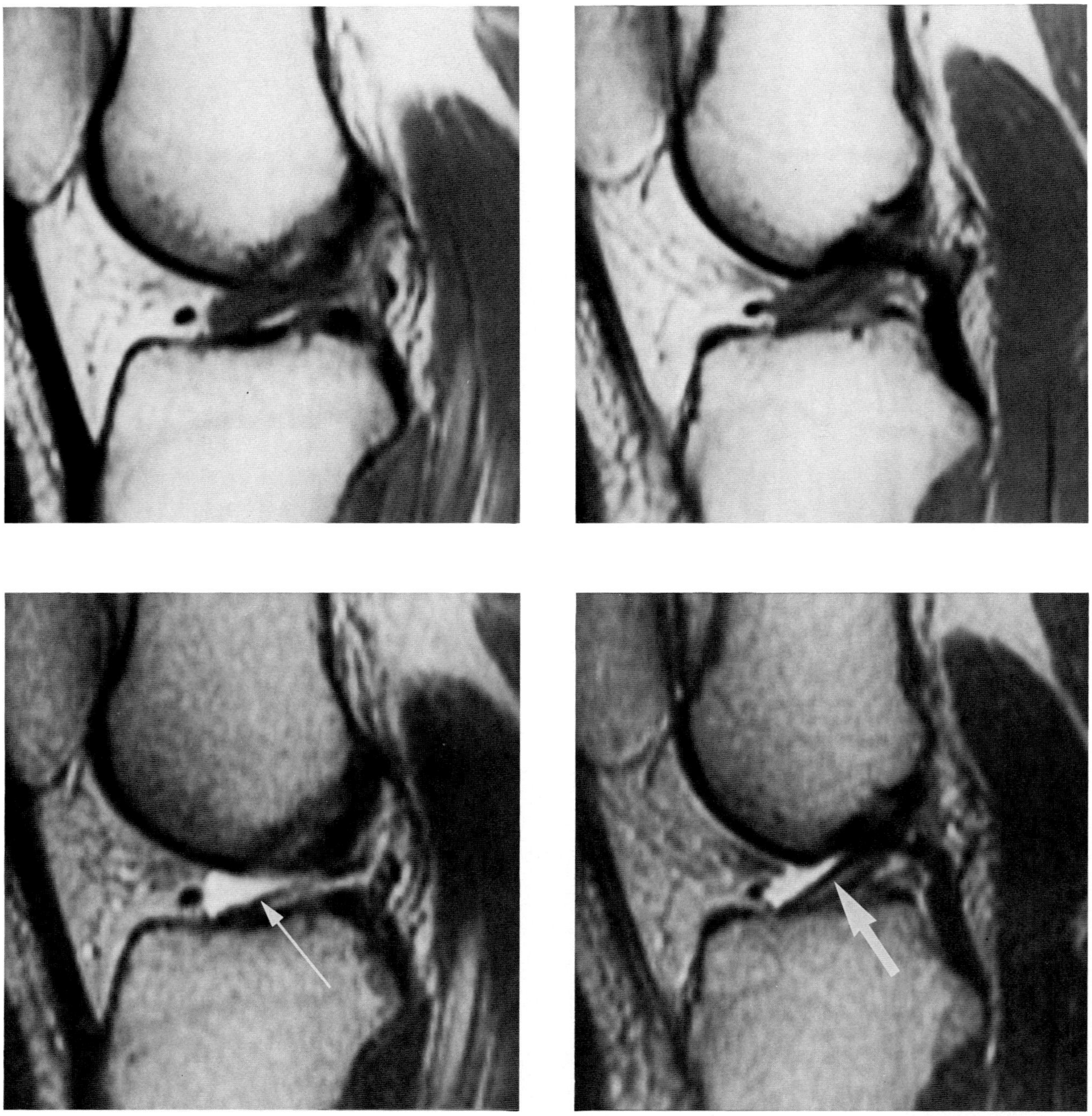

FIGURE 13–30

Consecutive T1-weighted images through the region of the anterior cruciate ligament (top), *with the corresponding T2-weighted images below, demonstrate an inflammatory mass in the region of the tibial insertion of the ligament with a torn strand* (small arrow, lower left) *oriented posteriorly, along with several intact strands* (large arrow, lower right).

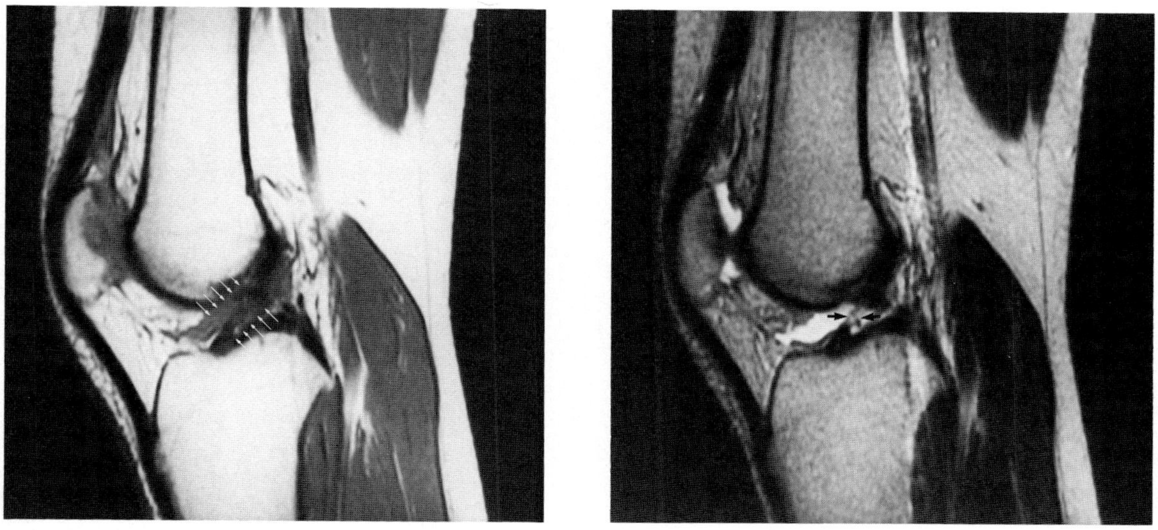

A B

FIGURE 13–31
A. *There is a large area of inflammation* (arrows) *in the region of the anterior cruciate ligament, which is incompletely visualized, although it is oriented properly.* B. *The T2-weighted images demonstrate a partial tear, with at least two thin strands still intact* (arrows).

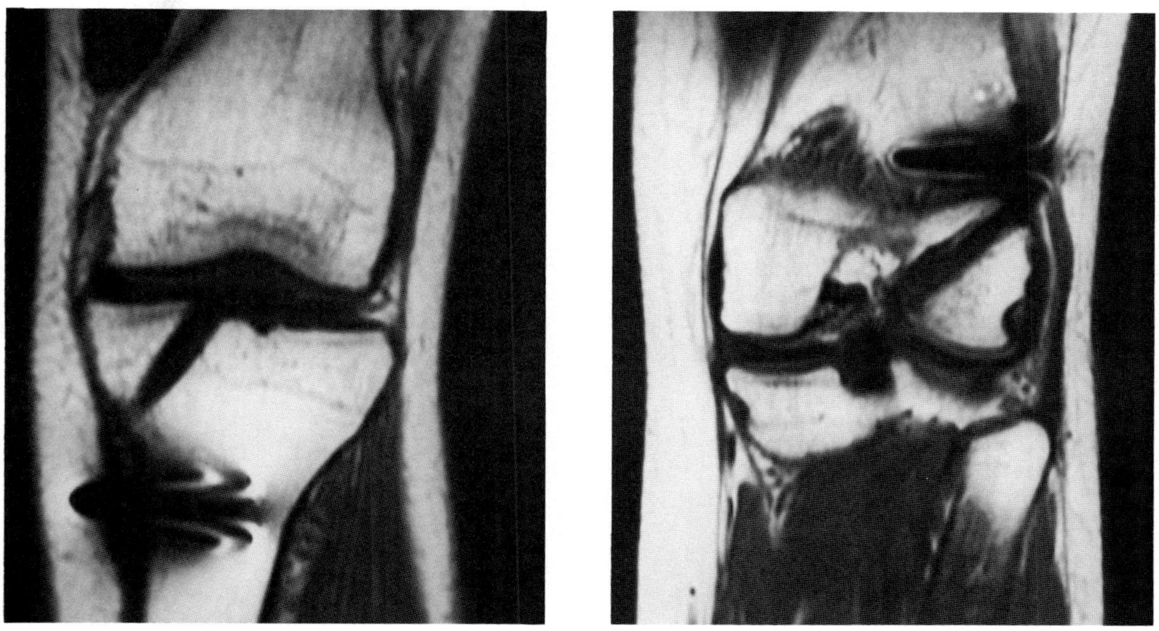

A B

FIGURE 13–32
A, B. *A ruptured anterior cruciate ligament can be repaired by replacing it with a natural (patellar, semitendinosus, or iliotibial band) or synthetic (Teflon, Gore-Tex, or Dacron) tendon. Channels are drilled inferomedially through the anterior part of the tibia (A) and superolaterally through the posterior lateral femoral condyle (B), with the replacement tendon secured with metal staples to the femoral and tibial shafts. These staples are responsible for the metallic artifacts observed.*

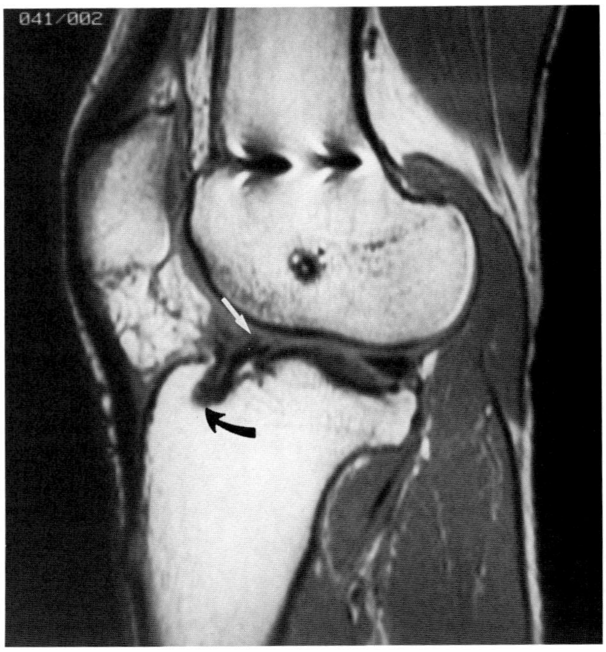

FIGURE 13–33
There has been a tear in a previously repaired tendon, 2 years after an anterior cruciate ligament repair. Torn replacement tendon (arrow) *can be observed emerging from the inferior (tibial) canal* (curved arrow).

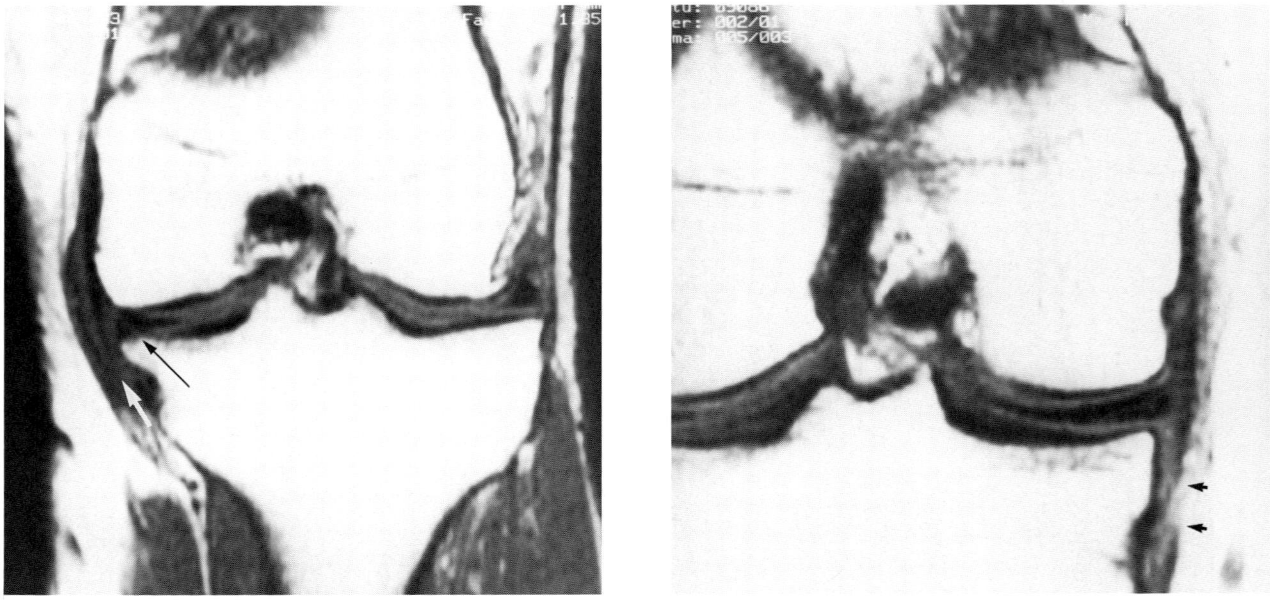

A B

FIGURE 13–34
A. *The superficial and deep components of the medial collateral ligament are severed* (white arrow) *in a patient with a large effusion. The black arrow indicates a meniscal tear.* B. *The arrows indicate interruption in the medial collateral ligament.*

sualized on spin-echo sequences. Metallic artifacts due to the pinning of the ligament are usually sufficiently removed from the area of interest so that they are not a problem.

MEDIAL AND LATERAL COLLATERAL LIGAMENTS

The medial and lateral collateral ligaments are best evaluated on the coronal images. As noted in Chapter 2, the medial collateral ligament is usually considered to consist of two parts—the more superficial tibial collateral ligament and the medial capsular ligament—with a bursa separating them. Both are dark, signal-poor structures on both T1- and T2-weighted images, and are well visualized in the coronal plane (Figure 13–34). A complete tear of the ligaments can easily be perceived as a lack of continuity, often with a widening of the joint space [32]. Partial tears often result in increased fluid within the bursa, further separating the two structures. On the lateral aspect, the iliotibial band, fibular collateral ligament, and biceps femoris tendon are evaluated for integrity by similarly checking for continuity (Figure 13–35). Rupture of the biceps femoris tendon often lead to a retraction of the muscle. Some caution must be exercised to make sure that an incompletely visualized ligament (on a single section) is not simply a relaxed, tortuous structure that is intact, but incompletely visualized in the imaging plane.

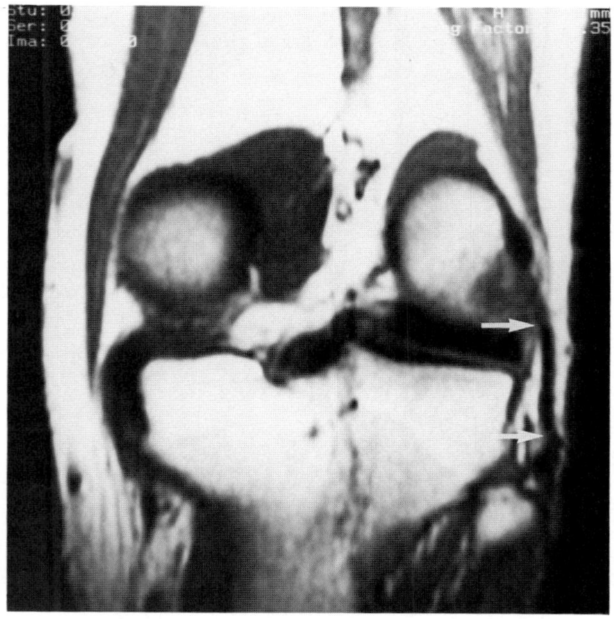

FIGURE 13–35
Incomplete tears in the fibular collateral ligament are visible (arrows) *in this coronal view.*

Fluid

Fluid collections, such as joint effusions, are easily detectable as regions of moderate signal intensity on T1-weighted images, becoming very bright on more T2-weighted images (Figure 13–36). The true joint space within the knee includes the suprapatellar bursa, but does not communicate with bursae medial and lateral to the joint, which are also often observed to contain fluid in acute pathologic conditions. Popliteal cysts exhibit similar signal characteristics, but are located posterior to the joint space, usually in the semimembranosus-gastrocnemius bursa, although they can be found in other bursae as well (Figures 13–37 and 13–38). On occasion, pop-

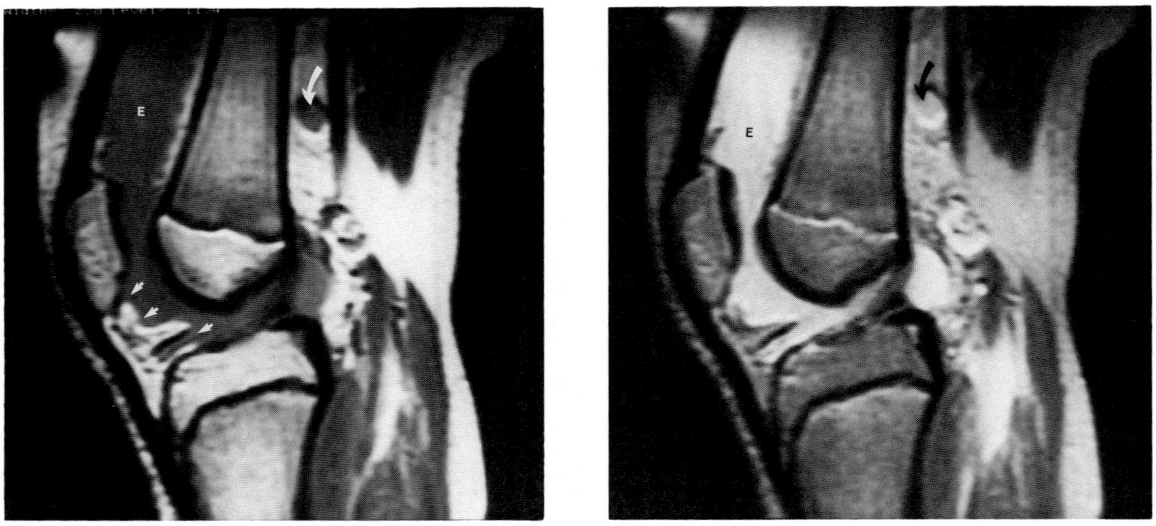

A B

FIGURE 13–36
T1-weighted and T2-weighted images illustrate a large effusion (E) in a 12-year-old boy with a 6-month history of recurrent effusions. No soft tissue or osseous abnormalities were observed on the scan. Two months after this study, the patient tested positive for Lyme's disease. Note the enlarged lymph node (curved arrow) *and irregularities in the contour of the infrapatellar fat pad* (arrow), *consistent with inflammation of the synovium.* A. *T1-weighted image.* B. *T2-weighted image.*

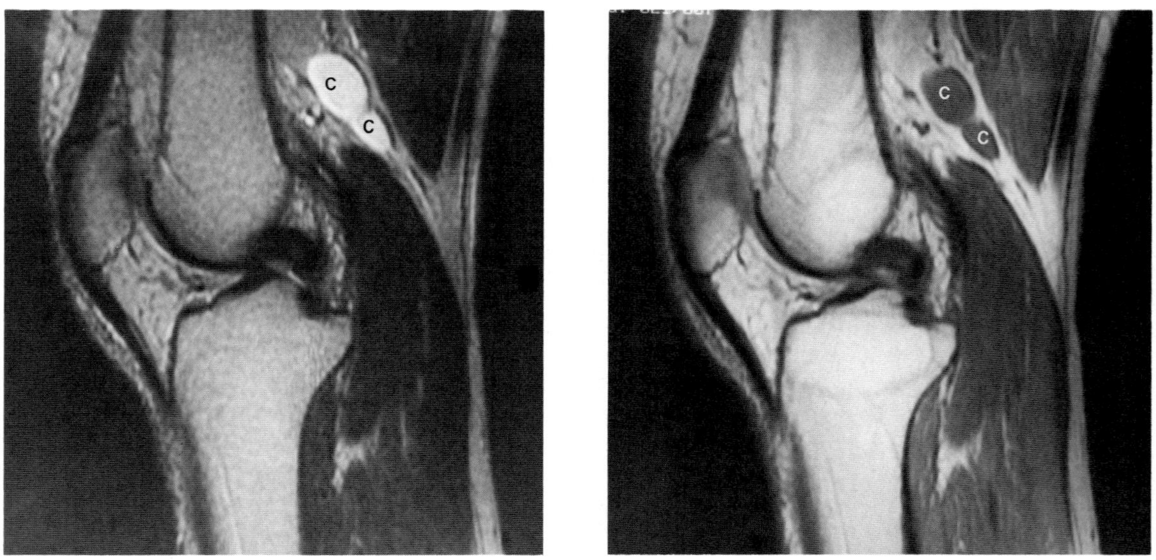

A B

FIGURE 13–37
Two small popliteal cysts (C) are shown within the popliteal fossa. A. *They are observed as medium-intensity structures on the proton density image.* B. *The signal intensity increases significantly on the T2-weighted image.*

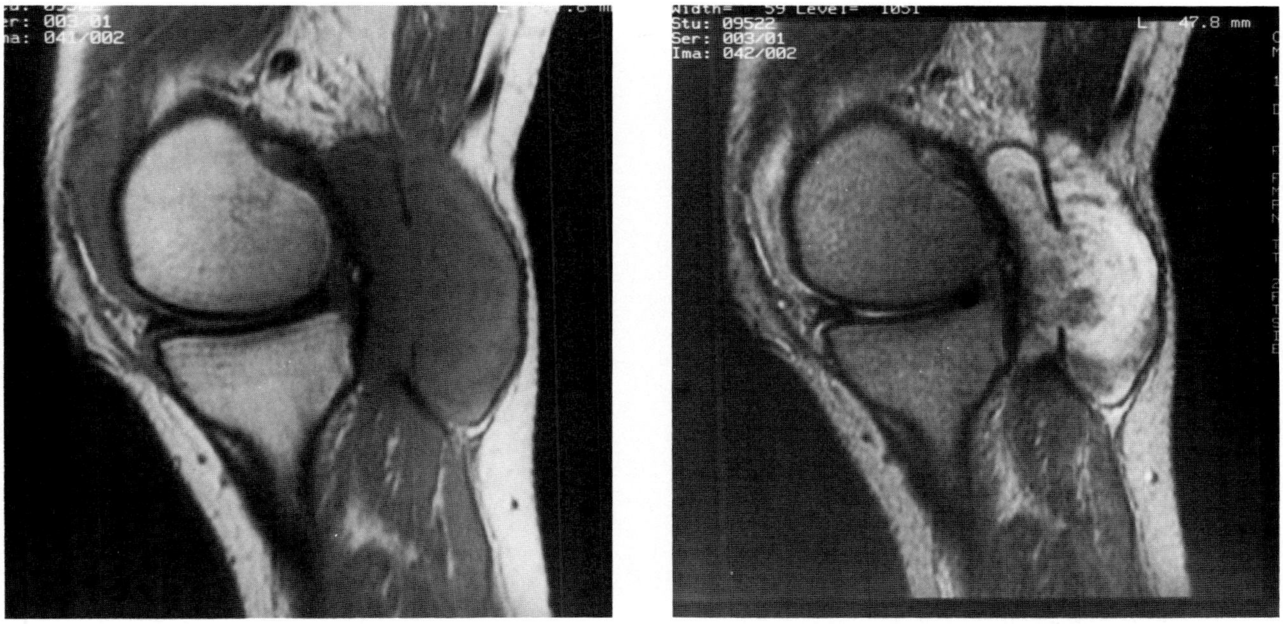

A B

FIGURE 13–38
A. *A large, septated popliteal cyst in the gastrocnemius-semimembranosus bursa.* B. *A T2-weighted image demonstrates areas of decreased signal within the cyst, consistent with pannus formation.*

liteal cysts can rupture and dissect through a muscle, mimicking pain from deep venous thrombosis (Figure 13–39) [33]. Ganglion cysts can be seen as homogeneous outpouchings of synovial fluid, which are completely outside of the joint space, arising from either synovium or a tendon sheath. These can be quite large, and appear isointense to synovial fluid on all echo sequences. Meniscal cysts are thought to be trauma-induced fluid collections, usually anteriorly located, and can be demonstrated to be in continuity with a meniscal tear (Figures 13–40 and 13–41). These require surgical treatment and should be distinguished from the more innocuous ganglion cysts, which do not communicate with a meniscal surface (Figure 13–42) [34].

Other Abnormalities

A wide variety of various abnormalities commonly found in the knee can be diagnosed by a careful examination of the MR images. In addition to detecting meniscal, tendon, and ligamentous tears, using a standard set of pulse sequences, MR images are able to provide a wealth of additional information that might otherwise require multiple and/or invasive studies [35–40].

Osteochondral bodies, caused by fragmentation of menisci or bone, can be well visualized by MRI as abnormal structures within the joint space. If they are meniscal fragments or are completely calcified, they appear as regions of absent signal on all pulse sequences (Figure 13–43). They may, however, contain yellow marrow, in which case they are seen as a region of fat intensity, surrounded by dark cortical bone (Figure 13–44). Osteochondral bodies are usually painful and accompanied by effusion. Their location within the joint space can be demonstrated by observing the fragment surrounded by fluid on the T2-weighted images. False-positive results, however, may arise from a failure to examine adjacent sections to be sure that the visualized abnormality is not simply part of the infrapatellar

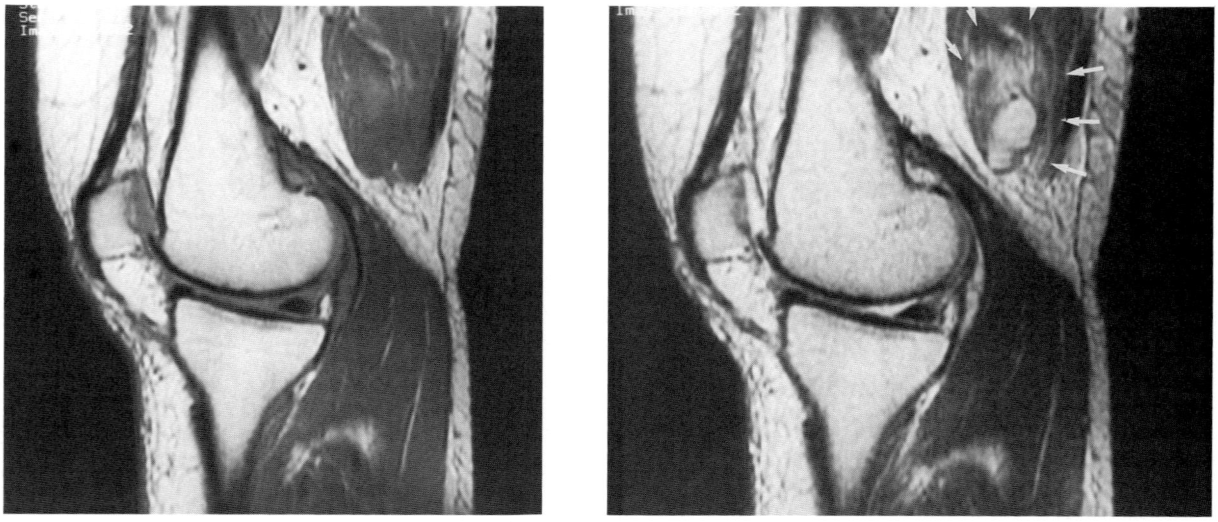

A B

FIGURE 13–39
A ruptured popliteal cyst, causing inflammation in the semimembranosus muscle, is illustrated by the increased signal within the muscle (arrows) *on the T2-weighted image compared to the proton density image.* A. *Proton density image.* B. *T2-weighted image.*

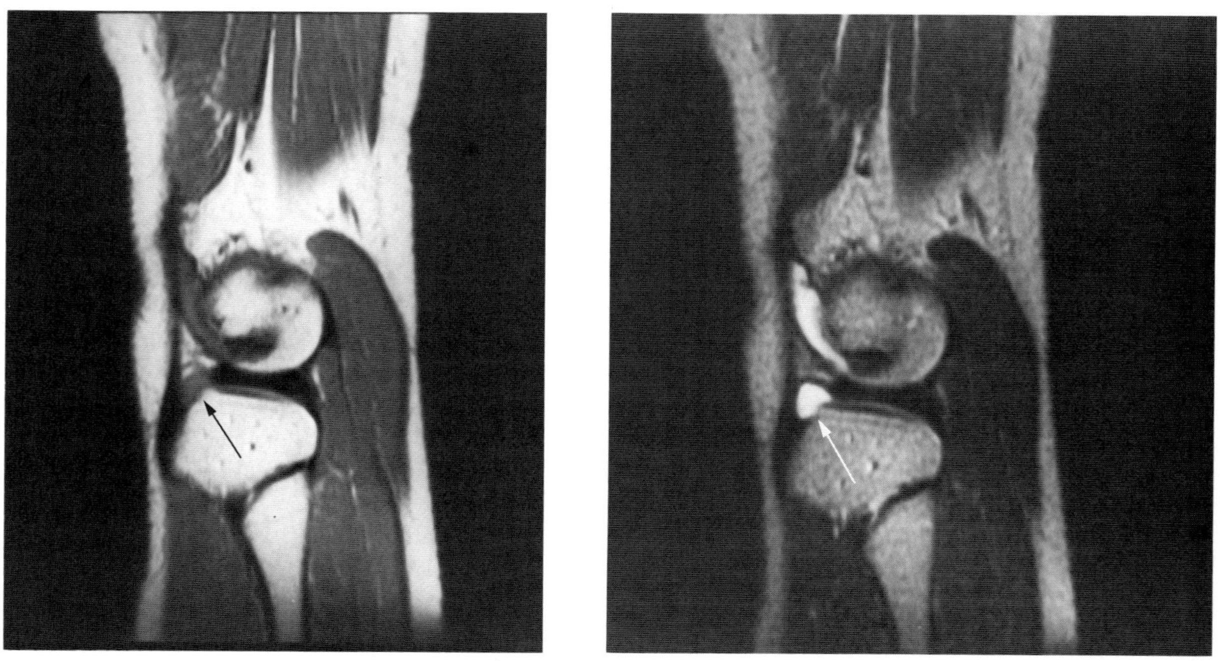

A B

FIGURE 13–40
A. *A small anterior meniscal cyst* (arrow) *extends from the lateral meniscus.* B. *It becomes very bright on the T2-weighted image. The cyst is isointense with the effusion (E).*

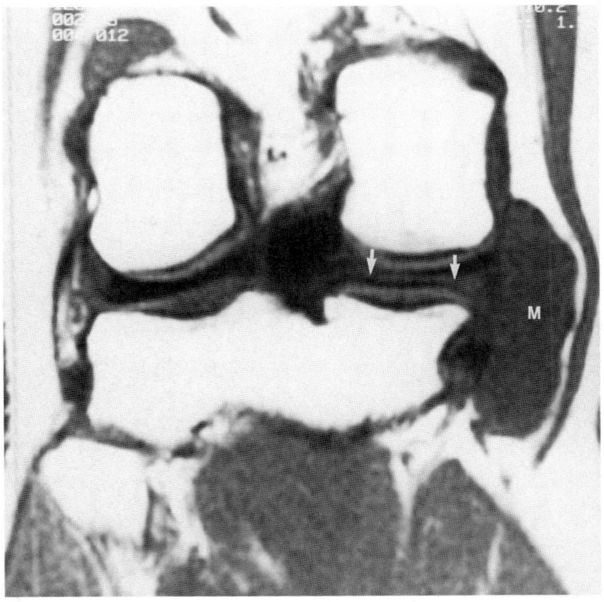

◀ *FIGURE 13–41*
A large meniscal cyst (M) communicates directly with a large tear (arrows) *in the posterior horn of the medial meniscus and displaces the distal sartorius muscle.*

FIGURE 13–42 ▶
A synovial cyst (S) is observed medial to the proximal tibia. A. T1-weighted sagittal image. B. T2-weighted sagittal image. C. T1-weighted coronal view.

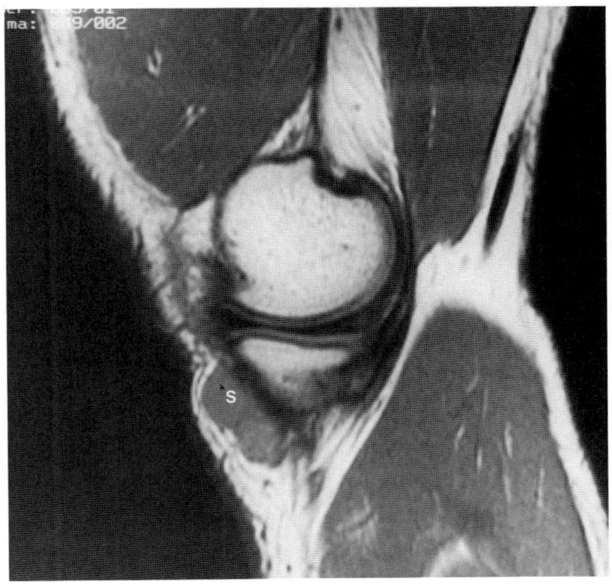

A

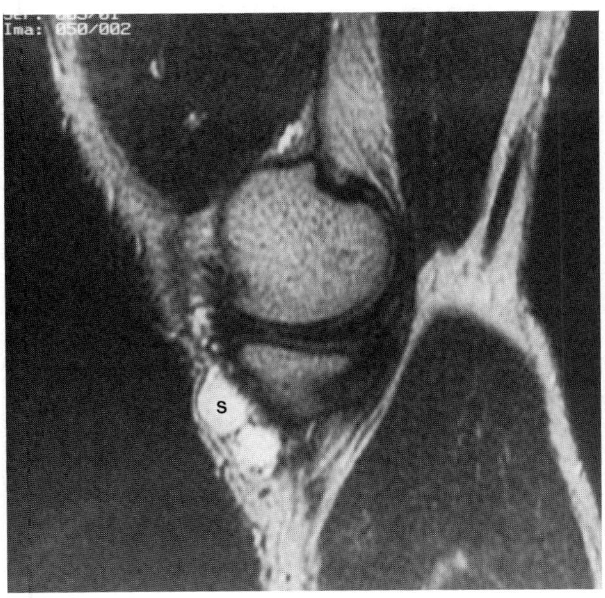

B

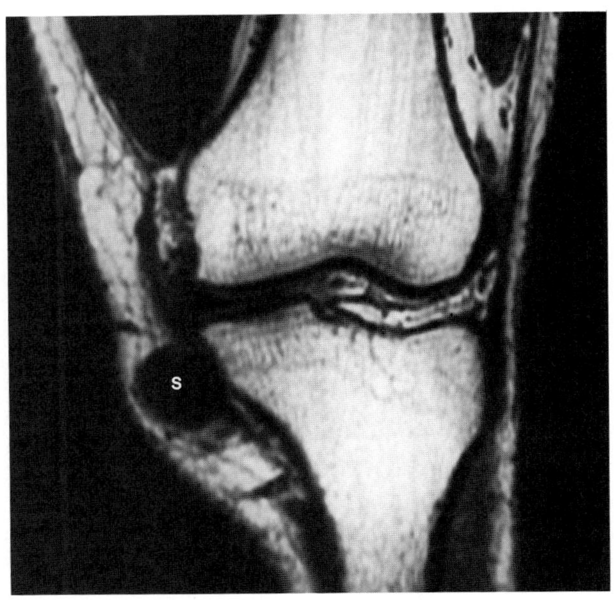

C

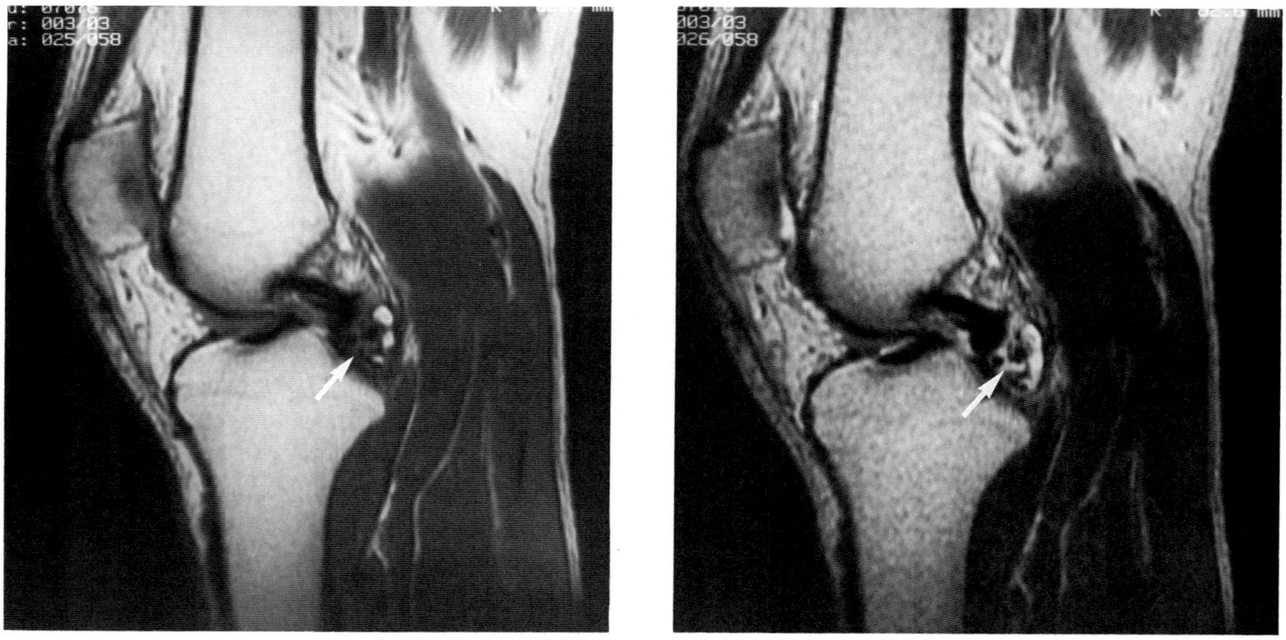

A B

FIGURE 13–43
A. Multiple osteochondral bodies (arrow) *are observed in the posterior joint space.*
B. They are accented on the T2-weighted image by the bright effusion.

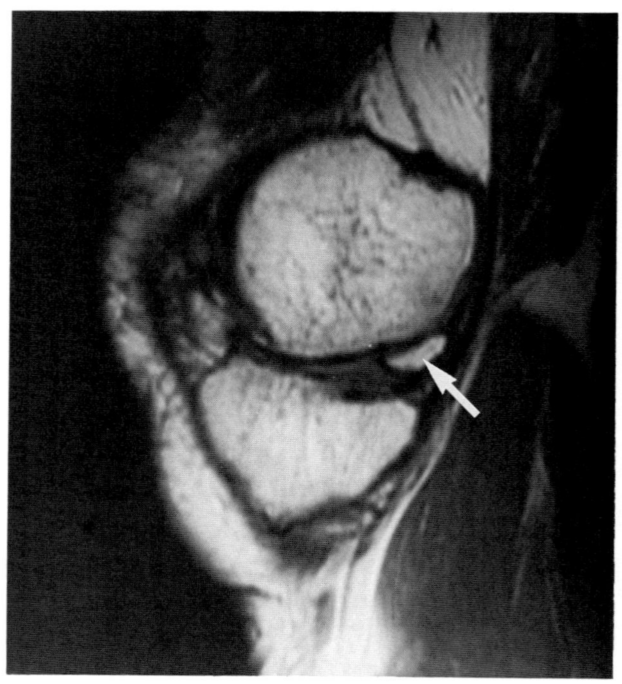

FIGURE 13–44
There is a large, well-corticated, marrow-containing bone fragment (arrow) *in the posterior joint space of a patient with severe degenerative joint disease.*

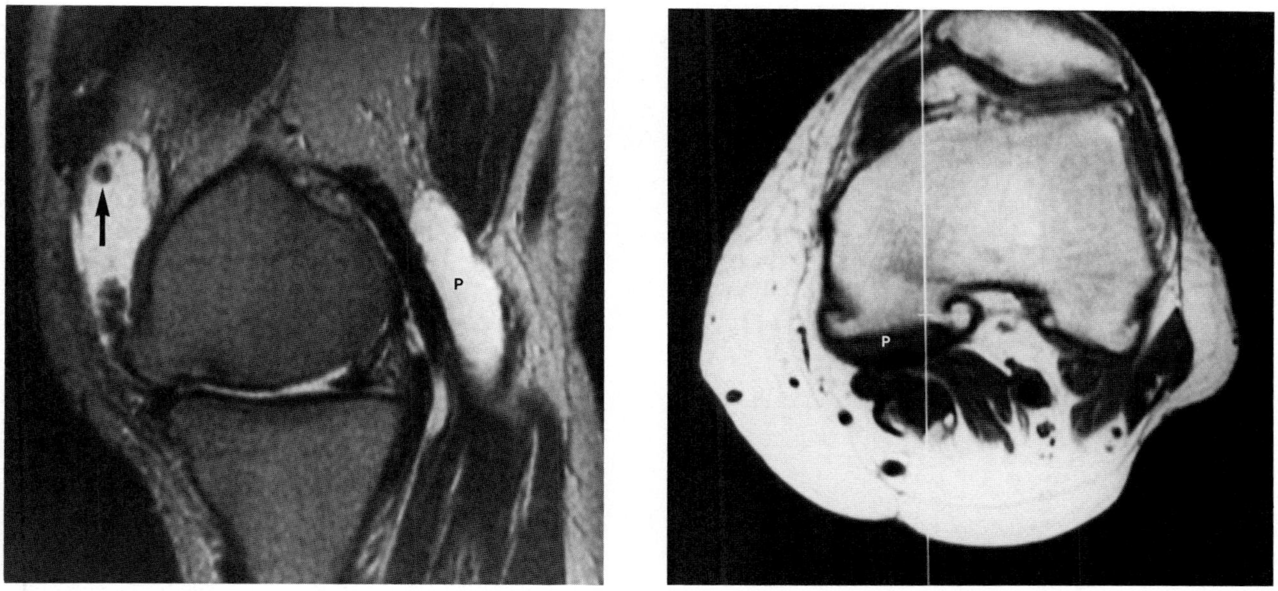

A B

FIGURE 13–45
A. *A "pseudofragment"* (arrow) *appears as an osteochondral body surrounded by joint fluid in this single T2-weighted image. Note the large popliteal cyst (P) in the gastrocnemius-semimembranosus bursa.* B. *The white line indicates the plane of the section of* A. *The tip of a large osteophyte, projecting into the joint effusion, gives the appearance of a free osteochondral body.*

fat pad, which appears as an isolated piece of marrow-containing bone fragment on a single section (Figure 13–45).

Osteochondritis dissecans is a disorder characterized by a fragmentation or separation of articular cartilage and subchondral bone from the primary osseous structure, most commonly the medial femoral condyle and less frequently the lateral condyle (Figures 13–46, 13–47, and 13–48) [41]. The talus and capitellum can occasionally be involved. This noninflammatory joint disease usually affects teenagers and young adults, and is of uncertain etiology, with acute trauma, chronic impingement, and avascular necrosis suggested as possible causes. Familial cases that have been reported suggest that it also may be transmitted as an autosomal dominant disorder [42]. Fragments that are not completely separated are found to be repairable with conservative therapy, while totally detached fragments are treated by internal fixation or surgical removal [43]. MRI has been found useful in determining the size, stability, and precise location of fragments [44,45].

Chondromalacia patellae is a syndrome primarily affecting young adults, in which a softening of the articular cartilage of the patella produces patellofemoral joint pain [46]. The softening of the articular cartilage most often begins over the medial femoral condyle, followed by gradual swelling, thinning and irregularity of the surface, and eventually complete erosion and exposure of the subchondral bone. Chondromalacia patellae may be idiopathic or induced by trauma, and arthritic changes occurring during adolescence may not be clinically manifested until adulthood [47–49]. MRI has the potential to be quite effective at diagnosing this entity, because of its ability to distinguish between articular cartilage, cortical bone, and surrounding structures (Figures 13–49 and 13–50). Excellent agreement has been found between MRI and surgical findings [50,51]; however, the practical value of MRI in the evaluation of this disease has

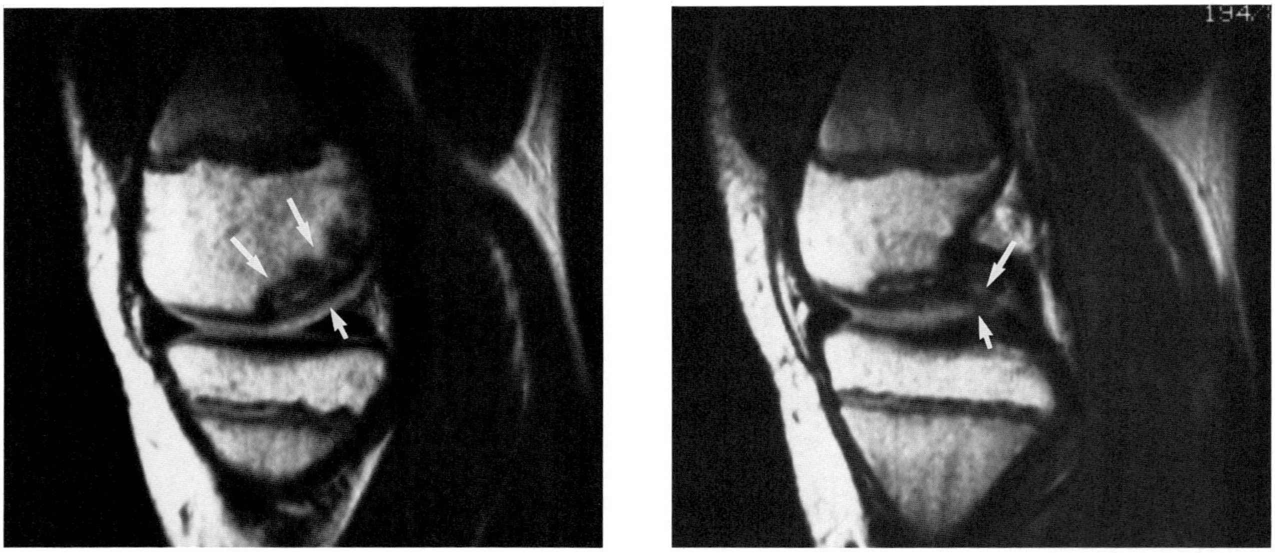

A B

FIGURE 13–46

A. *Sagittal section through the medial femoral condyle in a 14-year-old patient demonstrates a nondisplaced lesion* (large arrows). *Note the intact articular cartilage* (small arrows). B. *A more medial section approaching the intercondylar region shows a displaced fragment* (large arrow) *with a distinct break in the articular cartilage* (small arrow). *(Courtesy of Middletown Radiology Group, Middletown, NY.)*

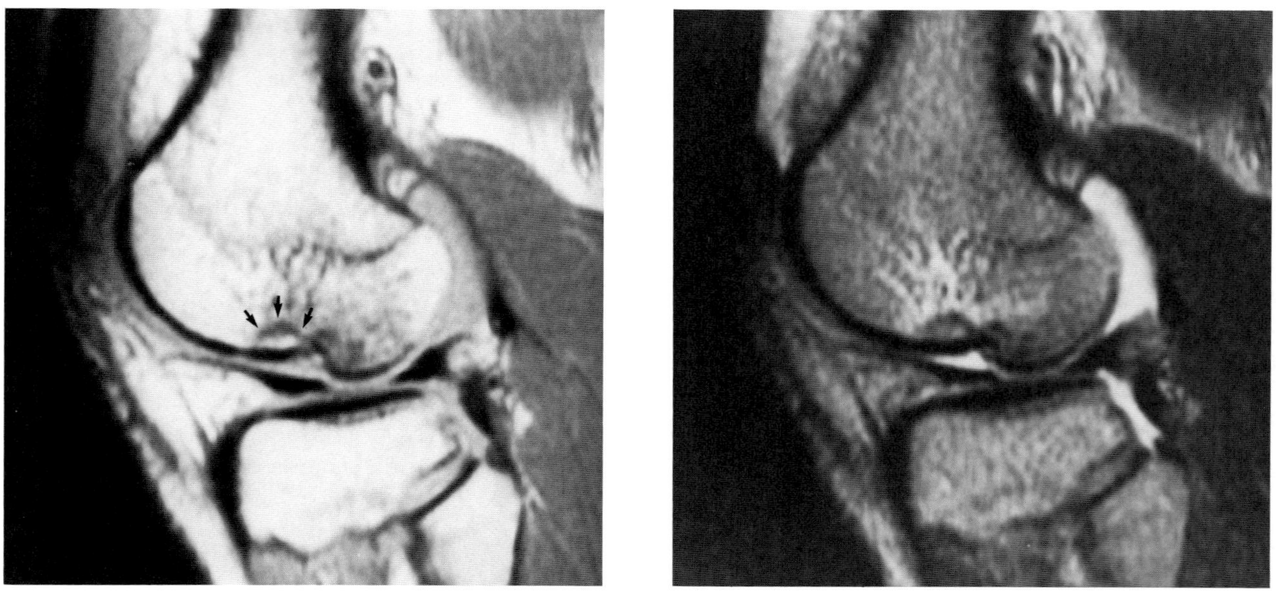

A B

FIGURE 13–47

A. *Nondisplaced subchondral fragment in the lateral femoral condyle in a 16-year-old patient with osteochondritis dissecans (T1-weighted image).* B. *Radial pattern of edema within the marrow cavity is evident in the T2-weighted image.*

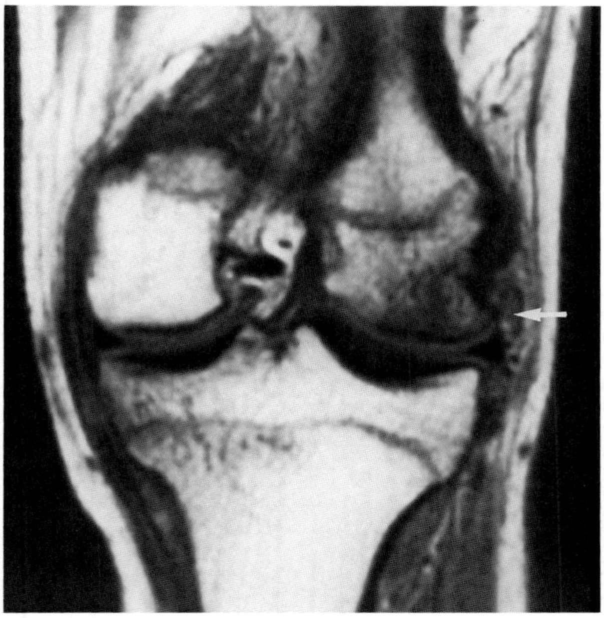

FIGURE 13–48
Osteochondritis dissecans is demonstrated by a large area of decreased signal in the medial femoral condyle and irregular region of absent bone in the medial aspect of the condyle. The arrow points to a medial osteochondral fragment.

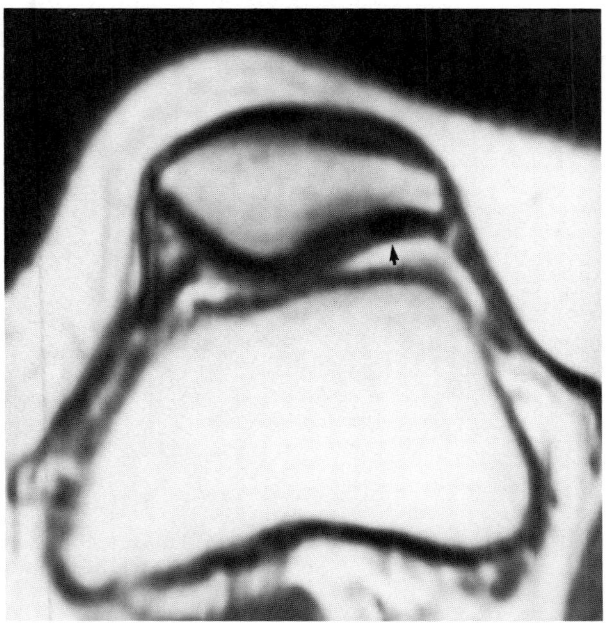

FIGURE 13–49
Erosive changes in the lateral facet of the patellar cartilage and cortex are noted in this axial image of a patient with chondromalacia patellae.

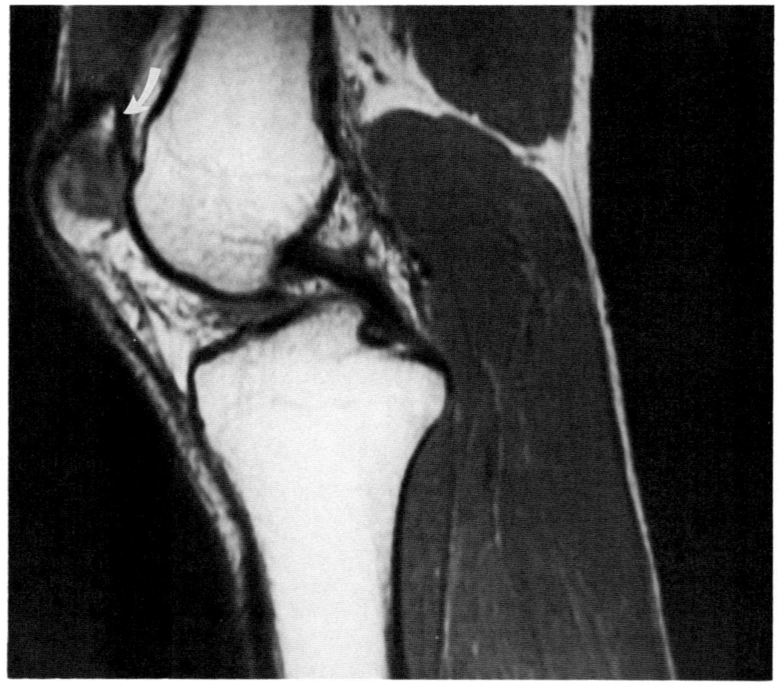

A

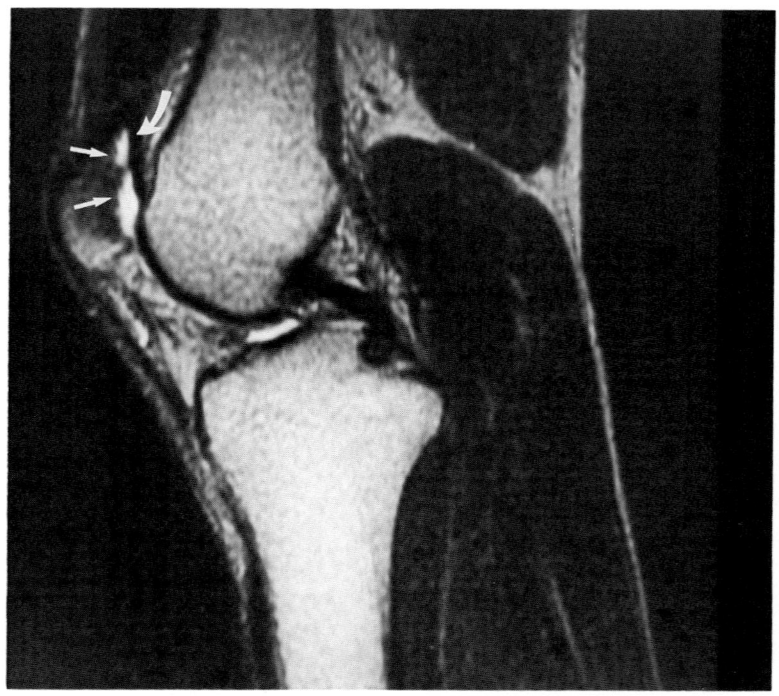

B

FIGURE 13–50
A. Chondromalacia patellae with a loss of definition of the patellar cartilage is shown. B. The small effusion on this T2-weighted image highlights an erosion of the articular and cortical surfaces (arrows). *Note the thickened medial patellar plica* (curved arrow).

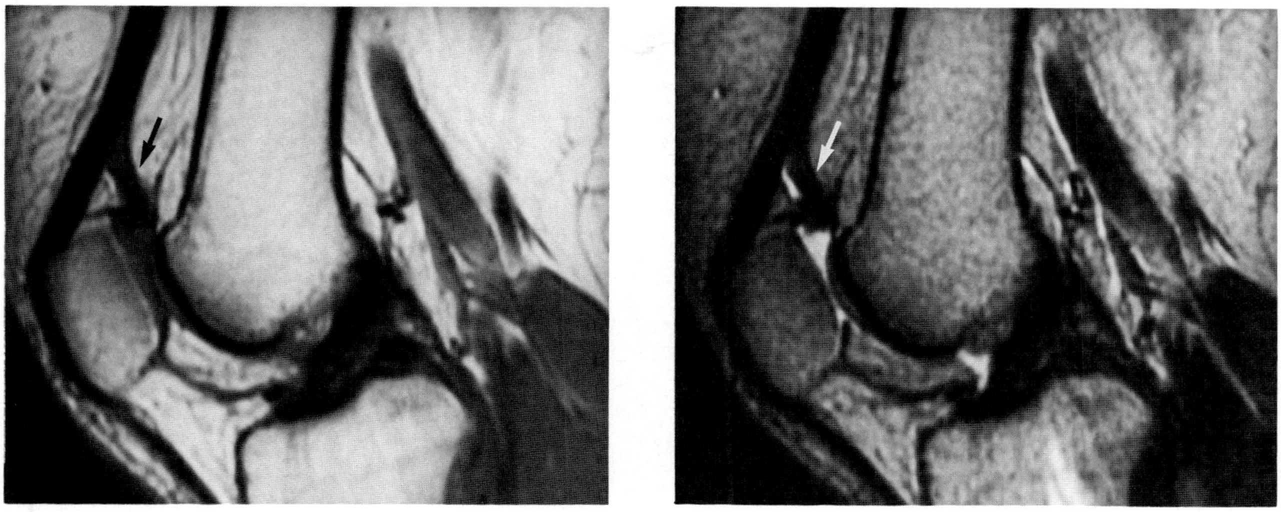

A B

FIGURE 13–51
A. *Hypertrophied medial patellar plica* (arrows) *can resemble a small suprapatellar effusion on a T1-weighted image.* B. *The T2-weighted image demonstrates only a small amount of fluid within the suprapatellar pouch, and a significantly thickened medial patellar plica. Increased linear signal in the proximal part of the tibia is due to an aliasing artifact.*

not been established in a controlled study. Both axial and sagittal planes permit visualization of the patellar surface, with the axial views better delineating the patellar facets. Thinning of the articular cartilage must be evaluated cautiously, so that artifacts due to chemical shift are not mistaken for pathologic features.

Patellar plica is a reduplication of synovial folds and represents remnants of embryologic synovial mesenchyme, which at one point divided the fetal knee into three compartments. These vestiges can be observed in up to one-half of the normal population [52]. Hypertrophy of the medial patella plica, secondary to inflammation or trauma, can be easily observed on axial and midline sagittal images. The appearance of a broad band of decreased intensity on both T1- and T2-weighted images, extending from the inferoposterior aspect of the patella, superiorly, is indicative of thickened medial patella plica (Figure 13–51). This abnormality can cause pain, effusion, and a clicking sound on extension of the knee, as the band hits against the anterior aspect of the medial femoral condyle [53,54]. Erosion of the condyle and eventual degenerative arthritic changes can follow [55]. Problems involving the superior or inferior plica are only rarely encountered [56].

Early synovitis can sometimes be detected by observing irregularities in the contour of the infrapatellar fat pad, even though the synovium itself is too thin to be visualized (Figure 13–52) [57]. This observation is nonspecific and can result from a wide variety of inflammatory disorders.

Joint pathologic conditions not unique or most common to the knee are discussed in general in Chapter 12, and a description of abnormalities of bone and soft tissue can be found in Chapters 9 and 10, respectively. Specific images illustrating some of the more common general problems associated with the knee caused by trauma and bone and soft tissue diseases are presented in Figures 13–53 through 13–56.

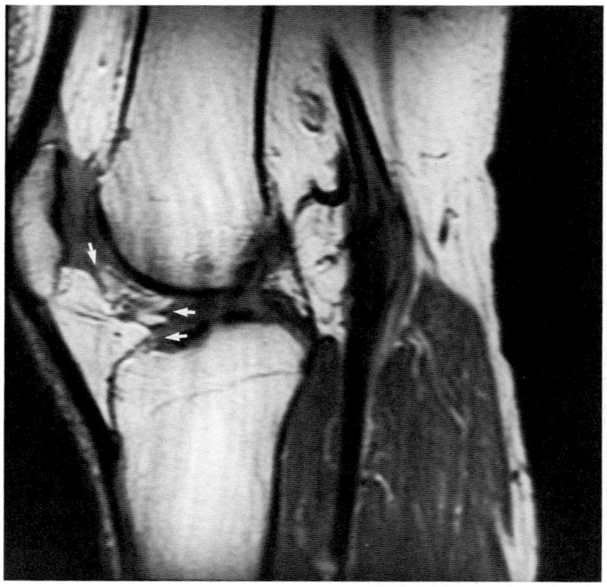

FIGURE 13–52
An irregular infrapatellar fat pad (arrows) *is an indication of chronic synovitis, in this case secondary to progressive degenerative changes.*

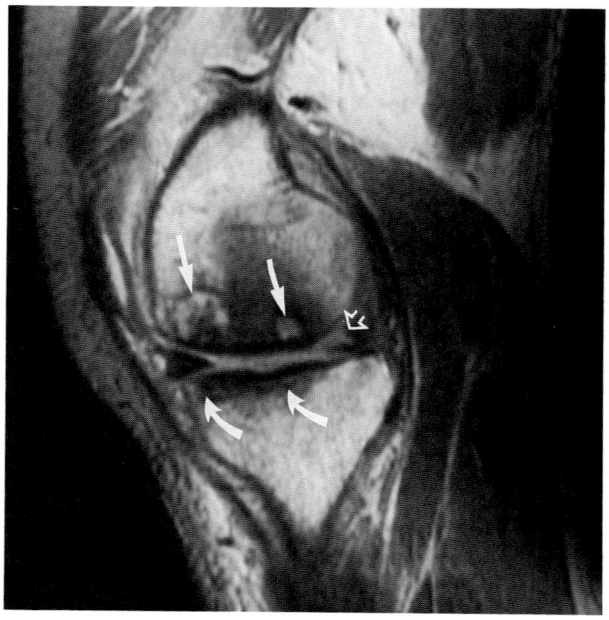

A

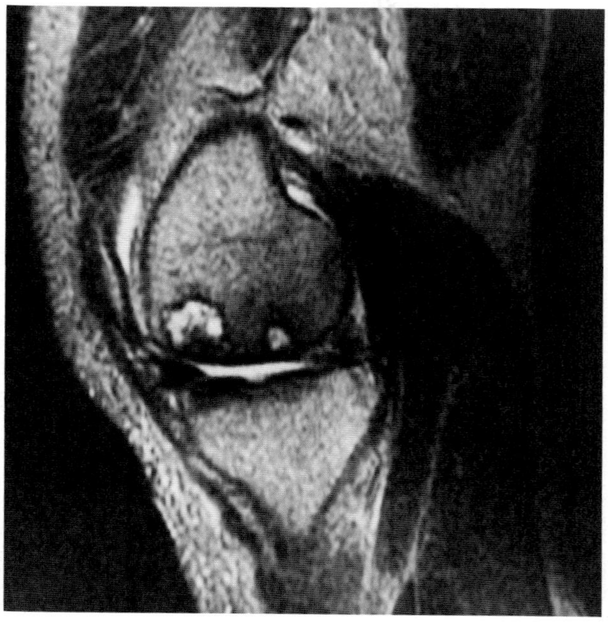

B

FIGURE 13–53
Marked degenerative changes include subchondral cyst formation (arrows), *sclerosis* (curved arrows), *and a large meniscal tear* (open arrow). A. *T1 image.* B. *T2 image.*

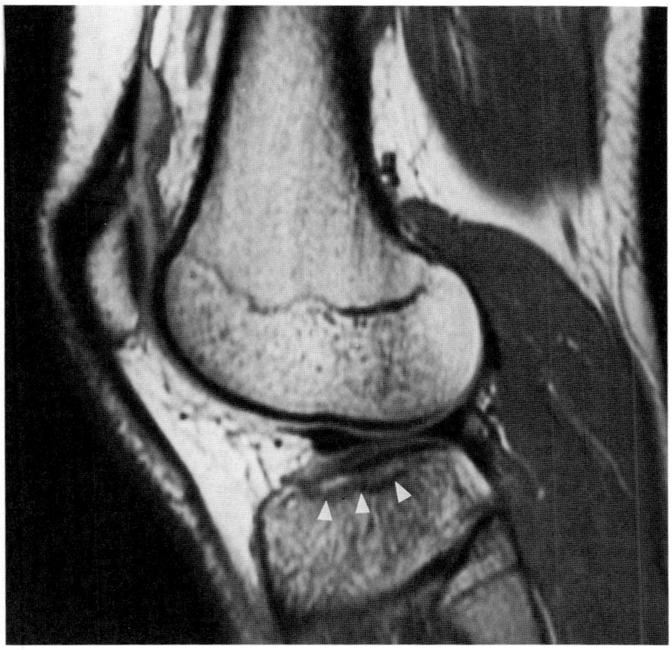

A

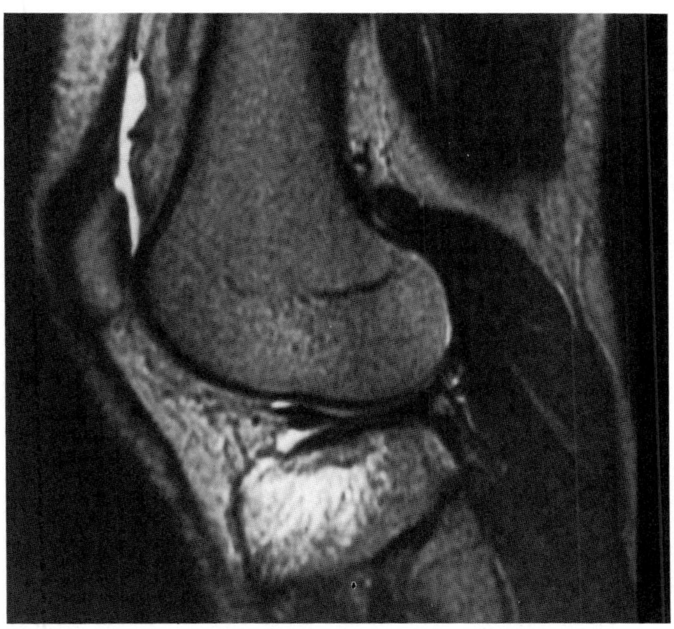

B

FIGURE 13–54
A. *An acute subchondral fracture* (arrowhead) *is clearly delineated on a T1-weighted image.*
B. *The T2-weighted images explicitly demonstrate the pattern of edema within the proximal part of the tibia.*

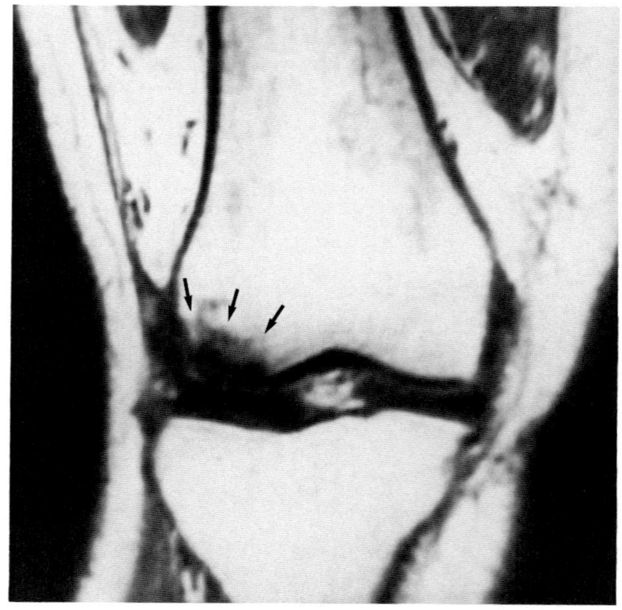

FIGURE 13–55
The area of decreased signal in the medial femoral condyle is due to osteonecrosis (arrows).

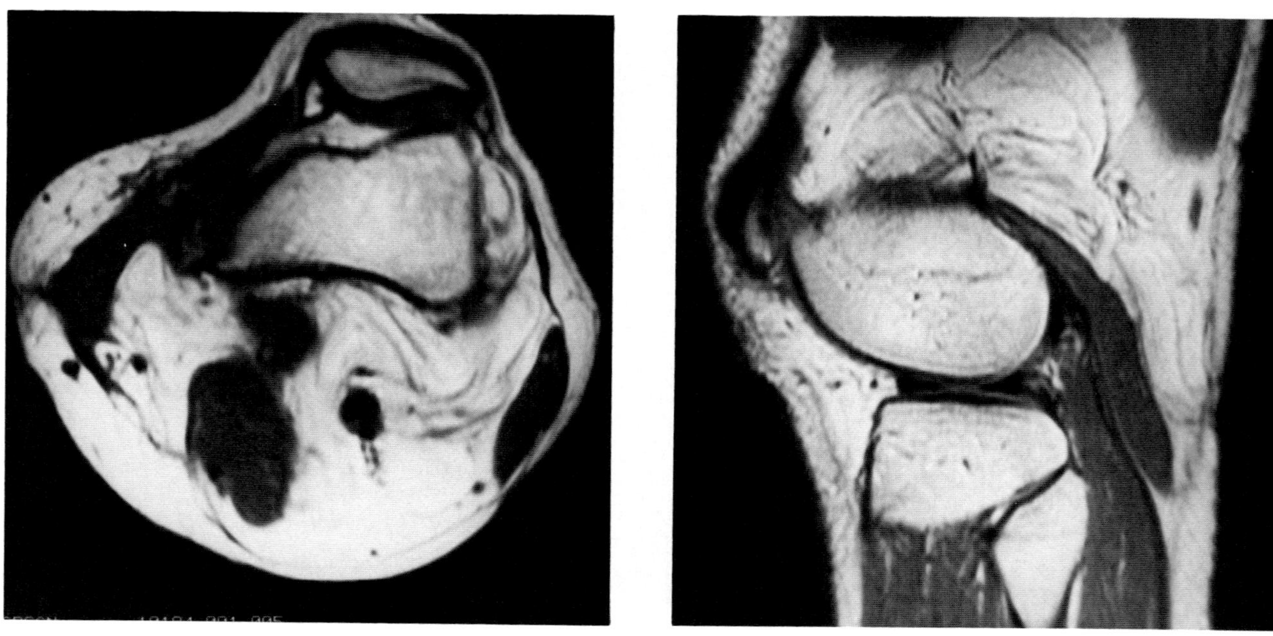

A B

FIGURE 13–56
A large multiseptated mass in the popliteal fossa is isointense with fat. This pattern is most consistent with a low-grade liposarcoma. A. *Axial view.* B. *Sagittal view.*

REFERENCES

1. Hajek PC, Baker LL, Sartoris DJ, Neumann CH, Resnick D. MR arthrography: anatomic-pathologic investigation. Radiology 1987; 163:141–7.
2. Beltran J, et al. Meniscal tears: MR demonstration of experimentally produced injuries. Radiology 1986; 158:691–3.
3. Gallimore GW Jr, Harms SE. Knee injuries: high-resolution MR imaging. Radiology 1986; 160:457–61.
4. Wright SM, Wright RM. Bilateral MR imaging with switched mutually coupled receiver coils. Radiology 1989; 170:249–55.
5. Mink JH, Reicher MA, Crues JV III. Magnetic resonance imaging of the knee. New York: Raven Press, 1987.
6. Beltran J, Noto AM, Mosure JC, Weiss KL, Zuelzer WA, Christofordis AJ. The knee: surface-coil MR imaging at 1.5 T. Radiology 1986; 159:747–51.
7. Mandelbaum BR, Hartzman S, Bassett LW, et al. Magnetic resonance imaging as a tool for evaluation of traumatic knee injuries: anatomical and pathoanatomical correlations. Am J Sports Med 1986; 11:436–8.
8. Reicher MA, Hartzman S, Bassett LW, Mandelbaum BR, Duckwiler G, Gold RH. MR imaging of the knee joint: Part I: traumatic disorders. Radiology 1987; 162:547–51.
9. Mink JH, Levy T, Crues JV. Tears of the anterior cruciate ligament and menisci of the knee: MR imaging evaluation. Radiology 1988; 167:769–74.
10. DeSmet AA. Meniscal tears on knee arthrography: patterns of arthrographic abnormalities. Skeletal Radiol 1985; 14:802–10.
11. Crues JV, Mink J, Levy TL, Lotysch M, Stoller DW. Meniscal tears of the knee: accuracy of MR imaging. Radiology 1987; 164:445–8.
12. Mink JH, Reicher MA, Crues JV III. Magnetic resonance imaging of the knee. New York: Raven Press, 1987; pp. 60–67.
13. Levensohn ME, Baker BE. Prearthrotomy diagnostic evaluation of the knee: evaluation of 100 cases diagnosed by arthrography and arthroscopy. AJR 1980; 134:107–11.
14. Watts I, Tasker T. Pitfalls in double contrast knee arthrography. Br J Radiol 1980; 53:754–9.
15. Kaplan EB. Discoid lateral meniscus of the knee joint. J Bone Joint Surg [Am] 1957; 39:77–87.
16. Herman LJ, Beltran J. Pitfalls in MR imaging of the knee. Radiology 1988; 167:775–81.
17. Mink JH, Reicher MA, Crues JV III. Magnetic resonance imaging of the knee. New York: Raven Press, 1987:149.
18. Burk DL, Kanal E, Brunberg JA, et al. 1.5 T surface-coil MRI of the knee. AJR 1986; 147:293–300.
19. Silverman JM, Mink JH, Deutsch AL. Discoid menisci of the knee: MR imaging appearance. Radiology 1989; 173:351–4.
20. Tyrrell RL, Gluckert K, Pathria M, Modic MT. Fast three-dimensional MR imaging of the knee: comparison with arthroscopy. Radiology 1988; 166:865–72.
21. Solomon SL, Totty WG, Lee JKT. MR imaging of the knee: comparison of three-dimensional FISP and two-dimensional spin-echo pulse sequences. Radiology 1989; 173:739–42.
22. Harms SE, Flamig DP, Fisher CF, Fulmer JM. New method for fast MR imaging of the knee. Radiology 1989; 173:743–50.
23. Konig H, Sauter R, Deimling M, Vogt M. Cartilage disorders: comparison of spin-echo, CHESS, and FLASH sequence of MR images. Radiology 1987; 164:753–8.
24. Solomon SL, Totty WG, Lee JKT. MR imaging of the knee: Comparison of three-dimensional FISP and two-dimensional spin-echo pulse sequences. Radiology 1989; 173:739–42.
25. Kennedy JC. The posterior cruciate ligament. J Trauma 1967; 7:367–77.
26. Torisu T. Isolated avulsion fracture of the tibial attachment of the posterior cruciate ligament. J Bone Joint Surg 1977; 59:68–72.
27. Kennedy JC, Weinberg HW, Wilson AS. The anatomy and function of the anterior cruciate ligament. J Bone Joint Surg 1974; 56:223–35.
28. Mink JH, Reicher MA, Crues JV. Magnetic resonance imaging of the knee. New York: Raven Press, 1987:29.
29. Lee JK, Yao L, Phelps CT, Wirth CR, Czajka J, Lozman J. Anterior cruciate ligament tears: MR imaging compared with arthroscopy and clinical tests. Radiology 1988; 166:861–4.
30. Butler DL, Noyes FR, Grood ES. Ligamentous restraints to anterior-posterior drawer in the human knee. J Bone Joint Surg 1980; 63:259–70.
31. Moeser P, Brechtold RE, Clark T, Rovere G, Karstaedt N, Wolfman N. MR imaging of anterior

cruciate ligament repair. J Comput Assist Tomogr 1989; 13:105–9.

32. Turner DA, Prodromos CC, Petasmick JP, Clark JW. Acute injury of the ligaments of the knee: magnetic resonance evaluation. Radiology 1985; 154:717–22.

33. Lieberman JM, Yulish BS, Bryan PJ, et al. Magnetic resonance imaging of ruptured baker's cyst. J Can Assoc Radiol 1988; 39:295–7.

34. Burk DL, Dalinka MK, Kanal E, et al. Meniscal and ganglion cysts of the knee: MR evaluation. AJR 1988; 150:331–6.

35. Lynch TP, Crues JV III, Morgan FW, Sheehan WE, Harter LP, Ryu R. Bone abnormalities of the knee: prevalence and significance at MR imaging. Radiology 1989; 171:761–6.

36. Mink JH, Deutsch AL. Occult cartilage and bone injuries of the knee: detection, classification and assessment with MR imaging. Radiology 1989; 170:823–9.

37. Shellock FG, Mink JH, Deutsch AL, Fox JM. Patellar tracking abnormalities: clinical experience with kinematic MR imaging in 130 patients. Radiology 1989; 172:799–804.

38. Deutsch AL, Mink JH, Rosenfelt FP, Waxman AD. Incidental detection of hematopoietic hyperplasia on routine knee MR imaging. AJR 1989; 152: 333–6.

39. Yao L, Lee JK. Avulsion of the posteromedial tibial plateau by the semimembranosus tendon: diagnosis with MR imaging. Radiology 1989; 172: 513–4.

40. Hartzman S, Reicher MA, Bassett LW, Duckwiler GR, Mandelbaum B, Gold RH. MR imaging of the knee. Part II. Chronic disorders. Radiology 1987; 162:553–7.

41. Aegerter E, Kirkpatrick JA Jr. Orthopedic diseases: physiology, pathology, radiology. 3rd ed. Philadelphia: WB Saunders, 1968:336–8.

42. Ribbing S. Hereditary multiple epiphyseal disturbance and its consequences for oetiogenesis of local malacias particularly osteochondritis dissecans. Acta Orthop Scand 1955; 24:286–97.

43. Hughston JC, Hergenroeder PT, Courtney BG. Osteochondritis dissecans of the femoral condyle. J Bone Joint Surg [Am] 1984; 66:1340–8.

44. Mesgarzadeh M, Sapega AA, Bondakdarpour A, et al. Osteochondritis dissecans: analysis of mechanical stability with radiography, scintigraphy, and MR imaging. Radiology 1987; 165:775–80.

45. Reicher MA, Bassett LW, Gold RH. High-resolution magnetic resonance imaging of the knee joint: pathologic correlations. AJR 1985; 145:903–9.

46. Freiberger RH, Kaye JJ. Arthrography. New York: Appleton-Century-Crofts, 1979.

47. Goodfellow J, Hungerford DS, Woods C. Patellofemoral joint mechanics and pathology. II. Chondromalacia patellae. J Bone Joint Surg [Am] 1976; 58:291–9.

48. Insall J, Falvo KA, Wise DW. Chondromalacia patellae. J Bone Joint Surg [Am] 1976; 58:1–8.

49. Bentley G. Articular cartilage changes in chondromalacia patellae. J Bone Joint Surg [Am] 1985; 67:769–75.

50. Yulish BS, Montanez J, Goodfellow DB, Bryan PJ, Mulopulos GP, Modic MT. Chondromalacia patellae: assessment with MR imaging. Radiology 1987; 164:763–6.

51. Mink JH, Reicher MA, Crues JV. Magnetic resonance imaging of the knee. New York: Raven Press, 1987:128.

52. Pipkin G. Knee injuries. The role of suprapatellar plica and suprapatellar bursa in simulating internal derangements. Clin Orthop 1971; 74:161–76.

53. Broom MJ, Fulkerson JP. The plica syndrome: a new perspective. Orthop Clin North Am 1986; 17:279–81.

54. Apple JS, Martinez S, Hardaker WT, Daffner RA, Gehweiler JA. Synovial plicae of the knee. Skeletal Radiol 1982; 7:251–4.

55. Deutsch AL, Resnick D, Dalinka MK, et al. Synovial plicae of the knee. Radiology 1981; 141: 627–34.

56. Patel D. Plica as a cause of anterior knee pain. Orthop Clin North Am 1986; 17:273–7.

57. Stoller DW, Genant HK. MR imaging of knee arthridites. Radiology 1987; 165(P):233.

Index

C